KV-335-120

Handbook of Cognitive Science

Previous books published in the Perspectives on Cognitive Science series:

Creativity, Cognition and Knowledge—An Interaction
Terry Dartnall

Language Universals and Variation
Mengistu Amberber and Peter Collins

Perspectives on Cognitive Science, Vol. 2—Theories, Experiments, and Foundations
Janet Wiles and Terry Dartnall

Perspectives on Cognitive Science, Vol. 1—Theories, Experiments, and Foundations
Peter Slezak, Terry Caelli and Richard Clark

Representation in Mind: New Approaches to Mental Representation
Hugh Clapin, Phillip Staines and Peter Slezak

Competition and Variation in Natural Languages: The Case for Case
Mengistu Amberber and Helen de Hoop

Aspects of Knowing: Epistemological Essays
Stephen Hetherington

Handbook of Cognitive Science: An Embodied Approach

Edited by

Paco Calvo and
Antoni Gomila

Amsterdam • Boston • Heidelberg • London • New York • Oxford • Paris
San Diego • San Francisco • Singapore • Sydney • Tokyo

Elsevier
525 B Street, Suite 1900, San Diego, CA 92101-4495, USA
Linacre House, Jordan Hill, Oxford OX2 8DP, UK
Radarweg 29, PO Box 211, 1000 AE Amsterdam, The Netherlands

First edition 2008

British Library Cataloguing in Publication Data
A catalogue record for this book is available from the British Library

Library of Congress Cataloging-in-Publication Data
A catalog record for this book is available from the Library of Congress

ISBN: 978-0-08-046616-3

For information on all Elsevier publications visit our web site at books.elsevier.com

Typeset by Charon Tec Ltd., A Macmillan Company. (www.macmillansolutions.com)

Transferred to digital print 2009
Printed and bound in Great Britain by CPI Antony Rowe, Chippenham and Eastbourne

Contents

3

EMBODIMENT AND EXPLANATION 41

ANDY CLARK

4

CAN A SWARM BE EMBODIED? 59

AMANDA J. C. SHARKEY AND NOEL SHARKEY

SECTION II

ROBOTICS AND AUTONOMOUS AGENTS 79

5

CAJUNBOT: A CASE STUDY IN EMBODIED COGNITION 81

ISTVAN S. N. BERKELEY

6

THE DYNAMICS OF BRAIN–BODY–ENVIRONMENT SYSTEMS: A STATUS REPORT 99

RANDALL D. BEER

SECTION III

PERCEIVING AND ACTING 159

9

ECOLOGICAL PSYCHOLOGY: SIX PRINCIPLES FOR AN EMBODIED–EMBEDDED APPROACH TO BEHAVIOR 161

MICHAEL J. RICHARDSON, KEVIN SHOCKLEY, BRETT R. FAJEN, MICHAEL A. RILEY, AND MICHAEL T. TURVEY

10

SEEING WHAT WE CAN DO: INSIGHTS INTO VISION AND ACTION THROUGH OBSERVATIONS OF NATURAL BEHAVIOR 189

JASON A. DROLL AND MARY M. HAYHOE

11

WHY WE DON'T MIND TO BE INCONSISTENT 207

JEROEN B. J. SMEETS AND ELI BRENNER

SECTION IV

A DYNAMIC BRAIN 219

12

NEURONAL AND CORTICAL DYNAMICAL MECHANISMS UNDERLYING BRAIN FUNCTIONS 221

ANJA STEMME AND GUSTAVO DECO

13

DYNAMIC FIELD THEORY AS A FRAMEWORK FOR UNDERSTANDING EMBODIED COGNITION 241

SEBASTIAN SCHNEEGANS AND GREGOR SCHÖNER

14

A LAZY BRAIN? EMBODIED EMBEDDED COGNITION AND COGNITIVE NEUROSCIENCE 273

PIM HASELAGER, JELLE VAN DIJK, AND IRIS VAN ROOIJ

SECTION V

EMBODIED MEANING 291

15

THE ROLE OF SENSORY AND MOTOR INFORMATION IN SEMANTIC REPRESENTATION: A REVIEW 293

LOTTE METEYARD AND GABRIELLA VIGLIOCCO

16

EMBODIED CONCEPT LEARNING 313

BENJAMIN BERGEN AND JEROME FELDMAN

17

MATHEMATICS, THE ULTIMATE CHALLENGE TO EMBODIMENT: TRUTH AND THE GROUNDING OF AXIOMATIC SYSTEMS 333

RAFAEL E. NÚÑEZ

18

EMBODIMENT FOR EDUCATION 355

ARTHUR M. GLENBERG

23

GETTING TO THE HEART OF EMOTIONS AND CONSCIOUSNESS 453

MAXINE SHEETS-JOHNSTONE

PREFACE

What has become of the cognitive sciences since the heyday of cognitivism? The information processing paradigm that dominated the field during the second half of the 20th century has given way to a diversity of *post-cognitivist* approaches. These, under the banner of "embodied cognitive science," have erupted with special force in the last decade. Unfortunately, the lack of a shared standpoint, together with the increasing number of the subdisciplines involved, has made progress difficult to gauge. In this *Handbook of Cognitive Science: An Embodied Approach* we aim to show—through the adoption of an interactive, embodied, and embedded view of cognition and behaviour—that a new more solid foundation is already in the making. This is not an easy project, but we believe that success depends on being able to generate the right intellectual climate and a common research agenda.

A panoramic view of the "embodied movement" was much needed. Embodiment has become fashionable in some quarters but these quarters are somewhat diverse and dispersed. Controversies and skepticism have come along, as the feeling that different lines of research were at times following dead ends for the lack of a better approach. Our goal is to suggest a path for an integrative approach, so that the different traditions dissatisfied with cognitivist orthodoxy realize the synergistic potential that they share. We are well aware that the whole project may well be jeopardized if one research program tried to become hegemonic. Progress depends on real synergistic convergence among post-cognitivist methodologies. This handbook will furnish the reader with a comprehensive picture of the ways in which the embodied approach is making progress on various topics. We introduce the different alternative research programs that constitute post-cognitivism and identify the problems that an embodied cognitive science must meet as a challenger candidate to cognitivism.

Our aim is also to inform all the mainstream Cognitive Sciences, not just those areas, such as cognitive linguistics, robotics, motor control, or cognitive neuroscience, where embodiment has taken root. In particular, the handbook addresses the way to scale up from the basics of sensorimotor coordination to the higher level cognitive processes, avoiding the big problems that have trapped

Cognitive Science until now: the frame problem, the grounding problem, the common code problem, and the problem of homuncular control. We do not claim that a new consensus already exists, but that the shift of attention called for by the embodiment movement is essential to understanding cognition and behaviour.

In this volume we have collected together a representative sample of the best work being done from an embodied perspective from disciplines such as psychology, computer science, computational and cognitive neuroscience, movement sciences, robotics, linguistics, and philosophy, among others. The contributions by outstanding specialists in their respective fields speak for themselves as to their explanatory power and richness and illustrate that the diversity of traditions and research programs can converge into a minimal common ground to propel Cognitive Science forward. Sections on (i) the conceptual grounds of embodiment, (ii) robotics and autonomous agency, (iii) perception and action, (iv) brain dynamics, (v) meaning and understanding, (vi) higher cognition, and (vii) emotion and social interaction all serve to structure the material herewith presented. We hope that the handbook will serve as a landmark in the establishment of an embodied cognitive science, understood as the beginning of an effort to unify the study of cognition by taking into account the more basic perceptuo-motor, bodily, structure, under an embracing post-cognitivist explanatory framework.

The idea of this handbook started as part of our research project "Cognition and Representation: non-classical alternatives," funded by the Spanish Ministry of Education BFF2003-129, and continued with research project HUM2006-11603-C02-01 (Spanish Ministry of Science and Education and European Union FEDER Funds). A key milestone involved a workshop in Palma de Mallorca, in December 2006, where about half of the authors met to present their work and reflect on grounds of the new, embodied, approach. Several of the chapters have grown out of those presentations and the discussions that followed. We also invited additional authors to contribute to this volume. We are grateful to all the contributors for their interest, commitment, and terrific work—and for trusting us and agreeing to become part of this enterprise. Special thanks go to Gregor Schöner, for his support and guidance, and to Peter Slezak, editor of the Cognitive Science series for Elsevier, for encouraging us to pursue this project. Last, but not least, to our families with love—especially as, during the course of this work, our embodied minds were not always *there*.

Rincón de Seca, and Sóller
May, 2008

List of Contributors

Idoia Alarcón (395), Technical University of Madrid, Autonomous Systems Laboratory, Madrid, Spain.

Michael L. Anderson (423), Institute for Advanced Computer Studies, Program in Neuroscience and Cognitive Science, University of Maryland, College Park, MD, USA.

Dana Ballard (139), Department of Computer Science, University of Texas at Austin, Austin, TX, USA.

Randall D. Beer (99), Cognitive Science Program, Department of Computer Science and Department of Informatics, Indiana University, Bloomington, IN, USA.

Benjamin Bergen (313), Department of Linguistics, University of Hawaii, Honolulu, HI, USA.

Istvan S. N. Berkeley (81), Philosophy and Cognitive Science, The University of Louisiana at Lafayette, Lafayette, LA, USA.

Mark H. Bickhard (29), Department of Philosophy, Lehigh University, Bethlehem, PA, USA.

Eli Brenner (207), Faculty of Human Movement Sciences, Vrije Universiteit, Amsterdam, The Netherlands.

Paco Calvo (1), Departamento de Filosofia, Edificio Luis Vives, Campus de Espinardo, Universidad de Murcia, Murcia, Spain.

Andy Clark (41), Department of Philosophy, University of Edinburgh, Edinburgh, Scotland, UK.

Gustavo Deco (219), Theoretical and Computational Neuroscience, Universitat Pompeu Fabra, Barcelona, Spain.

Jason A. Droll (189), Department of Psychology, University of California, Santa Barbara, CA, USA.

Brett R. Fajen (161), Department of Cognitive Science, Rensselaer Polytechnic Institute, Troy, NY, USA.

Jerome Feldman (313), Department of Electrical Engineering and Computer Science, University of California, Berkeley, Berkeley, CA, USA.

Shaun Gallagher (439), Department of Philosophy, University of Central Florida, Orlando, FL, USA.

Arthur M. Glenberg (355), Department of Psychology, Arizona State University, Tempe, AZ, USA.

Jaime Gómez (395), Technical University of Madrid, Autonomous Systems Laboratory, Madrid, Spain.

Toni Gomila (1), Departamento de Psicologia, Universitat Illes Balears, Edifici Beatriu de Pinos, Palma de Mallorca, Spain.

Pim Haselager (273), Artificial Intelligence/Cognitive Science, Nijmegen Institute for Cognition and Information, Radboud University Nijmegen, Nijmegen, The Netherlands.

Mary M. Hayhoe (189), Department of Psychology, University of Texas, Austin, TX, USA.

Carlos Hernández (395), Technical University of Madrid, Autonomous Systems Laboratory, Madrid, Spain.

Max Lungarella (121), Department of Informatics, University of Zurich, Zurich, Switzerland.

Lotte Meteyard (293), MRC Cognition and Brain Sciences Unit, Cambridge, London, UK.

Rafael E. Núñez (333), Department of Cognitive Science, University of California, San Diego, La Jolla, CA, USA.

Rolf Pfeifer (121), Department of Informatics, University of Zurich, Zurich, Switzerland.

Michael J. Richardson (161), Department of Psychology, Colby College, Mayflower Hill, Waterville, ME, USA.

Michael A. Riley (161), Department of Psychology, University of Cincinnati, Cincinnati, OH, USA.

Ricardo Sanz (395), Technical University of Madrid, Autonomous Systems Laboratory, Madrid, Spain.

Sebastian Schneegans (241), Institut fur Neuroinformatik, Ruhr-Universitat-Bochum, Bochum, Germany.

Gregor Schöner (241), Institut fur Neuroinformatik, Ruhr-Universitat-Bochum, Bochum, Germany.

Amanda J. C. Sharkey (59), Department of Computer Science, Regent Court, University of Sheffield, Sheffield, UK.

Noel Sharkey (59), Department of Computer Science, Regent Court, University of Sheffield, Sheffield, UK.

Maxine Sheets-Johnstone (453), Department of Psychology, University of Oregon, Yachats, OR, USA.

Kevin Shockley (161), Department of Psychology, University of Cincinnati, Cincinnati, OH, USA.

Jeroen B. J. Smeets (207), Faculty of Human Movement Sciences, Vrije Universiteit, Amsterdam, The Netherlands.

Olaf Sporns (121), Department of Psychological and Brain Sciences, Indiana University, Bloomington, IN, USA.

Anja Stemme (219), Institute for Biophysics, Computational Intelligence Group, University of Regensburg, Regenburg, Germany.

Michael T. Turvey (161), Department of Psychology, University of Connecticut, Storrs, CT, USA.

Jelle van Dijk (273), Mediatechnology, Utrecht University of Applied Sciences, Utrecht, The Netherlands.

Iris van Rooij (273), Nijmegen Institute for Cognition and Information, Radboud University Nijmegen, Nijmegen, The Netherlands.

Gabriella Vigliocco (291), Department of Psychology, University College London, London, UK.

Margaret Wilson (375), Department of Psychology, University of California, Santa Cruz, CA, USA.

1

Directions for an Embodied Cognitive Science: Toward an Integrated Approach

Toni Gomila[1] and Paco Calvo[2]

[1]*Departamento de Psicologia, Universitat Illes Balears, Edifici Beatriu de Pinos, Palma de Mallorca, Spain*
[2]*Departamento de Filosofia, Edificio Luis Vives, Campus de Espinardo, Universidad de Murcia, Murcia, Spain*

Cognitivism in a Blind Alley

Is cognitive activity more similar to a game of chess than to a game of pool? In order to answer this question we need to know first what are the relevant differences, cognitively speaking, between a game of pool and a game of chess. Questions of this sort (Kirsh, 1991; Haugeland, 1998) highlight the contraposition between those aspects of cognition where a rule-governed, formal approach appears to apply, and those other aspects of cognition that are deeply rooted in the physical nuts-and-bolts of the interacting agent. Whereas chess is a *formal* game that can be played regardless of details of physical implementation (think of Kasparov's legal defeat to Deep Blue), and where *rule-governed* manipulations of symbolic states suffice for the purposes of conforming to the rules of chess, in the case of pool, the actual striking of the cue ball with a nice solid hit cannot be dispensed with. Simply, pool is not a formal game. Rather than the rule-governed manipulation of inner states, real-time physical interactions need to be honored if a game that conforms to the rules of pool is to be played. However insofar as *digital* systems can be described by abstracting away from details of implementation, another way to emphasize the distinction (Haugeland, 1998) is by noting that whereas chess is digital, pool is not. In short, we may say that formal features identified at an algorithmic level of description are critical to a game of chess (even Kasparov's reliable offline mental rehearsal of the lost game would lack any relevant twists). However, an analogous formal level description would not work in the case of pool.[1]

The assumption that cognition is like a game of chess has been the driving force of research programs in cognitive science since the inception in the mid 1950s of the cognitivist paradigm, writ large (Chomsky, 1959; cf. the canonical history of the cognitive revolution: Gardner, 1985). Classical cognitivism takes as its starting point the concepts of representation and computation. Very roughly, models of the mind are likened to a von Neumann architecture in such a way that cognitive processing boils down to the computational manipulation of representational inner states. That the brain is a piece of biological hardware, and the mind is the software running on top, means that it can be modeled as a Turing machine. Cognition consists in the rule-governed manipulation of symbols that Newell and Simon's (1972) *Physical Symbol System* epitomized.

The computer metaphor with its hardware/software divide is held quite literally by proponents of classical cognitivism. As Fodor and Pylyshyn (1988) claim, "the symbol structures in a Classical model are assumed to correspond to real physical structures in the brain and the combinatorial structure of a representation is supposed to have a counterpart in structural relations among physical properties of the brain. For example, the relation 'part of,' which holds between a relatively simple symbol and a more complex one, is assumed to correspond to some physical relation among brain states. This is why Newell (1980) speaks of computational systems such as brains and Classical computers as 'physical symbol systems'" (p. 13). And we would add, this is why cognitivism is better at spelling out the algorithms and heuristics that Deep Blue deployed in its victory over Kasparov than at trying to cash out the physically constrained scenario of a game of pool. Information processing of discrete abstract symbols delivers the goods in the former case quite straightforwardly, or so the story goes (cf. Dreyfus, 1992), but not in the latter.

Connectionist theory, however, has traditionally been not as uneasy with less abstract, and more context-dependent tasks. Unfortunately, the way cognitivism has been learned and taught in the 1980s and 1990s has sometimes assumed a temporal sequence, with classical theories (Newell & Simon, 1972) as the pinnacle of cognitivism, followed in the 1980s by neural network theory as the alternative to cognitivism, courtesy of the *bible* of connectionism (McClelland,

[1] Of course, it goes without saying that you can play a virtual *game of pool* on a computer, but this shares with real pool nothing other than the commercial name. In fact, there cannot be a *Deep Pool* counterpart of Deep Blue. It would have to be a robot, with a vision system, and a cue mechanism, etc., and this would be different from Deep Blue. Put bluntly, the robot would actually have to *do* something.

Rumelhart et al., 1986; Rumelhart, McClelland et al., 1986), with hybrid models (Anderson, 1993) pulling together both ends, in an attempt to exploit the best of both worlds. However, from a historical point of view, all working hypotheses developed more or less at the same time. In fact, the basics of neurocomputation date back to the 1940s with McCulloch and Pitts (1943), and Hebb (1949). Unfortunately, although Rosenblatt's (1959) perception was a breath of fresh air, Minsky and Papert's (1969) devastating critique of two-layer neural networks and the delta rule pushed the field out of the spotlight. It was only in the 1980s with the deployment of the backpropagation learning algorithm (the generalized delta rule) in multilayered networks that connectionism re-emerged as a viable candidate to explain cognition.

Such a time line has fostered the illusion that, properly speaking, cognitivism *is* Newell and Simon (1972) and their physical symbol system hypothesis, Chomsky (1980) and his theory of rules and representations, Marr (1982) and his theory of vision paired with the well-known threefold distinction between computation, algorithm, and implementation, and Fodor and Pylyshyn (1988) with their seminal critique of connectionism. Connectionism was, as a matter of fact, believed for almost two decades to be *the* alternative to cognitivism. Certainly, Chomsky, Newell and Simon, Marr, and Fodor and Pylyshyn (among others, it goes without saying) represent orthodoxy in cognitive science,[2] but the discipline is on the move, and nowadays connectionist theory just represents the alternative to *classical* cognitivism, not to cognitivism. In fact, in a broader sense, as we shall show next, connectionism belongs to cognitivism.

We strongly believe that if we are to exploit scientifically the similarities between cognitive activity and a game of chess or a game of pool, cognitivism with a twist of neural networks is helpless. A more drastic change of focus is needed. Rather than between classicist and connectionist contenders, the critical contrast to be currently drawn is between cognitivism (both classicist and connectionist) and post-cognitivism: between a view of cognition as abstract computation versus a view of cognition as interactive, embodied and embedded. Hence, this *Handbook of Embodied Cognitive Science*, where we aim to provide a panoramic view of the richness, variety, and potential of current research programs working within this broad, post-cognitivist approach, and assess their potential convergence. We proceed in this introduction as follows: in the remainder of this section, we map out the blind alley into which, we believe, cognitivism has been driven. This allows us in Section 2 to introduce the different alternative research programs that feed post-cognitivism and recognize the problems that such an alternative must meet as a challenger candidate to explain cognition and behavior (another, superior, "game in town"). Section 3 discusses the commonalities among these

[2]It's ironic that neither Chomsky nor Fodor really espouse cognitivism as "the" theory of cognition; they keep it restricted to some narrow areas of the human mind. See Chomsky (1996) and Fodor (2000).

different programs that coincide in calling cognitivism into question, and highlights the milestones required in order to make progress toward an integrated approach. Finally, we address the question of how to scale up from bodily based cognitive interaction to higher cognition, within this alternative approach. As this set of issues unfolds, a short review of the contributions that shape the volume will be provided along the way, as the best way to show how they all cohere.

So, first of all, what do we make of the standard way of framing the debate of the post-behaviorism era? According to classical cognitivism, symbols are stored in memory which are retrieved and transformed by means of algorithms that specify how to compose them syntactically and how to transform them. The supposed manifest systematicity and inferential coherence of human thought, among other things, calls for this working hypothesis as an inference to the best explanation (Fodor & Pylyshyn, 1988). Underneath this conception is the idea that thought can be understood as some form of logic-like inferential processing—bluntly, what Deep Blue does. But the result of embracing classicism is the detachment of central cognitive processes from the perceptual and motor systems. The latter reduce to input and output modules that feed the system and output the system's response (say, 1 … e5 in response to white's opening 1 … e4), respectively. The propositions that cognitive psychology posits, or the Fodorian picture that results in the philosophy of psychology, have the same result, namely, the endorsement of the view that cognition is information-processing as conceived by the representational-computational view of the mind, or, as we may say, some form of symbol-crunching according to algebraic rules.

The "hundred step" constraint (Feldman & Ballard, 1982), graceful degradation, neurobiological plausibility, pattern recognition, and content-based retrieval of information were all interpreted as reasons to turn to connectionism (Rumelhart, McClelland et al., 1986). From the connectionist perspective, cognition was seen as the emergent outcome of the interconnectivity of numerous basic processing units connected in parallel within an allegedly biologically plausible neural network. Under this lens, the retrieval of information has as a consequence flow of inhibitions and excitations throughout an entire network of weighted connections, which are shaped as the weights are modified in response to the statistical regularities that the network is fed.

Unfortunately, things do not look that different from the connectionist perspective than they do from the classical position, despite the different technical jargon. If cognition amounted to some form of symbol crunching according to algebraic rules under classicism, connectionism now attempts to explain cognition in terms of the computational manipulation of subsymbols, according to statistical rules. Connectionism is in fact a form of cognitivism, in spite of the obvious architectural differences between symbol systems and connectionist networks (serial vs. parallel, discrete vs. distributed, etc.). Whereas orthodox classical cognitivism assigns symbolic content to the sort of physical entities that get stored in von Neumann architectures, connectionist cognitivism assigns subsymbolic content to the sort of physical entities that are fully distributed and superposed on the network's weight matrix.

Hurley (1998) makes a similar point when she warns against the "cognitive sandwich" metaphor implicitly endorsed by many cognitive scientists. Classicism is committed to a *sandwich* architecture insofar as it understands cognition "proper" as the filling in between a perception-action *bun*. But cognitive sandwiches need not be Fodorian. A feed forward connectionist network conforms equally to the sandwich metaphor. The input layer is identified with a perception module, the output layer with an action one, and hidden space serves to identify metrically, in terms of the distance relations among patterns of activation, the structural relations that obtain among concepts. The hidden layer this time contains the meat of the connectionist sandwich. In this way, we may say that, in the worst case, connectionism amounts to a hypothesis as to the implementation on top of which classical algebraic rules operate. In the best case, it amounts to an algebraic variation of a classical algorithm, insofar as symbols are incorporated into the neural network, either in the teaching pattern of the learning algorithm itself, in the case of supervised learning, or in the input encoding, as the network is fed with patterns of activation as training proceeds (Marcus, 2001). Of course, there is a lot to be said in response to Marcus' line of argument in terms of *implementational* versus *eliminative* connectionism. Marcus' criticism echoes Fodor and Pylyshyn's (1988) attack (connectionism can only account for the systematicity, productivity, etc., of thought insofar as it implements a classical model in doing so; otherwise, it becomes eliminativist, and fails to explain the character of thought). In doing so, it is subject to the same sort of criticisms that Fodor and Pylyshyn have encountered (Chalmers, 1990; Elman, 1998). However, although self-supervised learning algorithms of the sort employed by Elman (1990, 1998) may be less suspected of furnishing the network with symbols, insofar as the teaching pattern is this time the next input pattern in the training pool, and although fully distributed input encodings may reflect the statistical regularities of the environment in more subtle ways than localist encodings, we shall not pursue this line any further, since connectionist computations are also conceived as abstract, just as the inputs and outputs are codified quite apart of the real details of perception and action.

Granting this setting then, why is cognitivism in a blind alley? The recent exchanges in the empirical and modeling literature between classicist and connectionist cognitivists demonstrate the reason. Sympathizers of classicism (Marcus et al., 1999; Marcus, 2001) continue to search for cognitive abilities that, defying a statistical explanation under the Chomskian poverty of the stimulus lens, may embarrass their connectionist foes. In turn, connectionist rule-following skeptics (Seidenberg & Elman, 1999) rejoin by showing the informational richness of ecological data, that can be exploited statistically and allow connectionist networks to remain computationally adequate. In this way, the architecture of cognition is in dispute, but assumptions about its computational underpinnings remain unchallenged. The debate focuses upon whether cognition boils down to the manipulation of symbolic items according to explicit algebraic rules, as opposed to the manipulation of subsymbolic items according to implicit statistical rules. The dispute, however, is entirely internecine warfare among proponents of a generalized, representationalist information-processing paradigm. The past-tense debate (Pinker & Ullman,

2002; Ramscar, 2002), the systematicity debate (Fodor & Pylyshyn, 1988; Elman, 1998), the algebra versus statistics debate (Marcus et al., 1999; Seidenberg & Elman, 1999), and, more recently, the speech segmentation debate (Peña et al., 2002; Perruchet et al., 2004; Laakso & Calvo, 2008) have all re-enacted the "classical-connectionist, *within*-paradigm battle to win souls" (Calvo, 2005).

As the reader familiar with these debates knows, the connectionist's overall strategy is to show that stimuli are not that poor after all! Although things are never black and white, the debate has moved along these lines since the re-emergence of connectionism in the mid-1980s. The debate has been fruitful insofar as contributions have filled in empirical gaps at algorithmic levels of description. This is, however, a "cognitive decathlon" (Anderson & Lebiere, 2003) where we might never be able to declare a winner! Perhaps we are stuck in a never-ending dialectic of positing challenges to connectionism and then trying to account for them statistically, forever and ever (Calvo, 2003). We are not sure that this dialectic can deliver much more significant scientific progress. Worse still, focussing attention on this project becomes a way to ignore the deep roadblocks that cognitivism has stumbled upon (the frame problem, the grounding problem, the common code problem, etc.), and which only arise from a cognitivist standpoint. Progress on these problems, then, seems conditional on jettisoning cognitivist assumptions in the first place (Gomila, 2008).

The truth is, if one considers some of the conditions that a successful theory of cognition must satisfy according to Newell's (1980) classical paper (flexible behavior, real-time performance, adaptive behavior, vast knowledge base, dynamic behavior, knowledge integration, natural language, learning, development, evolution, and brain realization), it is easy to realize that little progress has been made within cognitivism, either classical or connectionist, on most of these problems (Anderson & Lebiere, 2003). Hence, the blind alley! We think it is high time to consider ways to make real progress in all these critical challenges, ways to get out of the blind alley, and to put those ways that already show the direction of progress in the foreground. Development, real-time performance, flexible, adaptive and dynamic behaviors, evolution and brain realization, to name but a few, are dimensions that post-cognitivist theories of cognition aim at accounting for, and where their successes, even at this early stage of development, clearly outperform their cognitivist competitors. The present volume will provide both evidence for this claim and reflection on how to make further progress.

The time is ripe indeed for a real alternative approach to cognitivism to establish itself on a firm basis. As the song goes, "The times they are a-changin'."

ALTERNATIVE APPROACHES TO COGNITIVISM

The philosophical interest in the notion of embodiment (Clark, 1997; Hurley, 1998; Haugeland, 1998; Noë, 2004; Shapiro, 2004; Gallagher, 2005; Wheeler, 2005; Rowlands, 2006; Chemero, in press) goes back to the move from a Cartesian framework into phenomenology, especially with Heidegger (1962) and

Merleau-Ponty (1962), and with Wittgenstein (1953). More generally, as a more or less radical scientific alternative to cognitivism, post-cognitivism (understood as the vindication of embodiment for the understanding of cognition), has existed for quite some time, but accelerated gradually during the last three decades, gaining visibility, influence and substantial momentum, since the 1990s.[3] A number of research programs clearly fall under the umbrella of post-cognitivism. These include ecological psychology (Gibson, 1966, 1979; Turvey & Carello, 1995), behavior-based AI (Brooks 1986, 1991, 1999; Beer, 1990; Pfeifer & Scheier, 1999), embodied cognition (Ballard, 1991; Varela et al., 1991; Clancey, 1997), distributed cognition (Hutchins, 1995), perceptual symbol systems (Glenberg, 1997; Barsalou, 1999), some forms of connectionism (Rolls & Treves, 1998; Freeman, 1999), interactivism (Bickhard & Terveen, 1995), and dynamical systems theory (Kelso, 1995; Port & van Gelder, 1995; Erlhangen & Schöner, 2002), to name but a few. In this handbook, we have tried to represent all these research programs, and to show the additional diversity in the area. We have also considered whether the various different programs are converging into a unified approach, and which conceptual and foundation issues would be required to facilitate this development.

At a minimum, all these approaches conceive of cognition and behavior in terms of the dynamical interaction (coupling) of an embodied system that is embedded into the surrounding environment. As a result of their embodied-embedded nature, cognition and behavior cannot be accounted for without taking into account the perceptual and motor apparatus that facilitates the agent's dealing with the external world in the first place, and to try to do so amounts to taking this external world also into account. This tells directly against the aforementioned cognitive sandwich and other forms of "methodological solipsism" (Fodor, 1980). Cognition is not a matter of crunching symbolically or subsymbolically, but of interacting, of coupling. To understand a cognitive system we need to take as the unit of analysis the"system" embedded into its surrounding environment—a kind of interaction in analogy with the biological notions of species and habitat. In a sense to be further spelt out below, understanding cognition involves understanding the coupled system as such, and not the mind/brain in itself.

Areas of research where post-cognitivist principles have been applied successfully include (cognitive) neuroscience (Skarda & Freeman, 1987; Damasio, 1994; Chiel and Beer, 1997; Jeannerod, 1997), AI and evolutionary robotics (Arkin, 1998; Murphy, 2000; Nolfi & Floreano, 2000), cognitive anthropology (Suchman, 1987; Hutchins, 1995), cognitive linguistics (Lakoff & Johnson, 1980, 1999; Langacker, 1987, 1991; Regier, 1996; Tomasello, 1998), motor control and learning (Thelen & Smith, 1994), enactivism in the philosophy of perception (Thompson, 1995; O'Regan & Noë, 2001; Noë, 2004), neurophenomenology (Hanna & Thompson, 2003), education (Resnick, 1994; Greeno, 1996), and even social psychology, a field traditionally less closed to post-cognitivist

[3] Anderson (2003) provides a good entry point to the reader unfamiliar with the literature.

methodologies that is attaining increasing attention (Semin & Smith, 2002). It is no exaggeration to say that virtually all of the connections in Gardner's (1985) well-known *cognitive hexagon* are at present being thoroughly explored via the same computational, linguistic, and behavioral methodologies that helped to shape the cognitive hexagon in the first place. Neuroimaging data, moreover, are helping to pile up a whole new set of evidence, although if a post-cognitivist framework is to be developed in all its consequences, a thorough revision in the cognitive neurosciences and the accompanying neuroimaging methodologies must be accomplished (Chapters 14 and 21).

Illustrations of insightful applications within post-cognitivism include using pen and paper (McClelland et al., 1986; Norman, 1993), counting with one's fingers, and drawing Venn diagrams. These actions permit us to off-load cognitive cargo into the world. Other examples include gesturing while speaking (Iverson & Goldin-Meadow, 1998), ballistic interception (Smeets & Brenner, 1995), the time course of motor response (Erlhangen & Schöner, 2002), epistemic actions (Kirsh & Maglio, 1994) such as helping yourself in a game of cards by laying the hand out in a particular order, body-based metaphors in thinking and reasoning (Lakoff & Johnson, 1980, 1999), and many others.

Accordingly, post-cognitivist approaches have applied and developed new formal instruments, such as the theory of dynamical systems, imported from the physical sciences. Limb movement is a classic example in the literature, as in the "HKB model" (Kelso et al., 1998) of finger coordination. Kelso (1995) studied the wagging of index fingers and a number of properties were successfully described and predicted dynamically. The phenomenon could be explained as a property of a non-linear dynamical system that achieves self-organization around certain points of instability. Post-cognitivist approaches lay the stress upon real world, time-pressured, situations as the context in which cognition and behavior take place and make sense. This ecological dimension is pivotal, for instance, in ecological psychology (Gibson, 1979). A Gibsonian approach fits nicely with post-cognitivism, insofar as affordances allow for a direct reach that avoids the exploitation of inner representational resources. Allegedly, no information-processing, no abstract symbol crunching, is required, but one can simply tune to environmental invariants through context-sensitive cue extraction and physical adjustment that do not involve a centralized process of control (Chapter 11). We cannot possibly account for a cognitive agent's behavior unless we treat it scientifically on a par with the environment in which the agent is acting.

Similarly, the interactivist developmental psychology of Piaget (1928, 1955) has also been a driving influence, with its processes of accommodation and assimilation that drive the reorganization of the system and the emergence of new cognitive abilities. This interactivist approach has been renewed by Thelen and Smith's dynamic systems approach to development, with its emphasis on decentralized motor development (discharged of the rationality, teleology, and systematicity of Piaget's approach). Remarkable contributions from this approach include the induction of steps in infancy, courtesy of a motorized treadmill (Thelen & Smith,

1994), and the well-known, although still highly controversial, explanation of the "A-not-B error" in infancy (Thelen et al., 2001). In the first case, a spring-like biomechanical response underlies stepping. In the second one, Thelen et al. (2001) go exhaustively over the literature on the A-not-B error, and offer yet another non-cognitivist explanation of motor control and development in that context. In their view, the A-not-B error can be perfectly explained in terms of the dynamical evolution of the coupling of perception, movement and memory, with no need to invoke information-processing concepts or operations (cf. Luo, 2007).

Several chapters in the handbook provide extensive reviews of remarkable research from an embodied view of cognition: Beer (Chapter 6) offers a state of the art status report on the field of robotics; Droll and Hayhoe (Chapter 10) document the interest in studying visually guided motor control and the results obtained; Meteyard and Vigliocco (Chapter 15) extensively review experimental and neurophysiological research on the sensorimotor and propioceptive involvement in semantic understanding; Bergen and Feldman (Chapter 16), Glenberg (Chapter 18), and Núñez (Chapter 17), offer a panorama of the ways in which abstract thought can be grounded in basic sensorimotor abilities.

It must be noted, though, that no single claim carries the full weight of the post-cognitivist research program. In fact, taxonomies and dimensions of embodiment and embeddedness abound in the literature. We need to be aware of the existing diversity, as witnessed by the plurality of ways of understanding the very notions of embodiment and embeddedness. In this regard, Wilson (2002) still provides a useful starting point by making the following distinctions:

(i) Cognition being situated (Thelen & Smith, 1994; Port & van Gelder, 1995; Steels & Brooks, 1995; Chiel & Beer, 1997; Clark, 1997; Pfeifer & Scheier, 1999; Beer, 2000). The emphasis is upon the maintenance of a competency as inputs and outputs relevant to the cognitive process keep impinging on the agent, as opposed to counterfactual thinking, the execution of an offline plan, etc.

But, (ii) cognition is also time-pressured (Brooks, 1991; Port & van Gelder, 1995; Pfeifer & Scheier, 1999). As Kirsh and Maglio's (1994) research on the game of Tetris (and Scrabble, see Maglio et al., 1999) nicely illustrates, players help themselves to the manipulated external environment ("epistemic actions," such as rotating a Tetris piece as it falls on the screen) in order to ease perceptual processing. In time-pressured tasks, the efficiency of rotating the piece, including over-rotations and corrections in real time, by contrast with *imagining* the potential fit of the piece in each specific context, is manifest.

As in the case of epistemic actions, cognitive agents also (iii) off-load cognitive work onto the environment. An example is Ballard et al.'s (1997) "minimal memory strategy" (Chapter 10), where subjects are asked to reproduce patterns of colored blocks, and where visual fixation and re-fixation serves to embody and approximate the experimental task with minimal demands in terms of storage. Glenberg and Robertson's (1999) compass-and-map task is another case in point.

However, laying the stress somewhere else, it has been argued that (iv) the environment itself is part and parcel of cognition (Greeno & Moore, 1993; Thelen &

Smith, 1994; Beer, 1995; see Clark, 1997, for discussion). In Clark and Chalmers' (1998) view, cognition spreads out into the world in a non-trivial way. As they argue, any worldly dimension that contributes to the achievement of a cognitive task and which would count as cognitive, had that contribution come from endogenous processes, should count as a cognitive input. Put bluntly, skin and skull are irrelevant to the identification of a cognitive process. Or as Richardson et al. (Chapter 9) point out, organism–environment systems, and not organisms as such, are the proper units of analysis.

Wilson (2002) distinguishes two more views: (v) Cognition as action, with the fields of vision (Churchland et al., 1994; Ballard, 1996) and memory (Glenberg, 1997) being actively explored. A representative illustration of "cognition as for action" is the identification of the dorsal and ventral visual pathways with "what"/"how" neural routes, instead of "what"/"where" ones (Goodale & Milner, 1992). However, as Smeets and Brenner (Chapter 11) convincingly argue, such a dichotomy is still a tributary of the cognitivist idea of perception as the construction of a visual scene representation (attributed to the "what" route). Instead, what the evidence suggests is that all vision is for action, while different cues may be useful for different tasks, even if they turn out to be inconsistent. One way or the other, the idea of cognition as for action explicitly drives a number of chapters in the handbook.

Finally, (vi) offline cognition is body based, as in mental imagery, and in general, sensorimotor functions are exploited for approximating offline competencies (Dennett, 1995; Glenberg, 1997; Barsalou, 1999; Grush, 2004). Several chapters in this volume also deal with such an idea mainly in relation with meaning and the "grounding problem." Bergen and Feldman (Chapter 16), Meteyard and Vigliocco (Chapter 15), Glenberg (Chapter 18), and Núñez (Chapter 17) all offer complementary approaches to explaining how abstract meaning gets its hold on sensorimotor interaction with the environment. Sanz et al. (Chapter 20) also tackle these issues through their discussion of the hierarchy of control and the "internal model" kind of control, with its use of an "efferent copy," which allows the system to have expectancies on its interaction in a fast way.

These six notions do not stand or fall together. In fact, Wilson (2002) takes (i), (ii), (iii), and (v) to be true, finds (iv) somewhat problematic, and considers (vi) to be the most interesting, although the less explored, conception. We very much agree that special care needs to be taken with (vi), if post-cognitivism is to scale up at all (see Section "Scaling Up: Higher Level Cognitive Processes"). But this sixfold taxonomy is not the only one available, and the emphasis can equally be laid upon another axis. For example, Berkeley (Chapter 5) adopts Ziemke's (2003) distinction between structural coupling, historical embodiment, physical embodiment, "organismoid" embodiment, and organismic embodiment. Sharkey and Sharkey (Chapter 4) also relate this classification of notions of embodiment to a "weak" and "strong embodiment" dichotomy, reminiscent of the "weak/strong AI" distinction, as well as a distinction between mechanistic and phenomenal embodiment. The latter is the stronger version, characteristic of the phenomenological tradition, where the focus is on how the body is felt from within (Chapters 22 and 23). Clark (Chapter 3) defends instead a functionalist understanding of embodiment, where the same cognitive organization may multiply and be realized across different triplets of bodies–brains–environments.

Richardson et al. (Chapter 9) offer yet another six principles central to the ecological perspective: (vii) Organism–environment systems as the proper units of analysis; (viii) a call for the definition of environmental realities at the ecological scale; (ix) behavior as emergent and self-organized; (x) perception and action as continuous and cyclic; (xi) information as specificational, and finally, (xii) perception as of affordances.

In any case, regardless of the minutiae, all these different notions of embodiment may roughly be described as incompatible with at least some central tenet of cognitivism, and *prima facie* reciprocally compatible. They are unified in rejecting the metaphor of cognition as a centralized, information-processing mechanism, but still in the business of interactive control. However, as Anderson (Chapter 21) reminds us, it cannot be taken for granted that they are somehow convergent, and different predictions of each have to be developed to find out how they fare.

Summing up, whereas an information-processing agent counts as a computational system insofar as its state-transitions can be accounted for in terms of manipulations on abstract, amodal, representations, with the related problems of framing, grounding, binding, and the like, the central idea that underlies post-cognitivist programs is a denial or radical transformation of the dogma that our minds must be described as *computing* and/or *representing*, understood as symbol/subsymbol-crunching. However, it remains to be determined whether they are just a heterogeneous cluster of approaches that happen to coincide in the rejection of cognitivism, or whether these different notions and approaches can converge into a unified view of cognition. In other words, whether they only allow for a negative characterization of cognition (what cognition is not), or whether they share a common positive ground. We now turn to this question.

POST-COGNITIVISM IN THE MAKING: COMMON GROUND AND CONCEPTUAL ISSUES

A post-cognitivist interactive and extended architecture is an empirical working hypothesis that needs to be made explicit in operational terms. It is only by looking at the details of what post-cognitivism has to offer that we can assess the extent to which we are confronting something truly different from classical cognitive science. We do think that there are reasons to answer this question in the affirmative, despite the obvious differences of emphasis, point of view, notions of embodiment, and areas of research among the different trends. In this section, we, first, show the common threads that shape the fabric of post-cognitivist programs, as illustrated by the chapters in the volume. Second, underline the conceptual problems that need to be addressed in order to develop this common ground into a well-founded research program.

Although not every author in this volume would agree with every item, this list, in our opinion, captures the central tendencies of the post-cognitivist approach:

- Rather than the topical emphasis on embodiment, even though the interaction is made possible by the body (Chapter 2), it seems to us that interactivism and dynamicism are the central postulates of post-cognitivism. What really unites post-cognitivist approaches is an interactivist and dynamic view of cognition, such that to understand the cognitive system attention has also needs to be paid to the context or environment in which it moves, evolves, develops, and the time course of the interaction, at the different time scales at which it unfolds. This sort of robust, but flexible, interaction is what the term "coupling" refers to.
- This dynamic interaction depends upon the body, in a way that has still to be made more precise and committed—not just physical interaction but social interaction as well (Chapters 4 and 22). Thus, interaction for coupling happens at all levels of physical aggregation.
- This emphasis on interaction brings sensorimotor aspects to the center of the study of cognition. As Pfeifer et al. (Chapter 7) and Richardson et al. (Chapter 9) forcefully argue, the informational structure a system can exploit depends on its bodily constitution in terms of sensors and effectors, materials, morphology, etc.
- Higher cognition is to be understood as constructed from this basic set of restrictions and allowances (plus maybe some new form of control: see next section).
- This standpoint breaks apart the "cognitive sandwich," and makes it clear that perception is active ("enactive"), and action is perceptually guided. Several chapters of this volume develop this theme: Ballard (Chapter 8) presents his latest work on perceptually guided action in a simulated environment; Smeets and Brenner (Chapter 11) show how in taking into account that perception is for action various puzzles dissolve, especially those derived from the idea that perception consists in building a visual representation; Droll and Hayhoe (Chapter 10) review research on visually guided action; and Beer (Chapter 6), and Pfeifer et al. (Chapter 7) review the progress made in robotics in this regard.
- There is also a growing consensus that proper explanation requires the simultaneous scientific understanding of neural, bodily, and environmental factors as they interact with each other in real time. The time course of the interaction and the activation turns out to be crucial to the explanation. New formal methods are called to deal with this requirement, and it is in this regard that the application of dynamical systems theory proves especially appropriate. A model of mental activity must respect the same principles of non-linearity, time dependence, and continuity that are generally invoked in explanations of bodily interactions and neural activity (Freeman, 2000).
- All of the strands also coincide in viewing cognition as an emergent, self-organizing phenomenon, arising out of the local activity of distributed units; no

global plan is required, and there is no single location in the system in control where everything comes together. The notion of criticality of dynamic systems theory helps in this respect by making the notion of emergence non-mysterious. As a matter of fact, it is a rather general natural phenomenon.

• The interactivism of the approach also involves an extended, situated, view of cognition; a clear way to make this point concrete is by saying that the unit of analysis is the system-cum-normal environment, and not the system in isolation.

• This set of basic assumptions naturally implies a transformation of the research questions: what should be studied shifts according to a naturalistic, ecological setting. Instead of artificial tasks in laboratory settings, post-cognitivism studies how the cognitive system deals with its contextual demands: grasping, reaching, interception, navigation, problem solving in the real world, etc. But attention also needs to be paid to consciousness. In doing so, post-cognitivism introduces more complex tasks and settings. In doing so, though, it also avoids artificial complexity, such as the requirement to fixate on a point in visual laboratory experiments, which requires inhibitory control of spontaneous saccades (Chapter 10).

• The formal toolkit of modeling and simulation has been deeply renovated. Logicist approaches recede and are replaced by formal and mathematical approaches that are more appropriate to deal with the interactivity and time dependence of the processes. It is no surprise that mathematical dynamic systems theory, and evolutionary algorithms, have been resorted to (Chapter 13), as well as connectionist modeling that allows for non-linearity and neurobiological plausibility (Chapter 12). An appeal to kinetics has also been useful in accounting for the nature of the forces and movements the body has to exert (Chapters 9 and 11). Explanation thus becomes understanding a behavior as a trajectory in a state space, rather than identifying its ballistic cause.

• Also, neurobiogical plausibility is a must (Chapter 12), as much as interest in bodily detail is, in the configuration of forces and torques (Chapter 7). Moreover, phenomenology, how the processes are experienced, constitutes part of the explananda, something to account for.

This basic set of common postulates, though, is not without problems. For instance, an approach such as the one outlined has to offer a viable notion of scientific explanation. Can explanation boil down to the mathematical description of the range of changes an extended system can experience over time? How does this state-space characterization of the coupled system relate to its mechanistic components; components that interact causally as well as informationally? How can an embodied cognitive science relate to an explanation of the inner mechanisms of sensorimotor coordination that give rise to higher level cognitive activity? It is important to emphasize that the answers to these questions have a direct bearing upon the epistemology of science, with consequences as far as methodologies and the generation of testable predictions are concerned.

Of course, there is disagreement over many other aspects as well. We have already taken stock of the plurality and ambiguity of the notion of embodiment, even though its different aspects need not be seen necessarily as contradictory. However, in some cases they may be in conflict, especially as regards the level of intrinsic dependence between cognition and body. Sheets-Johnstone (Chapter 23) opts for a more strict dependency of the mind on the specifics of the body, along the phenomenological tradition, whereas Clark (Chapter 3) defends a form of extended functionalism that departs from more radical readings of embodiment such as Noë's (2004) or Shapiro's (2004).

Another area where further clarification is required concerns the issue of whether the notion of representation still has a role to play in cognitive explanation. Someone might take away the wrong idea that, by eschewing the information-processing notions of computation and representation, we are throwing out the representational baby with the bathwater (Hayes et al., 1994). Nevertheless, as Bickhard (Chapter 2) points out,[4] embodiment is necessary for representation, and therefore, for cognition. His interactivist project derives representation from action and interaction, which only makes sense if the system is embodied to interact with its medium. In like vein, the reader can see how it bears upon the aforementioned problems. This takes us back to the concept of circular causal flow. Notice once again the contrast with the cognitive sandwich. As the output of a connectionist system exerts no influence upon the input patterns of activation, the system is engaged in no interaction whatsoever (Chapter 2).

Moreover, doubts may arise regarding the convergence of different methodologies. Thus, evolutionary considerations (Sheets-Johnstone, 1990; Nolfi & Floreano, 2000) are congenial with the post-cognitivist take on the aforementioned problems. The modeling of toy embodied, embedded systems by means of evolutionary algorithms constitutes a promising approach (Nolfi & Floreano, 2000) insofar as fitness is measured globally, and no *a priori* decisions as to what belongs to the (cognitive) system need to be made in advance (Beer, 2008). However, we must ask to what extent these modeling strategies add to the neurobiologically plausible models. Although, for obvious reasons, these models will be of little use in the generation of quantitative predictions (Beer, 2008), the reasons for concern run deeper. Insofar as computational neuroethology (Beer, 1990) honors critical biomechanical and ecological aspects, it is certainly a move in the right direction, but we need to know whether neurobiologically plausible artificial neural network architectures and algorithms (Rolls & Treves, 1998) will converge with the statistical analyses of these models. That is, it is *not* simply a question of being able to generate quantitative as opposed to qualitative testable predictions. Rather, it is also a question of methodological convergence with the fast-growing neurosciences.

In closing, it must be emphasized that post-cognitivism aims not merely at cashing out the posits of the information-processing paradigm in trendy mathematical terms, but rather at articulating a brand new way to understand cognition.

[4]For an insightful analysis of the problem of representation see Bickhard and Terveen (1995).

This is not an easy project, and we are well aware that it is certain to be jeopardized if one of these research programs tried to become hegemonic; in other words, the only way for that to happen is real synergistic convergence. Contributing to this goal is the aim of this handbook, and its success will depend upon being able to generate the right intellectual climate and a common agenda. This seems to be required because, although at first sight the articulation of ecological, dynamic, interactive, situated, and embodied approaches within one single framework may look pretty straightforward, one reason for the lack of progress on effective convergence seems to reside in the fact that conceptual issues are usually treated by philosophers, and empirical ones by the rest of the cognitive science community, separately. We need to put together conceptual analysis of the notions of representation, computation, emergence, embodiment, and the like, with empirical work that allows us to bring together ecological, dynamic, interactive, situated, and embodied approaches to the scientific study of cognition. The effort will be comprehensive insofar as it succeeds in unifying a conceptual/empirical framework for the cognitive sciences that allows for conceptual constraints upon the experimental paradigms and contrasting hypotheses, on the one hand, and whose empirical results inform further theoretical developments, on the other hand.

Thus, for this unified approach to consolidate, a systematic and forward-looking approach is also needed, beyond the temptation to just identify post-cognitivism as the alternative to cognitivism. In a controversial paper that appeared in *Science* in the 1960s, Platt (1964) asked what it is that allows some disciplines to make substantial progress in very little time (think of molecular biology, for instance), whereas other areas of research (think this time maybe (?) of cognitive psychology) advance at a slower pace. In Platt's view, it is not a matter of the intrinsic difficulty of the subject (theoretical physics) or of the money injected in the area (high-energy physics). It is instead an intellectual matter that makes up the divide. Whereas all scientific disciplines in their application of the scientific method accumulate inductive inferences in support of the working hypotheses, only in some fields this is done *systematically* ("formally and explicitly and regularly"). In our view, if we want cognitive science to be problem oriented, rather than method oriented, we must be willing to call into question the grounds of post-cognitivism itself in order to make progress. In fact, we need to be systematic, not just in the search for crucial sets of experiments to help us decide between cognitivist and post-cognitivist hypotheses, but rather in the empirical comparison of different post-cognitivist hypotheses (Chapter 21). This is a point that too often goes unnoticed, and unless it is focused upon more thoroughly, the question as to whether post-cognitivism is moving toward an integrated approach cannot be definitely answered in the positive.

It is in this context that the shift of paradigm that post-cognitivism represents should be submitted to critical scrutiny. In the next section, we thus review efforts at discounting post-cognitivism as just a form of cognitivism. In doing so, we also address the problem of scaling up—the ultimate challenge, if post-cognitivism is to present itself as a comprehensive and viable alternative.

SCALING UP: HIGHER LEVEL COGNITIVE PROCESSES

Despite the aforementioned momentum, and the degree of convergence in a unified approach, recent episodes in cognitive science suggest that cognitivism might still resist by assimilating some post-cognitivist methodologies and insights as complementary to its main thrust. We envisage two distinct strategies such a reaction might take. On one hand, it could be claimed that post-cognitivism is not so much an alternative view of cognition, but the right approach to deal with low-level cognitive processes, those involved in sensorimotor coordination, whereas cognitivism is still the right way to approach high-level, symbolic, cognition—even allowing for some kind of grounding of the latter on the former. On the other hand, it could be claimed that the sort of interactivism put forward by post-cognitivism still requires some sort of internal stand-in to account for the causal powers of the mind/brain states involved, so that, in the end, even at the basic sensorimotor level, the commitment to internal representations and computations over them is inescapable. Of course, these strategies are not equally challenging for post-cognitivism. The latter, in fact, tries to absorb the post-cognitivist principles into basic cognitivist architecture, whereas the former amounts to an acceptance of some form of hybridism. In what follows, we address these two issues in reverse order.

Just as an illustration of how this latter strategy may be carried out, consider the response by Vera and Simon (1993a, b) to the Gibsonian and to Brooks' (1991) challenges of "doing without representing" (Clark & Toribio, 1994). In Vera and Simon's view, cognitivism and post-cognitivism need not be antithetical. As a matter of fact, as Vera and Simon argue, Gibson's affordances and Brooks "navigation without representation" approaches should be seen as an illustration of "orthodox symbol systems." As Vera and Simon claim, the information-processing paradigm does not ignore the medium in which cognitive activity takes place. In their view, "the thing that corresponds to an affordance is a symbol stored in central memory denoting the encoding in functional terms of a complex visual display, the latter produced, in turn, by the actual physical scene that is being viewed" (p. 20). Gibsonian affordances, "far from removing the need for internal representations, are carefully and simply encoded internal representations of complex configurations of external objects, the encodings capturing the functional significance of the objects" (p. 41). Commenting on Brooks' line of research, Vera and Simon (1993a) assert that "[sensory] information is converted to symbols which are then processed and evaluated in order to determine the appropriate motor symbols that lead to behavior" (p. 34).

This very strategy of response, consisting of internalizing the relevant interactive relationships as symbolic states, has proliferated and might be recognized in several recent dismissive discussions of post-cognitivism as a real alternative (for instance, Markman & Dietrich, 2000). Although from these comments we cannot conclude much except that the debate is far from settled, it should be acknowledged that this "assimilationist" strategy is facilitated by the lack of a similar explanatory grip on the part of the post-cognitivist challenger. Unfortunately, it is not crystal clear what we mean when we say that post-cognitivist approaches require a *non-symbolic* interpretation of a cognitive system's ecological interactions (Winograd & Flores, 1986). Does for example "non-symbolic" mean "subsymbolic" or "non-representational" *tout court*?

This relates to the question of Representation with a capital "R." Intuitively speaking, online forms of co-variation do not amount to representation (think of the classical example of the sunflower's solar tracking behavior that is interpreted in purely reactive, non-cognitive, terms—Smith, 1996). The explanation would be that there is some exogenous feature that the sunflower manages to keep track of adaptively. But does the distinction between adaptive coupling and "other things" make sense in a full-blown post-cognitivist science? In case representations and mediating states do not vanish altogether, what properties do they have? Must they endure? Does it make sense to talk of enduring states at all? How can a representation be amodal? We are far from reaching a consensus here (Brooks, 1991; Vera & Simon, 1993a, b; Port & van Gelder, 1995; Clark 1997; Beer, 2000; Markman & Dietrich, 2000; Keijzer, 2002). We take it, anyway, that what is needed is rather a different notion of representation, relative to the sort of processes it sustains—so, non-syntactically individuated, not subsymbolically constituted—not as an internal reflection of an external feature, but as what allows for the coupling or interaction. It thus has normativity (Chapter 2), although it has none of the features of cognitivist representation.

This revision of the notion of representation may also be instrumental in foreclosing another interpretation of post-cognitivism as a sort of neo-behaviorism. Thus, someone may argue that the move from cognitivism to post-cognitivism involves a shift, indeed a U-turn, in the status of cognitive science itself (Ramsey, 2007). This claim presupposes both that (i) the emergence of cognitivism, as a reaction to behaviorism, capitalized on the concept of representation, and that (ii) the materialization of post-cognitivism involves a return to some form of pre-cognitivist behaviorism. As Ramsey (2007) puts it, after the cognitive revolution, a "revolution in reverse" (p. 223) is now taking place.[5] We believe that a premature endorsement of this reading may risk misinterpreting the new explanatory principles and models of post-cognitivism. The real issue is whether the traditional Western ways to think of mental representation up until now are the only way to conceive of such internal mediating states. For instance, if one takes Schneegans and Schöner's (Chapter 13) notion of "dynamic fields," the representational jargon adds little to its functioning.

At this point, guidance could be found in the way the parallel problem of how the genome codes the genetic information and controls its expression has been dealt with by development systems theory (Gray et al., 2001). In this area, as well, the standard view of representation drives the idea of a genetic coding of

[5]For an argument that contests Ramsey's analysis of where cognitive science is heading, to the effect that both (i) and (ii) may be called into question, see Calvo and García (submitted).

phenotypic characters, whereas when one realizes that the right unit of analysis is the genome-cum-environment (since the ontogenetic process depends upon the stability of the environment in supplying the chemicals needed), the idea that particular segments code particular characters by themselves loses its grip, because the information lies in the interaction between genetic sequences and robust enabling environment (which may involve the cell as well). To put in a more cognitive way, that the genetic information is context dependent. Of course, this does not diminish at all the reactive causal powers of the DNA sequences; it just underlines the fact that their informational interpretation depends upon the normal environment in which the epigenetic process takes place.

Anyway, the question of the representational nature of the mediating internal states is definitely key to understanding higher level cognition, dependent upon the offline activation of such internal states. It is in the context of higher level capacities that "representation-hungry" cases have to be addressed (Clark, 1999) and that representations appear to be explanatorily unavoidable. In the same vein, Wilson (2002) concluded that what is distinctive in human cognition is the possibility of offline, symbolic, processing, decoupled from current spatio-temporal context. Ballard (Chapter 8) and Berkeley (Chapter 5) also seem to assume that cognitivism may still be useful as regards symbolic, higher level processes, as far as postulating abstract, amodal, representations may keep its explanatory grip. The pressing question, then, is whether, and how, post-cognitivism can account for such a higher level cognition, whether it is able to go beyond simpler forms of adaptive coupling, which only involves some sort of online tracking, and for which it has already proved valid, to account as well for offline, higher level, cognition.

As Wilson (Chapter 19) points out, understanding our higher level abilities has to make evolutionary sense, and in this regard post-cognitivism is better placed than cognitivism. While cognitivism establishes a deep divide between animal cognition and human cognition (given the lesser degree of systematicity, productivity, and flexibility of the former), post-cognitivism tries to overcome all the big traditional dichotomies and thus stresses the elements of continuity with animal cognition, given the shared basic sensorimotor abilities. An evolutionary account is needed, then, to provide an account of how abstract, decoupled, symbolic, thought can emerge out of these basic abilities, and what special conditions of humans restrict the emergent level of cognition to our species. Haselager et al. (Chapter 14) further argue that effective control may be achieved by control systems that co-evolve in relation to constraints in terms of embodiment and embeddedness. Evolutionary considerations again allow for the "cognitive fit" of the extended system.

Wilson (Chapter 19) singles out motor control, analogy, and imitation as the key aspects from which such an account can be worked out. To take them in reverse order: it is clear that our ultrasociality has had something to do with our cognitive make-up, and it is beyond doubt that imitation may be a special kind of social learning in our species, allowing for the social scaffolding that introduces the new members of the species to the social ways of thinking of the group (Tomasello, 1999). The successful notion of a "mirror system" as the brain structure that supports such a competence may have something to do with how we are able to go along with others. But as Gallagher (Chapter 22) forcefully argues, the dominant interpretation of the workings of such a system, as simply the basis of imitation, is deeply contentious (suffice it to remember that macaques, where mirror neurons were first identified, using a single neuron paradigm, do not imitate, what seems to shortcut the link between such neurons and imitation). He proposes what we could term a post-cognitivist view of social interaction as a kind of bodily grounded intentional understanding, while reinterpreting the role of the mirror system in allowing it.

As regards analogy, Wilson refers to the pioneering work of Lakoff on conceptual metaphors, which has been successfully applied to areas as symbolic and abstract as mathematics (Chapters 17 and 18). Other contributors also work along this approach (Chapter 16). In a way, this approach could be seen to the compatible with the kind of symbolic, cognitivist processes, in that it offers a solution to the grounding problem for abstract, amodal representation, which, once constituted could then be worked according to syntactic processes (Barsalou, 1999; for a recent monograph on such an approach, de Vega et al., 2008). However, such an approach is ambiguous with regard to its commitment to cognitivism. We think it ill-advised to try to have it both ways, as an account of the grounding of cognitivist representations which are then submitted to computational processes. As we noted in the first section, such a project runs the risk of getting trapped in the blind alley. In addition, such a cognitivist interpretation fares poorly with the importance of imagination in such processes as depicted by these contributions (present even in the most abstract of problem solving; see Arp, 2008). In fact, what they suggest naturally is an analogical view of internal states, not as perceptual images, but as internal, dynamical, maps (Gomila, 2008).

A fashionable alternative reading would consist in placing such imaginative conceptual abilities in the context of simulation theories, as imagination is currently accounted for as simulation: visual imagination (Kosslyn, 1994), motor imagination (Jeannerod, 1997), empathic imagination (Goldman, 2006), etc. As a matter of fact, this is one of the ways abstract content is supposed to be grounded (Chapters 15–18, refer to this possibility). But as Anderson (Chapter 21) rightly points out, sensorimotor activations associated with higher level cognition cannot be viewed in simulationist terms without further ado. His "massive redeployment hypothesis" contends that this is an instance of a general phenomenon of re-use of structures for new functions. Maybe more important in this regard is to realize that such abilities have to be seen in the context of the third aspect Wilson mentions: a change in the nature of cognitive control, given that imagination involves precisely the sort of voluntary control that she views to be essential for detached, decoupled, abstract, thinking. Following Grush (2004), we think that such abilities are to be better conceived from the point of view of the internal model control architecture (Chapter 20; Gomila, 2007), which postulates that the controller sends an efferent copy to an internal emulator, an internal model of the interaction between systems and environment; such an internal model is what would support offline cognition, being accessed top-down, that is, non-stimulus driven. Of course, this is not the only possible control architecture, and Sanz et al. (Chapter 20) offer a wholesale view of control architectures in terms of levels of control, but the "internal model" architecture seems to be a good place to start an account of higher cognition. The question is how to conceive of such internal models. However, we think that neither classicism nor connectionism offer plausible suggestions, given that such internal models have to work in real time and along the same dimensions as the bodily interactions (sensory feedback, anticipation of propioceptive cues, etc.). On the contrary, proposals such as the dynamic field (Chapter 13) or neurobiologically plausible Hebbian networks (Chapter 12) provide an illustration of how such internal models can be conceived such that they can be viewed as representations, or thought to compute, in a non-contentious, mathematical way. It is clear anyway that complex cognition requires a complex integrated system, and that such a system requires forms of control that are not purely distributed and reactive (without being committed, for such reason, to postulating a "central executive" or anything homunculus-like).

Thus, instead of opting for a hybrid view of human cognition, the possibility exists to reinterpret higher level, decoupled, cognition, in terms of post-cognitivist principles. In our view, cognitivism, both in its classical version and in its connectionist form, is unable to deal fully with the dichotomy regarding cognition that the games of chess and pool example served to illustrate at the outset of this chapter. It is only when post-cognitivist models of cognition enter the picture that an answer to our opening question can begin to be given. Roughly speaking, the working hypothesis of post-cognitivism is that higher level cognitive activity never goes completely formal. Instead, it remains a "game of pool," we may say, in which non-formal perception-action activation patterns are ubiquitous.

As Kuhn taught us, though, it is only when a new paradigm is ready that the old one will begin to be overcome. It is tempting to ask whether there is anything here really deserving to be called a "new paradigm" yet. We would actually like to try to avoid too much talk in terms of paradigms and paradigm shifts. What really matters is the theoretical significance of the new framework. Our point here is: let us unify this research effort into a single framework. We want to assess the significance of an integrated, embodied cognitive science, and invite others to explore the path. Enjoy the ride.

ACKNOWLEDGMENTS

We are grateful to Mike Anderson, Aarre Laakso, Peter Slezak, and Meg Wilson for helpful comments and suggestions on a previous version of this chapter. Preparation of the manuscript was supported by DGICYT Project HUM2006-11603-C02-01 (Spanish Ministry of Science and Education and European Union FEDER Funds).

REFERENCES

Anderson, J. (1993). *Rules of the Mind.* Hillsdale, NJ: Erlbaum.

Anderson, J. & Lebiere, C. L. (2003). The Newell test for a theory of cognition. *Behavioral and Brain Sciences, 26,* 587–637.

Anderson, M. L. (2003). Embodied cognition: A field guide. *Artificial Intelligence, 149*(1), 91–130.

Arkin, R. C. (1998). *Behavior-Based Robotics.* Cambridge, MA: MIT Press.

Arp, R. (2008). *Scenario Visualization: An Evolutionary Account of Creative Problem Solving.* Cambridge, MA: MIT Press.

Ballard, D., Hayhoe, M., Pook, P., & Rao, R. (1997). Deictic codes for the embodiment of cognition. *Behavioral and Brain Sciences, 20,* 723–767.

Ballard, D. H. (1991). Animate vision. *Artificial Intelligence, 48,* 57–86.

Ballard, D. H. (1996). On the function of visual representation. In K. A. Akins (Ed.), *Perception* (pp. 111–131). Oxford: Oxford University Press.

Barsalou, L. W. (1999). Perceptual symbol systems. *Behavioral and Brain Sciences, 22,* 577–609.

Beer, R. D. (1990). *Intelligence as Adaptive Behavior: An Experiment in Computational Neuroethology.* Academic Press. Boston: Massachusetts.

Beer, R. D. (1995). A dynamical systems perspective on agent-environment interaction. *Artificial Intelligence, 72,* 173–215.

Beer, R. D. (2000). Dynamical approaches to cognitive science. *Trends in Cognitive Sciences, 4*(3), 91–99.

Beer, R. D. (2008). Beyond control: The dynamics of brain–body–environment interaction in motor systems. In D. Sternad (Ed.), *Progress in Motor Control: A Multidisciplinary Perspective.* Berlin: Springer.

Bickhard, M. H. & Terveen, L. (1995). *Foundational Issues in Artificial Intelligence and Cognitive Science: Impasse and Solution.* Amsterdam: Elsevier Scientific.

Brooks, R. A. (1986). A robust layered control system for a mobile robot. *IEEE Journal of Robotics and Automation, 2*(1), 14–23.

Brooks, R. A. (1991). Intelligence without representation. *Artificial Intelligence, 47,* 139–159.

Brooks, R. A. (1999). *Cambrian Intelligence.* Cambridge, MA: MIT Press.

Calvo, P. (2003). Non-classical connectionism should enter the decathlon. *Behavioral and Brain Sciences, 26,* 603–604.

Calvo, P. (2005). Rules, similarity, and the information-processing blind alley. *Behavioral and Brain Sciences, 28,* 17–18.

Calvo, P. & García, A. (submitted). Where is cognitive science heading?

Chalmers, D. J. (1990). Why Fodor and Pylyshyn were wrong: The simplest refutation, *Proceedings of the twelfth annual conference of the Cognitive Science Society* (pp. 340–347). Hillsdale, New Jersey: Lawrence Erlbaum Associates.

Chemero, A. (in press). *Radical Embodied Cognitive Science*, MIT Press.

Chiel, H. & Beer, R. (1997). The brain has a body: Adaptive behaviour emerges from interactions of nervous system, body and environment. *Trends in Neurosciences, 20,* 553–557.

Chomsky, N. (1959). Review of Skinner's verbal behavior. *Language, 35*(1), 26–57.

Chomsky, N. (1980). *Rules and Representations*, New York: Columbia University Press. Oxford: Basil Blackwell Publisher.

Chomsky, N. (1996). *Powers and Prospects. Reflections on Human Nature and the Social Order.* Boston: South End Press.

Churchland, P. S., Ramachandran, V. S., & Sejnowski, T. J. (1994). A critique of pure vision. In D.Koch (Ed.), *Large-Scale Neuronal Theories of the Brain* (pp. 23–60). Cambridge, MA: Bradford Book/MIT Press.

Clancey, W. J. (1997). *Situated Cognition.* Cambridge: Cambridge University Press.

Clark, A. (1997). *Being There: Putting Brain, Body and World Together Again.* Cambridge, MA: MIT Press.

Clark, A. (1999). An embodied cognitive science? *Trends in Cognitive Sciences, 3*(9), 345–351.

Clark, A. & Chalmers, D. (1998). The extended mind. *Analysis, 58*(1), 7–19.

Clark, A. & Toribio, J. (1994). Doing without representing? *Synthese, 101*, 401–431.

Damasio, A. (1994). *Descartes' Error: Emotion, Reason, and the Human Brain*. New York: Avon Books.

de Vega, M., Glenberg, A., & Graesser, A. (Eds.) (2008). *Symbols and Embodiment. Debates on Meaning and Cognition*. Oxford: Oxford University Press.

Dennett, D. (1995). *Darwin's Dangerous Idea: Evolution and the Meanings of Life*, New York: Simon & Schuster. London: Penguin.

Dreyfus, H. (1992). *What Computers Still Can't Do*. Cambridge, MA: MIT Press.

Elman, J. L. (1990). Finding structure in time. *Cognitive Science, 14*, 179–211.

Elman, J. L. (1998). Generalization, simple recurrent networks, and the emergence of structure. In M. A. Gernsbacher & S. J. Derry (Eds.), *Proceedings of the Twentieth Annual Conference of the Cognitive Science Society*. Mahwah, NJ: Lawrence Erlbaum Associates.

Erlhagen, W. & Schöner, G (2002). Dynamic field theory of movement preparation. *Psychological Review, 109*, 545–572. 200.

Feldman, J. A. & Ballard, D. H. (1982). Connectionist models and their properties. *Cognitive Science, 6*, 205–254.

Fodor, J. (1980). Methodological solipsism considered as a research strategy in cognitive psychology. *Behavioral and Brain Sciences, 3*, 63–110.

Fodor, J. (2000). *The Mind Doesn't Work that Way: The Scope and Limits of Computational Psychology*. Cambridge, MA: MIT Press.

Fodor, J. & Pylyshyn, Z. (1988). Connectionism and Cognitive Architecture. *Cognition, 28*, 3–71.

Ford, K. M., Hayes, P. J., & Agnew, N. (1994). On babies and bathwater: A cautionary tale. *AI Magazine, 15*(4), 14–26.

Freeman, W. J. (1999). *How Brains Make Up Their Minds*. London: Weidenfeld and Nicolson.

Freeman, W. J. (2000). *Neurodynamics: An Exploration of Mesoscopic Brain Dynamics*. London, UK: Springer-Verlag.

Gallagher, S. (2005). *How the Body Shapes the Mind*. New York: Oxford University Press.

Gardner, H. (1985). *The Mind's New Science: A History of the Cognitive Revolution*. New York: Basic Books.

Gibson, J. J. (1966). *The Senses Considered as Perceptual Systems*. Boston: Houghton Mifflin.

Gibson, J. J. (1979). *The Ecological Approach to Visual Perception*. Hillsdale, NJ: Lawrence Erlbaum.

Glenberg, A. M. (1997). What memory is for. *Behavioral and Brain Sciences, 20*(1), 1–55.

Glenberg, A. M. & Robertson, D. A. (1999). Indexical understanding of instructions. *Discourse Processes, 28*(1), 1–26.

Goldman, A. (2006). *Simulating Minds: The Philosophy, Psychology and Neuroscience of Mindreading*. Oxford: Oxford University Press.

Gomila, A. (2007). Executive Awareness: De-constructing the Homunculus. *Estudios de Psicología, 28*, 211–230.

Gomila, A. (2008). Mending or abandoning cognitivism? In M. de Vega, A. Glenberg & A. Graesser (Eds.), *Symbols and Embodiment. Debates on Meaning and Cognition* (pp. 789–834). Oxford: Oxford University Press.

Goodale, M. A. & Milner, A. D. (1992). Separate visual pathways for perception and action. *Trends in Neurosciences, 15*, 20–25.

Gray, R., Griffiths, P., & Oyama, S. (Eds.) (2001). *Cycles of Contingency*. Cambridge, MA: MIT Press.

Greeno, J. (1996). On claims that answer the wrong questions (Response to Anderson et al.). *Educational Researcher, 25*, 5–11.

Greeno, J. C. & Moore, J. L. (1993). Situativity and symbols: Response to Vera and Simon. *Cognitive Science, 17*(1), 49–59.

Grush, R. (2004). The emulation theory of representation: Motor control, imagery, and perception. *Behavioral and Brain Sciences, 27*, 377–442.

Hanna, R. & Thompson, E. (2003). The problem of consciousness: New essays in phenomenological philosophy of mind. In E. Thompson (Ed.), *Neurophenomenology and the Spontaneity of Consciousness* (pp. 133–161). Calgary, Alberta, Canada: University of Calgary Press.

Haugeland, J. (1998). *Having Thought: Essays in the Metaphysics of Mind.* Cambridge, MA: Harvard University Press.

Hebb, D. O. (1949). *The Organization of Behavior.* New York: John Wiley.

Heidegger, M. (1962). *Being and Time.* New York: Harper & Row.

Hurley, S. L. (1998). *Consciousness in Action.* Cambridge, MA: Harvard University Press.

Hutchins, E. (1995). *Cognition in the Wild.* Cambridge, MA: MIT Press.

Iverson, J. M. & Goldin-Meadow, S. (1998). Why people gesture as they speak. *Nature, 4,* 228–396.

Jeannerod, M. (1997). *The cognitive neuroscience of action.* London: Blackwell.

Keijzer, F. (2002). Representation in dynamical and embodied cognition. *Cognitive Systems Research, 3,* 275–288.

Kelso, S. (1995). *Dynamic Patterns.* Cambridge, MA: MIT Press.

Kelso, S., Fuchs, A., Lancaster, R., Holroyd, T., Cheyne, D., & Weinberg, H. (1998). Dynamic cortical activity in the human brian reveals motor equivalence. *Nature, 392,* 814–818.

Kirsh, D. (1991). Foundations of AI: The big issues. *Artificial Intelligence, 47*(1–3), 3–30.

Kirsh, D. & Maglio, P. (1994). On distinguishing epistemic from pragmatic actions. *Cognitive Science, 18*(4), 513–549.

Kosslyn, S. (1994). *Image and Brain: The Resolution of the Imagery Debate.* Cambridge, MA: MIT Press.

Laakso, A. & Calvo, P. (2008). A connectionist simulation of structural rule learning in language acquisition. *Proceedings of the 30th Annual Conference of the Cognitive Science Society.* Mahwah, NJ: Lawrence Erlbaum Associates.

Lakoff, G. & Johnson, M. (1980). *Metaphors We Live By.* Chicago, IL: University of Chicago Press.

Lakoff, G. & Johnson, M. (1999). *Philosophy in the Flesh.* Cambridge, MA: MIT Press.

Langacker, R. W. (1987). *Foundations of Cognitive Grammar: Theoretical Prerequisites.* Stanford, CA: Stanford University Press.

Langacker, R. W. (1991). Foundations of Cognitive Grammar. Descriptive Application. Stanford: Stanford University Press. Vol. II.

Luo, J. (2007). Object permanence as relational stability or how to get representation from the dynamics of embodiment, *Proceedings of the 29th Annual Conference of the Cognitive Science Society.* Mahwah, NJ: Lawrence Erlbaum Associates.

Maglio, P., Matlock, T., Raphaely, D., Chernicky, B., & Kirsh, D. (1999). Interactive skill in Scrabble, *Proceedings of Twenty-first Annual Conference of the Cognitive Science Society,* Mahwah, NJ: Lawrence Erlbaum.

Marcus, G. F. (2001). *The Algebraic Mind: Integrating Connectionism and Cognitive Science.* Cambridge, MA: MIT Press.

Marcus, G. F., Vijayan, S., Bandi Rao, S., & Vishton, P. M. (1999). Rule learning in seven-month-old infants. *Science, 283,* 77–80.

Markman, A. B. & Dietrich, E. (2000). Extending the classical view of representation. *Trends in Cognitive Sciences, 4,* 470–475.

Marr, D. (1982). *Vision.* New York: W.H. Freeman.

McClelland, J. L. & Rumelhart, D. E. and the PDP Research Group (1986). Parallel Distributed Processing: Explorations in the Microstucture of Cognition. Cambridge, MA: MIT Press. Vol. 2.

McCulloch, W. & Pitts, W. (1943). A logical calculus of the ideas immanent in nervous activity. *Bulletin of Mathematical Biophysics, 7,* 115–133.

Merleau-Ponty, M. (1962). A sensorimotor account of vision and visual consciousness. In C. Smith (Ed.), *Phenomenology of Perception.* London: Routledge & Kegan Paul.

Minsky, M. & Papert, S. (1969). *Perceptrons.* Cambridge, MA: MIT Press.

Murphy, R. R. (2000). *Introduction to AI robotics.* Cambridge, MA: The MIT Press.

Newell, A. (1980). Physical symbol systems. *Cognitive Science, 4,* 135–183.

Newell, A. & Simon, H. A. (1972). *Human Problem Solving*. Englewood Cliffs, NJ: Prentice Hall.

Noë, A. (2004). *Action in Perception*. Cambridge, MA: The MIT Press.

Nolfi, S. & Floreano, D. (2000). *Evolutionary Robotics: The Biology, Intelligence, and Technology of Self-Organizing Machines*. Cambridge, MA: MIT Press.

Norman, D. A. (1993). *Things That Make Us Smart*. New York: Addison-Wesley.

O'Regan, K. & Noë, A. (2001). A sensorimotor account of vision and visual consciousness. *Behavioral and Brain Sciences, 24*(5), 883–917.

Peña, M., Bonatti, L., Nespor, M., & Mehler, J. (2002). Signal-driven computations in speech processing. *Science, 298*, 604–607.

Perruchet, P., Tyler, M. D., Galland, N., & Peereman, R. (2004). Learning nonadjacent dependencies: No need for algebraic-like computations. *Journal of Experimental Psychology: General, 133*, 573–583.

Pfeifer, R. & Scheier, C. (1999). *Understanding Intelligence*. Cambridge, MA: MIT Press.

Piaget, J. (1928). *The Child's Conception of the World*. London: Routledge and Kegan Paul.

Piaget, J. (1955). *The Child's Construction of Reality*. London: Routledge and Kegan Paul.

Pinker, S. & Ullman, M. (2002). The past and future of the past tense. *Trends in Cognitive Sciences, 6*, 456–463.

Platt, J. (1964). Strong Inference: Certain systematic methods of scientific thinking may produce much more rapid progress than others. *Science, 146*, 3642. Comment: Number

Port, R. & Van Gelder, T. (1995). *Mind as Motion*. Cambridge, MA: MIT Press.

Ramscar, M. (2002). The role of meaning in inflection: Why the past tense doesn't require a rule. *Cognitive Psychology, 45*, 45–94.

Ramsey, W. M. (2007). *Representation Reconsidered*. New York: Cambridge University Press.

Regier, T. (1996). *The Human Semantic Potential: Spatial Language and Constrained Connectionism*. Cambridge, MA: MIT Press.

Resnick, M. (1994). *Turtles, Termites and Traffic Jams*. Cambridge, MA: MIT Press. Vol. 1.

Rolls, E. T. & Treves, A. (1998). *Neural Networks and Brain Function*. Oxford: Oxford University Press.

Rosenblatt, F. (1959). *Principles of Neurodynamics*. New York: Spartan Books.

Rowlands, M. (2006). *Body Language: Representation in Action*. Cambridge, MA: MIT Press.

Rumelhart, D. E. & McClelland, J. L. and the PDP Research Group (1986). Parallel Distributed Processing: Explorations in the Microstructure of Cognition. Cambridge, MA: MIT Press.

Seidenberg, M. & Elman, J. L. (1999). Networks are not hidden rules. *Trends in Cognitive Sciences, 3*, 288–289.

Semin, G. R. & Smith, E. R. (2002). Interfaces of social psychology with situated and embodied cognition. *Cognitive Systems Research, 3*, 385–396.

Shapiro, L. (2004). *The Mind Incarnate*. Cambridge, MA: MIT Press.

Sheets-Johnstone, J. & Maxine, G. (1990). *The Roots of Thinking*. Philadelphia: Temple University Press.

Skarda, C. A. & Freeman, W. J. (1987). How brains make chaos to make sense of the world. *Behavioral and Brain Sciences, 10*, 161–195.

Smeets, J. B. J. & Brenner, E. (1995). The visual guidance of ballistic arm movements. In T. Mergner & F. Hlavacka (Eds.), *Multisensori Control of Posture* (pp. 191–197). New York: Plenum Press.

Smith, B. C. (1996). *On the Origin of Objects*. Cambridge, MA: MIT Press.

Steels, L. & Brooks, R. A. (Eds.) (1995). *The Artificial Life Route to Artificial Intelligence: Building Embodied Situated Agents*. Hillsdale, NJ: Lawrence Erlbaum Associates Inc.

Suchman, A. (1987). *Plans and Situated Actions*. New York: Cambridge University Press.

Thelen, E. & Smith, L. (1994). *A Dynamic Systems Approach to the Development of Cognition and Action*. Cambridge, MA: MIT Press.

Thelen, E., Schöner, G., Scheier, C., & Smith, L. (2001). The dynamics of embodiment: A field theory of infant perseverative reaching. *Behavioral and Brain Sciences, 24*, 1–33.

Thompson, E. (1995). *Colour Vision: A study in Cognitive Science and the Philosophy of Perception.* London: Routledge.

Tomasello, M. (Ed.) (1998). *The New Psychology of Language: Cognitive and Functional Approaches to Language Structure*. Hillsdale, NJ: Lawrence Erlbaum.

Tomasello, M. (1999). *The Cultural Origins of Human Cognition.* Cambridge: Harvard University Press.

Turvey, M. & Carello, C. (1995). Some dynamical themes in perception and action. In R. Port & T. Van Gelder (Eds.), *Mind as Motion* (pp. 373–401). Cambridge, MA: MIT Press.

Varela, F., Rosch, E., & Thompson, E. (1991). *The Embodied Mind.* Cambridge, MA: MIT Press.

Vera, A. H. & Simon, H. A. (1993a). Situated action: A symbolic interpretation. *Cognitive Science, 17,* 7–48.

Vera, A. H. & Simon, H. A. (1993b). Situated action: Reply to William Clancey. *Cognitive Science, 17,* 117–133.

Wheeler, M. (2005). *Reconstructing the Cognitive World: The Next Step.* Cambridge, MA: MIT Press.

Wilson, M. (2002). Six views of embodied cognition. *Psychonomic Bulletin and Review, 9,* 625–636.

Winograd, T. & Flores, C. F. (1986). *Understanding Computers and Cognition. A New Foundation for Design*. Norwood: Ablex.

Wittgenstein, L. (1953). *Philosophical Investigations.* Oxford: Basil Blackwell. (Anscombe, G.E.M., trans.)

Ziemke, T. (2003). What's that thing called embodiment? *Proceedings of the 25th Annual Meeting of the Cognitive Science Society.* Philadelphia, PA: Lawrence Erlbaum.

SECTION I

THE INTERACTIVE ARCHITECTURE OF COGNITION: CONCEPTUAL ISSUES

2

Is Embodiment Necessary?

Mark H. Bickhard

Department of Philosophy, Lehigh University, Bethlehem, PA, USA

Every known instance of a genuine cognitive agent is embodied, and it is clear that embodiment has a major influence on, and can be a major help for, artificial agents (Pfeifer & Scheier, 1999). It is also clear that the notion of embodiment has multiple interpretations (Ziemke, 2001; Wilson, 2002; Svensson & Ziemke, 2005). But these points leave open a basic question: Is there a sense of embodiment in which being embodied is *necessary* to cognition? Is embodiment in some way necessary to the nature of cognition, or is it (merely) an important but secondary consideration for (most) implementations of cognitive agents (whether biological or artificial)? I will argue that embodiment is in fact necessary—it is essential in the nature of representation, and, therefore, of cognition.

There are several parts to this argument. First, there are considerations of approaches to the modeling of representation that do not have as consequences any such necessity of embodiment. I will argue that these approaches do not and cannot succeed. Second, there is the development of an alternative model of the nature of representation and cognition. This will be outlined, and it is clear that it requires embodiment, because it requires genuine interaction between a cognitive agent and the world. Finally, a word or two will be in order to look at the *kind* of embodiment that is involved in this *interactive* approach to representation and cognition.

CRITIQUES

Critiques of standard models of representation can be partially compressed because they are all heirs to an underlying error. This error has many manifestations, some of which have been known for millennia and some of which have been discovered relatively recently. Because the central error is held in common among the various models of representation on offer in the literature, only minimal particularization of the central critiques is required in order to demonstrate the applicability of those critiques to specific models.

This central error (or family of errors) has to do with the normativity of representation: the sense in which representation can be true or false. One criterion for a model of representation is that the model be able to account for the simple possibility that the representation is in error. This can be difficult if the representational relationship is purportedly constituted in some sort of factual relation between representation and represented—for example, a causal or informational or nomological relation—because the proper factual correspondence to constitute a representation cannot exist unless the environmental end of the correspondence exists, so that the representation of that existence, in such a view, *must* be correct. There have been multiple attempts to avoid this problem,[1] but I will focus primarily on a strengthened variant of it that has *not* been addressed.

This variant is the criterion of being able to account for *organism* (or *system*) *detectable* error—that the system can itself detect its own errors. Such possibilities of detection may be restricted to certain kinds of organism complexity, and may be quite fallible, but we know that they occur, so any model of representation that cannot account for the possibility of such detections is at best incomplete, and any model that precludes such detections is refuted. This criterion is, in fact, of central importance to any complex cognitive system because system detectable error is necessary in order for error guided behavior and learning to occur, and error guided behavior and learning underpin major portions of most species' cognitive world.

Nevertheless, there is no attempt to address this criterion in the major approaches to representation in the literature. One reason why, so I argue, is that there is no major approach that can possibly account for system detectable error.

Some sense of the depth of the problem posed by this criterion can be found by realizing that it is equivalent to the radical skeptical argument: we cannot check our own representations for truth or falsity because, in order to do so, we would have to step outside of ourselves to obtain independent epistemic access to what we are attempting to represent and then compare what we are trying to represent with our attempted representation of it (Rescher, 1980). We cannot step outside of ourselves, so this is impossible—therefore, so this argument goes, checking our representations for error is impossible. Again, however, we know that error guided behavior and learning occur, so there must be something wrong with this argument. It is not "merely" an armchair philosophical argument: it is a long-standing manifestation of an error in fundamental assumptions about representation.

[1] Without success, I argue elsewhere (Bickhard, 1993, 2004a, in press, in preparation; Bickhard & Terveen, 1995).

The radical skeptical argument also illustrates a second, related, criterion: we must be able to compare represented with representation, so we must have access to our own representational contents—we (or the organism, or system, or agent, or central nervous system, etc.) must have access not only to the represented (which is what the skeptical argument focuses on) but also to the *content* that we are applying to the represented. We must have access to this content in order to make the comparison in order to determine whether or not the content truly applies to the represented. So, in order to engage in any such comparison, we must have access to both sides that are to be compared. Most models fail this second criterion as well as the first.

FODOR

With this pair of criteria in mind, then, I will take a look at some of the central contemporary models of representation, and I begin with Jerry Fodor. Fodor's model is a version of an information semantics, with the crucial representation constituting relationship, in this case, being a nomological relationship between the represented and the representing state in the organism (Fodor, 1987, 1990, 1991, 1998, 2003). As such, the model encounters difficulties accounting for the possibility of representational error, and Fodor has proposed an ingenious attempted solution.

Fodor's model turns on the intuition that false evocations of a representation are dependent on correct evocations, but that there is no reverse dependency—the dependency is asymmetric. Thus, a cow representation may be evoked by cows, but also perhaps by a horse on a dark night. But the possibility of evocation by the horse is dependent on the possibility of evocation by cows, and this dependency is not reciprocated: evocations by cows could continue even if there were never any possibility of evocations by horses.

One problem with this proposal is that such asymmetric dependencies, even among nomological relationships, do not suffice to pick out representation at all. For example, a neurotransmitter docking with a receptor triggers ensuing activity in a nomological manner, and a mimicking poison molecule docking with the same receptor also evokes nomologically related activity—and the poison molecule's possibility of such evocation is asymmetrically dependent on the possibility of the neurotransmitter evoking such activity. Yet there is at best a biologically functional relationship here, not a representational relationship (Levine & Bickhard, 1999).

Setting this concern aside, however, we still find that the alleged representational error is characterizable as error only for an external observer who could (1) (supposedly) determine the counterfactual asymmetries involved among various families of nomological relationships in order to characterize what a representation is supposed to represent—that is, to characterize the content, (2) epistemically access the represented, and (3) compare the two in order to determine whether the content holds of the represented.

Note that this external observer is in precisely the position that the radical skeptical argument points out that no actual epistemic agent can be in for itself. First, no epistemic agent can have access to its own relevant counterfactual nomological relations in order to determine content, and, second, to access the represented in order to make the comparison with the content is precisely the representational problem all over again. This is the circularity that is at the center of the skeptical argument.

For Fodor, the consequence is that content is not accessible, the represented is not independently accessible, and system detectable error is therefore impossible. Consequently, error-guided behavior and learning are not possible. But error-guided behavior and learning occur, therefore the model is refuted.

MILLIKAN

In Millikan's etiological model, representing X is a particular kind of function that some things or conditions might have, and having such a function is constituted in having (or being properly derived from something that has) the right kind of evolutionary selection history (Millikan, 1984, 1993). The crucial selection history is one of undergoing a sufficient number of generations of selection for having the (functional) consequence in question.[2] In this sense, having a function is constituted, roughly, in being designed to have that function by evolutionary selection as the designer.

I would first note that this approach does not have the problem of accounting for the possibility of representational error *per se*: what something is supposed to represent—content—is determined by evolutionary history, while what is being represented is in the present. The two are thus pulled apart in a way that permits the possibility that the content will be incorrectly applied to a present entity or state of affairs—in a way that is false. This is a distinct advantage over information semantic approaches. The etiological approach, however, ultimately does not succeed either.

One specific problem with this approach is that it renders function, thus representation, causally epiphenomenal: having a function is constituted in the past, not in the present, and present state of a system is not specific to having that requisite past evolutionary history. This is illustrated by Millikan's example of a lion that pops into existence that is molecule by molecule identical to a lion in the zoo: the lion in the zoo has all the right evolutionary histories, therefore its organs and processes have functions, including functions of representing, but the science fiction lion has no evolutionary history, therefore no functions at all,

[2] The number of generations required is a matter of discussion and dispute (Godfrey-Smith, 1994). This literature cannot use the locution above of having a functional consequence, and must make the point much more indirectly, because no such consequence can be a functional consequence until the requisite number of generations of selections have transpired.

whether representational or not. Yet the two lions, by assumption, are causally, dynamically, identical: etiological function makes no causal difference.

This is a thought experiment, but an equivalent actual case occurs every time in evolutionary history that something occurs for the first time and is selected for. If sufficient generations of such selections ensue, then whatever organ is involved will come to have that consequence as its function, and it will serve that function when it produces that consequence (in the right conditions). But this first time and all subsequent times till the magic of constituting a function occurs are instances in which identical (or extremely similar) consequences occur but are not (etiologically) functional. So, again, we have some systems that are causally identical to others, yet do not have functions, whereas the others do have functions. Again we have that etiological function is causally epiphenomenal.

To return now to the general critique, no organism has access to the evolutionary histories of its parts, therefore no organism (or system) has access to its own representational contents. Therefore system detectable error is not possible.

Furthermore, the problem of access to what is being represented is identical in this case to the case for information semantics: that is the fundamental representational problem all over again. Again we encounter the fundamental circularity of the radical skeptical argument.[3]

THE SYMBOL SYSTEM HYPOTHESIS

Cognitive science was long dominated by computational approaches in which relevant processes were symbol manipulation processes (Franklin, 1995), and certainly such approaches are still very prevalent. There are, obviously, many powers and advantages afforded by such design approaches over, for example, simple associationistic approaches, but, regarding representation *per se*, they are hopeless. Basic representations in such models are taken to represent something in virtue of being in a correspondence with that something—a correspondence that somehow encodes its distal end—with the crucial correspondence variously taken to be one of a "stand-in" or perhaps a structural isomorphism (Newell, 1980; Vera & Simon, 1993). But such models cannot account for the bare possibility of representational error, and have no way to address the possibility of system detectable representational error (Bickhard, 2004a, in press, in preparation). If the crucial representational (encoding) relationship exists, then it is correct, and if it does not exist, then the representation does not exist, and there is no third possibility for modeling the representation existing but being incorrect.

[3]Dretske's approach to representation (Dretske, 1988) is also an etiological approach, though with a learning etiology instead of an evolutionary etiology, and, so suffers from similar problems. For more detailed analyses of Fodor's and Millikan's models, as well as those of Dretske (1988) and Cummins (1996), see Bickhard (2004a, in press, in preparation).

CONNECTIONIST REPRESENTATION

There was great hope for connectionist approaches in the 1980s because, among other reasons, they offered a seemingly natural way to realize learning processes. Certainly, connectionist models accommodate training in ways that symbol manipulation processes do not, but, again, with respect to representation *per se*, they do not offer any real advantages. In particular, a trained connectionist net establishes an informational correspondence with some class of input patterns, and, therefore, is a version of an information semantical approach.[4] Therefore, they suffer all of the same problems as do other versions, including the impossibility of system detectable error: no connectionist system has access to its own informational relationships, nor to what those relationships might be with, in order to be able to make any relevant comparisons regarding the possibility of error.[5]

AGENTIVE ANTI-REPRESENTATIONALISM

With the development of dynamic and agentive approaches in the late 1980s and 1990s, a strong anti-representationalism emerged. The very concept of representation was argued to be unnecessary and even misleading in designing and understanding complex organisms and systems (Brooks, 1991; van Gelder, 1995). One of the stronger *counter* arguments turned on what were called representation-hungry situations, in which representational tracking, such as of a hidden predator, was required (Clark & Toribio, 1995). These discussions were confused because different detailed conceptions of what constituted representations were often involved: for example does tracking *per se* constitute representation or is representation constituted only in manipulable symbols that track?

In either case, the underlying conceptions of representation were of the basic representational correspondence sort—something that encodes—and did nothing to avoid the fundamental problems of accounting for the possibility of error and of system detectable error. That is, my claim here is that both the anti-representationalists and the representationalists made equivalent fundamental assumptions about the nature of representation, and that both were wrong—neither could account for the normativities, the endogenous truth values, of representation.

If this is correct, then clearly the issues regarding the embodiment of cognitive systems cannot be properly addressed within such frameworks. In fact, issues of embodiment of cognitive systems cannot be properly addressed within

[4] Ironically, this model differs from Fodor's primarily in that Fodor's transductions are nomological, while connectionist nets are trained. In at least one of Fodor's incarnations, transduction is the evocation of representation via some process other than inference, and, by this definition, trained connectionist nets *are* Fodorian transducers (Bickhard, 1993; Bickhard & Terveen, 1995).

[5] For extensive discussion of connectionist systems, see Bickhard and Terveen (1995).

any of the frameworks that have dominated the history of cognitive science:[6] representation as correspondence or representation as encoding is a strictly input processing notion of representation. In any such view, there is no necessary involvement of any body, other than to house an input processing system. There is no necessity for action, thus none for a body that can act. In such a view, then, embodiment can at best constitute a source of constraint, quite possibly enabling constraint, on processing and on action, but can have no deep relationships with representation *per se*.

INTERACTIVE REPRESENTATION

This is in strong contrast to models of representation that derive representation from action and interaction, from general pragmatist principles (Rosenthal, 1983; Joas, 1993). If representation, thus cognition, is derived from (inter)action, then some sort of embodiment is required in order for such action and interaction to be possible: actions, thus embodiment, are not mere auxiliaries to representation, but, instead, are essential to it.

The task at this point, therefore, is to outline such an action-based model of representation, and show that it is a viable alternative to the passive, encoding correspondence, models addressed above. In particular, the task is to show that the model can account for the possibility of representational error and of system detectable error.

Consider a complex agent—animal or artificial. It faces an ongoing task of selecting what interactions to engage in and of guiding interactions underway. In order to select (or guide), there must be available in some functionally accessible way indications of what interactions are possible: it does no good to select opening the refrigerator to get something to eat if you are in the forest, miles from the nearest refrigerator. A frog, for another example, might have two possibilities for flicking its tongue and eating a fly, one possibility for eating a worm and one for jumping in the water. It does no good for the frog to select a tongue-flicking interaction for eating a fly when no detection of anything that might support such tongue-flicking-and-eating has occurred.

There are two aspects to this point: (1) some means for indicating interaction possibilities and (2) some way of selecting among those possibilities. Both aspects are of fundamental importance, but I will be focusing on the indications here.[7]

[6] There is a more general critique that underlies this one. If this more general critique is correct, then the fundamental problems extend much further back in time: to the pre-Socratics (Bickhard, 2006, in press, in preparation).

[7] The processes of selection underlie motivation (Bickhard, 2003), while, so I will argue, the indications underlie representation.

TRUTH VALUE

The crucial point about such indications of interaction possibilities for current purposes is that they constitute the emergence of the most fundamental aspect of representation: truth value. In particular, such indications, if selected, may turn out to be correct, in the sense that the interaction proceeds as indicated, or they may not, if the actual interaction violates the range of indicated possibilities.[8] The frog might flick its tongue, but not be able to proceed with eating because it was just a pebble, not a fly.

Note that the functional task of selecting among indicated interactive possibilities is inherent for any complex agent. The evolution of such indicative capabilities, therefore, is necessary for the evolution of complex agents. Consequently, the evolutionary emergence of representational truth value is inherent in the evolution of complex agents. And, therefore, the evolutionary emergence of representation is inherent in the evolution of complex agents.[9]

MORE COMPLEX REPRESENTATIONS

However, indications of interactive possibilities may possess truth values, but they do not look much like "standard" representations, such as of objects. Can this framework account for more complex representing?

At this point, I can borrow from one of the few extant models of representation in the literature that is based on action: that of Jean Piaget. In particular, Piaget has outlined a model of the representation of small manipulable objects in terms of organizations of potential actions that I can translate directly into the language of the interactive model (Piaget, 1954).[10]

Consider a child's wooden toy block: it offers (or affords)[11] multiple possible interactions, ranging from visual scans to manipulations to throwing, chewing, and so on. A fundamental manifestation of the persistence of objects is that, if any of these potential interactions are available, then they all are with, perhaps, intervening interactions. A particular visual scan, for example, might require an intermediate turning of the block so that what was the reverse side is now visible. In this sense, all visual and manipulation interactions are reachable from each other. Furthermore, this organization of internally reachable interaction possibilities has the property of being invariant under an important class of additional

[8]This point turns on the normativity of such indications being "correct" or "incorrect," and, therefore, on a more primitive, pragmatic, kind of normativity. Elsewhere, I develop a model of the emergence of normative function—a non-etiological model—that serves as the framework for this emergence of representational normativity (Bickhard, 1993, 2004a, in press, in preparation; Christensen & Bickhard, 2002).

[9]For discussion of how such anticipative indicating might occur in central nervous systems, see Bickhard and Terveen (1995), Bickhard (in preparation).

[10]For discussions of some of the differences between Piaget's model and the interactive model, see (Campbell & Bickhard, 1986; Bickhard, 1988; Bickhard & Campbell, 1989).

[11]For discussions of Gibson's model, see Bickhard and Richie (1983).

interactions: for example, if the block is left on the floor, it can (sometimes) be recovered by walking back into that room. The organization of internally reachable interaction potentialities, thus, is itself reachable given a reversal of locomotions and transportations—it is invariant under such locomotions and transportations. This is a further manifestation of the permanence of objects. Clearly, this organization is not invariant under transformations that destroy the block. For small children, these properties constitute what objects are.

For my current purposes, the important point is that an action- and interaction-based model of representation can address more complex forms of representation. There are many further kinds of representation (or purported representation) and cognition to address—for example abstract representation, perception, language, and so on[12]—but it is clear that the model is not restricted to just simple anticipations of single interaction possibilities. It is a candidate for capturing the basic ground of representation.

EMBODIMENT IS NECESSARY

There is a narrower point, however, that is sufficient for the point at issue here: Is embodiment necessary? The narrower point is that interactive representation clearly exists (whether or not it accounts for *all* representation), and it clearly requires interaction, and, therefore, requires some form of embodiment in order to be able to interact. It clearly exists because it is required for complex agents to function in the world, and it is representational because it manifests truth value. So embodiment is necessary for any form of cognition or representation that is of an anticipative interactive kind; such cognition and representation do exist, and therefore embodiment is necessary.

The case can be made stronger, however: if indicated interactions are engaged in and they fail to honor the indicated range of possibilities, then the indicated interactions are false, and they are falsified for the organism or system itself.[13] System detectable error can be accounted for within this model. Because no other model currently on offer can account for system detectable error, there is a strong claim that embodiment is required for *all* representation and cognition: absent an alternative model of representation that can satisfy the system detectable error criterion

[12] See, for example, Bickhard (2003, 2004b, 2005b, 2006, 2007a, in press, in preparation).

[13] It is important to note that detections of such failures of interaction anticipation do not require detecting any external, environmental, anticipated consequences. Such external anticipations would require being represented, and, thus, would not avoid the skeptical argument. Anticipated *internal* processes, however, do not require representation in order to be monitored: this can be done strictly internally and functionally. System detectable error, therefore, is possible. The skeptical argument is a valid argument, but it is based on an assumption that all representation has the form of encoding, and that assumption is false. The argument, then, is unsound, and constitutes a reductio of encoding models of representation.

without requiring interaction, thus embodiment, the hypothesis that all representation and cognition requires (interactive) embodiment is the only one left standing.[14]

WHAT KIND OF EMBODIMENT?

Embodiment is involved, and even required, for multiple phenomena. Any living being must be realized in some form that can maintain itself in its environment(s). Action and interaction are constrained and enabled not only by environments, but also by the specific forms of embodiment that must engage in those interactions—consider, for an extreme example, the differences between a caterpillar and a butterfly.

But perhaps these consequences and necessities of embodiment are different from those necessitated by representation and cognition. What can be said about the minimal form of embodiment that is necessitated for representation *per se*?

The central property involved in the interactive form of representation is the functional anticipation of future interactive processes (Bickhard, 2005a). I have been focusing on the necessity that there be involvement of the environment in determining the course of those interactive processes.[15] This involvement of the environment imposes a basic requirement on the nature of the embodiment of a representational system: the interactions must be of a sort that influences the environment in such ways as will, in turn, influence the internal interactive processes of the organism (or system). The interactions must be capable of being *full* interactions, not just inputs being processed into outputs—outputs that, in turn, have no influence on subsequent inputs.

A minimal embodiment requirement, therefore, is that the system or organism be capable of full interactions, with outputs having causal influence on inputs—with circular causal flow. This is actually a quite minimal requirement: many

[14] The claim here is not that all representation is directly interactive. Encodings do exist—Morse code is a perfectly good example—but genuine encodings are in fact representational stand-ins, and thus, they require something to be stood-in-for. Encodings cannot stand-in for other encodings unboundedly—that is the infamous regress of interpreters of symbolic encodings—so there must be some other form of representation that can ground any such encoding stand-in hierarchy. That other form of representation, thus the *basic* form of representation, is interactive representation. And that is what requires embodiment.

[15] There is also the requirement that those functional anticipations be normative, in the sense that they can be true or false. I have set aside the normativity requirement here, but it too is crucial. For an account of the emergence of basic normativity, see Bickhard (2004a, in press, in preparation). This emergent normativity, interestingly, imposes its own requirements on the nature of normative systems. In fact, they impose constraints on forms of embodiment—and thereby serve as the foundation for interactive representation. In particular, they constrain normativity to certain kinds of far from thermodynamic equilibrium systems, of which all living systems are examples. The nature of such systems, in turn, requires forms of interactivity, which constitute the ground for interactive representation—representation builds on normativity in a very natural way. Normativity, thus, imposes additional constraints on what kinds of systems can manifest representation and, thus, be cognitive systems, and these impose additional constraints on forms of embodiment. But those constraints on *forms* of embodiment do not add to the basic *necessity* that cognitive systems be embodied.

kinds of systems that have unusual embodiments—perhaps, for example, being constituted in some distributed network fashion over large, and perhaps even ongoingly changing, spatial regions—could satisfy it. What cannot satisfy it are passive input processors or throughput processors for which the outputs have no particular influence on the inputs. That is, passive computer systems, including passive connectionist systems, cannot be interactive in the required sense.

Animals and robots, clearly, *can* be interactive in the required sense. And, as mentioned at the beginning of this discussion, they are embodied. Although embodied and capable of being interactive, it is nevertheless an interesting question whether mechanical robots can capture full representation and cognition, whether they can manifest the normativities involved in interactive anticipations being true or false.[16]

CONCLUSION

Embodiment is necessary for representation and cognition. The minimal requirements on embodiment are, on the one hand, sufficient to eliminate most artificial systems in the world today as candidates for being genuinely cognitive, but, on the other hand, they are quite general in what they do impose on sufficient forms of embodiment. But it is not just interaction that is required for representation, it is also normative anticipation of interaction, and that involves its own further requirements.

REFERENCES

Bickhard, M. H. (1988). Piaget on variation and selection models: Structuralism, logical necessity, and interactivism. *Human Development, 31,* 274–312.

Bickhard, M. H. (1993). Representational content in humans and machines. *Journal of Experimental and Theoretical Artificial Intelligence, 5,* 285–333.

Bickhard, M. H. (2003). An Integration of motivation and cognition. In L. Smith, C. Rogers, & P. Tomlinson (Eds.), *Development and Motivation: Joint Perspectives* (pp. 41–56). Leicester: British Psychological Society, Monograph Series II.

Bickhard, M. H. (2004a). Process and emergence: Normative function and representation. *Axiomathes—An International Journal in Ontology and Cognitive Systems, 14,* 135–169. Reprinted from: Bickhard, M. H. (2003). Process and emergence: Normative function and representation. In J. Seibt (Ed.), *Process Theories: Crossdisciplinary Studies in Dynamic Categories* (pp. 121–155). Dordrecht: Kluwer Academic.

Bickhard, M. H. (2004b). The social ontology of persons. In J. I. M. Carpendale & U. Muller (Eds.), *Social Interaction and the Development of Knowledge* (pp. 111–132). Mahwah, NJ: Erlbaum.

Bickhard, M. H. (2005a). Anticipation and representation. In C. Castelfranchi (Ed.), *From Reactive to Anticipatory Cognitive Embodied Systems* (pp. 1–7). *AAAI Fall Symposium, Technical Report FS-05.05.* Menlo Park, CA: AAAI Press.

Bickhard, M. H. (2005b). Consciousness and reflective consciousness. *Philosophical Psychology, 18*(2), 205–218.

Bickhard, M. H. (2006). Developmental normativity and normative development. In L. Smith & J. Voneche (Eds.), *Norms in Human Development* (pp. 57–76). Cambridge: Cambridge University Press.

[16]Elsewhere I argue that 'Mechanism is not enough' (Bickhard, 2007b).

Bickhard, M. H. (2007a). Language as an interaction system. *New Ideas in Psychology, 25*(2), 171–187.
Bickhard, M. H. (2007b). Mechanism is not enough. In Q. Gonzalez, M. Eunice, W. F. G. Haselager, & I. E. Dror (Eds.), *Mechanicism and Autonomy: What Can Robotics Teach Us About Human Cognition and Action?* Special issue of *Pragmatics and Cognition, 15*(3), 573–585.
Bickhard, M. H. (in press). The interactivist model. *Synthese*.
Bickhard, M. H. (in preparation). The Whole Person: Toward a Naturalism of Persons—Contributions to an Ontological Psychology.
Bickhard, M. H. & Campbell, R. L. (1989). Interactivism and genetic epistemology. *Archives de Psychologie, 57*(221), 99–121.
Bickhard, M. H. & Richie, D. M. (1983). *On the Nature of Representation: A Case Study of James Gibson's Theory of Perception*. New York: Praeger Publishers.
Bickhard, M. H. & Terveen, L. (1995). *Foundational Issues in Artificial Intelligence and Cognitive Science: Impasse and Solution*. Amsterdam: Elsevier Scientific.
Brooks, R. A. (1991). Intelligence without representation. *Artificial Intelligence, 47*(1–3), 139–159.
Campbell, R. L. & Bickhard, M. H. (1986). *Knowing Levels and Developmental Stages*. Basel, Switzerland: Karger, Contributions to Human Development.
Christensen, W. D. & Bickhard, M. H. (2002). The Process Dynamics of Normative Function. *Monist, 85*(1), 3–28.
Clark, A. & Toribio, J. (1995). Doing without representing? *Synthese, 101*, 401–431.
Cummins, R. (1996). *Representations, Targets, and Attitudes*. Cambridge, MA: MIT Press.
Dretske, F. I. (1988). *Explaining Behavior*. Cambridge, MA: MIT Press.
Fodor, J. A. (1987). *Psychosemantics*. Cambridge, MA: MIT Press.
Fodor, J. A. (1990). *A Theory of Content and Other Essays*. Cambridge, MA: MIT Press.
Fodor, J. A. (1991). Replies. In B. Loewer & G. Rey (Eds.), *Meaning in Mind: Fodor and His Critics* (pp. 255–319). Oxford: Blackwell.
Fodor, J. A. (1998). *Concepts: Where Cognitive Science Went Wrong*. Oxford: Oxford University Press.
Fodor, J. A. (2003). *Hume Variations*. Oxford: Oxford University Press.
Franklin, S. (1995). *Artificial Minds*. Cambridge, MA: MIT Press.
Godfrey-Smith, P. (1994). A modern history theory of functions. *Nous, 28*(3), 344–362.
Joas, H. (1993). American pragmatism and German thought: A history of misunderstandings. In H. Joas (Ed.), *Pragmatism and Social Theory* (pp. 94–121). Chicago, IL: University of Chicago Press.
Levine, A. & Bickhard, M. H. (1999). Concepts: Where Fodor went wrong. *Philosophical Psychology, 12*(1), 5–23.
Millikan, R. G. (1984). *Language, Thought, and Other Biological Categories*. Cambridge, MA: MIT Press.
Millikan, R. G. (1993). *White Queen Psychology and Other Essays for Alice*. Cambridge, MA: MIT Press.
Newell, A. (1980). Physical symbol systems. *Cognitive Science, 4*, 135–183.
Pfeifer, R. & Scheier, C. (1999). *Understanding Intelligence*. Cambridge, MA: MIT Press.
Piaget, J. (1954). *The Construction of Reality in the Child*. New York: Basic Books Inc.
Rescher, N. (1980). *Scepticism*. Totowa, NJ: Rowman and Littlefield.
Rosenthal, S. B. (1983). Meaning as habit: Some systematic implications of Peirce's pragmatism. In E. Freeman (Ed.), *The Relevance of Charles Peirce* (pp. 312–327). LaSalle, IL: Monist.
Svensson, H. & Ziemke, T. (2005). Embodied representation: What are the issues. In B. Bara, L. Barsalou, & M. Buccarelli (Eds.), *Proceedings of the 27th Annual Meeting of the Cognitive Science Society* (pp. 2116–2121). Mahwah, NJ: Lawrence Erlbaum.
van Gelder, T. J. (1995). What might cognition be, if not computation? *The Journal of Philosophy, XCII*(7), 345–381.
Vera, A. H. & Simon, H. A. (1993). Situated action: A symbolic interpretation. *Cognitive Science, 17*(1), 7–48.
Wilson, M. (2002). Six views of embodied cognition. *Psychonomic Bulletin & Review, 9*(4), 625–636.
Ziemke, T. (2001). Are robots embodied? In C. Balkenius, J. Zlatev, C. Brezeal, K. Dautenhahn, & H. Kozima (Eds.), *Proceedings of the First International Workshop on Epigenetic Robotics: Modeling Cognitive Development in Robotic Systems* (pp. 75–93). Lund, Sweden: Lund University Cognitive Studies.

3

EMBODIMENT AND EXPLANATION

ANDY CLARK

Department of Philosophy, University of Edinburgh, Edinburgh, Scotland, UK

THREE THREADS

There are at least three (distinct but sometimes overlapping) ways in which embodiment seems to matter for mind and cognition. These are:

1. Spreading the load
   The body and the brain, thanks to evolution and learning, are adept at spreading the load. Bodily morphology, development, action, and biomechanics, as well as environmental structure and interventions, can reconfigure a wide variety of control and learning problems in ways that promote fluid and efficient problem-solving and adaptive response.
2. Self-structuring of information[1]
   The presence of an active, self-controlled, sensing body allows an agent to create or elicit appropriate inputs, generating good data (for themself and for others) by actively conjuring flows of multimodal, correlated, time-locked stimulation.
3. Supporting extended cognition
   The presence of an active, self-controlled, sensing body (a) provides a resource that can *itself* act as part of the problem-solving economy and (b) allows for the *co-opting* of bio-external resources into extended but deeply integrated cognitive and computational routines.

[1] The notion of information self-structuring can be found in Lungarella and Sporns (2005).

The three threads comport nicely with a supporting hypothesis (Gray & Fu, 2004; Gray et al., 2006) that Clark (2008a) dubs:

> Hypothesis of Cognitive Impartiality
>
> Our problem-solving performances take shape according to some cost-function or functions that, in the typical course of events, accords no special status or privilege to specific types of operation (motoric, perceptual, introspective) or modes of encoding (in-the-head or in-the-world).

Cognitive Impartiality explains the emergence of organizations (both long and short term) in which the storage, processing, and transformation of information are spread so indiscriminately between brain, body, and world.

Examples of all of these effects form the basis of the literature on "embodied, embedded cognitive science." Examples of (1) include work on passive dynamic walking (Collins et al., 2005), on sensor placement (Webb, 1996), and on the productive use of bodily and environmental structure (Pfeifer & Bongard, 2007). Examples of (2) include Ballard et al.'s (1997) work on just-in-time sensing and deictic pointers, and Yu et al.'s (2005) work on learning visually grounded meanings. In addition, studies of sensory substitution systems further underline the importance of self-controlled temporally nuanced cycles of sensor movement and (resultant) input in tuning bodily and sensory equipment in ways apt to support perception and action (Bach y Rita & Kercel, 2003). Examples of (3) include Clark and Chalmer's (1998) arguments for "the extended mind," empirical work on gesture-for-thought (Goldin-Meadow, 2003), and Sterelny's (2003) (with a different spin) take on cognitive niche construction or incremental epistemic engineering.

Rather than rehearsing all these findings here (see Clark, 2008a for such a review), I want to take something like the three threads story pretty much for granted and ask whether (despite some recent publicity) these kinds of appeal to embodiment, action, and cognitive extension are best understood as a revolutionary change or as fully continuous with computational, representational, and (broadly speaking) information-theoretic approaches to understanding mind and cognition. In defending the latter, more conservative, view I hope to display at least something of the likely shape of a mature science of the embodied mind.

THE SEPARABILITY THESIS[2]

Larry Shapiro (2007), in a recent review article on Embodied Cognition, glosses it in part as:

> ... an approach to cognition that departs from traditional cognitive science in its reluctance to conceive of cognition as computational.[3]

[2] Some of the material in this section is taken from Clark (2008b) by permission.

[3] See the entry for The Embodied Cognition Research Program in the online journal *Philosophy Compass* (http://www.blackwellcompass.com/subject/philosophy/).

Whereas Rohrer (2006, p. 2) claims[4] that:

> Unlike the computationalist-functionalist hypothesis, embodiment theorists ... argue that the specific details of how the brain and body embody the mind do matter to cognition.

Of course, even the most traditional of machine functionalists thought that cognitive processes needed to be implemented in physical stuff. The point was just that the physical stuff mattered only in virtue of what were broadly considered as its functional or organizational properties. Cognition, for the machine functionalist, was independent of its physical medium in the sense that if you could get the right set of abstract organizational features in place (typically, some set of input to internal-state-transitions to output functions), you would get the cognitive properties "for free." Importantly, as long as the right abstract organization could be instantiated, you would (as it was claimed) get the very same mental and cognitive properties regardless of the materials you were using (Cummins, 1983) and any details of gross physical shape or form. The traditional functionalist thus held that cognition was in *some* sense "platform-independent." The question that arises, then is does work in embodied cognition really cast doubt on such claims of platform-independence?

Shapiro (2004) seems to suggest that it does.[5] He presents an argument against one version of the claim of platform-independence that he dubs as the separability thesis (ST). According to ST a humanlike mind could perfectly well exist in a very non-humanlike body. Against ST, Shapiro urges us to embrace what he calls the embodied mind thesis (EMT) which holds that "minds profoundly reflect the bodies in which they are contained" (Shapiro, 2004, p. 167).

Why reject ST? One reason, Shapiro tells us, turns on quite basic facts about sensing and processing. Human vision, for example, involves a great deal of sensor movement. We move our heads to gain information about the relative distances of objects, since nearer objects will (courtesy of parallax effects) appear

[4]To be fair, Rohrer allows that the notions of functional and computational explanation might be broadened in many of the ways we have scouted. But such broadening, it seems to me, should not result in our putting the terms (as Rohrer then does) in scare quotes. Rather, to fail to recognize key events and processes as genuinely computational (because trafficking in representations, information, and information-based control) is to fail to account for what is special about minds: what distinguishes them from volcanos and other complex but non-cognitive phenomena (Clark, 1998).

[5]In a similar vein Alva Noë (2007, p. 537, emphasis in original) writes that "one deplorable legacy of functionalism is the idea that embodiment—the way we are put together, brains and body—is irrelevant to how our minds work. For functionalism, embodiment is just a matter of the way our mental functioning happens to be implemented." I shall try to show that while this is indeed true, it is by no means evidently "deplorable." Importantly, it is consistent with taking embodiment very seriously indeed. For insofar as certain key operations and encodings are accomplished by gross bodily (non-neural) means, features of embodiment (and action) turn out to provide the material means whereby minds like ours are realized. If embodiment thus turns out to be as important as (but no more important than) "embrainment," that would surely constitute a good reason to take embodiment seriously when pursuing the sciences of the mind.

to move the most. Such movements, Shapiro argues, are not simply an aid to vision. They are part and parcel of the visual processing itself. They are "as much a part of vision as the detection of disparity or the calculation of shape from shading" (Shapiro, 2004, p. 188). Similar points can be made about audition and the placement of the ears on the head. The idea is that:

> ...psychological processes are incomplete without the body's contributions. Vision for human beings is a process that includes features of the human body ... this means that a description of various perceptual capacities cannot maintain body-neutrality, and it also means that an organism with a non-human body will have non-human visual and auditory psychologies
>
> Shapiro (2004, p. 190)

Body-neutrality, for Shapiro is the idea that "characteristics of bodies make no difference to the kind of mind one possesses" and is further associated with the idea that "mind is a program that can be characterized in abstraction from the kind of body/brain that realizes it" (both quotes are from Shapiro, 2004, p. 175). According to Shapiro, work on the role of bodily movements in visual processing suggests that body-neutrality fails and that human-style vision requires a human-style body.

Another corpus of research that appears to contest claims of body-neutrality, at least regarding the contents of perceptual awareness is the so-called enactive approach to perception. Laying out this approach, Noë (2004, p. 25) comments that:

> If perception is in part constituted by our possession and exercise of bodily skills ... then it may also depend on our possession of the sorts of bodies that can encompass those skills, for only a creature with such a body could have those skills. To perceive like us, it follows, you must have a body like ours.

Another (through very different) way of rejecting ST appeals to the considerations of the role of the body in structuring human concepts. The *locus classicus* here is Lakoff and Johnson's (1980, 1999) work on the role of body-based metaphors in human thought and reason. Many of our basic concepts, they argue, are quite evidently body-based, for example concepts like front and back, up and down, inside and outside:

> If all beings on the planet were uniform stationary spheres floating in some medium and perceiving equally in all directions, they would have no concepts of *front* and *back*.
>
> Lakoff and Johnson (1999, p. 34)

But these basic concepts, they go on to argue, end up structuring our understandings (and our inferences) in more rarefied domains. Happiness and sadness, to take the standard example, are humanly conceived in terms of upness and downness. The specifics of embodiment thus shape the basic concepts that in turn inform (so it is argued) the rest. Summing up the Lakoff and Johnson line, Shapiro (2004, p. 201) suggests that:

> Organisms that didn't have bodies like our own would develop other metaphors to characterize happiness and sadness. *Happy* and *sad* would be structured in other ways and would thus assume different meanings.[6]

The common upshot of all these arguments, then, is a kind of principled body-centrism, according to which the presence of humanlike mind depends quite directly upon the possession of a humanlike body.

BEYOND FLESH-EATING FUNCTIONALISM

It is revealing, I think, that Shapiro's spirited defense of profound bodily[7] involvement in the mental processes comes in the larger context of a series of arguments aimed at a different, logically independent but thematically related target. That target was the thesis of multiple realizability: a staple of non-reductionist Philosophy of Mind ever since the heady days of early Machine Functionalism. At about that time, the notion that minds like ours might be directly identified with their specific *neural* underpinnings was widely cast as a kind of unacceptable meat- or species-chauvinism, to be replaced by the identification of mind as a functional kind: a kind capable in principle of being realized by many different physical substrates (Putnam, 1975; see also Putnam, 1960, 1967). In this new regime, mindware stood to neural hardware, as software stood to the physical device. Just as the same software could run on different bedrock machines, the same kinds of mind might, it was supposed, turn up in various kinds of material form. What mattered was not the bedrock physical forms but rather the abstract patterns (of input to internal-state-transitions to output) that the material structures were able to support. Sameness at this rather abstract level was meant to guarantee sameness at the mental level. Or at any rate, any remaining slack was to be taken up by rather arcane details of history and/or distal environmental embedding. As far as the *local machinery* of mind itself was concerned, functional identity fully fixed any contribution to mentality.

[6] That there is something problematic about this argument is evident in the tension between the easy use of a common notion of happiness and sadness in the first quoted sentence and the subsequent assertion that happy and sad would then "assume different meanings." But the point, in any case, is simply that arguments stressing the pervasive influence of embodiment on conceptualization look to be arguments against ST, since they assert the ineliminable involvement of bodily details in an account of mental states.

[7] I use "bodily" here to refer to the gross physical body rather than to the (of course equally bodily) brain.

Shapiro's appeal to work in embodied, embedded cognitive science depicts it as in spirit rather inimical[8] to the platform-neutral machine functionalist model of mind. But the notion of platform-neutrality is a slippery beast. For as we saw, even the standard machine functionalist need not (and should not) deny that properties of the bodily "platform" *matter* to mind and cognition. All she need claim is that insofar as the bodily platform matters, it matters in virtue of the suite of abstract opportunities (encodings, operations) that it makes available[9], and in contrast to the suite of encodings and operations that it makes unnecessary. Thus the machine functionalist, to take a simple example, need not (and should not) ignore the potent effects of passive dynamics (Collins et al., 2005) on the requirements for a control system supporting powered goal-driven locomotion. For the presence of rich passive dynamics reconfigures the problem space so as to enable biological organisms to produce and control locomotion in amazingly efficient ways. Moreover, we should not be misled into thinking that the kinds of operation and encoding at stake must be restricted to the familiar (digital, discrete, typically local, often temporally impoverished) suite of possibilities explored by classical artificial intelligence. Instead, human intelligent performance may be best understood by approaches that recognize the role of analog elements that change continuously with time or that exploit continuous state (of coupled unfoldings that criss-cross brain, body and world), of motor-loop involving self-stimulating routines, and of the active self-structuring of the flow of information.[10]

Thus, consider the claim (Ballard et al., 1997) that the brain creates its programs to minimize the amount of working memory that is required, and that eye motions are recruited for just-in-time retrieval of information from the environment. Ballard et al. were able to systematically alter the particular mixes of biological memory and active, embodied retrieval recruited to solve different versions of the problem, concluding that at least in this kind of task at least, "eye movements, head movements, and memory load trade off against each other

[8] Inimical to, but not inconsistent with, ST is said to be logically independent of MRT (Multiple Realizability Thesis) since "it is logically possible that a mind could be realized in a number of different kinds of structure, but that all of these structures are contained in similar sorts of bodies (and) it is logically possible that there is only one or a few ways of realizing a humanlike mind but that these few types of realizations can exist in many different sorts of bodies" (Shapiro, 2004, p. 167). Such concessions make the intended force of the earlier arguments depicting physical structures as proper parts of psychological processes unclear, though Shapiro does add that he is willing to bet that "if there are but a few ways to realize a humanlike mind, probably there are but a few kinds of bodies that could contain such a mind" (Shapiro, 2004, p. 167).

[9] It is also compatible even with traditional forms of machine functionalism that, just as it happens, only one kind of stuff in the universe might be capable of implementing a given functional profile.

[10] Taken to the extreme, one may here discern the possibility of what Wheeler (in press) describes as a form of "non-computational functionalism" that is nonetheless compatible with the multiple realizability of cognitive mechanisms.

in a flexible way" Ballard et al. (1997, p. 732). Ballard et al.'s work is thus an example of the kind of approach that Clark (2008a) dubs Distributed Functional Decomposition (DFD). Such an approach analyzes a cognitive task as a sequence of less intelligent subtasks (in this case using recognizable computational and information processing concepts), but it does so relative to a larger (not merely neural) organizational whole. Such approaches recognize the profound contributions that embodiment and environmental embedding make to the solution of the problem, and display those contributions clearly and distinctly. They do this by identifying the information processing role of specific (both gross-bodily and neural) operations in our performance of the task. Bodily actions and worldly encodings and transformations might thus emerge as some of the means by which certain key operations are implemented. In this way bodily and worldly elements emerge as genuine parts of extended problem-solving regimes apt for formal description in either (or both) dynamical and information processing terms.

Or, To cite one final example, consider the role of bodily gesture in the unfolding of thought. An intriguing suggestion found in McNeill (2005) and Goldin-Meadow (2003) is that actual spatially extended physical gestures sometimes act as cognitive elements in their own right, so that speech, gesture, and neural activities unite to form a single integrated cognitive system. If that were indeed the case then, for humanbeings like us, the body might thus provide for a kind of cognitive functionality that neural unfoldings alone do not typically support. But viewed from a greater distance this merely represents one way of implementing a much more abstract routine whose essence was seen to lie in the productive tension between two forms of loosely coupled encoding: one visuospatial, the other verbal. The increasingly popular image of functional, computational, and information processing approaches to mind as flesh-eating demons is thus subtly misplaced. For rather than *necessarily* ignoring the body, such approaches may instead help target larger organizational wholes in ways that help reveal where, why, how, and even how much (see section "Quantifying Embodiment" following) embodiment and environmental embedding really matter[11] for the construction of mind and experience.

ADA, ADDER, AND ODDER

We are now in a position to reconsider Shapiro's opposition (mounted in the name of Embodied Cognition) to the idea that "the same kind of mind can exist in bodies with very distinct properties" (Shapiro, 2004 p. 175). On the basis of

[11] Shapiro (personal communication) notes that on the account I favor, bodies matter because they can play certain roles in the processing cycles that constitute cognition, but (in another sense) bodies do not matter since what matters is the resulting overall processing profile, not the presence of any specific bodily features *per se*, nor the precise way that various operations are distributed between brain, body, and world. Shapiro fears that this robs the embodied approach of much of its distinctive appeal. I fear that the alternative buys bodily appeal at the price of scientific mystery.

the kinds of evidence described above, Shapiro rejects the idea that "snakelike organisms and creatures of science fiction" (Shapiro, 2004, p. 174) might share our kind of mind. Shapiro suggests that if the theorists of embodied cognition are correct, Body Neutrality [the idea that "characteristics of bodies make no difference to the kind of mind one possesses" (Shapiro, 2004, p. 175)] is false.

It should now be clear that something has gone by rather too swiftly. For imagine now a case in which we have two intelligent beings. One of them is a snake-like creature lying on top of an advanced touch-screen like environment. In this flat screen setting every little wriggle of the snake can cause specific external symbolic tokens to appear elsewhere on the screen: tokens that are themselves apt for perceptual uptake (perhaps via a kind of Braille). The snake like creature being (call it Adder) uses this setup, let us suppose, to carry out the same complex accounting as a standard, pen-and-paper using accountant (let us call her Ada). As far as the DFD story goes, there is no reason to suppose (from anything we have said so far) that the accounting-relevant states of Ada and Adder need differ in any respect. Each implements the same extended computational process. They even (we may suppose) divide the biological and non-biological contributions in the same way, making use of external storage and notations at exactly the same points in their distributed problem-solving routines.

More radically, however, we may next imagine a case where there are differences at the level of what gets done where. Odder enters, performs certain computations internally so that Ada and Adder both perform using action and perception routines in the non-biological arena. Here too, the DFD theorist is at liberty to assert that the very same cognitive routines are being implemented, with nothing distinguishing the cases apart from some non-essential matters of location. Just as, on a standard internalist model, we need not care exactly where *within* the brain a given operation is performed, likewise (it might now be urged) we should not care whether, in some extended computational process, a certain operation or encoding occurs inside or outside some particular membrane or metabolic boundary.

DFD-style work in embodied, embedded cognition thus lends, it seems to me, no support to the idea that minds like ours require bodies like ours, even though it insists that bodily and worldly operations can be active and crucial participants in extended information processing routines. What matters, the DFD theorist insists, is just the full suite of encodings and operations made available by the some combination of neural, gross-bodily, and worldly opportunities. Creatures with radically different bodies, brains, and worlds might thus contrive to use their varying resources to implement many of the very same cognitive and information processing routines.

A TENSION REVEALED

All this reveals a tension at the heart of the program that is sometimes so easily (so unitarily) glossed as the study of "embodied, embedded cognition." It is the tension between seeing body (and world) as expanding the palate of opportunities for the realization of cognitive processes and mental states, and something more fundamentally—though mysteriously—fleshy; the idea that embodiment vastly restricts the space of "minds like ours," tying human thought and reason inextricably (and non-trivially) to the details of human bodily form.[12] (Non-trivially in that, of course, the encountered (seen, touched) shape and propioceptively sensed unfolding activity of the body will be part of what is given in conscious experience, and is thus apt to impact and inform our self-image and attitudes in many well-understood ways. To that extent, the details of specific forms of embodiment clearly *make* a difference. The question is, must all differences in bodily form make differences that go beyond these direct and as it were instrumental effects?)

Thus consider Shapiro's (2004, p. 188) observation that:

> The instructions by which the human brain computes relative depth do not work in creatures with eye configurations other than those in a human being. This is the sense in which depth perception is embodied. The procedures by which human beings perceive depth — a fact about human psychology — are contingent on a fact about human bodies.

Recall that from facts such as these, Shapiro concludes that "human vision needs a human body" (Shapiro, 2004, p. 189). Such a claim is, however, quiet ambiguous. It might mean only that the brain's algorithms factor in the bodily structures and opportunities. This is surely correct and (as we saw) fully compatible with platform-flexible forms of DFD. Or it might mean that being able to make the kinds of gross visual discrimination that we can make *requires* having exactly the same kind of body (in respect of eye configuration at least) as we do.[13] But this claim is surely false, since an alternative distribution of the very same information processing steps, in some differently brained and differently bodied being, would be capable of implementing that same algorithm.[14] Or it might (finally) mean that any such alternative implementation need not preserve the qualitative feel of human depth perception: a qualitative feel that is somehow non-trivially tied, not to the abstract algorithm but to the use of two gross physical eyes of such-and-such shape and character, located a certain distance apart.

[12]That something might be awry with this latter picture is perhaps indicated by the simple fact that human bodies already come in a wide variety of shapes and forms. Just what then is "human embodiment" that it might so cleanly limn the space of "human mentality"?

[13]Shapiro (personal communication) clarifies that the intended meaning was indeed the former (that the brain's algorithms factor in contingencies about the body). Given this reading, however, it seems unclear why facts about embodiment are taken to work against the separability thesis.

[14]Thus consider FLICKER. Flicker is a creature with just one eye that moves very rapidly from side to side of its face, sending signals only while at the two locations that happen to match those of the human eyes. With some canny tweaks of the neural control and downstream sensory post-processing circuitry, such a being could implement much the same basic stereo depth perception algorithm as ourselves. The situation would be not unlike the use of a fast serial computer to simulate a parallel processing device.

The wild card in this debate is thus our old friend phenomenal experience itself. Might the body be making some special kind of contribution, one that cannot help but impact (in non-trivial ways) certain qualitative aspects of our mental life? This is probably the best way to understand Noë's previously quoted assertion that "the character of our experience depends on … idiosyncratic aspects of our sensory implementation" (Shapiro, 2004, p. 26). If you think that the "sensory implementation" plays a unique (supra-functional) role that contributes directly and non-trivially to experiential content, you may very well think that every difference in implementation makes a real (though perhaps vanishingly small) difference to the felt nature of the experience itself.[15]

It is by no means obvious, however, that we should endorse, even where conscious experience is concerned, any such full and principled sensitivity to the fine details of a being's embodiment and/or sensory apparatus. From a mechanistic standpoint it seems compelling that two beings could be very different in respect of gross sensory apparatus and embodiment and yet, courtesy perhaps of compensatory differences in key aspects of downstream processing, end up realizing the same set of experience determining operations and state transitions. Noë (2004) and also O'Regan and Noë (2001) seem to leave no room for this even as a bare possibility. Noë is explicit that "to see *as we do*, you must … have a sensory organ and a body like ours" (Noë, 2004, p.112, italics in original).[16]

Perhaps this is right, and experience is non-trivially permeated by the full details of biological embodiment. My own view is that this is most unlikely to be true. By simply *identifying* the contents of experience with implicit knowledge of the full suite of contingencies defined at the sensorimotor surfaces, this kind of strong sensorimotor account leaves no room for compensatory downstream adjustments to yield identical experiences despite surface dissimilarities.[17] Nor does it leave room for small differences at the sensorimotor surfaces to be such as to make no experiential difference, as a result of failing to deliver any *salient*

[15]Noë may actually have an even more pervasive role for the body in mind. In his book (2004, p. 25) he writes that:

> In general it is a mistake to think that we can sharply distinguish visual processing at the highly abstract algorithmic level, on the one hand, from processing at the concrete implementational level, on the other. The point is not that algorithms are constrained by their implementation, although that is true. The point, rather, is that the algorithms are actually, at least in part, formulated in terms of items at the implementational level. You might actually need to mention hands and eyes in the algorithms!

It is unclear, though, just what Noë here has in mind. For some useful discussion, see Shapiro (in press).

[16]Such an account of course makes it in principle impossible for a differently embodied being to fully share human perceptual experiences.

[17]Thus Noë (personal communication) does indeed assert that "you couldn't have the very same experience unless you have the same underlying sensorimotor exercise." This may turn out to be true, but it is not yet obvious to me why it should be true, or how we can at this time know it to be true.

differences in signals to downstream processors. Perhaps, that is to say, downstream processing provides a kind of grid relative to which certain differences at the level of the sensory inputs (and associated contingencies) simply fail to *make* a difference.

A related worry threatens at least the strongest versions of Lakoff and Johnson's claims concerning the tight links between forms of embodiment and basic conceptual repertoires. This is because, what embodied experience actually delivers as the baseline for learning and metaphorical thought surely depends on some complex mixture of bodily form, environmental structure, and (possibly innate) downstream internal processing. Here too, compensatory adjustments in either of the two non-bodily arenas look likely to make forms of thought and reason available, which are not tethered in any simple way to the gross bodily bedrock.

PARTICIPANT MACHINERY AND MORPHOLOGICAL COMPUTATION

A simple illustration of this interchangeability of resources is provided by Chandana Paul's (2004, 2006) demonstration "that a robot body can be used for computation in addition to merely acting as an effector for the controller." The backdrop to the demonstration involves a very simple class of neural networks known as perceptrons (Rosenblatt, 1962). It is well known that a perceptron, if given two inputs A and B, can compute OR and AND functions (in fact, all linearly separable functions), but not linearly inseparable ones such as exclusive-or. Exclusive-or, normally written XOR, is true if either *but not both* disjuncts are true: that is, if (A or B) is true but (A and B) is false. Paul's demonstration involves a simple "vehicle" of the kind made famous by Braitenberg (1984)whose behavior is determined by the activity of two perceptrons. Perceptron 1 computes OR and controls M1, a forward drive delivered to the single central front wheel of a front-wheel drive vehicle. This means that power is delivered to the single central front wheel if either or both inputs are active (it is thus computing the standard INCLUSIVE OR function). Perceptron 2 computes the standard AND function and controls M2, a lifting device that will raise the single front wheel of the forward drive vehicle off the ground only if both inputs are active.

You can probably see where this is going. When A and B are both reading OFF (zero, false), both nets output zero, the wheel is on the ground, but no power is delivered so the robot stays stationary. When only A is ON, the AND net delivers zero, the wheel stays grounded, and the OR net outputs a one. The wheel turns and the robot goes forward. The same type of scenario occurs when only B is ON. But (and this is the crucial case) when A and B are both ON, the OR net causes M1 to move but the AND net lifts the wheel from the ground so the robot stays stationary. The embodied system's response profile to the different possible values of the A and B inputs thus has the form of the standard XOR

TABLE 3.1 Behavioral profile of the XOR robot.

A	B	Behavior
F	F	Stationary
F	T	Moving
T	F	Moving
T	T	Stationary

From Paul (2004) with permission.

truth table, *despite* the fact that the computational controllers are perceptrons, congenitally unable to compute non-linearly separable functions such as XOR. Lifting the front wheel in response to the conjunction of the two inputs now stands in for the "missing line" of the XOR truth table. In this way, the physical vehicle, despite having only perceptrons for controllers, is able to behave exactly as if it were controlled by an XOR net. For it now behaves in the way displayed in Table 3.1.

The active body of the robot is here providing the functional equivalent of the missing second layer of neural processing: the extra processing that would be required to solve the linearly inseparable problem of computing XOR. The overall embodied system thus provides the missing functionality, equivalent to performing a NOT on the first input, followed by an AND. In this way "the example shows that through its configuration a robot body can perform a quantifiable operation on its inputs" (Paul, 2004, p. 33).

At this point a skeptic might argue that the XOR computation is in some way unreal: more in the eye of the observer than a true resource for a reasoning robot? And there is (as things stand) some truth in this. For what the robot currently displays is what Paul nicely dubs "latent morphological computation": computation that is visibly (to us) implicit is the response profile of the overall physical device, but not yet available *to the device itself* as a general-purpose problem-solving resource. A simple (and, as we shall see, biologically unexceptional) tweak, however, makes the new functionality available to the device itself. Thus Paul next describes a "vacuum cleaning robot" (the precise details of which need not concern us here). The vaccum cleaning robot is like the XOR robot except that this time it is augmented with a sensor informing it of the behavioral consequences of its own action. Thus augmented, the robot can learn (or be programmed) so as to incorporate the body involving XOR circuit into an open-ended set of other routines, routing various A, B signals through the body circuit and reading the XOR result off from a rapid, self-perceived bodily "twitch" of the front wheel: a twitch that need not even persist long enough to cause actual forward motion. The body involving XOR computation (that may previously have appeared to be merely in the eye of the beholder) is now a general-purpose resource that can be invoked much like a regular logic gate. Quite generally then:

> when a robot with latent morphological computation is augmented with a sensor which can sense the behavioral consequences, it makes the computational function defined by the morphology explicit, such that it can be used as a standard computational sub-unit at any stage of the processing
>
> (Paul, 2004, p. 36)

It might seem that this is all just a clumsy trick: why use the robot body to perform a computation that would be so cheaply and easily handled using a simple three layer neural network? To think this is, however, to miss the point and force of the demonstration. For the idea is that evolved biological intelligences, unlike the more neatly engineered solutions with which we are still most familiar as designers, are perfectly able to find and exploit unexpected forms of *multiple functionality*.[18] That is to say, they may find and exploit solutions in which a single element (such as a bodily routine or motion) plays many roles, some of them merely practical, others more "epistemic" (Kirsh & Maglio, 1994) in nature. The clean division between mechanical (body) design and controller design that characterizes many humanly engineered solutions looks quite unimportant (indeed, often counter-productive) if what we seek is efficiency and maximal exploitation of resources. Paul's demonstration may be compared to Thompson's (1998) and Thompson et al.'s (1996) work using genetic algorithms to evolve real electronic circuits. The evolved circuits turned out to exploit all manner of physical properties, usually ignored or deliberately suppressed by human engineers.[19] The lesson, according to the authors, was that:

> It can be expected that all of the detailed physics of the hardware will be brought to bear on the problem at hand: time delays, parasitic capacitances, cross-talk, meta-stability constraints and other low-level characteristics might all be used in generating the evolved behavior
>
> Thompson et al. (1998, p. 21)

What thus goes for the brain (the hardware chip) goes too for the rest of the physical body. It too may be exploited, in all manner of unexpected ways, as an essential part of an information processing organization. In real-world cases, Paul goes on to suggest, we should expect to find that the computational roles being played by bodily acts are much more complex than the computation of a common binary function, perhaps involving analog functions of quite unexpected degrees of complexity. The case of gesture-for-thought may be an example of just this kind, in which actual hand and arm motions help to implement encoding and processing operations that are (as McNeill himself suggests) holistic and analog rather than local, symbolic, and discrete.

[18] For an excellent account of multiple functionality in evolved systems, see Anderson (2007).

[19] For discussion, see Clark (2001) chapter 5.

QUANTIFYING EMBODIMENT

I would like to end by briefly broaching a topic whose very label causes raised eyebrows among some of the more radical friends of embodied cognition The topic is quantifying embodiment, that is to say, measuring exactly *how much* difference embodiment makes with regard to some behavior, capacity, or ability.

At first the question sounds peculiar. What can it mean to quantify the effects of embodiment? Relative to *what* might we measure them? The question sounds less peculiar, though, once we begin to view embodiment through a broadly speaking information-theoretic lens. That is to say, once we attempt to understand the cognitive roles of body, action,[20] and environment by understanding their roles in the elicitation, storage, transformation, and processing of information, and in securing its poise for use in the control of intelligent action. It is, in fact, quite a small step from viewing body, world, and action as elements in extended dynamical–computational routines to attempts at quantification. As early as 1995 we read that:

> ...it is necessary to understand the way various external actions fit into an overall strategy of computation. This requires identifying mental functions served by external actions and changes, and enumerating the resources saved in specific cognitive components such as visual memory, articulatory loop, attention, and perceptual control
>
> Kirsh (1995, p. 60)

In the same paper, Kirsh measured the performance benefits gained by the "cognitive use" (as he puts it) of hands, fingers, and surrounding material objects in a variety of tasks. More recently, Maglio et al. (2003) plotted the increase in the value of the so-called hazard function that results from information self-structuring during expert TETRIS play. (The hazard function is the instantaneous probability of completing a process in the next move, and serves as a rough measure of information processing pay-off.)

Such attempts at quantifying the benefits of embodiment and action remain in their infancy. But there is for optimism. Lungarella et al. (2005) describe a variety of methods for quantifying increases in the information present in raw sensory experience as a result of coordinated sensorimotor activity ("information self-structuring" as described in section "Three Threads" above). The experimental setup involved a robot able to deploy active vision (in the form of a robot-controlled camera) so as to detect informational structure in a video data stream. The study investigated the extent to which the ability to produce self-generated motor activity (activity that actively structures the sensory input that guides the ongoing motor activity itself) increases the information structures present in the sensory signals used to guide learning and response. The results were unambiguous. The presence of coordinated self-generated motor activity (when compared to a control condition), resulted in a suite of measurable differences in the information

[20]Notice that "action" here must play a dual role, both as practical action *per se* and as part of the information-processing routines that select (other) actions.

structure implicit in the sensory array. For example, there were measurable increases in mutual information (the statistical dependence of one variable in the simple experiment, the state of an individual pixel in the visual array on another), in integration (the total amount of statistical dependence among the variables, hence the degree to which they share information) and in complexity[21] (the degree to which elements manage to be specialized, reporting statistically independent events, while also sharing information). Such increases in the information structure present in the sensory signal provide, the authors argue, a clear functional rationale for the evolution and use of coordinated sensorimotor behavior as a means of actively structuring our own sensory experience.

In a neat inversion, these informational measures can also be used to drive the evolution of artificial agents. By using the measures as part of a fitness function, Sporns and Lungarella (2006) were able to investigate the morphologies and behaviors that result from direct pressure to maximize the information structure in the sensory signal. The idea was tested in simulation using a simple creature and environment. The creature was provided with vision and touch, in the form of an "eye" (a 25 × 25 pixel moving window with a 5 × 5 central "fovea") that could sweep the environment and an arm-hand-and-touch-pad appendage that could also move across the environment. The environment itself was just a 100 × 100 pixel area, where each pixel and each time step displayed a randomly generated color, either red, green, or blue. Across this little world a single colored object (5 × 5 pixels) moved at a constant speed in a random path. The object, unlike the rest of the environment, had tactile features too, either ridges or knobs. When the touch pad encounters the object, the object stops, thus allowing the pad to sweep the surface to detect tactile properties. Once touch is broken, the object resumes its random walk. Controlling the simple body was a neural system appraised of the visual and tactile inputs, and provided with an attention system involving the use of a saliency map to drive eye and arm activity. Using a mixture of behavioral and information-theoretic cost functions, Sporns and Lungarella were able to evolve agents capable of coordinated visuomotor action. Before evolution, the accidental touch of the target object did not yield foveation, tracking, or prolonged object "capture". After evolution, arm and eye worked together to acquire and scan the objects. Maximizing specific forms of information structure was thus seen to lead to the emergence of key adaptive strategies including visual foveation, tracking, reaching, and tactile exploration of objects. In this way, actively maximizing key parameters relating to the self-structuring of information flows helps explain the emergence of coordinated sensorimotor activity in embodied agents, and provides a new design tool for evolving artificial agents able to profit from various forms of embodied intervention and (hence) information self-structuring.

[21]This is an especially interesting measure. It captures the degree to which a system is both functionally specialized and functionally integrated, a property that delivers maximum information-processing power. See Sporns (2002) for an accessible discussion.

CONCLUSIONS

The contemporary tendency to speak of mind as embodied is, according to one recent writer, just "a lexical band-aid covering a 350 year old wound generated and kept suppurating by a schizoid metaphysics."[22] Where some see a band-aid, others see a panacea, finding in the appeal to embodiment and environmental embedding a sweeping radical alternative to standard forms of cognitive scientific exploration and understanding. Neither view should compel our assent. To take embodiment seriously is simply to embrace a more balanced view of our cognitive (indeed, our human) nature. We are thinking beings whose nature qua thinking is not accidentally but profoundly and continuously informed by our existence as physically embodied, and as socially and technologically embedded organisms.

To understand how this is so, where it is so, how much it is so, and just what kinds of difference it makes, we will need all the tools currently at our disposal, and probably several more besides. We will need to combine a dynamic sensibility to the importance of action, timing, and closely coupled unfolding with (I predict) the use of a variety of more familiar tools and constructs. These will include the various computational, representational, and information-theoretic lenses that currently seem to provide our best understanding of the rich and complex space of adaptive trade-offs between neural, bodily, and environmental contributions and operations. But despite the use of some familiar (and some unfamiliar) tools, the object of study here is not the same as before. Our target is not just a neural control system but a complex cognitive economy spanning brain, body, and world.

ACKNOWLEDGMENTS

This chapter is a based on material drawn from Clark (2008a). Thanks to the press for permission to reuse this material here. This version was prepared thanks to support from the AHRC, under the ESF Eurocores CNCC scheme, for the CONTACT (Consciousness in Interaction) project, AH/E511139/1.

REFERENCES

Anderson, M. (2007). The massive redeployment hypothesis and the functional topography of the brain. *Philosophical Psychology, 20*(2), 143–174.

Bach y Rita, P. & Kercel, S. W. (2003). Sensory substitution and the human–machine interface. *Trends in Cognitive Sciences, 7*(12), 541–546.

Ballard, D., Hayhoe, M., Pook, P., & Rao, R. (1997). Deictic codes for the embodiment of cognition. *Behavioral and Brain Sciences, 20*, 723–767.

Braitenberg, V. (1984). *Vehicles: Experiments in Synthetic Psychology*. Cambrige, MA: MIT Press.

[22]This quote is from Maxine Sheets-Johnstone (1999, p. 275).

Clark, A. (1998). Time and mind. *Journal of Philosophy, 95*(7), 354–376.

Clark, A. (2001). *Mindware: An Introduction to the Philosophy of Cognitive Science*. New York: Oxford University Press.

Clark, A. (2008a). *Supersizing the Mind: Embodiment, Action, and Cognitive Extension*. New York: Oxford University Press.

Clark, A. (2008b). Pressing the flesh: A tension in the study of the embodied, embedded mind? *Philosophy and Phenomenological Research, 76*(1), 37–59.

Clark, A. & Chalmers, D. (1998). The extended mind. *Analysis, 58*(1), 7–19.

Collins, S. H., Ruina, A. L., Tedrake, R., & Wisse, M. (2005). Efficient bipedal robots based on passive-dynamic Walkers. *Science, 307*, 1082–1085.

Cummins, Robert (1983). *Psychological Explanation*, Cambridge Mass. and London: England MIT Press.

Goldin-Meadow, S. (2003). *Hearing Gesture: How Our Hands Help Us Think*. Cambridge, MA: Harvard University Press.

Gray, W. D. & Fu, W.-T. (2004). Soft constraints in interactive behavior: The case of ignoring perfect knowledge in the world for imperfect knowledge in the head. *Cognitive Science, 28*(3), 359–382.

Gray, W. D., Sims, C. R., Fu, W.-T., & Schoelles, M. J. (2006). The soft constraints hypothesis: A rational analysis approach to resource allocation for interactive behavior. *Psychological Review, 113*(3), 461–482.

Kirsh, D. (1995). Complementary strategies: Why we use our hands when we think. *Proceedings of the Seventeenth Annual Conference of the Cognitive Science Society*. Lawrence Erlbaum Associates.

Kirsh, D. & Maglio, P. (1994). On distinguishing epistemic from pragmatic action. *Cognitive Science, 18*, 513–549.

Lakoff, G. & Johnson, M. (1999). *Philosophy in the flesh: The embodied mind and its challenge to Western thought*. New York: Basic Books.

Lakoff, G. & Johnson, M. (1980). *Metaphors We Live By*. Chicago, IL: University of Chicago Press.

Lungarella, M., Pegors, T., Bulwinkle, D., & Sporns, O. (2005). Methods for quantifying the information structure of sensory and motor data. *Neuroinformatics, 3*(3), 243–262.

Lungarella, M., & Sporns, O. (2005). Information self-structuring: key principles for learning and development. *Proceedings 2005 IEEE Intern. Conf. Development and Learning*, pp. 25–30.

Maglio, P. P., Wenger, M. J., Copeland, A. M. (2003). The benefits of epistemic action outweigh the costs. In *Proceedings of the Twenty Fifth Annual Conference of the Cognitive Science Society*.

McNeill, D. (2005). *Gesture and Thought*. Chicago, IL: University of Chicago Press.

Noë, A. (2004). *Action in Perception*. Cambridge, MA: The MIT Press.

Noë, A. (2007). Understanding *Action in Perception*: Reply to Hickerson and Keijzer. *Philosophical Psychology, 20*(4), 531–538.

O'Regan, J. K. & Noë, A. (2001). A sensorimotor approach to vision and visual consciousness. *Behavioral and Brain Sciences, 24*(5), 883–975.

Paul, C. (2004). Morphology and computation. In S. Schaal, A. J. Ijspeert, A. Billard, S. Vijayakumar, J. Hallam, & J.-A., Meyer (Eds.), *From Animals to Animats. Proceedings of the 8th International Conference on the Simulation of Adaptive Behaviour.* (pp. 33–38). Cambridge. MA: MIT Press.

Paul, C. (2006). Morphological Computation: A Basis for the Analysis of Morphology and Control Requirements. In *Robotics and Autonomous Systems*, *54*, 619–630.

Pfeifer, R. & Bongard, J. (2007). *How the Body Shapes the Way We Think*. Cambridge, MA: MIT Press.

Putnam, H. (1960). Minds and machines. In S. Hook (Ed.), *Dimensions of Mind*. New York: New York University Press.

Putnam, H. (1967). Psychological predicates. In W. Capitan & D. Merrill (Eds.), *Art, Mind and Religion* (pp. 37–48). Pittsburgh, PA: Pittsburgh University Press.

Putnam, H. (1975). Philosophy and our mental life. In H. Putnam (Ed.), *Mind, Language, and Reality* (pp. 291–303). Cambridge, UK: Cambridge University Press.

Rohrer, T. (2006). The body in space: Embodiment, experientialism and linguistic conceptualization. In J. Zlatev, T. Ziemke, R. Frank, & R. Dirven (Eds.), *Body, Language and Mind* (Vol. 2, pp. 89–150). Berlin: Mouton de Gruyter.

Rosenblatt, F. (1962). *Principles of Neurodynamics*. New York: Spartan Books.

Shapiro, L. (2004). *The mind Incarnate*. Cambridge, MA: MIT Press.

Shapiro, L. (2007). *The Embodied Cognition Research Program*. Philosophy Compass: 2(2), 338–346.

Shapiro, L. (in press). Reductionism, embodiment, and the generality of psychology. In H. Looren de Jong & M. Schouten (Eds.), *Reductionism*. Blackwell.

Sheets-Johnstone, M. (1999). Emotion and movement: A beginning empirical–phenomenological analysis of their relationship. *Journal of Consciousness Studies, 6*(11–12), 259–277.

Sporns, O. (2002). Network analysis, complexity and brain function. *Complexity, 8*, 56–60.

Sporns, O. & Lungarella, M. (2006). Evolving coordinated behavior by maximizing information structure. *Procedings of 10th International Conference on Artificial Life*. Cambridge, MA: MIT Press.

Sterelny, K. (2003). *Thought in a Hostile World: The Evolution of Human Cognition*. Oxford: Blackwell.

Thompson, A. (1998). *Hardware Evolution: Automatic Design of Electronic Circuits in Reconfigurable Hardware by Artificial Evolution*. Distinguished dissertation series. Springer-Verlag. Berlin.

Thompson, A., Harvey, I., & Husbands, P. (1996). Unconstrained evolution and hard consequences. In E. Sanchez & M. Tomassini (Eds.), *Towards Evolvable Hardware*. Notes in Computer Science, Springer-Verlag, Berlin.

Webb, B. (1996). A Cricket robot. *Scientific American, 275*, 62–67.

Wheeler (in press). Minds, things and materiality. In C. Renfrew & L. Malafouris (Eds.), *The Cognitive Life of Things*. McDonald Institute for Archaelogical Research: Cambridge.

Yu, C., Ballard, D., & Aslin, R. (2005). The role of embodied intention in early lexical acquisition. *Cognitive Science, 29*(6), 961–1005.

4

Can a Swarm Be Embodied?

Amanda J.C. Sharkey and Noel Sharkey

Department of Computer Science, University of Sheffield, Sheffield, UK

> *The most general organismal character of the ant-colony is its individuality. Like the cell or the person, it behaves as a unitary whole, maintaining its identity in space, resisting dissolution and, as a general rule, any fusion with other colonies of the same or alien species.*
>
> *Wheeler (1911) essay on "The Ant Colony as an Organism"*

INTRODUCTION

The current enthusiasm for swarms prompts us to consider how they are related to questions about embodiment. We begin by identifying three examples of self-organized swarms: swarms of (i) social insects, (ii) physical robots, and (iii) software agents. We consider the way in which the individuals within each of these swarms are embodied, and also examine the idea that a swarm as a whole could be embodied. In doing so, our concern is to determine whether they are embodied or not, and the form that their embodiment takes.

Little attention has been paid to the embodiment of swarms and swarm members before. Questions about embodiment are more often pursued with reference to humans, and the manner in which their bodily experiences provide a grounding for cognition and language (Gibbs, 2005), or with reference to artificial agents and the way that embodied cognition provides both a solution to the symbol grounding problem (Harnad, 1990; Anderson, 2003) and an alternative to the traditional cognitivist emphasis on representation. Swarm research, on the other hand, is inspired by the self-organization shown in biological systems such as swarms, or colonies, of social insects and usually involves simple, or even reactive, individuals. It can be assumed that social insects have more limited representational, communicative, and symbol processing abilities than humans (even though evidence of their information processing abilities is gradually accumulating, cf. Detrain et al., 1999). The typically greater simplicity of the individuals in a swarm provides a different perspective on embodiment and its relationship to cognition than that offered by an anthropocentric view. In addition, although embodiment is usually considered in terms of the relationship between an individual and its physical environment, swarms encourage an awareness that that environment also consists of the other members of the swarm.

In these discussions of swarms and embodiment, we are particularly interested in determining the presence, or absence, of *strong* embodiment. We use the word "strong" by analogy to the debates about strong and weak Artificial Intelligence. The concept of strong Artificial Intelligence, or strong AI, was introduced by Searle (1980, 1997), who distinguished it from weak AI: the more cautious view that the computer is a useful tool for simulating the human mind. As Searle puts it, "*according to strong AI, the computer is not merely a tool in the study of mind; rather, the appropriately programmed computer really is a mind, in the sense that computers given the right programs can be literally said to understand and have other cognitive states*" (Searle, 1980). Debates about strong embodiment are similarly about the extent to which an artificial robot or agent can be said to be embodied in the same way as living creatures. The alternatives for artificial robots or agents are weak embodiment, or the conclusion that the agent is not embodied at all. Weak embodiment, like weak AI, refers to the situation where the effects of embodiment are modeled, and explored, but true embodiment is not achieved. However, it is also possible to identify different versions of weak embodiment (Ziemke, 2001; Chrisley & Ziemke, 2002). A further distinction between the two forms, or interpretations, of strong embodiment—mechanistic and phenomenal embodiment—(Sharkey & Ziemke, 1998, 2000, 2001; Ziemke & Sharkey, 2001) is also found to be useful.

THREE EXAMPLES OF SWARMS

Before discussing swarms and their embodiment, we need to establish definitions for the terms "swarm", "swarm intelligence", and "swarm robotics", and to explain the idea of self-organization on which they are based. Both swarm robotics and swarm intelligence are inspired by biological self-organized systems. Beni and Wang (1989) used the word "swarm" to describe their simulated cellular robots because of the features that they shared with social insects, namely, "decentralized control, lack of synchronicity, simple and (quasi) identical members" (Beni, 2005). Bonabeau et al. (1999) define swarm intelligence as "any attempt to design algorithms or distributed problem-solving devices inspired by the collective behaviour of social insect colonies and other animal societies" (Bonabeau et al., 1999, p. 7). They emphasize that self-organization is one of the key features of such societies. Although there has been some debate about its defining features (Dorigo & Sahin, 2004), swarm robotics has been described as the application of swarm intelligence principles to collective robotics (Sharkey & Sharkey, 2006) and clearly the concepts of swarm robotics, swarm intelligence, and self-organization are closely related.

Camazine et al. (2001, p. 8) offer the following definition of self-organization, "*Self-organization is a process in which pattern at the global level of a system emerges solely from numerous interactions among the lower-level components of the system. Moreover, the rules specifying interactions among the system's components are executed using only local information, without reference to the global pattern*". Social insect colonies are prototypical examples of biological self-organized systems, or swarms. In such colonies, sophisticated behavior emerges as a consequence of interactions between a large number of individuals who communicate locally, and who seem to operate independently and autonomously without knowledge of what has to be done to keep the colony in operation. Insect societies are able to maintain themselves as a collective, to accomplish the coordinated action needed to construct nests, to feed and raise their young, and to react to invasion or other interference. They are able to do so in the absence of centralized control. As Gordon, a prominent ant researcher puts it, "*the basic mystery about ant colonies is that there is no management*" (Gordon, 1999).

Self-organization in biological systems differs from its occurrence in physical systems. In physical systems, patterns can result from interactions based on physical laws. For instance, a pattern of regularly spaced ripples in sand can occur as a consequence of forces attributable to gravity and wind (Anderson, 1990; Forrest & Haff, 1992). Self-organizing biological systems also obey the laws of physics, but because they are subject to natural selection, they have an added dimension. The subunits of biological systems (individual insects in the case of social insects) can "*acquire information about the local properties of the system and behave according to particular genetic programs that have been subjected to natural selection*" (Camazine et al., 2001). The interactions between the subunits can involve the transmission of information, and since natural selection tunes the rules, the emerging behaviors are adapted to the environment. The adaptive nature of self-organization in biological systems underlies the sophisticated coordination of social insect societies.

Social insects include all ants, all termites, and some bees and wasps (some bees and wasps live solitary lives). The most advanced social (or eusocial) insects share three biological traits: the adults care for the young; two or more generations of adults live in the same nest; and the members of each colony are divided into a reproductive caste and a non-reproductive worker caste. The self-organization of such biological living swarms depends on their local communicative abilities and interactions, and on their caste and behavioral specializations. Their

local communication mechanisms include antennation, stridulation, regurgitation, and indirect stigmergic communication by leaving chemical traces in the environment. Pheromones, or "semiochemicals", can be used for a variety of purposes, from foraging recruitment to alarm communication in case of attack, and recruitment for nest construction, nest defense, and migration. These individual interactions and behaviors seem to be pursued blindly. As E.O Wilson describes it: *An important first rule concerning mass action is that it usually results from conflicting actions of many workers. The individual workers pay only limited attention to the behavior of nestmates near them, and they are largely unaware of the moment-by-moment condition of the colony as a whole* (Wilson, 1971, p. 224).

Both swarm intelligence algorithms and swarm robotics are based on the biological inspiration of social insects. Both emphasize the principles of decentralized control, local communication, and redundancy. Martinoli (2001) sees these principles as central, and identifies three main advantages to what he describes as taking a swarm intelligence approach to robotics: (i) the resulting collective systems will be scalable because the same control architecture is used for both a few and for thousands of units; (ii) such systems are flexible, because the individual units can be dynamically added or removed without the need for explicit reorganization; and (iii) the systems will be robust due to the reliance on both unit redundancy and minimalist unit design. Bonabeau et al. (1999) also list flexibility, cheapness, and fault tolerance as potential advantages, and the possibility that due to self-organization, the collective behavior of swarm-based robots could result "in patterns that are qualitatively different from those that could be obtained with a single agent or robot". The promise of swarm intelligence systems is that of developing "*artificial distributed problem-solving devices that self-organize to solve problems*" (Bonabeau et al., 1999).

We can briefly describe typical manifestations of swarm intelligence algorithms and swarm robotics. The ant colony optimization algorithms are prime examples of swarm intelligence algorithms. They depend on software agents that mark the route that they take with the equivalent of a pheromone trace that decays over time. The algorithms depend on simultaneous exploration of different routes by a collection of virtual ants that are influenced by the pheromone trails left by others. Ant-based algorithms have been shown to be useful for the approximate solution of combinatorial optimization problems. In particular, the ant colony optimization meta-heuristic has been successfully applied to problems such as the Traveling Salesman problem (Dorigo & Gambardella, 1997) and routing and load balancing in telecommunication networks (Schoonderwoerd et al., 1996; Di Caro & Dorigo, 1998). The typical characteristics of many swarm robotics studies are represented by a series of studies (Beckers et al., 1994; Holland & Melhuish. 1999; Wilson et al., 2004). These studies investigated the emergence of sorting and clustering behaviors as a consequence of the interactions between reactive robots that responded reflexively to pucks that they encountered in their environment; picking them up as they came across them, and dropping them when they bumped into other pucks. The interest here is in the emergence of solutions as a consequence of self-organized interactions between the agents, each other and the environment.

Swarm intelligence and many (but not all) examples of swarm robotics share a commitment to decentralization, local communication and redundancy (for a discussion of examples that do not have a commitment to these features. See Sharkey, 2007). The two approaches can also be distinguished from each other, particularly in the extent to which they are physically realized. Swarm robotics more often involves the use of actual physical agents or, at least more realistic simulations of physical agents than is the case in swarm intelligence algorithms. The use of real physical robots can result in some simpler solutions, as aspects of the environment can be exploited to achieve some tasks, but their use also creates extra problems, as mechanical and sensorial challenges have to be solved. The two artificial swarm approaches also differ in terms of the closeness of their relationship to the biological swarms that inspired them. Swarm intelligence researchers are usually not concerned to model the biological systems that originally inspired the approach very closely. Bonabeau et al. (1999) point out that swarm intelligence algorithms do not need to be accurate models of biological systems, and that, "*efficiency, robustness and flexibility are the driving criteria, not biological accuracy*". Not all swarm robotics researchers are interested in modeling biological systems, but at least a subset of them are concerned to constrain the control, communicative and representational abilities of their robots to those plausible for the social insects that inspired their approach (e.g., Krieger & Billeter, 2000; Parker & Zhang, 2004a, b, 2005), and swarm robotics often involves a greater commitment to biological modelling than swarm intelligence.

In summary, we have outlined the main properties of three different examples of swarms: biological swarms or colonies of social insects, swarm intelligence algorithms, and swarm robotics. All three share the features of being decentralized, and dependent on local communication, and of the self-organized emergence of global or collective behaviors as a consequence of interactions between their subunits. They also depend on the advantages of redundancy and expendability—individual insects, software agents, or robots may be removed, or replaced without the need to update the methods of control or communication used. At the same time, the three approaches can be distinguished from each other. They can be in different ways: the living example can be distinguished from the artificial ones; and the two that exist in tangible physical form can be distinguished from the software example. In addition, the two artificial swarms differ both in terms of their physical manifestation, and also in the extent of their interest in modeling or adopting constraints from biological swarms of insects.

ARTIFICIAL SWARMS AND STRONG EMBODIMENT

Having outlined the main characteristics of these swarms, we can now consider their embodiment. We will consider the artificial examples first, and the biological example in the following section. To begin with, we focus on questions about the embodiment of the individuals in a swarm. Is a software agent in a swarm intelligence system embodied, or is a physical body, of some kind, essential? Can a physical robot in a swarm robotic system be seen as embodied, and if so what form does that embodiment take?

It is sometimes suggested that software agents, without physical bodies, can be embodied, in the sense of being sufficiently connected to their environment. For instance, Franklin (1997) writes, "*Software systems with no body in the usual physical sense can be intelligent. But they must be embodied in the situated sense of being autonomous agents structurally coupled with their environment*". Pfeifer & Scheier (1999) also suggest that although intelligence requires a body, that body could be a simulated one: "*Embodiment: A term used to refer to the fact that intelligence cannot merely exist in the form of an abstract algorithm but requires a physical instantiation, a body. In artificial systems, the term refers to the fact that a particular agent is realised as a physical robot or as a simulated agent*". The underlying idea here is that intelligent behavior requires an agent to be situated in an environment, such that the environment can affect it, and it is able to affect the environment. This view of embodiment makes sense when contrasted to the Good Old Fashioned Approach to AI (GOFAI) that emphasized internal representation and reasoning about the world, at the same time as it ignored the issues and complexities involved in interacting with the environment. GOFAI assumed that information and data from the world was somehow available to the system, and was uninterested in the details of how it became available. When viewed in the context of providing an alternative to GOFAI, systems that are coupled to, and capable of interacting with, their environment would be considered to be embodied.

Dautenhahn and her colleagues made a related but different argument about embodiment. For them, there is no such thing as an unembodied agent— under their argument everything is embodied to some degree. They complain about "*simplistic binary distinctions between systems that are embodied and those that are not*", and suggest that instead it is possible to quantify their "*degree of embodiment*". In their arguments, Dautenhahn et al. (2002) make use of a definition that emphasizes the relationship between a system, and its environment: "*A system **S** is embodied in an environment **E** if perturbatory channels exist between the two. That is, **S** is embodied in **E** if for every time **t** at which both **S** and **E** exist, some subset of **E**'s possible states with respect to **S** have the capacity to perturb **S**'s state, and some subset of **S**'s possible states with respect to **E** have the capacity to perturb **E**'s state*" (Quick & Dautenhahn, 1999). They claim that this minimal definition of embodiment provides an opportunity to quantify embodiment without specifying the metric to be used.

The primary candidate for such measurement that they discuss is the perturbatory bandwidth, or the range of events that the system and its environment can produce that the other is sensitive to (Quick & Dautenhahn, 1999). As they point out, some perturbation can be found even amongst inanimate material objects, where climactic and geological forces can impact on the geographical features. Thus a granite outcrop is weathered and perturbed by the wind, and also perturbs the air currents around it. In Fong et al. (2003), it is argued that AIBO (Sony) with 20 actuators has more perturbatory channels than the Khepera with two actuators and can therefore be considered to be more strongly embodied than a Khepera. They also argue that a social robot can be embodied without a physical body and a conversational agent could be embodied to the same extent as robots with limited actuation. Dautenhahn et al. (2002) compare the degree of embodiment, construed as perturbatory bandwidth, of the *E. coli* bacteria, and that of the beaver. The *E.coli* has "*limited and minimal sensory and effector surfaces (that) provide a low bandwidth coupling between organism and environment*", whereas the beaver "has a highly developed sensory system, and through behaviors such as dam-building, significantly perturbs its environment and its own ongoing relationship with that environment". They do not explicitly contrast living with non-living examples—but their definition implies that they would consider a Sony AIBO to be more embodied than an *E.coli* bacteria.

According to the definitions quoted above by Pfeifer & Scheier (1999) and Franklin (1997), and the degrees of embodiment account (Dautenhahn et al. 2002) all of our swarm examples are made up of embodied individuals. However, describing both the artificial and biological examples as embodied overlooks what we see as a fundamental distinction between living and non-living bodies. Our contention is that although artificial agents that interact directly with their environment can be viewed as embodied, their embodiment is a *weak* as opposed to a *strong* embodiment. As examples of weak embodiment, they can provide useful models, and increase understanding of the ways in which the environment and the body affect intelligence. However, there remains an insurmountable difference between examples of the strong embodiment, and the weak embodiment of living systems. A living system is part of the environment in a way that an artificial system cannot be. At the same time, a living system is an integrated whole in the way that an artificial system, cannot be.

A distinction between weak and strong embodiment is also implied by the different views of embodiment identified by Ziemke (2001) and Chrisley & Ziemke (2002). They identify four views of embodiment: the first three are forms of weak embodiment, and the last corresponds to what we are calling strong embodiment:

1. **Physical realization**: realized in some physical substrate, as even "a web-prowling virtual agent" must be (Chrisley & Ziemke, 2002). This view was described as "structural coupling" by Ziemke (2001), and can be seen to correspond also to Franklin's (1997) definition quoted above.
2. **Physical embodiment**: where the realizing physical system is a coherent integral system that persists over time—an account of embodiment that includes conventional robots.
3. **Organismoid embodiment**: where the localized physical realization of the system shares some characteristics with the bodies of natural organisms,

but need not be alive. The examples they give are of humanoid robots such as Cog (Brooks & Stein, 1993) and Kismet (Breazeal & Sassellati, 2000), and the cricket robot complete with ears developed by Webb (Webb, 1995; Webb & Scutt, 2000).

4. **Organismal embodiment**: "The body must not only be organism like, but actually organic and alive" (Chrisley & Ziemke, 2002).

If we consider these four versions of embodiment in terms of our artificial swarm examples, it becomes apparent that the software agents in swarm intelligence algorithms correspond to view (1) since they are physically realized, or structurally coupled, whereas physical robots are physically embodied (view 2). Neither can be said to be organismally, or strongly embodied.

Organismal embodiment corresponds to strong embodiment. There are however, two different possible interpretations of strong embodiment, and of what such embodiment implies in terms of cognition: mechanistic and phenomenal embodiment. These two versions of strong embodiment were identified by Sharkey & Ziemke (1998, 2000, 2001; Ziemke & Sharkey, 2001). We introduce them here to reinforce our claim that artificial swarms are not strongly embodied, and consider them again in relation to natural swarms in Section 4.

Under a mechanistic interpretation of embodiment, it is assumed that there is no separate cognitive or representational apparatus apart from the mechanism itself. The behavior of an organism is grounded in the interaction between the agent and the environment. The mechanistic viewpoint can be traced back to the work of Loeb (1918) and Sherrington (1906), and the idea of the environment determining the response of the organism. Sherrington was concerned to provide an account of the solidarity of multi-cellular animals, and how their bodily reactions are integrated. He noted the mechanical combination of cells into a single mass, the solidarity that results from chemical communication between cells and body parts, and saw the reflex arc as the elementary unit of integration and behavior. Loeb was interested in the response of the whole animals in reaction to its environment. He developed a theory of *tropisms*, the forced movement of higher organisms in response to stimuli such as light or gravity (using the term tropism as opposed to "taxis" in acknowledgement of the similarity to the tropic responses of plants).

In contrast, the phenomenal interpretation, refers to the embodiment of a mental or subjective world, and stresses the embedding of the organism in the physical environment, and in its own phenomenal world, and the close coupling between the two. The idea of phenomenal embodiment is based on the work of von Uexküll, who objected to the mechanistic doctrine, and Loeb's work in particular, claiming, "*We no longer regard animals as mere machines, but as subjects whose essential activity consists of perceiving and acting. We thus unlock the gates that lead to other realms, for all that a subject perceives becomes his perceptual world and all that he does, his effector world. Perceptual and effector worlds together form a closed unit, the Umwelt*". (von Uexküll, 1957, p. 6).

Phenomenal embodiment is also consistent with the views of Maturana and Varela, who argue that living systems cannot be properly analyzed at the level of physics, but require a biological phenomenology: "autopoietic unities specify biological phenomenology" (Maturana & Varela, 1987).

We claim that the individual members of artificial swarms are not strongly embodied in either the mechanistic or the phenomenal sense. The difference between living bodies, and artificial robot bodies, can be made clearer with reference to Maturana and Varela, and the concepts of autopoietic and allopoietic organization. The term, "autopoiesis" was introduced by Maturana & Varela (1973) and means "auto (self)-creation" from the Greek *auto* for self and *poiesis* for creation or production. An autopoietic machine is a homeostatic machine that maintains its own organization. It has a unity that distinguishes it from its environment, a unity that depends on the autonomy of its individual cells, or "first-order autopoietic units". The solidarity of these individual cells constitute the organism as an integrated behavioral entity and second-order autopoietic unit", due to the fact that "the structural changes that each cell undergoes in its history of interactions with other cells are complementary to each other, within the constraints of their participation in the metacellular unity they comprise". Maturana & Varela (1987)

An allopoietic machine such as a car or a robot is not self-sustaining, but depends on components produced by other processes that are independent of the organization of the machine. In addition, the chemical, mechanical and integrating mechanisms of living things are missing from robots. New sensors or body parts could be added to or removed from a robot without affecting its "multicellular solidarity". No pain could be felt by a robot as it is dismantled, for there is no integrated "self" there to feel. Nor could any physiological evidence of stress be found in the robot, in the way that Bradshaw & Bateson (2000), for instance, found evidence of stress in the bloodstream and muscles of deer that had been hunted, for again there is no integrated mechanism that connects the body parts that could be so affected. The question of whether or not an insect can feel pain or stress is undecided (Eisemann et al., 1984), but nonetheless, the point is that an insect's body forms an integrated whole, in the way that a robot's does not. As well as being an autopoietic system, an insect also has a nervous system and a brain. The physical instantiation of a robot appears to give it a body, in the sense that when it moves, it takes all its parts with it. But such a body is of a very different kind to that possessed by living organisms.

If the individual members of artificial swarms are not strongly embodied, in either a mechanistic or a phenomenal sense, then it follows that the behaviors that emerge from artificial swarms do so in a different way than from biological swarms. Emergent effects do occur in artificial swarms: the sorting that results from the interactions of robots with pucks, the environment, and each other is emergent in the sense of being more than the sum of the individual behaviors of the robots. However, the major difference between the emergent effects obtained in swarm robotic and swarm intelligent studies, and those exhibited by natural swarms as they regulate and balance the needs of their colony or nest, is the necessary involvement of human researchers. In the artificial examples, researchers make a number of decisions about the robot behaviors, the situation and the environment they are placed in; all of which affect the solutions that subsequently "emerge". The puck sorting robots did not spontaneously begin to pick up the pucks that they encountered: they were constructed with this behavior in mind. There is no reason intrinsic to the robots for undertaking the task. Although sorting and clustering is achieved without any explicit planning or programming, these emergent behaviors were indirectly programmed by humans when they set up the initial conditions for the robots.

The same is true even if computational evolutionary methods are used to design behaviors—when the bodies involved are not strongly embodied, they cannot be subject to the effects of the environment in the same way that living organisms can be. The solutions that "emerge" are still the result of decisions made by the researchers about the situations the robot will encounter, the fitness function, the behaviors required and so on. The emergent self-organized behavior of an insect colony, on the other hand, *is* the product of evolutionary adaptation. Insects adapt to environmental pressures, and this adaptation is dependent on the intimate relationship between their bodies and the environment. Their living bodies have evolved with their environment, and, in the case of social insects they have also co-evolved to form a mutually dependent swarm.

In summary, we have considered the embodiment of both swarm intelligence algorithms and swarm robotics. Both examples fit with current zeitgeist and its emphasis on embodied Artificial Intelligence: in terms of the views of embodiment identified by Ziemke (2001) and Chrisley & Ziemke (2002), swarm intelligence algorithms can be seen as being structurally coupled to their environment, and *physically realized*, whilst robots in a swarm are *physically embodied*. However, it is concluded that the robots and software agents in these swarms are not organismally, or strongly embodied, in either a mechanistic, or a phenomenal sense. Self-organized effects do emerge as a consequence of artificial individuals interacting in a swarm, but these effects emerge in a different way than they do from biological swarms, and do not change their designation from weak embodiment to strong. Because the individual members are not strongly embodied, the collective behaviors that emerge from artificial swarms depend on the interventions and decisions of researchers, and do not result purely from the adaptive pressures of the environment.

IS A LIVING SWARM AN EMBODIED ENTITY?

In this section, we turn to a consideration of biological swarms and strong embodiment, and the idea that a swarm as a whole could be strongly embodied. The notion of collective embodiment has not been discussed before, although it is implied by Wheeler's concept of the superorganism (Wheeler, 1911). In considering the idea of a swarm as an embodied entity we again make use of the concepts of mechanistic and phenomenal embodiment introduced in the preceding section. However, before considering the natural swarm as a whole, we should first consider the embodiment of the individual insect within it.

We can assume that as living organisms, insects are strongly embodied: but should they be viewed as mechanistically or phenomenally embodied? If insects are phenomenally embodied, this would imply that they were capable of cognition and experience. Indeed von Uexküll's original explication of the *Umwelt* used the example of the subjective experience of the tick, an insect that exists in an Umwelt that consists of the odor of butyric acid that emanates from the sebaceous follicles of mammals, the temperature of 37°C, and the hairiness of surfaces:

> We are not concerned with the chemical stimulus of butyric acid, any more than with the mechanical stimulus (released by the hairs), or the temperature stimulus of the skin. We are solely concerned with the fact that, out of the hundreds of stimuli radiating from the qualities of the mammal's body, only three become the bearers of receptor cues for the tick ... What we are dealing with is not an exchange of forces between two objects, but the relations between a living subject and its object... The whole rich world around the tick shrinks and changes into a scanty framework consisting, in essence of three receptor cues and three effector cues—her Umwelt.
>
> von Uexküll (1957)

Maturana and Varela would also assume insects to be capable of cognition. For them, "autopoiesis is necessary and sufficient to characterize the organization of living systems" (Maturana & Varela, 1973), whilst "living systems are cognitive systems, and living as a process is a process of cognition" (Maturana, 1970). The alternative to their position is to view insects as being mechanistically embodied, and hence not capable of thinking about, or experiencing the world.

Of course, barring an unexpected metamorphosis (Kafka, 1968), we will never know what an insect's experience of the world is. In part, a decision about whether or not an insect should be viewed as being mechanistically or phenomenally embodied depends on one's views about the possibility of insect cognition. Certainly the suggestion that insects might be capable of some form of cognition used to be far more controversial than it is now. For instance, in his paper, "What is it like to be a bat?" Nagel justified his choice of bats thus: "*I have chosen bats instead of wasps ... because if one travels too far down the phylogenetic tree, people gradually shed their faith in there being any experience at all*" (Nagel, 1981). Similarly, Moser (1970) seems to be describing a form of mechanistic embodiment when he wrote, "*Insects function like tiny robots programmed to do specific jobs. Their nervous systems act like biological computers; they are activated, as if by punch cards, when their receptors are stimulated. The external receptors respond to pressure, sound, light, heat and chemicals.*" However, there is now a greater interest in the notion of minimal cognition (van Dujin, 2006), and in a biogenenic view of cognition which takes simpler forms of cognition as its starting point (Lyon, 2006) and sees cognition in terms of a continuum, that stretches from insects, or even bacteria, to human cognition. There is also an accumulating body of evidence which illustrates insects' abilities to learn, remember, and process information (Papaj & Lewis, 1992; Detrain et al., 1999).

We conclude then that the individuals in a living swarm are strongly embodied— and that there are reasons to believe their embodiment is a phenomenal one. But can a biological swarm as a whole be embodied? The idea seems counter-intuitive at first—surely a swarm is not contained within a body, but is made up of many individual bodies?

It can be argued that a living swarm is an example of an autopoietic, and not an allopoietic system. As biological self-organized systems, swarms seem coherent with Maturana and Varela's definition of autopoietic systems: "*An autopoietic machine is a machine organized (defined as a unity) as a network of processes of production (transformation and destruction) of components that produces the components which: (i) through their interactions and transformations continuously regenerate and realize the network of processes (relations) that produced them; and (ii) constitute it (the machine) as a concrete unity in the space in which they (the components) exist by specifying the topological domain of its realization as such a network*" (Maturana & Varela, 1973).

In addition, an autopoietic machine "*continuously generates and specifies its own organisation through its operation as a system of production of its own components*" (Maturana & Varela, 1973), and so does a living swarm of ants or bees. Such swarms have evolved as self-organized systems as the consequence of their unmediated interactions with the environments in which they occur. They are structurally coupled to the environments in which they exist: "*In these interactions, the structure of the environment only triggers structural changes in the autopoietic unities (it does not specify or direct them), and vice versa for the environment. The result will be a history of mutual congruent structural changes as long as the autopoietic unity and its containing environment do not disintegrate: there will be a structural coupling*" (Maturana & Varela, 1987).

A biological swarm should presumably be viewed as a higher order autopoietic entity, since it is composed of individual autopoietic systems, or insects, which are themselves made up of autopoietic cells. Maturana & Varela (1973) discuss how multi-cellular organisms are autopoietic entities of a second order: providing the following definition, "*if the autopoiesis of the component unities of a composite autopoietic system conforms to allopoietic roles that through the production of relations of constitution, specification and order define an autopoietic space, the new system becomes in its own right an autopoietic entity of second order*" and claim that "*if the higher order autopoietic system undergoes self-reproduction ... an evolutionary process begins in which the evolution of the manner of realization of the component autopoietic systems is necessarily subordinated to the evolution of the manner of realization of the composite unity*" (Maturana & Varela, 1973).

MECHANISTIC EMBODIMENT AND LIVING SWARMS

If a swarm can be viewed as an autopoietic entity, it should also be possible to consider it to be mechanistically embodied. It is the case that a living swarm does not require the intervention of outside agencies. A biological swarm is directly affected by its environment—indeed, the acceptance of the idea that it has evolved presupposes that its whole development is a consequence of such effects. Biological swarms of social insects evolved as a consequence of their direct and unmediated interactions with the environment—and they are able to respond directly to changes and events in the environment, at the same time as maintaining their own homeostasis and equilibrium. It can be shown, for example, that a honeybee colony is very good at thermoregulation—maintaining a constant internal temperature between 34.5°C and 35.5°C: Gates (1914) recorded a situation in which the temperature of adult bee clusters was held at 31°C even though the outside temperature was –28°C.

A biological swarm is capable of many other forms of social homeostasis. Examples include the way in which the number of workers of fire ants attracted to food increases linearly as the amount of trail pheromone laid down increases, and the amount of pheromone increases with the number of workers laying trails, such that the number of workers approaching a food source is a linear function of its size (E.O. Wilson, 1971). Another (slightly disturbing) example reported by Wilson (1971) is of the way in which colonies of the termite genus *Zootermopsis*, when kept on a diet lacking in nitrogen, are able to restore the balance by cannibalism—which ceases as soon as more protein is added to their diet.

The suggestion that a biological swarm is strongly embodied in a mechanistic sense is related to the once popular concept of the swarm, or ant colony as a superorganism (Wheeler, 1911). Wheeler compared the tight organization of such swarms of eusocial insects to that of an organism, and argued that they behave as a unit that "*possesses distinctive properties of size, behavior, and organisation that are transmitted from colony to colony and from one generation to the next. The queen is the reproductive organ, the workers the supporting brain, heart, gut and other tissues. The exchange of liquid food among the colony members is the equivalent of the circulation of blood and lymph*" (Hölldobler & Wilson, 1994). The notion of the superorganism was much discussed at the time, but gradually abandoned as biologists turned their attention to more detailed studies of the communication and caste formation that underlies colony organization. But "*old ideas in science ... never really die*" (Hölldobler & Wilson, 1994), and there is a renewed interest in the idea of sociogenesis, and the way in which, by means of evolution, individuals change and specialize as they form a society. Sociogenesis is the next level up from morphogenesis, which is the process by which cells change their shape and chemistry as they form an organism.

ARE LIVING SWARMS PHENOMENALLY EMBODIED?

It seems that it is possible to make an argument that a biological swarm is mechanistically embodied. Can a convincing argument also be made that a living swarm is phenomenally embodied? Maturana and Varela argue, or assume, that all living systems are capable of cognition. On this basis, accepting a living swarm as an autopoietic system would imply that it is also capable of cognition. The same implication could be drawn from the use of the term "mind" to refer to emergent swarm effects. For example, the "hive mind" referred to by Kelly (1994), or the "collective mind" (Couzin, 2007), or "swarm cognition" (Passino et al., 2008). Such suggestions are backed up by pointing out the similarity between the emergence of a mind from the collection of neurons that makes up a brain, and the emergence of behaviors as a consequence of the interactions between insects in a living swarm. Could the combination of individual insects in a living swarm similarly give rise to the ability to think, reason, and plan?

The similarity between brains and swarms is considered explicitly in a recent paper on honeybees (Passino et al., 2008) entitled, "Swarm cognition in honey bees". The authors argue that there are key features of cognition in "the neuron-based brains of vertebrates" that are also to be found in swarms of honeybees. These include "*the existence of interconnected subunits, parallel processing of information, a spatially distributed memory, layered processing of information, lateral inhibition, and mechanisms of focusing attention on critical stimuli*". The focus of their consideration is the ability of bees to make collective decisions about nest site selection, something that was once found surprising, as evidenced by the following quote, (borrowed from Crist, 2004), "*Surely these are some of the most astonishing things that have yet been discovered in the whole realm of bee behaviour? How can bees which one supposes have no powers of reasoning, reach what amounts to an agreement on one of several possible nesting sites?*" (Butler, 1954, p. 166).

The abilities of ant colonies and swarms of bees to come to collective decisions about a new nest site may appear to support the notion that a swarm is phenomenally embodied. However, when their decision-making processes are examined in more detail, this seems less likely. Visscher & Camazine (1999) for instance, report a detailed study of swarming honeybees, and their ability to make a collective decision about the new site to move to when reproductive swarming occurs. Although few individual bees have the opportunity to visit and compare multiple nest sites, as a group the bees are able to evaluate a number of sites and reach unanimous agreement on a single site. Visscher and Camazine's explanation of the decision-making mechanism is that it depends on the local communication of finds, and positive feedback as bee scouts dance, and on attrition, as scouts stop dancing for the less good sites. Ants can also make effective collective decisions about nests, although in their case the relative advantages of different nests are indicated by chemical recruitment trails and tandem running, decisions based on quorum sensing, and the number of ants visiting a site (Pratt et al., 2002).

The organization of a swarm, as was emphasized in Section 2, is inherently decentralized. Ants and bees can make collective decisions, but these seem to be the result of the individual members of the swarm being influenced by their neighbors and by the environment. The problem for the notion of phenomenal embodiment of living swarms is that it can be argued that some form of centralized organization is a prerequisite for conscious experience.

The organization of a human body is inherently centralized, in that much of its operation is dependent on and coordinated by the brain. The same is true for simpler organisms also—individual insects within a swarm for example. An argument can be derived from Damasio's neurophysiological account of consciousness (Damasio, 2000) to the effect that that conscious experience, and thereby phenomenal embodiment, depends on the existence of centralizing structures and organization that appear to be absent in the swarm.

Damasio distinguishes between three kinds of self: the "proto self"; the "core self" ; and the "autobiographical self". These can be thought of as levels, with each depending on its predecessor. The "proto self" depends on centralizing structures (the brain stem, the hypothalamus, and the basal forebrain) that receive chemical signals from the body, and cause the release of neurotransmitters in the central cortex, thalamus, and basal ganglia. The proto self provides a model of the body and its state, and is, Damasio argues, a prerequisite for the development of the next level of self: "*I have come to conclude that the organism, as represented inside its own brain, is a likely biological forerunner for what eventually becomes the elusive sense of self. The deep roots for the self, including the elaborate self which encompasses identity and personhood, are to be found in the ensemble of brain devices which continuously and nonconsciously maintain the body state within the narrow range and relative stability required for survival. These devices continually represent, nonconsciously, the state of the living body along its many dimensions. I call the state of activity within the ensemble of such devices the proto-self, the nonconscious forerunner for the levels of self which appear in our minds as the conscious protagonists of consciousness: core self and autobiographical self*" (Damasio, 2000). For Damasio, "core consciousness", or "the feeling of knowing" depends on the proto self. It occurs "*when the brain's representational devices generate an imaged, non-verbal account of how the organism's own state is affected by the organism's processing of an object*", and for our purposes, the point is that such consciousness depends on the representation of the organism's state provided by the proto self.

Damasio's account of the prerequisites for even the simplest form of consciousness suggests that in order for something to experience the world, or in a further step, to be aware of itself experiencing the world, it needs centralizing organization and structures. However, as we have seen, the predominant characteristic of swarms is their decentralized organization. Effects like decision making, and thermoregulation do emerge from swarms, but it is usually accepted that they are the result of decentralized local interactions. Other examples of emergent effects include the "effective" extended perception that can result for instance, from individuals in a flock or swarm responding to their neighbors. Thus, as Couzin (2007) points out, individual fish in a school can behave as though their perceptual range was greater than it was, because they turn as their neighbors turn, even though they themselves have not detected the predator. There is a range of mechanisms by which such individual effects can be transmitted and amplified as a result of interactions between individuals, but it is difficult to see how the sense data received by one fish, or one insect, could be directly shared with another. Similarly the decision making that occurs as ants, or bees, choose a nest site is an emergent effect, and no individual, or group, has access to all the factors that influenced the decision. It seems that creatures with nervous systems and brains are organized and structured in a way that a swarm of insects is not.

Suggesting that a biological swarm is a strongly embodied autopoietic system, but that this embodiment is mechanistic rather than phenomenal, means going against Maturana and Varela's assumption that all living systems are capable of cognition. Other authors have expressed reservations about equating life with cognition: suggesting for instance that there are reasons for wanting to be able to "differentiate between the cognitive processes of the rabbit and those of the carrot" (van Dujin et al., 2006). Some have argued that cognition requires a nervous system (Moreno et al., 1997), since this enables an organism to go beyond merely metabolic functions. Van Dujin et al. (2006) question the need for a nervous system, and argue that "the core of cognition revolves around sensorimotor coupling". They make an interesting case for the behavior of the *E.coli* bacterium, and its sensorimotor system as representing an instance of minimal cognition. However, such discussions have been phrased in the context of individual organisms and not in terms of swarms. The interesting point about the present discussion is that it raises the possibility of a different set of prerequisites for minimal cognition—suggesting that cognition requires more than collective information processing and decision making, but also some rudimentary form of self-hood and experience.

In summary, we argue that it is possible to view a biological swarm as being strongly embodied. Because it is composed of living bodies, it is part of, and directly affected by the environment without the intervention of outside agencies. The tight interdependent organization of eusocial insect colonies means that they can be viewed as higher order autopoietic systems. However, it is easier to make the case that such living systems are mechanistically rather than phenomenally embodied. A living swarm can be viewed as being mechanistically embodied because its self-organized decentralized organization has developed in response to the unmediated evolutionary pressures of the environment. However, there are differences between the decentralized organization of biological swarms, and the centralized and hierarchically structured organization of organisms with nervous systems. If such organization is held to be a prerequisite for conscious experience, then it follows that a swarm is unlikely to be phenomenally embodied.

Addressing the idea of embodiment and swarms, particularly biological swarms, opens up a new set of questions and issues, and this chapter represents a first stab at addressing them. Distinguishing between mechanistic and phenomenal embodiment helps to characterize the embodiment of a living swarm. Describing a biological swarm as being strongly (mechanistically) embodied captures the interdependence and unity of its members. At the same time, the idea that phenomenal embodiment might require centralizing organization and structure, highlights the differences between the decentralized way that emergent effects occur in a swarm, and the way that conscious experience may occur in individual living organisms. The same distinction between mechanistic and phenomenal embodiment may prove to be useful in the further development of the notion of minimal cognition. In the future, we plan to develop these considerations further, and to extend them to other collective phenomena, particularly the examples of multi-cellular aggregation found in slime mold, and bacteria (Camazine et al., 2001).

CONCLUSION

We have explored the concept of embodiment from the perspective of biological and artificial swarms, with the aim of determining the presence or absence of weak and strong embodiment, and the form that each takes. We began by describing the main characteristics of three examples of swarms: swarm intelligence algorithms, swarm robotics, and biological swarms; before exploring questions about their embodiment. This exploration has been pursued in two parts—first in terms of artificial swarms, and secondly in terms of biological swarms. A consideration of the individuals in artificial swarms led to the conclusion that they were not strongly embodied, even though collective effects emerged as a result of their interactions. Unlike living organisms, they are not autopoietic systems, and they are neither mechanistically nor phenomenally embodied.

In Section 4, the embodiment of biological swarms was considered. It was argued that the individual insects in such swarms are strongly embodied: whether they should be viewed as mechanistically or phenomenally embodied could be seen as a matter of personal preference, although we side with the latter. The question of whether a biological swarm could itself be viewed as strongly embodied was also addressed. It was argued that if it is accepted as a higher order autopoietic system, then a biological swarm can be considered to be strongly embodied—such swarms have evolved as the result of the unmediated interaction between themselves and the environment. The question of whether such strong embodiment should be interpreted as being mechanistic or phenomenal was also discussed. It was concluded that it is possible to argue that they are mechanistically embodied.

The question of strong phenomenal embodiment for a living swarm was difficult to address. There is no evidential basis for or against it and even knowing what would constitute an evidential basis is problematic. A swarm may be a strongly embodied superorganism (Wheeler, 1928) that is capable of making decisions, but it is not necessarily capable of conscious thought and experience. A swarm's decisions arise from decentralized self-organization. If centralizing structures are a necessary condition for conscious experience, then phenomenal embodiment of living swarms is not possible.

REFERENCES

Anderson, R. S. (1990). Eolian ripples as examples of self-organisation in geomorphological systems. *Earth-Science Reviews, 29*, 77–96.

Anderson, M. L. (2003). Embodied cognition: A field guide. *Artificial Intelligence, 149*(1), 91–130.

Beckers, R., Holland, O. E., & Deneubourg, J. L. (1994). From local actions to global tasks: Stigmergy and collective robotics. *Proceedings of A-Life IV*. MIT Press.

Beni, G. (2005). From swarm intelligence to swarm robotics, In E. Sahin & W.M. Spears (Eds.), *Swarm Robotics*, Lecture Notes in Computer Science, 3342 (pp. 1–9). Springer: Berlin Heidilberg New York.

Beni, G. & Wang, J. (1989). Swarm intelligence. *Proceedings of the Seventh Annual Meeting of the Robotics Society of Japan*, Tokyo, Japan, pp. 425–428.

Bonabeau, E., Dorigo, M., & Theraulaz, G. (1999). *Swarm Intelligence: From Natural to Artificial Systems*. Oxford University Press.

Butler, C. G. (1954). *The World of the Honeybee*. London: Collins.

Bradshaw, E. L. & Bateson, P. (2000). Welfare implications of culling red deer (*Cervus elaphus*). *Animal Welfare, 9*, 3–24.

Breazeal, C. & Scassellati, B. (2000). Infant-like social interactions between a robot and a human caretaker. *Adaptive Behavior, 8*, 1.

Brooks, R. A. & Stein, L. A. (1993). *Building Brains for Bodies*. Cambridge: MIT Artificial Intelligence Laboratory, (AI Memo no. 1439).

Camazine, S., Deneubourg, J.-L., Franks, N. R., Sneyd, J., Theraulaz, G., & Bonabeau, E. (2001). *Self-Organisation in Biological Systems*. Princeton and Oxford: Princeton University Press.

Chrisley, R. & Ziemke, T. (2002). Embodiment. In *Macmillan Encyclopedia of Cognitive Science*. ISBN 0-333-79261-0

Couzin, I. D. (2007). Collective minds. *Nature, 455*, 715.

Crist, E. (2004). Can an insect speak? The case of honeybee dance language. *Social Studies of Science, 34*(1), 7–43.

Damasio, A. (2000). *The Feeling of What Happens: Body, Emotion and the Making of Consciousness*. London: Vintage.

Dautenhahn, K., Ogden, B. & Quick, T. (2002). From embodied to socially embedded agents – Implications for interaction-aware robots. *Cognitive Systems Research 3*(3), 397–428. Special issue on Situated and Embodied Cognition: Tom Ziemke (guest-editor), Elsevier.

Detrain, C., Deneubourg, J. L., & Pasteels, J. M. (Eds.) (1999). *Information Processing in Social Insects*. Basel: Birkhauser Verlag.

Di Caro, G. & Dorigo, M. (1998). AntNet: Distributed stigmergic control for communications networks. *Journal of Artificial Intelligence Research (JAIR), 9*, 317–365.

Dorigo, M. & Gambardella, L. M. (1997a). Ant Colony System: A cooperative learning approach to the travelling salesman problem. *IEEE Transactions on Evolutionary Computation, 1*, 53–66.

Dorigo, M. & Sahin, E. (2004). Guest editorial: Swarm robotics. *Autonomous Robots, 17*(2–3), 111–113.

Van Dujin, M., Keijzer, F., & Franken, D. (2006). Principles of minimal cognition: Casting cognition as sensorimotor coordination. *Adaptive Behaviour, 14*(2), 157–170.

Eisemann, C. H., Jorgensen, W. K., Merritt, D. J., Rice, M. J., Cribb, B. W., Webb, P. D., & Zalucki, M. P. (1984). Do insects feel pain? – A biological view. *Cellular and Molecular Life Sciences, 40*(2), 164–167.

Fong, T., Nourbakhsh, I., & Dautenhahn, K. (2003). A survey of socially interactive robots. *Robotics and Autonomous Systems, 42*, 143–166.

Forrest, S. B. & Haff, P. K. (1992). Mechanics of wind ripple stratigraphy. *Science, 255*, 1240–1243.

Franklin, S. A. (1997). Autonomous agents as embodied AI. *Cybernetics and Systems, 25*(8), 499–520.

Gates, B. N. (1914). The temperature of the bee colony. *Bulletin, United States Department of Agriculture, 96*, 1–29.

Gibbs, R. W. (2005). *Embodiment and Cognitive Science*. New York: Cambridge University Press.

Gordon, D. (1999). *Ants at work: How an insect society is organised*. London: W.W. Norton and Company, Inc.

Harnad, S. (1990). The symbol grounding problem. *Physica D, 42*, 335–346.

Hölldobler, B. & Wilson, W. O. (1994). *Journey to the Ants: A Story of Scientific Exploration*. Cambridge, MA, London, UK: The Belknap Press of Harvard University Press.

Holland, O. & Melhuish, C. (1999). Stigmergy, self-organisation, and sorting in collective robotics. *Artificial Life, 5*, 173–202.

Kefka, F. (1968). *Metamorphosis and Other Stories*. . Penguin Modern Classics.

Kelly, K. (1994). *Out of Control: The New Biology of Machines, Social Systems, and the Economic World*. New York: Addison-Wesley.

Krieger, M. J. B. & Billeter, J. B. (2000). The call of duty: Self-organised task allocation in a population of up to twelve mobile robots. *Robotics and Autonomous Systems, 30*, 65–84.

Loeb, J. (1918). *Forced Movements, Tropisms and Animal Conduct*. Philadelphia: Lippincott Company.

Lyon, P. (2006). The biogenic approach to cognition. *Cognitive Processes, 7*, 11–29.

Martinoli, A. (2001). Collective complexity out of individual simplicity. Invited book review on "Swarm Intelligence: From Natural to Artificial Systems" by E. Bonabeau, M. Dorigo, and G. Theraulaz, *Artificial Life*, *7*(3), pp. 315–319.

Maturana, H. R. (1970). Biology of cognition. In H. R. Maturana & F. J. Varela (Eds.), *Autopoiesis and Cognition: The Realization of the Living*. Boston: Reidel, 1980.

Maturana, H. R. & Varela, F. J. (1973). Autopoiesis: The organisation of the living. In H. R. Maturana & F. J. Varela (Eds.), *Autopoiesis and Cognition: The Realization of the Living*. Boston: Reidel.

Maturana, H. R. & Varela, F. J. (1987). *The Tree of Knowledge – The Biological Roots of Human Understanding*. Boston, MA: Shambhala.

Moreno, A., Umerez, J., & Ibanez, J. (1997). Cognition and life: the autonomy of cognition. *Brain and Cognition, 34*(1), 107–129, New york: Academic Press.

Moser, J. C. (1970). Pheromones of social insects. In D. Wood, R. Silverstein, & M. Nakajima (Eds.), *Control of Insect Behavior by Natural Products* (pp. 161–178). Academic Press.

Nagel, (1981). What is it like to be a bat? In D. R. Hofstadter & D. C. Dennett (Eds.), *The Mind's I*. New York: Basic Books.

Papaj, D. R. & Lewis, A. C. (Eds.) (1992). *Insect Learning: Ecological and Evolutionary Perspectives*. London, New York: Chapman and Hall.

Parker, Chris A.C. & Zhang, H. (2005) Active versus passive expression of preference in the control of multiple-robot decision-making. In *Proceedings of the 2005 IEEE/RSJ Int'l Conference on Intelligent Robots and Systems (IROS 2005)*.

Parker, Chris A.C. & Zhang, H. (2004a) Biologically inspired decision making for collective robotic systems. In *Proceedings of the 2004 IEEE/RSJ Int'l Conference on Intelligent Robots and Systems (IROS 2004)*.

Parker, Chris A.C. & Zhang, H. (2004b) Collective decision making: A biologically inspired approach to making up all of your minds. In *Proceedings of 1st IEEE Int'l Conference on Robotics and Biomimetics (ROBIO 2004)*.

Passino, K. M., Seeley, T. D., & Visscher, P. K. (2008). Swarm cognition in honey bees. *Behavioural Ecology and Sociobiology, 62*, 401–414.

Pfeifer, R. & Scheier, C. (1999). *Understanding Intelligence*. Cambridge, MA: MIT Press.

Pratt, S. C., Mallon, E. B., Sumpter, D. J. T., & Franks, N. R. (2002). Quorum sensing, recruitment, and collective decision-making during colony emigration by the ant Leptothorax albipennis. *Behavioral Ecology and Sociobiology, 52*, 117–127.

Quick, T. & Dautenhahn, K. (1999). Making embodiment measurable, Workshop Contribution "Embodied Mind/Artificial Life" at 4. Fachtagung der Gesellschaft für Kognitionswissenschaft, 28 September to 1 October 1999 in Bielefeld (KogWis99).

Schoonderwoerd, R., Holland, O.E., Bruten, J., & Rothkrantz, L. (1996) Ant-based load balancing in telecommunications networks. *HP Labs Technical Report*, HPL-96-76, May 21.

Searle, J. R. (1980). Minds, brains and programs. *Behavioural and Brain Sciences, 3*, 417–457.

Searle, J. R. (1997). *The Mystery of Consciousness*. London: Granta Books.

Sharkey, A. J. C. (2007). Swarm robotics and minimalism. *Connection Science, 19*(3), 245–260.

Sharkey, A. J. C. & Sharkey, N. E. (2006). The application of swarm intelligence to collective robotics. In J. Fulcher (Ed.), *Advances in Applied Artificial Intelligence* (pp. 157–185). Hershey, PA: Information Science Publishing.

Sharkey, N. E. & Ziemke, T. (1998). A consideration of the biological and psychological foundations of autonomous robotics. *Connection Science, 10*(3–4), 361–391.

Sharkey, N. E. & Ziemke, T. (2000). Life, mind and robots – The ins and outs of embodied cognition. In S. Wermter & R. Sun (Eds.), *Hybrid Neural Systems*. Heidelberg, Germany: Springer Verlag.

Sharkey, N. E. & Ziemke, T. (2001). Mechanistic vs phenomenal embodiment: Can robot embodiment lead to strong AI? *Cognitive Systems Research, 2*(4), 251–262.

Sherrington, C. S. (1906). *The Integrative Action of the Nervous System*. New York: C. Scribner's Sons.

Von Uexküll, J. (1957). A stroll through the world of animals and men – A picture book of invisible worlds. In C. H. Schiller (Ed.), *Instinctive Behaviour – The Development of a Modern Concept* (pp. 5–80).

Visscher, P. K. & Camazine, S. (1999). The mystery of swarming honeybees: From individual behaviours to collective decisions. In C. Detrain, J. L. Deneubourg, & J. M. Pasteels (Eds.), *Information Processing in Social Insects*. Basel: Birkhauser Verlag.

Webb, B. (1995). Using robots to model animals: A cricket test. *Robotics and Autonomous Systems, 16*, 117–134.

Webb, B. & Scutt, T. (2000). A simple latency dependent spiking neuron model of cricket phonotaxis. *Biological Cybernetics, 82*, 247–269.

Wheeler, (1911). The ant-colony as an organism. *Journal of Morphology, 22*(2), 307–325.

Wheeler, W. M. (1928). *The Social Insects: Their Origin and Evolution*. London: Kegan Paul, Trench, Trubner and Co. Ltd.

Wilson, M., Melhuish, C., Sendova-Franks, A. B., & Scholes, S. (2004). Algorithms for building annular structures with minimalist robots inspired by brood sorting in ant colonies. *Autonomous Robots, 17*, 115–136.

Wilson, E. O. (1971). *The Insect Societies*. Cambridge, MA, London, UK: The Belknap Press of Harvard University Press.

Ziemke, T. & Sharkey, N. E. (2001). A stroll through the worlds of robots and animals: Applying Jakob von Uexküll's theory of meaning to adaptive robots and artificial life. *Semiotica, 134*(1–4), 701–746.

Ziemke, T. (2001). Are robots embodied? In C. Balkenius, J. Zlatev, H. Kozima, K. Dautenhahn, & C. Brezeal (Eds.), *Proceedings of the First International Workshop on Epigenetic Robotics-Modeling Cognitive Development in Robotic Systems*. Lund University Cognitive Studies, *85*, 75–83.

SECTION II

ROBOTICS AND AUTONOMOUS AGENTS

5

CAJUNBOT: A CASE STUDY IN EMBODIED COGNITION

ISTVAN S.N. BERKELEY

The Institute of Cognitive Science, University of Louisiana at Lafayette, Lafayette, LA, USA

INTRODUCTION

One of the difficulties with writing about so-called embodied cognition, is that the precise meaning and scope of the idea is a little unclear. For instance, Ziemke (2003) has identified several distinct conceptions of "embodied cognition." The most important of these distinct conceptions of embodiment can be summarized as follows:

1. *Structural Coupling*: A situation in which a structurally coupled relationship exists between the agent and the environment is required for a system to count as being embodied.
2. *Historical Embodiment*: A situation in which a history of Structural Coupling is required for a system to count as being embodied.
3. *Physical Embodiment*: A situation in which a system requires "physical instantiation" to count as being embodied.
4. *"Organismoid" Embodiment*: A situation in which a system requires an organism-like body to count as being embodied.
5. *Organismic Embodiment*: A situation in which only (biological) living bodies count as being embodied.

It should be clear that Structural Coupling is the least restrictive notion of embodiment, whereas Organismic Embodiment is the most restrictive. However, what all the versions of embodiment that Ziemke describes share in common is a close and essential link between the agent or system and the environment. As such, these conceptions of embodiment might be referred to as instances of "environmental embodiment.". The reason for raising this conception in the current context is because the bulk of the discussion here will concern a robotic device, known as "CajunBot.". The version of CajunBot (there are now several) that will be concentrated upon here was the autonomous robotic system that competed in the 2005 Defense Advanced Research Projects Agency (DARPA) Grand Challenge. As such, this system clearly counts as being "environmentally embodied" under at least the first three notions just described. This notion of embodiment, though, does not completely exhaust the notion of "embodied cognition," for reasons that will be clarified later.

The issue that will be addressed in this chapter is the status of embodied cognition, as compared with more traditional strategies in the study of cognition. On this issue, Cowart (2006) has identified two main positions. The so-called Compatibilist Approach, for example, is associated with Clark (1997). This approach assumes that embodied strategies and tools for studying cognition can be used in conjunction with more traditional strategies and tools for such study. This approach stands in stark contrast to the so-called Purist Approach. The latter rests upon the more radical thesis that traditional strategies and tools are inherently flawed, and thus must be replaced wholesale with tools and strategies inspired by an embodied perspective on cognition. This kind of position, for example, is advocated by Varela et al. (1991).

The main position that the fans of the Purist Approach to embodied cognition wish to take issue with is the so-called Classicist/Cognitivist view of cognitive science. Cowart (2006) characterizes the main differences between the Classicist/Cognitivist view and the Embodied Cognitivist view as consisting of a number of important contrasts.

While the Classicist/Cognitivist view is grounded upon a computer-based metaphor of the mind, the Embodied Cognitivist view instead advocates a coupling metaphor for the mind. The first view focuses primarily upon rule-based systems which are significantly logic driven, whereas the second takes the position that cognitive processes are importantly constrained by forms of embodiment, and the environment and actions within that environment. Cowart also characterizes the Classicist/Cognitivist view as being isolationist, with a primary focus on an organism's internal processes. This stands in contrast to the Embodied Cognitivist view which concentrates upon attempting relational analyses of the interplay among mind, body, and environment. Another important contrast is between computation being the primary interest for the Classical view, as opposed to goal-directed action unfolding in real time, being the primary focus for the Embodied view. On this characterization, the Classical view takes passive retrieval as being a paradigmatic cognitive activity, whereas the Embodied view sees cognition as active construction, based upon an organism's embodied, goal-directed actions, as a paradigmatic activity. Finally, under the Classical view, representations are conceptualized as being symbolic encodings (cf. Berkeley,

2008). These contrast with the supposedly sensorimotor representations of the Embodied view.

Although there are certain "family resemblances" between aspects of Cowart's description and the environmentally embodied conception of embodied cognition that Ziemke (2003) describes, this apparent similarity actually masks a considerably more complex situation. What Cowart describes has an architectural, or systematic (in the sense of "properties of the system"), or methodological component that goes beyond Ziemke's environmental conception of embodiment. This additional component involves commitments to certain (contingently) representational strategies and certain ways of computing functions. For clarity, these additional commitments will be termed "methodological embodiment." Although this terminology is not ideal, it does serve to identify notions associated with embodied cognition which lie outside the scope of simple environmental embodiment.

At this point, it is natural to ask about the relationship between environmental and methodological embodiment. There is a sense in which methodological embodiment and environmental embodiment are related, or allied, though independent of one another. This is because, for example, systems such as the famous Stanford Cart (Moravec, 1990) would count as being environmentally embodied, without being methodologically embodied. Furthermore, there are examples that are embodied, presumably in a methodological sense, which are not embodied in the environmental sense. Some of the work of Beer (2004, 2006) comes to mind in this context. However, the relationship between the two notions of embodiment is far from unproblematic.

One of the reasons that it is a little tricky to precisely describe the relationship between the environmental and methodological conceptions of embodiment has to do with the wide variety of positions that fall into the scope of the latter term. Although there is a general consensus that there should be a commitment to "non-Classical" representations and processes, there is little consensus on precisely what is taken to be appropriate. For example, some fans of methodological embodiment, such as Chemero (2000), embrace a broadly anti-representational (Brooks, 1991) position. Other theorists, such as Clark (2005), advocate a less radical and controversial position. Although a detailed taxonomy of the various versions of methodological embodiment would be both useful and an interesting undertaking, the development of such a taxonomy would not be appropriate here, as it would take us too far from the main focus of this chapter. For current purposes, as a general guideline, both environmentally embodied and methodologically embodied cases will be taken as instances of embodied cognition.

There is a certain tension generated by taking this approach, however, that is worth commenting upon. In so far as methodological embodiment defines itself as being non-Classical, there are two distinct worries that immediately come to mind. First, if the non-Classical status of a system, or a subsystem, is taken as being definitional for the system or the subsystem to have (methodologically) embodied status, then adopting this stance just simply begs the question with respect to whether hybrid systems are possible (thereby rendering an adjudication between the Compatibilist Approach and the Purist Approach impossible). As such, this stance is surely defective, at least in the current context. Second, given that having symbolic status is one of the definitional features of a Classical system, the fact that there is a good deal of confusion over what exactly is to count as a symbol (Berkeley, 2008) means that whether any particular system should be considered as methodologically embodied will always be debatable. For these reasons then, it will be left to the reader to determine whether or not a particular system or subsystem should count as methodologically embodied, when it comes to the detailed discussion of the CajunBot architecture. Although this leaves the methodologically embodied status of aspects of CajunBot pretty much in the proverbial "eye of the beholder," helpful comments will be added, where appropriate.

It is also worth noting that Cowart's characterization of the contrasts between the Classical and the Embodied views may not be perfect for analogous reasons. Not all theorists will necessarily agree with all points. However, this characterization is offered here as a handy, broad way of understanding the kinds of issues that can arise between the two positions. More importantly, this characterization gives a series of points of reference with which the debate between the Compatibilist Approach to embodied cognitive science, as opposed to the Purist Approach, can be judged. To put it plainly, if the Purists are correct, then we should not find cases of actual embodied systems, regardless of the kind of embodiment involved, that contain any classical elements. As will be illustrated below, the actual facts of the matter, with respect to real-world systems, are a good deal more complicated than this.

The strategy in the rest of this chapter will be as follows: First, the CajunBot system will be introduced. Next, various subsystems of the robot's architecture will be examined in some detail. Through this process of careful scrutiny, it should become clear that the computational strategies used in CajunBot contain examples which are highly consistent with the Embodied view, as well as cases which are much more obviously Classical in nature. All these examples, in combination, form a powerful argument in favor of the Compatibilist Approach to embodied cognitive science and against the Purist Approach. Hopefully, along the way some useful insights into the strengths, advantages, and weaknesses of both the Embodied and Classical views of cognitive science will be generated.

CAJUNBOT AND THE DARPA GRAND CHALLENGE, 2005

In June 2004, DARPA announced their Grand Challenge for the following year. The goal of the Grand Challenge was for teams to construct vehicles that could autonomously navigate a challenging course over varying terrain. In October 2004, DARPA issued the detailed rules for the Challenge. These rules stipulated that the course for the Challenge would be no more than 175 mile long (282 km). It was also announced that the terrain over which the vehicles would have to travel would consist of roads, trails, and off-road desert areas, containing a variety of obstacles.

A team from the University of Louisiana at Lafayette elected to enter the 2005 Grand Challenge. They decided to compete under the name "Team CajunBot," as this was the name that they had used when competing in the 2004 Grand Challenge. It is for analogous reasons that they decided to call their autonomous vehicle "CajunBot."

CAJUNBOT: HARDWARE

The CajunBot vehicle was based upon a Recreative Industries MAX IV All Terrain Vehicle (ATV), as a machine of this type was available to the team at no cost due to a donation. It is worth noting that cost savings were an important consideration to team CajunBot at the time, as the State of Louisiana was in a perilous financial state due to the impacts of Hurricanes Katrina and Rita in 2005. This meant that there were extremely limited resources available for the CajunBot project. This fact significantly influenced a number of the design decisions that were taken with respect to the CajunBot vehicle.

The ATV was very suitable for the project as it had all-wheel drive through each of its six wheels. This enabled the vehicle to navigate rough terrain with a minimal risk of getting stuck. Maneuvering the vehicle was effected through a skid steering system. The whole system was powered by a 28 hp Kohler engine. The basic ATV chassis also had to be augmented in a number of ways to accommodate the additional computational and sensory systems that the vehicle needed for autonomous operation. In particular, additional power was supplied from a pair of Honda generators.

Sensor input to the CajunBot vehicle came from two main sources. The first of these was an Oxford Technical RT3000 Inertial Navigation System. This was then enhanced by Starfire differential Global Positioning System (GPS), from a C & C Technology C-NAV receiver. The second source of input came from five LIDAR laser obstacle sensors. Output from the system was handled by a series of actuators that interfaced with the ATVs control systems (for further details, see Lakhotia et al., 2006). In between the sensors and the actuators was a computational architecture that is illustrated in Figure 5.1. This architecture provided CajunBot with the abilities necessary to sensibly navigate through the terrain of the Grand Challenge as well as other test terrains.

THE "EMBODIED" NATURE OF THE CIRCUMSTANCE

The reason why CajunBot is of particular interest in the current context is due to the nature of the task that it had to undertake. CajunBot had to operate autonomously in a real-world environment in real time. These factors served to constrain the kinds of solutions that could be used with CajunBot when it came to solving problems. For example, the inputs received from the sensor system could in no sense be idealized.

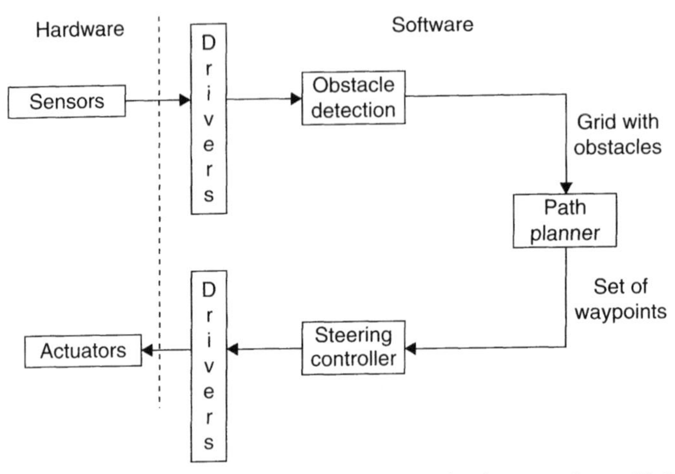

FIGURE 5.1 An architectural overview of the computational systems that enabled CajunBot to function autonomously.

The underlying software had to be able to handle any corruptions in the data stream in a manner that would have facilitated operation while minimizing the potential for catastrophic failure. It was also the case that the actual maneuverability of the vehicle had to be factored in, when computing commands to be sent to the control systems. Failure to take these factors into consideration would have rendered the system incapable of completing the task at hand. Finally, control system processes had to operate quickly enough to prevent mishaps. It is for these reasons, then, that CajunBot is an interesting case of a truly embodied system, at least in an environmentally embodied sense.

It turned out that these real-world constraints had a significant influence on the way that particular problems were solved by the software systems run by CajunBot. Indeed, examining the various subsystems of CajunBot reveals an interesting interaction between solutions that are in the spirit of embodied approaches to cognition, in addition to solutions that had a distinctly more traditional flavor. It is due to these influences that CajunBot is also a plausible candidate for an embodied system in the methodological sense. At this point, it is time to begin looking at the details of the way that CajunBot functioned, to illustrate these claims with concrete examples.

CAJUNBOT SENSOR SYSTEMS

LIDAR LASER SENSORS

In the words of CajunBot Team member, Anthony Maida (personal communication, 2006), "... the vehicle moves like a brick on wheels." This lack of

FIGURE 5.2 The effect of terrain upon the detection of obstacles by the CajunBot LIDAR laser sensors. When CajunBot was on an uphill slope, this would have the effect of focusing the sensors further away. When CajunBot was on a downhill slope, this would have the effect of focusing the sensors closer.

maneuverability, in conjunction with the fact that CajunBot could travel at speeds of up to 28 miles per hour (45 kph) over rough terrain, meant that there were some real issues when it came to integrating the sensor inputs from the LIDAR laser sensors. These sensors were CajunBot's primary obstacle detection system. As such, they were a crucial subsystem that had to function correctly and reliably, if CajunBot was to succeed in operating autonomously. A particular difficulty arose over the fact that where obstacles were detected depended in an important manner, upon the vertical angle of the vehicle. The problem is illustrated in Figure 5.2.

When CajunBot was on an uphill slope, the focus of the sensors would be further away than when it was on a flat surface. Similarly, when CajunBot was on a downhill slope, the focus of the sensors would be closer than when on a flat surface. The rough nature of the terrain over which CajunBot had to operate made this quite a serious issue that needed to be handled in an appropriate manner. This was because determining the distances between CajunBot and obstacles in its path was crucial to the prevention of collisions.

There were basically three strategies that could have been adopted to resolve this issue. The first strategy would be to ensure that the LIDAR sensors were mounted on a vehicle with very good suspension systems, such that it damped the movements of the sensors caused by the terrain. However, this option was unavailable to Team CajunBot, as they only had access to the ATV hardware that they had at hand. The second strategy would be to mount the LIDAR sensors on a platform that was stabilized by a Gimble, or some other mechanical system, that would ensure that the sensors remain focused on a constant distance ahead. This strategy was, in fact, used by other teams that competed in the 2005 Grand Challenge. However, such an approach was unavailable to Team CajunBot due to cost considerations. Such mechanical stabilization systems could cost around $70,000 which was far more than Team CajunBot's limited budget allowed. The third strategy for handling this issue was thus the only one available to Team CajunBot. This strategy involved just mounting all the LIDAR sensors on a single, extremely rigid, platform. However, it turned out that adopting this strategy, in practice, gave rise to an unanticipated and surprising advantage.

The way this problem was handled was to take the outputs from the LIDAR sensors and integrate it with a stream of data that was available from the inertial navigational system (INS). One of the things that the INS provided was data that encoded the "pose" of the vehicle. Thus, by determining where the system was, with respect to a horizontal plane, it became possible to compute the exact location that the LIDAR system was reading from. In fact, it turned out that adopting this strategy provided an unanticipated advantage, in so much as it increased the effective range of the LIDAR sensors. This was because the previously problematic up and down motion of the LIDAR sensors was able to be utilized as something akin to a poor man's vertical sweep system, thereby increasing the effectiveness of this sensor system. Thus, this is a case in which a challenge presented by the embodied nature of the task that CajunBot had to solve, actually gave rise to an unforeseen advantage for the system. Despite this though, the representations used by the system and the kinds of computations performed by the system were highly classical. This then is a case in which the Compatibilist position would seem to have the upper hand.

THE SPIKES AND Z-DRIFT PROBLEMS

The issue with the LIDAR system just described was not the only problem that arose with the sensor systems due to the real-world nature of the task. Two further problems arose with the INS/GPS system. The first problem was that, from time to time, the INS/GPS system would produce "spikes" in the data stream due to glitches in communications with the satellites. This "spike" data needed to be detected, filtered out, and discarded, to prevent corrupt information from entering the downstream systems. The second problem arose due to a phenomenon known as "Z-drift," that the INS/GPS system suffered from. It turned out that the value of the *Z* component (i.e., the vertical component) of the data produced by the INS/GPS system would exhibit a certain drift in values reported over time. Even when the vehicle was entirely stationary, the *Z* values reported by the INS/GPS system could drift by between 10 and 25 cm. This had the effect of making an otherwise perfectly flat surface appear uneven.

The solution to the spikes problem was handled by a clever use of the multiple data streams available through the various CajunBot sensory systems. The sensor system had inputs from both the INS/GPS system and from the LIDAR sensors. The INS/GPS system was filtered to remove the data spikes caused by satellite communication problems. This filtered data was then used to compute global coordinates of the LIDAR scans. This had the effect of preventing any corrupt data entering the data stream. When the information from the GPS system encountered an issue, due to a spike appearing, there was still data available from the INS sub-component of the system. Thus, by carefully using the multiple data streams available via the CajunBot sensor systems, it was possible to retain data integrity despite the problem presented by intermittent satellite communication systems.

The Z-drift issue was also handled by employing additional information available to the system. Whenever a global point reading was taken, a time relative to the internal system time of CajunBot was associated with it. Thus, each global point reading was represented to the system as a 4D value, with the format X, Y, Z, time-of-measurement. This enhanced representation was then used to ensure that only points that had a temporal distance between them of under 3 seconds were used in further computations. By doing this, it was possible to effectively overcome the Z-drift problem.

The spike and Z-drift problems could have been ignored in a purely abstract model of the task at hand. However, in the real-world context in which the CajunBot system had to operate, effective means of handling these issues had to be developed, in order for the system to be able to correctly handle the mission presented in the Grand Challenge. Thus, the environmentally embodied nature of the task at hand had a significant influence upon the technical details of the way that the CajunBot system operated. This serves to illustrate the important differences between embodied circumstances, as opposed to those assumed by more traditional approaches.

DATA INTEGRATION

Another problem that arose with the CajunBot sensor system came about due to the fact that the various different sensor subsystems produced data at different rates. For example, the INS system generated data at 100 Hz, producing data at 10 ms intervals. In contrast, the LIDAR laser systems operated at 75 Hz, producing data at 13 ms intervals. As a result of this, the most recent INS reading could be up to 9 ms old when a LIDAR scan was read. Given the potential speed of the CajunBot system, it was important that this discrepancy was allowed for, in order to avoid out-of-date information entering into the system, thereby potentially leading to the incorrect identification of the location of obstacles.

The data integration problem was solved by the CajunBot system by ensuring that the different modules in the system communicated with one another via a "blackboard" system. The blackboard system was conceived as being analogous to the working memory of more traditional cognitive systems (Englemore & Morgan, 1989; Craig, 1995). Instead of just fusing the most recent data from the LIDAR sensors and the INS/GPS systems, the blackboard system solved the data integration problem by computing global points. These global points were computed by interpolating the state immediately before and after each LIDAR scan was read. This approach, although highly traditional in many ways, served to solve the data integration problem and enabled the CajunBot system to operate effectively in the task environment.

The examples discussed above serve to show that the CajunBot system used a variety of strategies to solve the problems that were presented by the challenge of operating in a real-world, real-time environment. Some of these strategies, for example, those that were used to solve the difficulties presented by the up and down movement of the LIDAR laser sensors, were clearly highly embodied in spirit, at least in the environmental sense of embodiment. However, other strategies, most notably that adopted to solve the data integration problem, were far more traditional in their approach. There are also reasons to suspect that at least some of the problem solutions described above were such that they would count as being methodologically embodied, at least in the view of some theorists. Overall though, these facts, in combination, suggest that the Compatibilist Approach to embodied cognition is much more consistent with actual practice than the Purist Approach. However, it turns out that the sensor systems are not the only part of the overall CajunBot system architecture where interesting conclusions such as this can be drawn from. For this reason, it is also worthwhile looking at the CajunBot path planning system subsystem, in a little detail.

PATH PLANNING

Path planning, as a type of search, is a venerable topic in traditional Artificial Intelligence research (see for example, Rich & Knight, 1990; Sharples et al., 1989). Generally speaking, planning systems can be characterized as being either "deliberative," or "reactive." Blythe and Reilly (1993) describe the key strengths and weaknesses of the two kinds of systems as follows

> Deliberative systems that embody powerful techniques for reasoning about actions and their consequences often fail to guarantee a timely response in time-critical situations. Reactive systems that respond well in time-critical situations typically do not provide a reasonable response in situations unforeseen by the designer.

For CajunBot to perform successfully on the task presented by the Grand Challenge, it is clear that features of both kinds of systems would be required. A deliberative strategy would be extremely helpful in ensuring that CajunBot managed to reliably reach goals along the specified route. However, a reactive strategy would be vital in ensuring that CajunBot managed to successfully avoid obstacles that appeared along the way. Also of interest, in the current context, is that the deliberative type systems are broadly more consistent with traditional approaches to the study of cognitive systems, whereas the reactive type systems are broadly more consistent with methodologically embodied approaches to the study of cognitive systems. Path planning for the CajunBot system was handled by two distinct subsystems, the G-Nav and the L-Nav systems.

The long range planning system used with CajunBot was known as the G-Nav system. This system functioned based upon a sequence of static GPS waypoints that were held in a route description file. As such, the G-Nav system pretty clearly falls into the class of deliberative systems.

The local planning system was known as the L-Nav system. This system provided subgoals that enabled the navigation between the static waypoints. The L-Nav system was able to take into account the presence of local obstacles and make the appropriate adjustments to the relevant sub-goals. The status of the L-Nav system, with respect to the deliberative versus reactive classes of planning systems, is less clear than the case of the G-Nav system. The reason why this is the case should become clear by looking a little more closely at the details of the L-Nav system.

The path planning method used by the L-Nav system rested upon a metaphor of charged particles. The idea is described in Maida et al. (2006) as follows:

> ... the robot is (say) positively charged and a desired goal is negatively charged. Obstacles are given the same charge as the vehicle. The simulated force vectors can control the steering of the actual robot in the actual world so that the robot approaches the goal while avoiding obstacles.

The fact that the L-Nav system enabled CajunBot to escape from hazards, like dead end canyons, is suggestive that the system has at least some deliberative properties. In contrast, the real-time obstacle avoidance capacity of the L-Nav system appears to be suggestive that this system also exhibits reactive properties. In fact, this issue was one which generated a good deal of debate (although not too much agreement) between members of the CajunBot team, according to Anthony Maida (personal communication, 2006). On balance though, L-Nav is probably best thought of as being a blended system. As such, it is also a candidate for methodological embodiment.

Obstacles were represented in the L-Nav system as having an "expansion region" around them. This expansion region effectively made obstacles larger than they really were. The purpose of the expansion region was to provide a margin of safety, so as to allow for "... imperfect steering or other unanticipated physical event[s]." according to Maida et al. (2006). Given the potentially catastrophic consequences of a collision between CajunBot and an obstacle, this was a prudent and necessary affordance. However, this once again is a case where the environmentally embodied nature of the Grand Challenge task clearly made an augmentation to the system necessary to take into account real-world constraints and variables.

The final issue that is relevant in the current context, with respect to the path planner system, concerns waypoint filtering. This was necessary to take into account the limits of maneuverability of CajunBot. Particularly while traveling at speed, CajunBot was simply unable to make extremely rapid turns or execute sharp changes of direction, due to factors such as the momentum of the system. Also, attempting such extreme maneuvers would put the system at risk of toppling over. To take account of these limitations, the path planning system had to undertake the filtering of potential waypoints, to avoid adopting problematic trajectories. This was handled by first identifying places in a proposed path that would involve a change of direction. The potential waypoints in a proposed path were then filtered to ensure that none of them involved a change of direction that lay beyond the operational capabilities of CajunBot. Only the paths that were consistent with the capabilities of CajunBot were then selected. This then was another instance in which the real-world nature of the task at hand influenced the functioning of the CajunBot system.

From the above, it should be clear that the G-Nav system was deliberative. As such, it was very traditional in the way it operated. However, the same was not the case for the rest of the path planning system of CajunBot. The L-Nav system had both deliberative and reactive features. The expansion region approach adopted to effect obstacle avoidance and the waypoint filtering method of the path planning system explicitly were designed to take into account the real-world nature of the context in which CajunBot was designed to operate. As such, as with the sensor systems described in the previous section, the path planning system is much more consistent with the Compatibilist Approach to embodied cognition than the Purist Approach.

In the next section, a brief examination of the steering control system will be undertaken. It turns out that the conclusions that can be drawn from this system are similar to the conclusions that follow from looking at the sensor system and the path planning system.

STEERING CONTROL

Steering the CajunBot vehicle presented some interesting challenges. This was, in part, because the steering system presented a classic control system problem. As CajunBot turned toward a new heading, it was necessary to stop turning before the precise desired new heading was reached, to prevent oversteering. This problem arose, in large part, due to the real-time nature of the problem. Both software systems and hardware systems suffered from temporal lag. The solution deployed with CajunBot was a typical engineering solution, based upon a proportional integral differential (PID) control system (Sellars, 2001). In fact, the CajunBot steering system only used the proportional and differential terms, as will become clear below. However, as in the previously discussed CajunBot subsystems, these environmental constraints led to software approaches that at least potentially are consistent with the methodological version of embodiment as should become clear below.

The way that the steering controller worked was as follows. The controller had a number of inputs. One important input was a value that encoded the *current_heading*; another value encoded the *desired_heading*. From these it was then possible to compute a value for *error* by subtracting the value of *current_heading* from *desired_heading*. Once *error* had been computed, it was possible to compute a value for *error_rate* by subtracting *error* from *previous_error*, where *previous_error* was the value of *error* on the previous control loop execution.

The next step was to compute the proportional term (P-term) and the differential term (D-term). The P-term was computed by multiplying the value of *error* by a constant K_p. The D-term was computed by multiplying *error_rate* by a constant K_d. The values of the constants K_p and K_d were determined empirically in field trials. The P-term provided a measure of how much error needed to be corrected, whereas the D-term provided a metric of the rate of increase or decrease in *error.* The P-term and D-term were arranged such that they were generally of opposite signs, such that they could cancel each other out.

Once these computations were complete, it became possible to compute a value for the steering command as follows:

$$Steering = K_p\,(Error) + K_d\,(Error\_rate)$$

Using this method, it was found that the steering problem could be solved satisfactorily for CajunBot. When the results of this system were passed to the actuators, CajunBot was able to navigate successfully, without running into problems of oversteer, understeer, or falling into oscillatory states.

There were some highly dynamic elements that had to be taken into account in this method of solving the steering problem. These dynamic elements are the parts that are consistent with methodological embodiment, given that dynamical systems (van Gelder, 1997) are taken by some (Chemero, 2000) to be paradigmatic of methodologically embodied systems. However, in this case, the dynamic elements were included as a direct consequence of the environmentally embodied nature of the CajunBot steering task. These dynamic elements could have been abstracted away from, or just ignored, if the system just had to operate in a highly abstract domain that could idealize the environment (i.e., if this system was deployed in a purely classical domain). Unlike the previous cases though, in the case of the CajunBot steering control system, this is a system that is strongly consistent with both the environmental and methodological approaches to embodied cognition. So, this is the only part of the CajunBot system examined thus far which would be consistent with the Purist approach. All the other evidence is much more consistent with the Compatibilist Approach.

SIMULATIONS

Simulation environments were used extensively in the development of CajunBot and its software systems. Given that simulated environments have no direct contact with the real world, they appear to fall unambiguously into the Classical/Cognitive Approach far more than they do into the environmentally Embodied Approach. Despite this fact, they were absolutely crucial to refining the final functional architecture of CajunBot.

There were two main strategies that were used with respect to simulations, in the development of CajunBot. The first strategy involved the use of targeted simulations that were used to get answers to questions about particular problems. The second strategy involved the use of more comprehensive simulations to solve full system integration and testing problems.

TARGETED SIMULATIONS

Targeted simulations were of particular importance to the development of the CajunBot navigation system. For example, potential field visualizations were used extensively in the development of the L-Nav module. Different methods of generating field flow maps were tested extensively, using simulations, to determine which methods would produce the best results. Indeed, it was this testing that led to an abandonment of the neural network-based potential field generation strategy that was initially used in favor of using simple linear potential fields. Thus, this is a nice example of a case where simulation testing had a crucial influence upon the final CajunBot system architecture (for a detailed discussion, see Maida et al., 2006).

Targeted simulations were also important for developing the final methods by which the G-Nav and L-Nav systems interacted with each other. Initially, the G-Nav system was designed to invoke the L-Nav system whenever an object which was a potential obstacle was detected in the range of the sensors. However, through extensive simulation testing it was eventually determined that the G-Nav and L-Nav systems should run as concurrent processes. The integration of these two processes was also perfected through the use of simulations.

COMPREHENSIVE SIMULATIONS

The more comprehensive simulations of the CajunBot systems can be roughly divided into early and later phases. In the early phases, the simulations were constructed incrementally, with more and more realistic features being added, both with respect to CajunBot and the environment. As this process progressed, the addition of realistic steering delays led to something of a crisis. This was because it became clear that, even at low speeds, CajunBots direction of travel would oscillate, leading to crashes. Fortunately, it was also discovered from these simulations that the adoption of the waypoint interaction systems between the G-Nav and L-Nav systems significantly improved these steering issues.

In the later stages of comprehensive CajunBot simulations, the process continued to provide useful information and to highlight bugs. For example, there were coordinate transformation bugs that this process revealed. A bug concerning a failure to translate between centimeters and meters when reading from the blackboard communication system was discovered. A number of waypoint extracting bugs were also discovered through this process. In addition, the process of testing richer simulation environments revealed that the L-Nav system gave uninformative error messages when it encountered unanticipated types of data found in the richer simulation environments. This suggested that an enhanced simulation environment that used a broader spectrum of data should have been used in earlier simulations.

The simulation environment also made it possible to test CajunBot in various situations. For example, it was possible to determine how the system would behave when faced with an obstacle (e.g., a van) that was located exactly on top of a G-Nav global waypoint. Fortunately, under this set of circumstances, CajunBot performed perfectly. It was also possible to determine how the system would perform under circumstances, such as when CajunBot found itself in a dead-end canyon, with the next waypoint directly beyond the end of the canyon. The simulations showed that CajunBot would handle this situation effectively for the most part.

It is interesting, and perhaps a little ironic, to note that the use of simulations was motivated, in large part, by real-world factors, thus suggesting that embodiment may not really be quite as far removed from these otherwise classical approaches as might be initially supposed. Actual system testing had the drawbacks of being expensive and time consuming to conduct. It also carried with it the risk of damage to the hardware components of CajunBot. Given the very limited funds available to the CajunBot team, hardware damage had to be avoided at all costs. Thus, simulation testing mitigated against these drawbacks. This serves to show that practical, embodied considerations could directly motivate the otherwise abstract, disembodied, and more traditional simulation strategies.

Thus far, it is clear that simulation testing is antithetical to environmental embodiment. However, nothing has been said with respect to where such simulations stand with respect to methodological embodiment. It is perfectly possible for systems to make extensive use of simulations, as is the case, for example, with Grush's (2004) emulators, while still being unequivocally embodied in the methodological sense. In the case of the simulations used to test the CajunBot subsystems, it may well have been the case that techniques which are consistent with the positions of at least some methodologically embodied theorists were used. However, as much of the detailed data on this testing is not available, it is not possible to say with any degree of certainty. So, although the simulations were not embodied in the environmental sense, they still could have been embodied in the methodological sense.

Before turning to the drawing of final conclusions, for the sake of historical completeness, it is worth pausing briefly to record what happened to the CajunBot team when they competed in the Grand Challenge.

CAJUNBOT PERFORMANCE AND RESULTS

Team CajunBot made it through the 2005 Grand Challenge qualifying rounds. This enabled them, with 23 other teams, to compete in the final which was held in Primm, Nevada. CajunBot ran well for the first 17 miles of the course. CajunBot was then ordered to pause, to provide a safe distance between it and other competitors. Unfortunately, after this pause, CajunBot never moved again and was eventually eliminated from the competition.

Naturally, the team was disappointed and immediately investigated the cause of this mishap. After some investigation, Arun Lakhotia, the team leader, described what happened as follows (Lakhotia, 2005):

> CajunBot was put in pause mode for about fifty minutes to allow other oncoming bots on the track to clear. In the pause mode CajunBot pulls its breaks [*sic*] fully, which means the motors are engaged to their maximum capacity. Normally at this state the motor should lock and not use power. But for some reason, the motor continued to drain power, [at] too very high amperage [*sic*]. A sustain draw of that level of power for fifty minutes fried the motor.

A full analysis of the failure traced the underlying cause to a technical mishap that had arisen a few days before the actual Grand Challenge final run. A few days before the Challenge, CajunBot's transmission had failed and had to be replaced. Unfortunately, during the replacement process, the new transmission was installed half an inch out of alignment. It was this alignment error that caused the actuator motor to stay powered up during the pause period, and thus to burn out.

Although this was a great shame for the members of Team CajunBot, it also throws light upon an additional peril that needs to be associated with environmental embodiment. This is the decidedly low tech peril of simple engineering failures! All environmentally embodied systems have to contend with this kind of problem, in addition to the other rigors that embodiment throws upon such systems.

Now, it may be objected that this failure of CajunBot could actually be attributed to the fact that it was an insufficiently embodied system in some sense. Perhaps if CajunBot had been provided with internal sensors, furnishing it with a better "proprioceptive sense," then the problem could have been detected and avoided. However, such an objection would rather miss the point. The kind of issue that gave rise to this failure is sufficiently unusual so as to make furnishing the system with the appropriate sensors an implausible improvement. An analogous situation can arise with the lungs of human divers, when they rapidly ascend. Human lungs are in danger of getting over expanded, leading to pulmonary barotraumas, due to the fact that there is no associated discomfort when they are getting over-expanded (NAUI, 2000). However, this fact is not indicative of a "design fault" of human lungs. The CajunBot case is entirely similar.

CONCLUSION

The detailed examination of CajunBot and its component subsystems should make it clear that, at least in the case of this system, techniques that are consistent with the Classical/Cognitive view were used, as well as techniques that were more consistent with both the environmental and the methodological embodied views. This suggests that the Embodied view and the Classical/Cognitive views should not be thought of as much as being competitors with one another, but rather they should be seen as being complimentary. This should be clear from an examination of Table 5.1.

It is clear that both classical and embodied strategies are employed in the overall functional architecture of the CajunBot system. This conclusion also suggests that the Compatibilist Approach with respect to embodied cognition is much more plausible than the Purist Approach.

If this assessment is correct, then embodied cognition offers the promise of broadening the range of methods and techniques for studying cognition rather than offering a wholesale replacement for traditional methods and techniques. Furthermore, the detailed study of CajunBot suggests that the interrelation between the two kinds of strategy is much more complex and rich than it might

TABLE 5.1 Various subcomponents of the CajunBot system and overview of the way which these components should be classified with respect to whether they are classical, environmentally embodied, or methodologically embodied.

	Classical	Environmentally embodied	Methodologically embodied
Sensors	In parts	Yes	Maybe
Path planning (G-Nav)	Yes	Yes	No
Path planning (L-Nav)	No	Yes	Probably
Steering control	No	Yes	Yes
Simulations	Not known	No	Not known

previously have been assumed. The current discussion thus illustrates the importance of looking at the details of actual embodied systems rather than just engaging in abstract theorizing.

ACKNOWLEDGMENTS

Special thanks go to Anthony Maida, Suresh Golconda, Arun Lakhotia, and the rest of the CajunBot Team for all their assistance in preparing this chapter. Details of the efforts of this team, including their most recent activities, can be found at http://www.cajunbot.com.

REFERENCES

Beer, R. (2004). Autopoiesis and cognition in the game of life. *Artificial Life, 10,* 309–326.

Beer, R. (2006). Parameter space structure of continuous-time recurrent neural networks. *Neural Computation, 18,* 3009–3051.

Berkeley, I. (2008). What the <0.70, 1.17, 0.99, 1.07> is a symbol? *Minds and Machines, 18,* 93–105.

Blythe, J. & Reilly, W. (1993). Integrating reactive and deliberative planning for agents. *Technical Report*, CMU-CS-93-135, School of Computer Science, Carnegie Mellon University, May 1993.

Brooks, R. (1991). Intelligence without reason. *Artificial Intelligence, 47,* 139–159.

Chemero, A. (2000). Anti-representationalism and the dynamic stance. *Philosophy of Science, 67,* 625–647.

Clark, A. (1997). *Being There: Putting Brain Body and World Together Again.* Cambridge, MA: MIT Press.

Clark, A. (2005). Beyond the flesh some lessons from a mole cricket. *Artificial Life, 11,* 233–244.

Cowart, M. (2006). Embodied cognition. In *The Internet Encyclopedia of Philosophy* [web]. Martin, TN: University of Tennessee. Retrieved December 30, 2007 from http://www.iep.utm.edu/e/embodcog.htm.

Cox, I. & Wilfong, G. (Eds.) (1990). *Autonomous Robot Vehicles.* New York, NY: Springer-Verlag.

Craig, I. (1995). *Blackboard Systems.* Bristol, UK: Intellect Books.

Englemore, R. & Morgan, T. (Eds.) (1989). *Blackboard Systems.* Reading, MA: Addison-Wesley Publishing Company.

Grush, R. (2004). The emulation theory of representation: Motor control, imagery and perception. *Behavioural and Brain Sciences, 27,* 377–442.

Haugeland, J. (Ed.) (1997). *Mind Design II.* Cambridge, MA: MIT Press.

Lakhotia, A. (2005). October 9, 2005: CB actuator motor burned due to long pause. In *CajunBot Blog* [web]. Lafayette, LA. Retrieved December 30, 2007 from http://64.233.169.104/search?q=cache:J6XofftT6XwJ:www.cajunbot.com/site.php%3FpageID%3D24%26newsID%3D235+%22A+sustain+draw+of+that+level+of+power%22&hl=en&ct=clnk&cd=1&gl=us.

Lakhotia, A., Golconda, S., Maida, A., Mejiay, P., Puntambekar, A., & Seetharaman, G. (2006). CajunBot: Architecture and algorithms. *Journal of Field Robotics, 23,* 555–578.

Maida, A., Golconda, S., Mejia, P., Lakhotia, A., & Cavanaugh, C. (2006). Subgoal-based local navigation and obstacle avoidance using a grid-distance field. *International Journal of Vehicle Autonomous Systems (IJVAS), 4,* 122–142.

Moravec, H. (1990). The Stanford Cart and the CMU Rover. In I. J. Cox & G. T. Wilfong (Eds.), *Autonomous Robot Vehicles* (pp. 407–441). Springer-Verlag: New York.

NAUI (2000). *NAUI Scuba Diver Manual.* Tampa, FL: The National Association of Underwater Instructors.

Rich, E. & Knight, K. (1990). *Artificial Intelligence*, 2nd ed. Columbus, OH: McGraw-Hill.

Sellars, D. (2001). *An Overview of Proportional Plus Integral Plus Derivative Control and Suggestions for Its Successful Application and Implementation.* From Portland Energy Conservation Inc. Resource Library Website: Retrieved June 19, 2008 from http://www.peci.org/Library/PECI_ControlOverview1_1002.pdf.

Sharples, M., Hogg, D., Hutchison, C., Torrance, S., & Young, D. (1989). *Computers and Thought: A Practical Introduction to Artificial Intelligence*. Cambridge, MA: MIT Press.

van Gelder, T. (1997). Dynamics and cognition. In J. Haugeland (Ed.), *Mind Design II* (pp. 421–450). Cambridge, MA: MIT Press.

Varela, F., Thompson, E., & Rosch, E. (Eds.) (1991). *The Embodied Mind: Cognitive Science and Human Experience*. Cambridge, MA: MIT Press.

Ziemke, T. (2003). What's that thing called embodiment? *Proceedings of the 25th Annual Meeting of the Cognitive Science Society.* Philadelphia, PA: Lawrence Erlbaum.

6

The Dynamics of Brain–Body–Environment Systems: A Status Report

RANDALL D. BEER

Cognitive Science Program, Department of Computer Science and Department of Informatics, Indiana University, Bloomington, IN, USA

INTRODUCTION

The history of science can often be characterized as a sequence of revolutions and reactions. The birth of cognitive science can be traced to the cognitive revolution of the mid-20th century, which was a reaction against the behaviorist revolution of the early 20th century. Behaviorism in turn was a reaction to the introspectionist tradition of the late 19th and early 20th centuries. More recently, the connectionist revolution was a reaction to some of the symbolic assumptions of the computational core of cognitive science.

In the mid-1980s, just as mainstream cognitive science was becoming aware of connectionism, two new ideas appeared on the intellectual landscape: situatedness and embodiment. These were quickly followed in the early 1990s by a third: dynamics. Broadly speaking, situatedness concerns the role played in an agent's behavior by its ongoing interactions with its immediate environment. Embodiment, in contrast, concerns the role of the physical properties of an agent's body in its behavior. Finally, dynamical approaches emphasize the temporal dimension of behavior, seeking to apply the concepts and tools of dynamical systems theory to the analysis of agents. Of course, none of these ideas are really new. As is often the case in science, they each had important historical precedents, including cybernetics (Walter, 1953; Ashby, 1960; Braitenburg, 1984), phenomenology (Heideggar, 1962; Merleau-Ponty, 1962; Dreyfus, 1992), and ecological psychology (Gibson, 1979).

Historically, these three ideas entered cognitive science somewhat independently (Beer, in press). Situatedness arose primarily as a reaction against the classical AI planning view of action (Agre & Chapman, 1987; Suchman, 1987). In contrast, embodiment arose primarily from a dissatisfaction with the inability of symbolic AI approaches to cope with the sorts of problems encountered by real robots moving around in real environments (Brooks, 1991). Finally, dynamics arose from a rejection of the discreteness (in both time and state) of classical computationalism (Thelen & Smith, 1994; Van Gelder, 1995). Even today, there are people who hold each of these positions individually without necessarily committing themselves to the others.

However, it is becoming increasingly clear that situatedness, embodiment, and dynamics work much better as a unit. Combining these three ideas leads to the notion of a brain–body–environment system, wherein an agent's nervous system, its body, and its environment are each conceptualized as dynamical systems that are in continuous interaction (Beer, 1992, 1995a; Figure 6.1). Taking such a perspective seriously has fundamental implications across the cognitive, behavioral, and brain sciences, but it also raises many difficult empirical and theoretical challenges. Exploring these implications and addressing these challenges has been a major focus of my research program for almost 20 years (Beer, 1990, 1992, 1995a, b, 1997, 2003). In this chapter, I review both the experimental and the theoretical accomplishments of this research program to date, and then discuss some of the major challenges that remain.

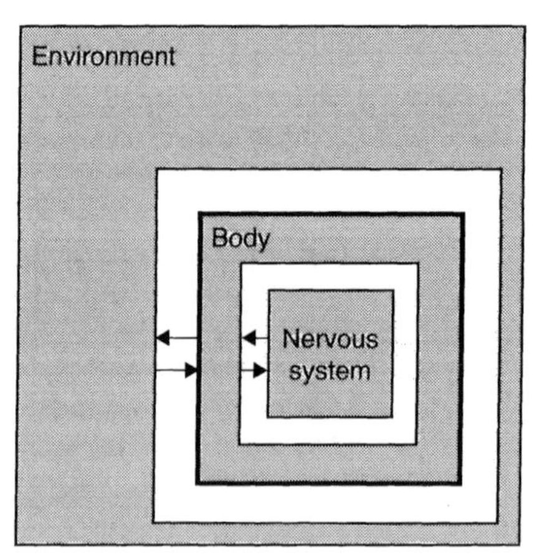

FIGURE 6.1 An agent and its environment as coupled dynamical systems. The agent in turn is composed of coupled nervous system and body dynamical systems.

EXPERIMENTAL ACCOMPLISHMENTS

The brain–body–environment perspective raises severe challenges for the experimental sciences. Studying any one component of a brain–body–environment system is difficult enough, but studying all three components and their interactions in any animal is currently beyond our experimental capabilities. We currently lack the technology to monitor and manipulate the activity of all the relevant neurons within the nervous systems of intact, freely behaving animals, let alone the key biomechanical and environmental properties as well. For this reason, most work in this area has utilized simpler idealized models of brain–body–environment systems. This section describes an evolutionary approach to the construction of such models and surveys the range of behaviors that have been successfully evolved to date. These models make no attempt to account for the behavioral or neurophysiological data from any particular animal. Rather, their goal is to clarify the general nature of brain–body–environment systems and to develop the tools we need to understand such systems.

AN EVOLUTIONARY APPROACH

Animals are evolved, not designed. Among other things, this means that they are selected for their overall behavioral efficacy, not for their understandability. Evolution can fully exploit the freedom to partition its solutions across the brain–body–environment boundaries in ways that will not necessarily align with our preconceptions about how such systems should work. Thus, if we wish to generate model brain–body–environment systems that exhibit the essential characteristics of animals, we can do no better than attempt to mimic the evolutionary process by which these natural brain–body–environment systems are produced (Beer & Gallagher, 1992; Cliff et al., 1993; Nolfi & Floreano, 2000).

Evolutionary algorithms (EVAs) are a by now standard class of search techniques whose operation is loosely based on biological evolution (Goldberg, 1989). EVAs maintain an initially random population of genetic strings that encode the relevant phenotypic characteristics. These strings are repeatedly subjected to evaluation, selection, and reproduction using mutation and crossover genetic operators. Many variations of this basic technique are possible, including different genetic string encodings, different mutation and crossover operators, different evaluation and selection procedures, etc. For our purposes, the key advantage of an evolutionary approach is that it allows the construction of model brain–body–environment systems that are unencumbered by a priori assumptions on our part about how such systems ought to work.

We use EVAs to evolve the parameters of model "nervous systems" coupled to model bodies in model environments so that the entire coupled system exhibits some behavior of interest. It is also possible to evolve body properties, although we will not consider this option here. Our neural model of choice is continuous-time recurrent neural networks or CTRNNs (Beer, 1995c). CTRNNs were selected for several reasons. First, they are arguably the simplest continuous nonlinear dynamical network models. Second, they can be interpreted neurobiologically either as a mean firing-rate model or as a model of nonspiking neurons with synaptic nonlinearites. Third, they are known to be universal approximators of smooth dynamics (Kimura & Nakano, 1998). Thus, they can also be interpreted as just a convenient basis dynamics for building arbitrary nonlinear dynamical systems. Finally, CTRNNs have been the target of considerable mathematical analysis, so a great deal is known about their dynamics.

EVOLUTION OF SENSORIMOTOR BEHAVIOR

Even basic sensorimotor behavior engages the full panoply of issues associated with the brain–body–environment perspective. Any situated and embodied agent must utilize its neural dynamics to coordinate the actions of its body with the spatiotemporal structure of its environment so as to accomplish the tasks necessary to its survival and reproduction. Accordingly, the first model agents that we evolved concerned orientation and locomotion (Beer & Gallagher, 1992).

In chemotaxis, an agent must orient to some source of chemical stimuli. In our experiments, the environment consisted of a square box with a chemical source in the center whose intensity fell off as the inverse square of the distance. The agent had a circular body with a pair of chemical sensors symmetrically placed about its midline (Figure 6.2). The agent's nervous system consisted of a bilaterally symmetric, 6-neuron fully interconnected CTRNN, for a total of 24 free parameters. Two neurons were sensory neurons that responded to the strength of the chemical signal at their location, two neurons were motor neurons that specified the amount of forward force generated at the left and right edges of the body, and the remaining two neurons were interneurons. Fitness was evaluated by averaging the separation between the agent and the patch at the end of a set of trials.

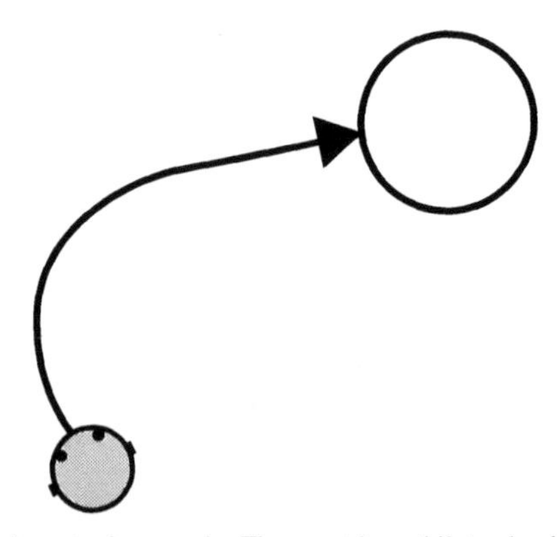

FIGURE 6.2 The chemotaxis scenario. The agent has a bilateral pairs of chemosensors (black disks) and motors (black rectangles). Its task is to navigate to the chemical source whose intensity falls off as the inverse square of distance.

Two different chemotactic strategies evolved. By far the most common strategy was the obvious one of moving forward while turning toward the side on which the chemical signal is stronger by an amount proportional to the gradient of the odor field. This is a spatial strategy, because it relies solely on the difference in signal intensity at the two chemical sensors. In contrast, a small number of runs evolved temporal strategies, in which agents moved from side to side with oscillations whose amplitude was biased toward the side on which the odor signal was stronger, causing the agent's path to curve toward the chemical source. Interestingly, once these agents neared the source, they switched to a spatial strategy. Agents differed in their behavior at the source. Some came to a stop, some spun in circles across the edge of the patch, some orbited around the patch, and some repeatedly crossed it. Additional experiments were run with an agent possessing a single chemosensor. In this case, agents evolved to loop through the environment, with the radius of curvature of the loop proportional to the intensity of the chemical stimulus. The range of different strategies that evolved on even this simple task illustrates the ability of an evolutionary approach to explore the space of possibilities in a relatively unbiased way.

We next evolved locomotion behavior in a legged agent. This is a considerably more complicated motor behavior than orientation, since it requires the coordination of multiple effectors to simultaneously solve the twin problems of support and progression. In our experiments, we utilized an insect-like body model with 12 active degrees of freedom actuated by 18 effectors (Figure 6.3). Each of the six legs could swing back and forth relative to the body and the foot could grasp the substrate. Each swing degree of freedom was actuated by a pair of opposing "muscles," while the foot was actuated by a single binary grasp/release effector. Each leg also possessed an angle sensor whose signed output was proportional to the deviation of the leg from perpendicularity to the long axis of the body. The agent's nervous system consisted of a 30-neuron CTRNN, with three motor neurons and two interneurons per leg and interleg coupling connections both across and along the body. Bilateral and front/back symmetries imposed on this circuit reduced the number of free parameters to 50. Fitness was evaluated by measuring the total forward movement of the body over a fixed length of time. Note that this fitness measure does not directly reward correct leg movement. Rather, the problems of support and coordination are implicit in the "physics" of the body.

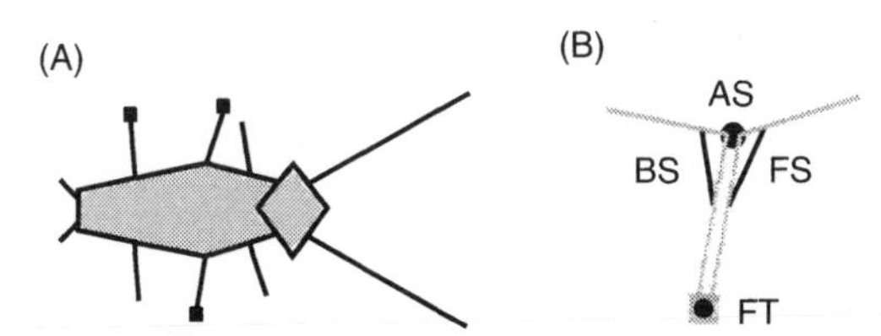

FIGURE 6.3 The walking scenario. (A) The body model. (B) Leg detail. Each leg possesses a binary foot effector (FT) and an antagonistic pair of effectors for swinging the leg: backward swing (BS) and forward swing (FS). In some experiments, an angle sensor (AS) was also utilized.

We evolved walking under three different sensory conditions: (1) sensory feedback was always available during evolution, (2) sensory feedback was never available, and (3) sensory feedback was available on half of the trials and unavailable the other half. In all cases, the best walking agents evolved to use a tripod gait, in which movements of the front and back legs on each side of the body were synchronized with the movement of the middle leg on the opposite side. However, sensory lesions revealed important differences between the agents evolved under the three different conditions. Agents evolved with reliable sensory feedback were capable of adjusting their walking pattern to environmental perturbations, but failed to walk in the absence of such feedback (reflexive pattern generators). In contrast, agents evolved without sensory feedback could intrinsically generate a walking pattern, but could not adjust their movements to external perturbations (central pattern generators, CPGs). Finally, agents evolved under conditions of unreliable sensory feedback were able to use such feedback when it was available to compensate for perturbations, but could intrinsically generate an adequate walking pattern in its absence (mixed pattern generators). Interestingly, such agents could also adapt their walking pattern to a growing leg simply as a consequence of the entrainment of this intrinsic oscillation by rhythmic sensory feedback.

Sensorimotor behavior has been a major focus of work in evolutionary robotics, and no brief survey can really do justice to it. The range of behaviors that have been successfully evolved include not only locomotion, but also obstacle avoidance, wall-following, goal-seeking, navigation, foraging, predator–prey interactions, etc. Typically, this work has focused on the use of evolutionary methods to design controllers for robots. However, some of this work has been designed specifically to target scientific questions. For example, evolutionary approaches have been used to study the role of neuromodulation in rhythmic neural circuits in *Aplysia* (Deodhar & Kupfermann, 2000), the transition from swimming to walking gaits in salamanders (Ijspeert, 2001), and the mechanisms of path integration for homing behavior in ants (Vickerstaff & Di Paolo, 2005).

EVOLUTION OF LEARNING BEHAVIOR

A fundamental characteristic of realistic environments is that they change over a wide range of timescales. Thus, any agent trying to survive in such environments must likewise be capable of changing its behavior over multiple timescales. For this reason, we next explored the evolution of learning behavior in model agents (Yamauchi & Beer, 1994a, b; Phattanasri et al., 2007). An unusual feature of our work in this area is that we do not assume a learning mechanism (e.g., a specific synaptic plasticity rule) a priori. Rather, we take full advantage of the universal dynamics approximation capabilities of CTRNNs and the ability of an evolutionary approach to explore the space of possible brain–body–environment dynamics that can generate the changes of behavior required to perform a given task.

Our initial work in this area involved landmark learning and learning to make sequences of decisions (Yamauchi & Beer, 1994a, b). In the landmark learning task, an agent with very limited sensory capabilities must move to a goal object as efficiently as possible in a one-dimensional environment which also contains a landmark (Figure 6.4A). In some environments, this landmark was placed on the same side of the agent as the goal and in other environments it was placed on the opposite side. The task of the agent was to discover which type of environment it was currently in and to adjust its behavior accordingly. In the sequential decision-making task, an agent must learn to make a particular sequence of binary decisions in response to an environmental trigger using reinforcement

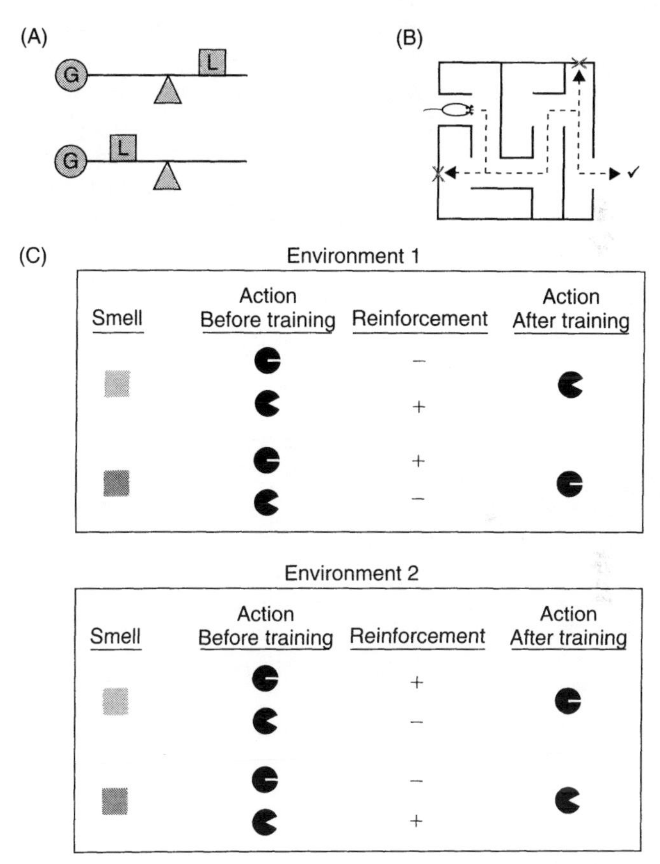

FIGURE 6.4 Various learning scenarios. (A) A simple landmark learning task. An agent (triangle) with extremely limited sensory capabilities must find the goal G. In some cases, the landmark L is on the same side of the agent as the goal, and in other cases it is on the opposite side. Based on its experiences in a given environment, the agent must learn how best to use the landmark to find the goal. (B) A sequential decision-making and learning task based on maze learning. At each T intersection, the agent is presented with a choice to turn either left or right. The agent has to learn the proper sequence of actions based on reinforcement from its environment. (C) A food edibility learning task. An agent must bite at edible food while not biting at inedible food. Because the relationship between food smell and edibility changes over time, the agent must be able to learn (or relearn) this relationship based on feedback from its actions in the current environment.

from its environment. This task was meant to be an abstraction of maze learning, in which each T-junction triggers a left/right decision and successful maze traversal depends on the correctness of the entire sequence of decisions (Figure 6.4B). CTRNNs without plastic synapses were successfully evolved for both the landmark learning and the sequential decision-making tasks.

More recently, we extended this work to a food edibility learning task (Phattanasri, 2002; Phattanasri et al., 2007). In this task, an agent equipped with a mouth, a smell sensor and a gut sensor (serving as a reinforcement signal) must learn to bite only edible food and avoid biting inedible substances (Figure 6.4C). Because which substance was edible and which was inedible varied from trial to trial, the agent had to learn to associate the smells of each substance with their edibility based on its experiences in that environment, and had to be able to relearn when this relationship changed. We first successfully evolved CTRNNs with nonplastic synapses to solve this task. We then explored the evolution of CTRNNs with Hebbian synapses. Interestingly, we found that, although successful plastic circuits could be evolved, the weights changed on the same timescale as the neural dynamics rather than on longer timescales as might be expected. When we restricted the learning rates to be slower than the neuronal time constants, successful circuits failed to evolve.

The evolution of learning in model agents has become an increasingly active area of research. Initially, most work in this area assumed a specific model of synaptic plasticity a priori (e.g., Chalmers, 1991; Miller & Todd, 1991; Floreano & Mondada, 1996). However, more recently, a number of researchers have been exploring the evolution of nonplastic CTRNNs. For example, Tuci et al. (2002) extended the landmark learning task described above, whereas Blynel and Floreano (2003) evolved agents that could learn to traverse a simple T maze. Finally, Izquierdo-Torres and Harvey (2006, 2007a, b) have recently explored the evolution of nonplastic CTRNNs to learn to associate stimuli from a continuum and to mimic Hebbian synapses.

EVOLUTION OF MINIMALLY COGNITIVE BEHAVIOR

How far can the evolutionary agents approach be taken? Given that there is considerable interest in exploring the implications of a situated, embodied and dynamical perspective on cognition, it would be especially useful to extend the evolutionary agent approach to cognitive behavior. However, it is difficult to imagine how the entirety of human cognition could be evolved or dynamically analyzed at this point. For this reason, I suggested the idea of minimally cognitive behavior, in which model agents exhibiting the simplest behaviors that are of cognitive interest are studied (Beer, 1996). Specifically, I proposed a model agent with simple "visual," locomotory, and manipulatory capabilities that could be used to explore a wide range of cognitively interesting behavior, building on earlier work on the evolution of visually guided behavior (Cliff et al., 1993).

We have studied a variety of tasks with this agent (Beer, 1996; Slocum et al., 2000; Beer, 2003; Goldenberg et al., 2004). For example, we evolved agents that

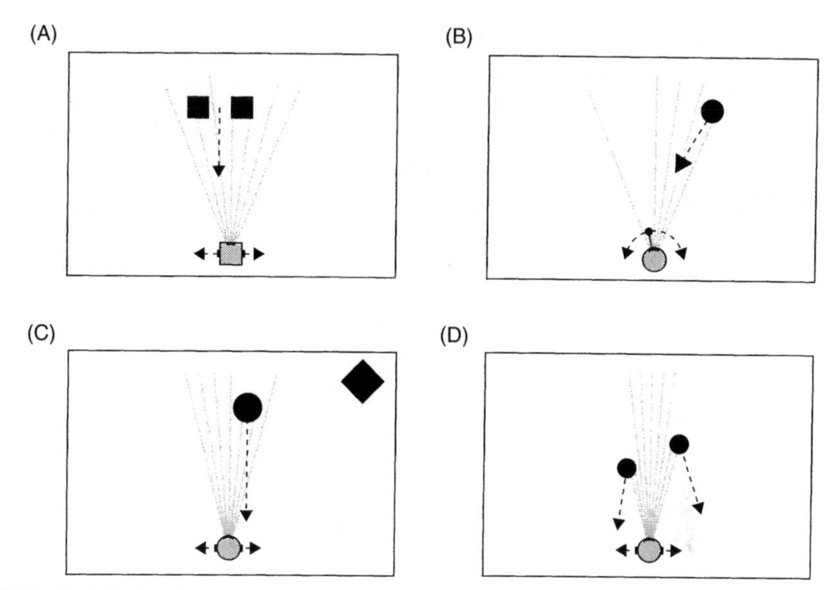

FIGURE 6.5 Various minimally cognitive behavior scenarios. (A) An affordance perception task. The agent must align itself with openings through which its body can pass while avoiding apertures that are too small. (B) A catching task. The agent must catch falling objects with its opaque hand. (C) An object discrimination task. The agent must move so as to catch all circles and avoid all diamonds. (D) A selective attention task. The agent must move so as to catch all objects within its field of view.

could discriminate between circles and diamonds by catching the former while avoiding the latter (Figure 6.5C). Interestingly, the best agent evolved an active perceptual strategy in which objects were foveated and scanned before they were caught or avoided. We also evolved agents that could discriminate between passable and impassable openings by judging the width of each aperture relative to its own body size (Figure 6.5A). In another set of experiments, we evolved agents that could catch falling objects with a simple manipulator (Figure 6.5B). An interesting feature of this experiment was that the "hand" of the manipulator was opaque, and therefore it appeared within the agent's visual field like any other object. Thus, an important problem that this agent had to solve was distinguishing between an object in its environment that was part of itself (and which it could directly control) from other objects in its environment (over which it had no direct control). Finally, we evolved agents that could catch multiple objects (Figure 6.5D). This is the richest task that we have so far evolved, since it involves attention (once the agent has decided to catch one object, it must not become distracted by other objects in its field of view), memory (if the agent loses sight of one object while pursuing another, it must return to catch the first object after the second object has been caught), and prediction (since objects move at different speeds, the agent's decision as to which object to catch first cannot be based on which is currently the closest, but must be based on a prediction of which object will reach it first).

In recent years, there has been a substantial increase in work on the evolution of minimally cognitive behavior. For example, Ward and Ward (2006) used the object discrimination task described above to test and ultimately reject the idea that resolving cognitive conflict requires explicit conflict monitoring (Botvinick et al., 2004). These same authors also compared the performance of evolved model agents and human subjects on the selective attention task described above and identified a common reliance on reactive inhibition (Ward & Ward, in press). Likewise, Di Paolo and colleagues have recently evolved a variety of minimally cognitive behavior, including perceptual crossing and agency detection (Di Paolo et al., 2008), the A-not-B error in infant reaching (Wood & Di Paolo, 2007), and behavioral preference (Iizuka & Di Paolo, 2007). Finally, there has been a long line of work on the evolution of communication between model agents which is directly relevant to minimally cognitive behavior (Quinn, 2001; Marocco et al., 2003; Steels, 2003).

THEORETICAL ACCOMPLISHMENTS

For the purposes of our research program, the ability to successfully evolve brain–body–environment systems is only half the story. The central goal of the research program is not merely to produce model brain–body–environment systems that exhibit a wide variety of interesting behavior, but to *understand* them. Through the analysis of such systems, we seek to develop the concepts, theoretical framework, and mathematical and computational tools necessary for the vastly more complex brain–body–environment systems that occur in animals. Unfortunately, most work in evolutionary robotics does not include any significant mathematical analysis of the agents that are evolved (with a few notable exceptions, e.g., Husbands et al., 1995; Pasemann, 2002). This section shows some of what can be accomplished by applying the standard tools of dynamical systems theory to evolved brain–body–environment systems.

CTRNN DYNAMICS

The model nervous systems that we utilize clearly play an extremely important role in any brain–body–environment dynamics that evolve. It would thus be worthwhile to understand as much as possible the general dynamics of CTRNNs as a foundation for the analysis of evolved agents. To the extent that CTRNNs have plausible neurobiological interpretations, such an analysis also serves as a nice "warm-up" exercise for the study of more biologically realistic circuits. In addition, given the universal approximation capabilities of CTRNNs, such a study is also of considerable mathematical interest. For all of these reasons, we have undertaken a long-term study of the general dynamical characteristics of CTRNNs.

Our initial work in this area began with a detailed analysis of the phase portraits, bifurcations, and parameter charts of small CTRNNs (Beer, 1995c). The primary purpose of this work was to build intuition for tackling the more general case. This work complemented and extended previous analyses of related recurrent neural network models. For individual neurons, we were able to derive exact expressions for the bifurcation boundaries between monostable and bistable behavior. We also derived approximate expressions for the locations of equilibrium points as a function of parameters. For 2-neuron circuits, we enumerated a preliminary catalog of 11 distinct phase portraits; this catalog has now been extended to a total of 16 entries (Beer, 1995c; Ermentrout, 1998; Beer, unpublished). We also derived expressions for some bifurcation boundaries. In addition, we identified a special class of 2-neuron CTRNNs known as center-crossing circuits, in which the nullclines of each neuron intersect at their exact centers of symmetry. For these circuits, some bifurcation boundaries could be calculated exactly in a way that extended to larger circuits. Based on our analysis, we hypothesized that an evolutionary search seeded with random center-crossing circuits would evolve oscillatory dynamics much faster than one seeded with a completely random initial population, a prediction that was later confirmed (Mathayomchan & Beer, 2002). Finally, we identified a small subset of the large number of possible 3-neuron phase portraits, including a 3-neuron CTRNN exhibiting chaotic dynamics.

More recently, we have moved from studying the dynamics of 1-, 2-, and 3-neuron CTRNNs to attempting to understand the overall structure of the infinite-dimensional space of all possible CTRNNs (Beer, 2006). Since this space is stratified by circuit size, in practice we seek to characterize the structure of this space using expressions in which the circuit size appears as a free parameter. Although such analysis is extremely difficult, a surprising amount of progress can be made. Building on the work of Haschke and Steil (2005), we first developed tools to explicitly calculate and visualize all local bifurcation manifolds of small CTRNNs. These visualizations revealed a set of extremal saddle-node bifurcation manifolds that subdivide the net input space into regions with dynamics of different effective dimensionality. We then completely characterized the combinatorics and geometry of an approximation to these regions for CTRNNs of arbitrary size. Finally, we derived estimates of both the probability of finding regions of parameter space with dynamics of different effective dimensionality and the probability of selected phase portraits.

DYNAMICAL ANALYSIS OF WALKING

The evolved walking agents were the first that we subjected to detailed dynamical analysis. For simplicity, we concentrated our analysis on agents with a single leg. While this task might seem too trivial to be of any broader interest, we have found it to be an extraordinarily rich source of insights into the fundamental nature of brain–body–environment systems. This section briefly reviews a sample of the results that have been obtained.

After some preliminary analysis of the dynamics of the best reflexive, central, and mixed pattern generators (Beer, 1995a), we focused on a population of over three hundred 3- to 5-neuron central pattern generators (CPGs) (Beer et al., 1999; Chiel et al., 1999). We found that the mean best walking performance increased significantly from 3- to 4-neuron CPGs, but only slightly from 4- to 5-neuron CPGs. We also found that there seemed to be a maximum achievable performance. Finally, a detailed study of the top ten 3-neuron CPGs revealed significant parametric variability despite only small differences in walking performance. In addition, the nonlinear dependence on parameters led to a failure of CPGs constructed by averaging the top circuits to even oscillate. It also led to the CPGs being highly sensitive to some combinations of parameter variations and insensitive to others. Multiple instantiability, failure of averaging, and sensitivity and robustness to parameter perturbations were subsequently observed in more biophysically realistic models of the lobster stomatogastric ganglion (Goldman et al., 2001; Golowasch et al., 2002; Prinz et al., 2004).

In order to achieve a more general understanding of the operation of the best CPGs, we used separation of timescales to decompose them into *dynamical modules* (sets of neurons that simultaneously made a transition from one quasistable state to another while the outputs of the other neurons remained relatively constant). Each module operated as a bistable element that was switched from one configuration to the other by other modules in a closed chain that produced the rhythmic pattern. We described how the steady-state input–output curves of each module varied as a function of synaptic input from other modules. This allowed us to quantitatively characterize constraints on circuit architecture, explain the duration of the different phases of the walking cycle, and predict the effects of parameter changes. Finally, the notion of a dynamical module made it possible to classify the evolved CPGs, to enumerate the possible dynamical modular architectures, and to quantitatively account for the observed parameter variability.

We next undertook a detailed biomechanical analysis of the model body. Due to the simplicity of the model, it was possible to derive exactly the optimal motor pattern and calculate its properties. For example, we found that we could quantitatively explain the maximum fitness that we observed in our evolutionary searches in terms of the maximum velocity achievable by this model body driven by the optimal motor pattern. We also found that, as the mechanical advantage of the leg decreased near the end of stance phase, neural outputs had a decreasing effect on the body's motion. This allowed us to explain why the evolved motor patterns were highly variable at this point while being tightly constrained at others. Finally, we have recently begun to characterize the structure of walking fitness space by combining our knowledge of CTRNN parameter space with our biomechanical analysis of the legged body in order to characterize the maps from neural and synaptic parameters to motor pattern to behavior to fitness.

Finally, we examined the impact of network architecture on locomotion performance (Psujek et al., 2006). Because of the very large number of distinct architectures that can occur in even small circuits, we evolved over 2 million different pattern generation circuits with nearly 10,000 different architectures. We then studied the relationship between locomotion performance and network architecture in these circuits, identifying particular circuit motifs that were strongly correlated with high performance. In addition, by comparing the best locomotion performance that could be obtained by a given architecture with the performance achieved on average, we found that some architectures of equivalent best performance nevertheless differed significantly in their evolvability. Furthermore, we were able to relate these differences in the evolvability of different architectures to the differing statistical structures of their parameter spaces.

DYNAMICAL ANALYSIS OF FOOD EDIBILITY LEARNING

The main goal of the analysis described in this section was to understand how CTRNNs lacking plastic synapses were nevertheless able to solve the food edibility learning task (Phattanasri et al., 2007). Since these circuits receive time-varying input from the chemical and gut sensors, they are nonautonomous dynamical systems. However, because the chemosensor was binary and we could idealize the gut sensor as either positive or negative in high fitness circuits, there were only five possible input patterns to consider. In each case, we determined the complete autonomous phase portrait for the circuit under each input pattern. We then studied the transient dynamics as the sensory inputs switched between their different possible values and the circuit state was attracted toward subsequent equilibrium points. By "strobing" the system state at the falling edges of each input signal, we found that these strobe points fell into fairly compact and distinct clusters. We then showed that each of these clusters could be associated with a state of a finite-state machine that captured the combinatorial structure of the food edibility learning problem. More recently, we have shown that nonplastic CTRNNs evolved to learn stimuli from a continuum instantiate manifolds of finite-state machines (Izquierdo-Torres et al., in press).

DYNAMICAL ANALYSIS OF CATEGORICAL PERCEPTION

The visually guided object discrimination task is the most sophisticated one that we have analyzed to date (Beer, 2003). We began our analysis by characterizing the performance and behavior of the best evolved agent when discriminating circles from diamonds. Although the agent's discrimination performance was generally quite good, we identified narrow regions of its field of view where mistakes were made. These regions provided a clue as to how the discrimination was being made. As mentioned earlier, we also found that this agent employed a scanning strategy, presumably to accentuate temporally the small spatial differences between circles and diamonds. We demonstrated that these scanning motions were actively generated by the agent in response to objects within its field of view, and that they varied systematically according to the type of object and its distance from the agent.

We next performed a series of psychophysical studies. In support of these studies, we defined a set of hybrid objects that smoothly interpolated between a circle and a diamond. We first demonstrated that the agent exhibited a sigmoidal labeling curve and a bell-shaped discrimination curve as a function of the "mixing" parameter. These curves are characteristic of categorical perception. We also found an anomalous peak at an intermediate value of the mixing parameter, which provided another clue as to how this agent operated. Next, we showed that object width was the primary stimulus feature on which the discrimination is based. Interestingly, subsequent experiments on this task have shown that randomizing over object size during evolution produces agents that are sensitive to object shape rather than width (Di Paolo & Harvey, 2003). Finally, we attempted to characterize when the discrimination was made by switching the object identity at various points during the interaction and determining at what point the agent had irreversibly committed to either a catch or avoid response. We found that, rather than a discrete event, the "decision" was a temporally extended process that became irreversible only after the behavioral expression of its decision was well underway.

Finally, we performed a detailed dynamical analysis of the behavior of this evolved brain–body–environment system. Dynamical analyses can be performed at many different levels (Figure 6.6). For this agent, we focused on three different levels of analysis.

First, we characterized the dynamics of the coupled brain–body–environment system. By plotting various projections of the trajectories of the coupled system, we found that the dynamics develops into two distinct bundles of trajectories. For a circle, these bundles wound around each other three times before colliding at the midline, resulting in catches. For a diamond, the bundles wound around one another only twice before diverging, resulting in avoidances. As the mixing parameter moved from a circle to a diamond and the pair of bundles unwound from three to two crossings, we found an intermediate value at which bundles collided. Interestingly, this corresponded exactly to the value at which the anomalous catch peak occurred in the labeling curve. We were also able to successfully predict the occurrence of two additional anomalous peaks as the trajectory bundles continued to unwind for larger object sizes.

Second, we examined the agent–environment interactions that give rise to this coupled system dynamics. Our primary tool for this analysis was the agent's steady-state velocity fields, which help us to understand how object motion influences agent motion and how agent motion influences object motion. Specifically, these fields portray the agent's equilibrium velocity in response to different objects fixed at all possible locations in its field of view. By superimposing the agent's trajectories of instantaneous velocity over these steady-state

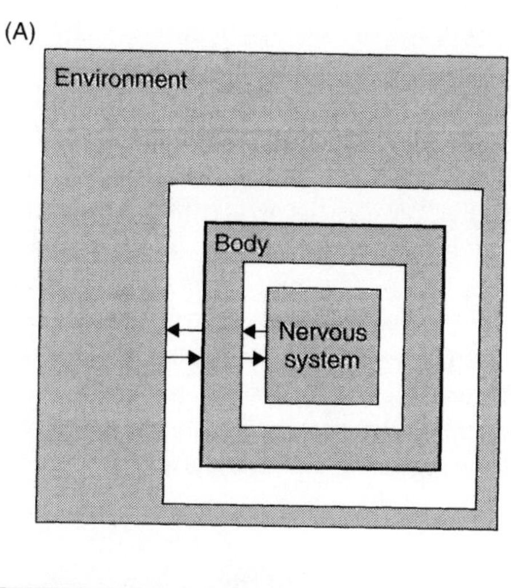

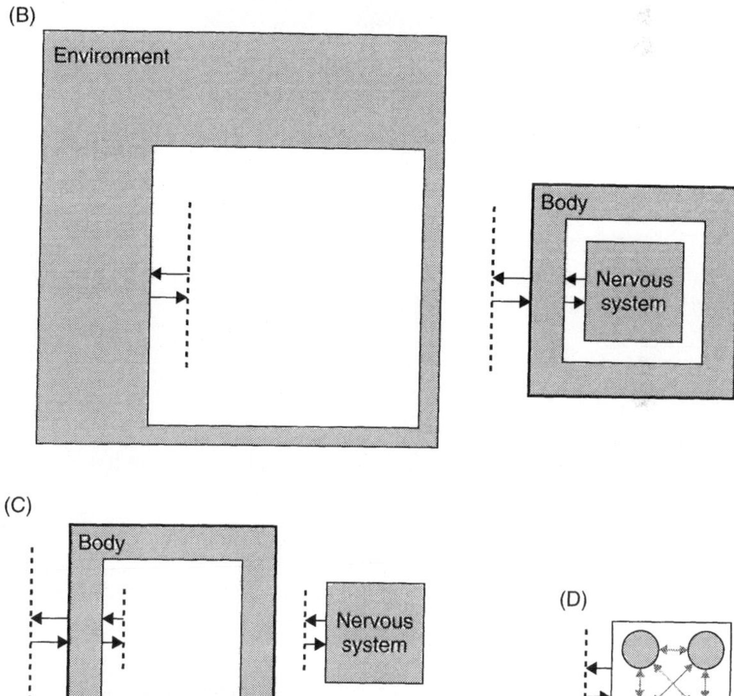

FIGURE 6.6 Dynamical analysis can be applied at several different levels. (A) The complete brain–body–environment system. (B) The agent-environment interactions that give rise to the coupled dynamics in part (A). Note that both subsystems become nonautonomous when we separate them. (C) The neuromechanical interactions underlying the agent dynamics in part (B). (D) The neural interactions underlying the neural dynamics in part (C).

velocity fields, we were able to account for both the normal catch and avoidance responses and the mistaken classifications that the agent sometimes made. This analysis also allowed us to successfully predict that all sufficiently peripheral objects would be avoided. A crucial component of this account was the lag that occurred between an agent's instantaneous and steady-state velocities, demonstrating the key role that transient dynamics plays in this agent's behavior.

Third, we studied the neural instantiation of the agent dynamics. In particular, we demonstrated how the specific neural and synaptic parameters set by the EVA implemented the observed steady-state velocity fields. We also examined some aspects of the neural basis of transient dynamics in the agent. This allowed us to predict that transiently lesioning a particular connection at a particular point in the agent's interaction with its environment would cause the agent to avoid a circle rather than catching it. This prediction was subsequently verified.

OUTSTANDING CHALLENGES

This chapter has reviewed the current status of a long-term research program aimed at elucidating the mechanisms of behavior and cognition through the evolution and analysis of model brain–body–environment systems. Looking back over the past 20 years, one cannot help but be struck by the progress that has been made both in the kinds of behaviors that can now be evolved and in the level of mathematical understanding of the evolved agents that can now be achieved. Nevertheless, substantial challenges remain. Some of the challenges that I find most pressing are briefly outlined in this final section.

EXPERIMENTAL CHALLENGES

The principal experimental challenge facing this research program is to expand its domain of applicability by enlarging the range of behaviors that can be evolved. We would like to address a wider range of perceptual, motor, learning, and minimally cognitive tasks. In addition, we would like to evolve agents that combine nontrivial perceptual, motor, learning, and cognitive capabilities in the service of multiple, sometimes conflicting, goals. This will require richer model bodies for our agents and richer model environments with which they can interact. Finally, it is very important to complement the idealized tasks that are often explored with tasks drawn from actual neuroscientific, ethological, and psychological applications in order to begin to make contact with the experimental data in these fields.

The major impediment to expanding the range of behavior that can be evolved is the scaling properties of EVAs. Despite the tremendous growth in computing power, there is growing anecdotal evidence that we are reaching a complexity ceiling for simple EVAs. For this reason, many people have turned to more complex neural models, network architecture evolution, and developmental approaches. However, I know of no clear-cut example where such augmented approaches have succeeded in evolving behavior substantially more complicated than that evolvable by standard approaches. It is not difficult to understand this failure. Take, for example, developmental processes. Although development is obviously a crucial component of biological evolution, the simple fact of the matter is that we do not yet understand it very well. Indeed, developmental processes themselves would greatly benefit from the kind of simpler idealized modeling that this chapter has advocated for the mechanisms of behavior. Until our understanding of development is significantly more advanced, the likelihood that simply adding another level of complexity we do not understand would allow us to circumvent the present limitations of EVAs seems rather low.

To my mind, the most promising approach to scaling EVAs to more sophisticated behavior, at least in the short term, is incremental shaping. Because EVAs start from completely random genomes, they operate in a mode that has more in common with the origin of life than it does biological evolution. Unfortunately, the gap from random agents to highly fit ones simply becomes too large for more sophisticated behavior. In contrast, biological evolution makes incremental changes to an already functioning organism. The simplest way to capture this property is to begin with a relatively simple version of the task and then incrementally complicate it as the evolutionary search progresses until the desired behavior is achieved. This is a strategy that we and others have applied quite successfully to some of the most sophisticated behaviors that have so far been evolved.

One difficulty with the incremental approach is that finding the right sequence of complexifications can sometimes be a challenge. If the population becomes stuck at an intermediate stage of evolution, sometimes adding more neurons (typically with very small initial connection weights so as not to disrupt the operation of existing circuits) can increase the dimensionality of the parameter space sufficiently to circumvent the barrier. Another issue is that the particular sequence of complexifications chosen might unduly bias the final result.

THEORETICAL CHALLENGES

Perhaps the most pressing theoretical challenge is the dynamical analysis of many more evolved agents beyond the small number that have so far been studied. Analyzing many examples of agents evolved on the same task is crucial for identifying commonalities and differences between them. It is also the only way to obtain a more general understanding of the operation of the evolved brain–body–environment systems that abstract over the particular details of any given instantiation. To date, we have successfully performed this abstraction only once, using the concept of dynamical modules to characterize the population of evolved walking agents (Beer et al., 1999; Chiel et al., 1999). In addition to analyzing multiple examples of agents for tasks that we can already reliably evolve, it will also be essential to attempt to analyze the most sophisticated agents that can be produced at any given time. This will push the further development of the techniques and tools for dynamical analysis. Finally, further development of a theory of the dynamics of CTRNNs and related neural models will continue to provide a crucial foundation for these analyses.

The dynamical analysis of brain-body-environment systems faces many mathematical obstacles. For example, as more sophisticated agents are analyzed, it will become increasingly difficult to carry out a complete microdynamical analysis of the brain–body–environment system. Thus, simplified macrodynamical descriptions will need to be considered, just as they are for natural systems. However, although some techniques do exist for reducing the dimensionality of dynamical systems under specific circumstances (e.g., near limit sets and bifurcations), systematic procedures for dimensionality reduction of arbitrary dynamical systems under arbitrary conditions is still very much an open research topic. Furthermore, there is a fundamental trade-off between simplicity and scope, since macrodynamical descriptions gain their simplicity by restricting their scope, whereas microdynamical descriptions gain their generality at the expense of simplicity.

Another mathematical challenge is analyzing brain–body–environment systems that include stochastic components. Such stochasticity typically arises as a simple model of variations whose dynamics are unknown in detail. Once again, although techniques exist for the analysis of stochastic dynamical systems (Lasota & Mackey, 1994), this is very much an open research area. One of the chief difficulties is that one must replace a consideration of the flow of individual values of the state variables with a consideration of the flow of distributions over these states, exchanging a finite-dimensional dynamical system for an infinite-dimensional one. Perhaps the best way to proceed is to first fully analyze the deterministic dynamics of such a system and then try to understand the impact of stochasticity.

Finally, probably the most urgent theoretical challenge is to come to grips fully with the multiple timescale transient nature of the dynamics of brain–body–environment systems. Such properties come to the fore either when the system has dynamics on timescales that are long relative to the lifetime of the agent (so the interaction never reaches an attractor during the agent's lifetime) or when we decompose the coupled system into an interacting agent and environment (so that the agent receives time-varying sensory signals from its environment). However, the mathematical tools of dynamical systems theory are most highly developed for autonomous dynamical systems. Although techniques exist for the rigorous analysis of nonautonomous dynamical systems when either the timescale of input variation is well-separated from the timescales of intrinsic dynamics or the time variation of input takes a particularly simple form (e.g., periodic), the analysis of general nonautonomous systems is a wide open research problem. Unfortunately, we have often found in our analysis of evolved agents that such transient dynamics plays a central role (e.g., in the food edibility learning and categorical perception tasks) and new mathematical tools will be necessary to deal with it. In addition, taking transient dynamics seriously may force us to reconceive the role of internal state in a dynamical agent, from representing its current situation and goals to setting a context in which subsequent sensory stimuli are "interpreted" (in terms of how the trajectory of stimuli influence subsequent behavior).

EDUCATIONAL CHALLENGES

When computational psychology first appeared on the scene in the late 1950s, a lot of silly things were said by people who did not really understand computation. This problem was only overcome when computational ideas became a core part of the educational curriculum of psychologists and philosophers of mind. Likewise, there is a need for the perspective and tools of dynamical systems theory to become a standard topic in the education of brain, behavioral and cognitive scientists. Only then can an informed critical assessment of these ideas take place. Even within the evolutionary agents community, where complex systems ideas are second nature, there is often still a strong reluctance to analyze how evolved agents work. The availability of better open source software tools for dynamical analysis would substantially lower the barrier to such analyses, as well as support educational initiatives in dynamical systems theory. Finally, we need to collectively do a better job of bringing the power and utility of the evolutionary synthesis and dynamical analysis approach to the attention of our colleagues in neuroscience and psychology as an important addition to the existing tools in their scientific toolbox.

REFERENCES

Agre, P. E. & Chapman, D. (1987). Pengi: An implementation of a theory of activity, *Proceedings of the Sixth National Conference on AI* (pp. 268–272). Seattle, WA: Morgan Kaufmann.

Ashby, W. R. (1960). *Design for a Brain*, 2nd ed. New York: John Wiley & Sons.

Beer, R. D. (1990). *Intelligence as Adaptive Behavior: An Experiment in Computational Neuroethology*. New York: Academic Press.

Beer, R. D. (1992). A dynamical systems perspective on autonomous agents. *Technical Report CES-92-11*, Department of Computer Engineering and Science, Case Western Reserve University, Cleveland, OH.

Beer, R. D. (1995a). A dynamical systems perspective on agent–environment interaction. *Artificial Intelligence, 72,* 173–215.

Beer, R. D. (1995b). Computational and dynamical languages for autonomous agents. In R. Port & T. van Gelder (Eds.), *Mind as Motion* (pp. 121–147). Cambridge, MA: MIT Press.

Beer, R. D. (1995c). On the dynamics of small continuous-time recurrent neural networks. *Adaptive Behavior, 3,* 471–511.

Beer, R. D. (1996). Toward the evolution of dynamical neural networks for minimally cognitive behavior. In P. Maes, M. Mataric, J. Meyer, J. Pollack, & S. Wilson (Eds.), *From Animals to Animats 4: Proceedings of the Fourth International Conference on Simulation of Adaptive Behavior* (pp. 421–429). Cambridge, MA: MIT Press.

Beer, R. D. (1997). The dynamics of adaptive behavior: A research program. *Robotics and Autonomous Systems, 20,* 257–289.

Beer, R. D. (2003). The dynamics of active categorical perception in an evolved model agent (with commentary and response). *Adaptive Behavior, 11*(4), 209–243.

Beer, R. D. (2006). Parameter space structure of continuous-time recurrent neural networks. *Neural Computation, 18,* 3009–3051.

Beer, R. D. (in press). Dynamical systems and embedded cognition. In K. Frankish & W. Ramsey (Eds.), *The Cambridge Handbook of Artificial Intelligence.* Cambridge University Press.

Beer, R. D. & Gallagher, J. C. (1992). Evolving dynamical neural networks for adaptive behavior. *Adaptive Behavior, 1,* 81–122.

Beer, R. D., Chiel, H. J., & Gallagher, J. C. (1999). Evolution and analysis of model CPGs for walking II. General principles and individual variability. *Journal of Computational Neuroscience, 7,* 119–147.

Blynel, J. & Floreano, D. (2003). Exploring the T-maze: Evolving learning-like robot behaviors using CTRNNs. In C. Ryan, T. Soule, M. Keijzer, E. Tsang, R. Poli, & E. Costa (Eds.), *Applications of Evolutionary Computing* (pp. 593–604). Heidelberg: Springer-Verlag, (Lecture Notes in Computer Science 2611).

Botvinick, M. M., Cohen, J. D., & Carter, C. S. (2004). Conflict monitoring and anterior cingulate cortex: An update. *Trends in Cognitive Sciences, 8,* 539–546.

Braitenburg, V. (1984). *Vehicles: Experiments in Synthetic Psychology.* Cambridge, MA: MIT Press.

Brooks, R. A. (1991). New approaches to robotics. *Science, 253,* 1227–1232.

Chalmers, D. J. (1991). The evolution of learning: An experiment in genetic connectionism. In D. S. Touretzky, J. L. Elman, T. J. Sejnowski, & G. E. Hinton (Eds.), *Connectionist Models: Proceedings of the 1990 Summer School* (pp. 81–90). San Mateo, CA: Morgan Kaufmann.

Chiel, H. J., Beer, R. D., & Gallagher, J. C. (1999). Evolution and analysis of model CPGs for walking I. Dynamical modules. *Journal of Computational Neuroscience, 7,* 99–118.

Cliff, D., Harvey, I., & Husbands, P. (1993). Explorations in evolutionary robotics. *Adaptive Behavior, 2,* 73–110.

Deodhar, D. & Kupfermann, I. (2000). Studies of neuromodulation of oscillatory systems in *Aplysia* by means of genetic algorithms. *Adaptive Behavior, 8,* 267–296.

Di Paolo, E. A. & Harvey, I. (2003). Decisions and noise: The scope of evolutionary synthesis and dynamical analysis. *Adaptive Behavior, 11,* 289–293.

Di Paolo, E. A., Rohde, M., & Iizuka, H. (2008). Sensitivity to social contingency or stability of interaction? Modelling the dynamics of perceptual crossing. *New Ideas in Psychology 26,* 278–294.

Dreyfus, H. L. (1992). *What Computers Still Can't Do.* Cambridge, MA: MIT Press, Original edition published in 1972.

Ermentrout, B. (1998). Neural networks as spatio-temporal pattern-forming systems. *Reports on Progress in Physics, 61,* 353–430.

Floreano, D. & Mondada, F. (1996). Evolution of plastic neurocontrollers for situated agents. In P. Maes, M. Mataric, J. Meyer, J. Pollack, & S. Wilson (Eds.), *From Animals to Animats 4: Proceedings of the Fourth International Conference on Simulation of Adaptive Behavior* (pp. 402–410). Cambridge, MA: MIT Press.

Gibson, J. J. (1979). *The Ecological Approach to Visual Perception.* Boston, MA: Houghton Mifflin.

Goldberg, D. E. (1989). *Genetic Algorithms in Search, Optimization, and Machine Learning.* Reading, MA: Addison-Wesley.

Goldenberg, E., Garcowski, J., & Beer, R. D. (2004). May we have your attention: Analysis of a selective attention task. In S. Schaal, A. Ijspeert, A. Billard, S. Vijayakumar, J. Hallam, & J.-A. Meyer (Eds.), *From Animals to Animats 8: Proceedings of the Eighth International Conference on the Simulation of Adaptive Behavior* (pp. 49–56). Cambridge, MA: MIT Press.

Goldman, M. S., Golowasch, J., Marder, E., & Abbott, L. F. (2001). Global structure, robustness and modulation of neuronal models. *Journal of Neuroscience, 21,* 5229–5238.

Golowasch, J., Goldman, M. S., Abbott, L. F., & Marder, E. (2002). Failure of averaging in the construction of a conductance-based neural model. *Journal of Neurophysiology, 87,* 1129–1131.

Haschke, R. & Steil, J. J. (2005). Input space bifurcation manifolds of recurrent neural networks. *Neurocomputing, 64C,* 25–38.

Heideggar, M. (1962). *Being and Time*. New York and Evanston: Harper and Row, Originally published in 1927.

Husbands, P., Harvey, I., & Cliff, D. (1995). Circle in the round: State space attractors for evolved sighted robots. *Robotics and Autonomous Systems, 15,* 83–106.

Iizuka, H. & Di Paolo, E. A. (2007). Toward Spinozist robotics: Exploring the minimal dynamics of behavioral preference. *Adaptive Behavior, 15,* 359–376.

Ijspeert, A. J. (2001). A connectionist central pattern generator for the aquatic and terrestrial gaits of a simulated salamander. *Biological Cybernetics, 84,* 331–348.

Izquierdo, E. Harvey, I. & Beer, R.D. (in press). Associative learning on a continuum in evolved dynamical neural networks. To appear in Adaptive Behavior.

Izquierdo, E. Harvey, I. & Harvey, I. (2006). Learning on a continuum in evolved dynamical node networks. In L. Rocha et al. (Eds.), *Artificial Life X: Proceedings of the Tenth International Conference on the Simulation and Synthesis of Living Systems* (pp. 507–512). Cambridge, MA: MIT Press.

Izquierdo-Torres, E. & Harvey, I. (2006). Learning on a continuum in evolved dynamical node networks. In L. Rocha et al. (Eds.), *Artificial Life X: Proceedings of the Tenth International Conference on the Simulation and Synthesis of Living Systems* (pp. 507–512). Cambridge, MA: MIT Press.

Izquierdo-Torres, E. & Harvey, I. (2007a). Hebbian learning using fixed weight evolved dynamical "neural" networks. In H. A. Abbass et al. (Eds.), *Proceedings of the First IEEE Symposium on Artificial Life* (pp. 394–401). IEEE Press.

Izquierdo-Torres, E. & Harvey, I. (2007b). The dynamics of associative learning in an evolved situated agent. In F. Almeida, E. Costa et al. (Eds.), *Advances in Artificial Life: Proceedings ECAL* (pp. 365–374). Springer.

Kimura, M. & Nakano, R. (1998). Learning dynamical systems by recurrent neural networks from orbits. *Neural Networks, 11,* 1589–1599.

Lasota, A. & Mackey, M. C. (1994). *Chaos, Fractals and Noise: Stochastic Aspects of Dynamics*. New York: Springer-Verlag.

Marocco, D., Cangelosi, A., & Nolfi, S. (2003). The emergence of communication in evolutionary robots. *Philosophical Transactions of the Royal Society of London. Series A, 361,* 2397–2421.

Mathayomchan, B. & Beer, R. D. (2002). Center-crossing recurrent neural networks for the evolution of rhythmic behavior. *Neural Computation, 14,* 2043–2051.

Merleau-Ponty, M. (1962). *Phenomenology of Perception*. New York: Humanities Press.

Miller, G. F. & Todd, P. M. (1991). Exploring adaptive agency: I. Theory and methods for simulating the evolution of learning. In D. S. Touretzky, J. L. Elman, T. J. Sejnowski, & G. E. Hinton (Eds.), *Connectionist Models: Proceedings of the 1990 Summer School* (pp. 65–80). New York: Morgan Kaufmann.

Nolfi, S. & Floreano, D. (2000). *Evolutionary Robotics*. Cambridge, MA: MIT Press.

Pasemann, F. (2002). Complex dynamics and the structure of small neural networks. *Network: Computation in Neural Systems, 13,* 195–216.

Phattanasri, P. (2002). *Associative Learning in Evolved Dynamical Neural Networks*. Ph.D. Dissertation, Department of Electrical Engineering and Computer Science, Case Western Reserve University, Cleveland, OH.

Phattanasri, P., Chiel, H. J., & Beer, R. D. (2007). The dynamics of associative learning in evolved model circuits. *Adaptive Behavior, 15*(4), 377–396.

Prinz, A. A., Billimoria, C. P., & Marder, E. (2003). Alternative to hand-tuning conductance-based models: Construction and analysis of databases of model neurons. *Journal of Neurophysiology, 90,* 3998–4015.

Psujek, S., Ames, J., & Beer, R. D. (2006). Connection and coordination: The interplay between architecture and dynamics in evolved model pattern generators. *Neural Computation, 18,* 729–747.

Quinn, M. (2001). Evolving communication without dedicated communication channels. In J. Kelemen & P. Sosík (Eds.), *Advances in Artificial Life: Proceedings ECAL 2001* (pp. 357–366). Springer.

Slocum, A. C., Downey, D. C., & Beer, R. D. (2000). Further experiments in the evolution of minimally cognitive behavior: From perceiving affordances to selective attention. In J. Meyer, A. Berthoz, D. Floreano, H. Roitblat, & S. Wilson (Eds.), *From Animals to Animats 6: Proceedings of the Sixth International Conference on Simulation of Adaptive Behavior* (pp. 430–439). Cambridge, MA: MIT Press.

Steels, L. (2003). Evolving grounded communication for robots. *Trends in Cognitive Sciences, 7,* 308–312.

Suchman, L. A. (1987). *Plans and Situated Actions*. New York: Cambridge University Press.

Thelen, E. & Smith, L. B. (1994). *A Dynamic Systems Approach to the Development of Cognition and Action*. Cambridge, MA: MIT Press.

Tuci, E., Quinn, M., & Harvey, I. (2002). An evolutionary ecological approach to the study of learning behavior using a robot-based model. *Adaptive Behavior, 10*(3–4), 201–221.

Van Gelder, T. (1995). What might cognition be if not computation? *Journal of Philosophy, 91,* 345–381.

Vickerstaff, R. J. & Di Paolo, E. A. (2005). Evolving neural models of path integration. *Journal of Experimental Biology, 208,* 3349–3366.

Walter, W. G. (1953). *The Living Brain*. W.W. Norton.

Ward, R. & Ward, R. (2006). Cognitive conflict without explicit conflict monitoring in a dynamical agent. *Neural Networks, 19,* 1430–1436.

Ward, R. & Ward, R. (in press). Selective attention and the control of action: Comparative psychology of artificial, evolved agents and people. *Journal of Experimental Psychology: Human Perception and Performance*.

Wood, R. & Di Paolo, E. A. (2007). New models for old questions: Evolutionary robotics and the "A not B" error. In F. Almeida, E. Costa et al. (Eds.), *Advances in Artificial Life: Proceedings ECAL 2007* (pp. 1141–1150). Springer.

Yamauchi, B. & Beer, R. D. (1994a). Integrating reactive, sequential and learning behavior using dynamical neural networks. In D. Cliff, P. Husbands, J. Meyer, & S. Wilson (Eds.), *From Animals to Animats 3: Proceedings of the Third International Conference on Simulation of Adaptive Behavior (SAB 94)* (pp. 382–391). Cambridge, MA: MIT Press.

Yamauchi, B. & Beer, R. D. (1994b). Sequential behavior and learning in evolved dynamical neural networks. *Adaptive Behavior, 2*(3), 219–246.

7

The Synthetic Approach to Embodied Cognition: A Primer

Rolf Pfeifer[1], Max Lungarella[1] and Olaf Sporns[2]

[1]*Department of Informatics, University of Zurich, Zurich, Switzerland*
[2]*Department of Psychological and Brain Sciences and Programs in Cognitive and Neural Science, Indiana University, Bloomington, IN, USA*

Introduction

In recent years, there has been a considerable amount of research showing that cognition is embodied and best understood as a situated activity (Brooks, 1991; Chiel & Beer, 1997; Clark, 1999; Lakoff & Johnson, 1999; Pfeifer & Scheier, 1999; Thompson & Varela, 2001; Wilson, 2002; Anderson, 2003; Pfeifer & Bongard, 2007). Building on this body of empirical and theoretical work this chapter addresses a specific set of issues surrounding the link between embodiment and information processing. Our main thesis is that the interaction between physical and information processes is central for the emergence and development of intelligence. Specifically, for agents in the real world, information is not just "out there," an infinite tape ready to be loaded and processed by the cognitive machinery of the brain. Instead, through physical (embodied) interactions with the environment, embodied agents actively induce information structure in their sensory inputs (e.g., spatio–temporal correlations in a visual input stream, redundancies between different perceptual modalities, or regularities in sensory patterns that are invariant with respect to changes in illumination, size, or orientation). In the context of this chapter, we will use the term information structure to refer to the organization of the sensory data typically induced by and meaningful with respect to some purposive or intended action such as grasping or walking. As suggested here, the presence of such structure might be essential for the acquisition of a broad range of cognitive and motor abilities such as multimodal sensory integration, cross-modal learning, perceptual categorization, reaching, object manipulation, language, and locomotion.

Because of the tutorial nature of this chapter, we start by introducing some basic concepts that we will use throughout: synthetic methodology, frame-of-reference, and self-organization. We then look at how morphology and the intrinsic dynamics of the body promote the self-organization of a repertoire of preferred movements that can greatly simplify the learning of complex movements. Subsequently, we expand on the notions of information structure and information self-structuring, and show how quantitative measures can be used to corroborate and theoretically underpin our claims. We then elaborate on the role of these ideas in learning and development, and use two case studies, categorization and intelligent prosthetic devices, to illustrate the main concepts. Before concluding, we discuss the implications of our ideas for theories of cognition and cognitive development.

BASICS

If properly applied, ideas from embodied cognition do not only lead to surprising theoretical insights but can also have great practical value, for example, for the design of autonomous adaptive systems. An embodied perspective, because it distributes control and processing to all aspects of the agent (its central nervous system, the material properties of its musculoskeletal system, the sensor morphology, and the interaction with the environment), provides an alternative avenue for tackling the challenges faced by robotics. The tasks performed by the controller in the classical approach are now partially taken over by morphology and materials in a process of self-organization (Box 7.1); for example, skin properties support the functionality of hands: grasping a glass with soft, compliant, slightly humid fingertips is much easier than grasping a glass with thimbles, because the deformation of the tissue on the fingertips, which is entirely passive, increases surface contact and friction. Clearly, the embodied view suggests that the actual behavior emerges from the interaction dynamics of agent and environment through a continuous and dynamic interplay of physical and information processes (Figure 7.1).

In this context, we also point out a distinct advantage of using robots rather than working with humans or animals (Figure 7.2). Robotic platforms allow for comprehensive recording and analysis of complete histories of sensory stimulation and motor activity, and enable us to conduct precisely controlled experiments while introducing systematic changes in body morphology, materials, and control architectures. They are the essential tools of the synthetic methodology which advocates "building in order to understand" (Box 7.2). Moreover, robots allow

BOX 7.1 Self-organization, self-stabilization, and emergence

Biological systems and consequently bio-inspired robots display self-organization and emergence at multiple levels: at the levels of movement generation, induction of sensory stimulation, exploitation of morphological and material properties, and interaction between individual modules and entire agents.

For example, a loosely hanging swinging arm will self-organize into a particular trajectory suited for exploration. This process is self-organizing because it is the result of the synergistic interaction of the muscle activity in the shoulder, the global structure of the musculo-skeletal system (which includes the muscle tone that stabilizes the upper body), and the fact that the arm acts like a pendulum that exploits gravitational forces. The elbow joint and the wrist joint are passive and not actuated, but they do change in desired ways, especially the elbow joint. The resulting trajectory of the hand turns out to be very useful to the agent because it leads to a high probability of something interesting happening. The word interesting in this context means an event that leads not only to sensory stimulation, but also to stimulation that is rich and contains a lot of information structure. The task of controlling the joint angles to achieve the desired hand trajectory is partially taken over by morphology and materials in a process of self-organization.

Other examples of self-organization include self-stabilization of a mechanical system, as in the paradigmatic passive dynamic walker (see text) or self-regulation as in the Yokoi robot hand or human hands for that matter (Yokoi et al., 2004). Mathematical analysis of self-stabilizing systems shows that the periodic gate of the passive dynamic walker corresponds to an attractor of a complex dynamical system. The basin of attraction, that is, the regions of stable walking, can be extended by adding just a bit of actuation, but again, without specifically controlling the joint angles. If we put pressure sensors on the feet of any of these robots, periodic patterns of sensory stimulation that reflect that natural dynamics of the system will be induced, because the stimulation is the result of self-organization. Additional examples are provided in the text.

Another level of self-organization occurs when individuals interact using local rules, as in swarm behavior, or ant trail formation. In biological development and in modular robotics, many components—the cells or the individual modules—self-organize into functional collectives, limbs, organs, wheels, or entire robots in the modular robotics case. Ensembles assembled from small components can be shown to display emergent functionality such as locomotion, rotation, or wall-following. The focus of the chapter is on self-organization in individual behavior which is why we do not pursue collective intelligence further, here.

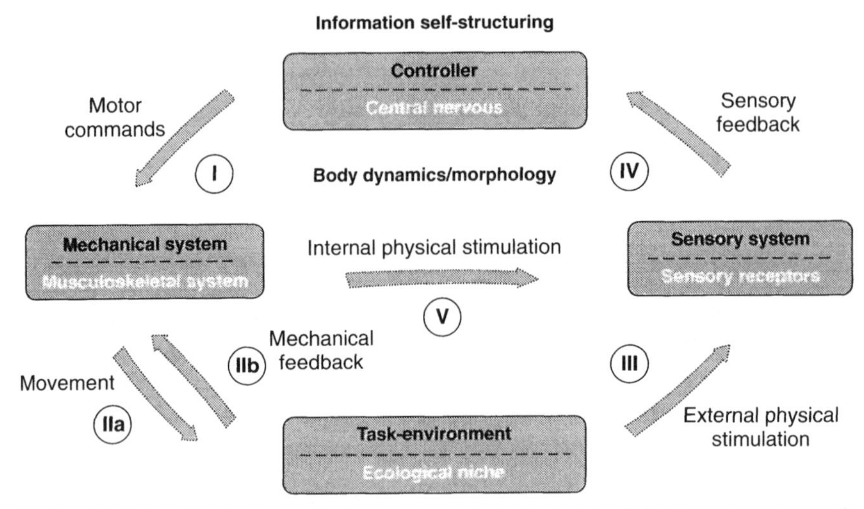

FIGURE 7.1 The interplay of information and physical processes. Driven by motor commands (I), the musculoskeletal system (mechanical system) of the agent acts on the external environment (task environment or ecological niche) (IIa). The action leads to rapid mechanical feedback (IIb) characterized by pressure on the bones, torques in the joints, and passive deformation of skin tissue. In parallel, external stimuli (III) (pressure, temperature, and electromagnetic fields) and internal physical stimuli (IV) (forces and torques developed in the muscles and joint-supporting ligaments, as well as accelerations) impinge on the sensory receptors (V) (sensory systems). The patterns induced thus depend on the physical characteristics and morphology of the sensory systems and on the motor commands.

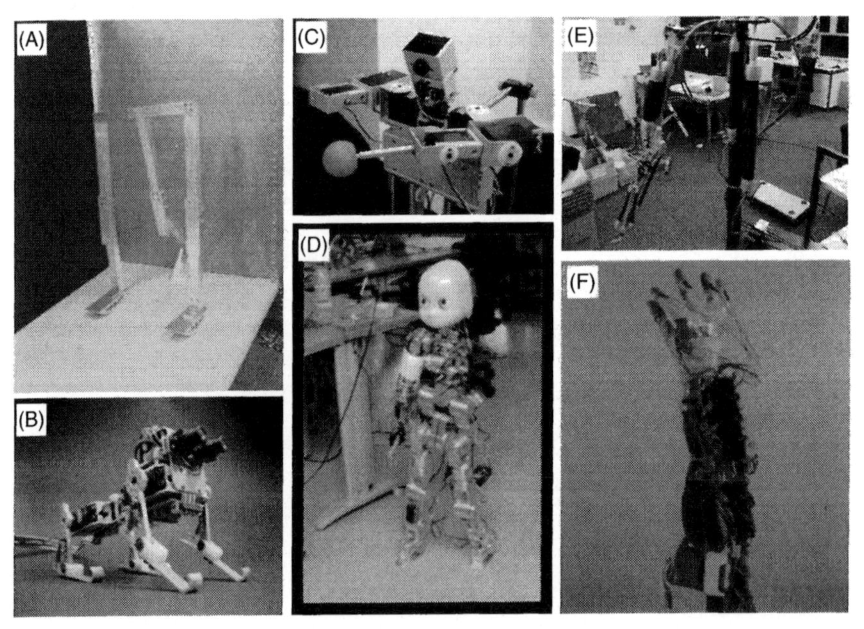

FIGURE 7.2 Collection of robots used in the synthetic approach to embodied intelligence. (A) Passive dynamics walker. (B) Quadruped used for experiments on the influence of morphology and control on behavior. (C) Humanoid used for experiments on information self-structuring. (D) Developmental humanoids iCub. (E) Anthropomorphic arm with pneumatic actuators mimicking, among other things, the loosely swinging arm described in the text. (F) Prosthetic hand–arm complex. (See color plate)

BOX 7.2 The synthetic methodology

The synthetic methodology can be characterized by the slogan "understanding by building." Given a phenomenon of interest—how we recognize a face in the crowd, how ants find their way back to the nest when returning from a food source, how humans walk or how they learn to make distinctions in the real world, how birds manage to fly in swarms, and how rats learn to navigate in a maze—we then design and construct a system that mimics certain aspects of this phenomenon. While studying embodiment and embodied cognition, it is essential to build actual physical systems, which, because we are interested in intelligent systems, will most likely be robots. For example, if we are trying to understand human walking, the synthetic methodology requires that we build an actual walking robot. Of course, simulations can also be employed, but they have to replicate the actual physical processes of walking to tell us something about walking in general. And there is always the question of the accuracy of a simulation. Experience has shown that building a real physical system always yields the most new insights. It is easy to "cheat" with simulations: a real-world walking agent, such as a human or a physical robot, has to somehow deal with bumps in the ground, while this problem can be ignored in a simulation (where each problem has to be explicitly programmed in). Moreover, compliant under-actuated systems often have a highly complex dynamics that is very hard to model accurately, which implies that although simulations seem to be easier to construct, building the actual system is often quicker and yields more interesting results. Of course, when working with artificial evolution or developmental systems, simulations are—given the current level of technology—the only feasible tool. Thus, when employing the synthetic methodology, the question is not whether to use simulation or to build real robots, but when to use which, and very often, it is best to employ both.

The synthetic methodology contrasts with the more classical analytical ways of proceeding as in biology, psychology, or neuroscience, where an animal or human is analyzed in detail by performing experiments on it. Having said that, it is interesting to note that the sciences in general have become more synthetic lately, as the brisk rise of the computational sciences demonstrates: physicists increasingly prepare experiments in simulation, surgeons prepare operations in simulation, and pharmacologists test the effects of drugs in simulation. If these simulations are to be useful, they of course have to be as accurate as possible. But even if there is a high level of simulation accuracy, eventually it will always be necessary to perform experiments in the real world.

us to clearly separate behavior (which is always the result of an embodied interaction of the agent with the real world) and the internal mechanisms underlying it—providing a means to avoid getting entangled in the frame-of-reference problem (Box 7.3).

BOX 7.3 The frame-of-reference problem

The frame-of-reference issue, that is concerned with the perspectives we can adopt when observing or designing agents, implies that we be very clear about what we are observing and how we interpret what we observe. The initial inspiration for this line of thought comes from Herbert Simon's seminal book *The Sciences of the Artificial*, in which he introduced the anecdote of an ant walking along a beach (Simon, 1976). He argued that from an observer's point of view, the ant describes a complex path because it walks around puddles, rocks, twigs, and pebbles. However, from the point of view of the ant, the mechanisms that bring about this behavior might in fact be quite simple, such as "if obstacle on right then turn left" or "if obstacle on left then turn right," and "go straight." The final path of the ant emerges from its interaction with the environment; in this case, the beach. The ant knows nothing about puddles, pebbles, and twigs but still manages to find its way around quite well (see also Pfeifer & Scheier, 1999).

From an external observer's point of view the ant is finding a path back to the nest, whereas from the ant's perspective, it may simply be reacting to sensory stimulation. This is the perspectives issue. Behavior is, by definition, always emergent from the interaction of an agent with the real world and thus cannot be predicted by looking at the internal control or brain mechanisms alone; we always have to know how the control is embedded into the physical system (which includes its connection to the sensors and the actuators), and the kinds of interactions with the environment. This is the behavior versus mechanism issue. Finally, from the apparent complexity of the behavior, we cannot draw firm conclusions about the complexity of the underlying mechanisms, as illustrated by Simon's ant on the beach. Note that in this context, emergence is precisely defined and can be rationally understood and explained—there is nothing mystical about it.

Here is another illustration. Imagine a loosely swinging arm. If you look at the hand, you will find that it describes a highly complex trajectory in 3D space. However, the control for this movement is very simple because gravity is doing part of the work (the arm is a bit like a pendulum), and the anatomy and the body tissue constrain the movement to certain preferred trajectories. Even if there was random neural stimulation, the arm movement would be highly constrained. Moreover, the elbow and wrist joints are largely passive and only minimally controlled: from the fact that the joints move, we cannot conclude that they are directly steered. Why this arm movement is not only easy to control, but is in fact highly useful is explained in the text.

BODY DYNAMICS AND MORPHOLOGY

Several studies with robots indicate that the computational processes involved in control can be partially taken over by the morphological properties of the agent (Collins et al., 2005; Tedrake et al., 2004; Pfeifer et al., 2007; Iida et al., 2008). A paradigmatic example is provided by passive dynamic walkers which are robots—or rather mechanical structures without microprocessors or motors—that walk down a slope without control and actuation (Collins et al., 2005; Figure 7.2A). The walker's morphology (center of mass, length of the limbs, and the shape of the feet) and its materials are carefully designed so as to exploit the physical constraints present in its ecological niche for locomotion (friction, gravity, inclination of the slope). To get the robot to learn to walk on level surfaces, one can use the mechanical design obtained during passive dynamic walking and endow it with actuators (e.g. located in the ankles or hips) (Tedrake et al., 2004). The natural dynamics of the body–environment system provides the target for learning the control policy for the actuators by stabilizing the limit cycle trajectory that the robot follows—the dynamics structures the output of the angle sensors located in the joints, so to speak—and the robot learns to walk adaptively on flat ground within a relatively short period of time.

It is interesting to observe that as a consequence of the different data distributions resulting from different sensory morphologies a dependency exists between morphology, dynamics, and learning speed (Lichtensteiger & Pfeifer, 2002; Tedrake et al., 2004). For example, by exploiting the non-homogenous arrangement of facets in the insect eye (denser in the front than on the side), the phenomenon of motion parallax can be "compensated away" and the adaptability of neural controller can be greatly improved (Lichtensteiger & Pfeifer, 2002). We infer that the design of controller and morphology are, in a sense, inseparable, because the structure of both impacts information processing. However, while some progress has been made to optimize the design of robot controllers, robot morphology largely remains a matter of heuristics. Future progress in the design of intelligent robots will require analytical tools and methodologies to exploit the interaction between morphology and computation (Lungarella et al., 2005; Lungarella & Sporns, 2006; Pfeifer et al., 2007).

The specific morphology of the body and the interaction of body and environment dynamics also shape the repertoire of preferred movements: a loosely hanging swinging arm moves in a complex trajectory but its control is extremely simple (the knowledge of how to move the limb seems to reside in the limb itself), whereas moving the hand in a straight path—a seemingly simple trajectory—requires a lot of control. It follows that part of the "processing" is done by the dynamics of the agent–environment interaction, and only sparse neural control needs to be exerted when the self-regulating and self-stabilizing properties of the natural dynamics can be exploited. The idea that brain, morphology, materials, and environment share responsibility in generating information structure has been called the "principle of ecological balance" (Pfeifer & Scheier, 1999) because there is a "balance" or task distribution between the different aspects of an embodied agent.

INFORMATION SELF-STRUCTURING

In the previous section, we have shown how the interaction of a given morphology with the environment can impose consistent and invariant (i.e., learnable) structure on sensory stimulation. Working in parallel with the specific body morphology, sensory–motor coordinated interaction is crucial in shaping the resulting information structure. This idea finds support in work on direct and active perception, as well as animate, interactive, and enactive vision (Gibson, 1979; Bajcsy, 1988; Ballard, 1991; Churchland et al., 1994). From an information theoretic point of view, embodied agents generate information structure in their sensory stimulation as they—actively—interact with the environment. It is important to note that in this process, the specific morphology and the materials used unavoidably determine the resulting information structure. For instance, because of the high density of touch sensors on the fingertips and because of the shape of the hand, grasping automatically leads to rich and structured tactile stimulation. The coordinated sensory–motor action of grasping induces stable patterns of stimulation characterized by correlations between the activities of neurons within a single sensor modality, as well as correlations between neurons across different modalities (vision, touch, audition, and proprioception). Such statistical dependencies (which are instances of information structure) create redundancy across sensory channels, which may help to reduce the effective dimensionality of the input. Given the typically large dimensionality of the state space formed by the sensory input, such redundancy can significantly simplify perception.

Theoretical studies and robot models provide quantitative evidence for the notion that self-generated motor activity can create information structure in sensory–motor data—an idea which has been called "information self-structuring" (Lungarella & Sporns, 2005, 2006; Nehaniv et al., 2008). For instance, in Lungarella & Sporns (2005) it is demonstrated how a simple robot capable of saliency-based attentional behavior—an instance of an active vision system—"self-structures" the information present in its sensory and motor channels. The results of the study also show that sensory–motor coordination yields a better embedding of the visual input into a low-dimensional space, as compared to uncoordinated behavior (for details, see Box 7.4). Traditionally, such dimensionality reduction has been studied in the context of internal processing of a neural architecture, for example, through mechanisms in early visual processing that lead to efficient low-dimensional (sparse) encoding by exploiting input redundancies and regularities (Barlow, 2001; Simoncelli & Olshausen, 2001). We suggest that the generation of structured information through embodied interaction provides an additional mechanism contributing to efficient neural coding.

BOX 7.4 How the body induces information structure

Sensory inputs are often incomplete and ambiguous, and pose, for example, significant challenges for traditional approaches to machine vision, which often require visual inputs segmented into objects, preprocessed, or brought into canonical formats for efficient processing (Palmeri & Gauthier, 2004). In an elegant set of robot experiments, Metta & Fitzpatrick, 2003 demonstrated how embodied interaction can be exploited to disambiguate and segment a complex visual scene. Working with the humanoid robot "Cog," they investigated the potential role of experimental manipulation (e.g., reaching for, touching and displacing objects within the visual field) in generating visual information about object boundaries and affordances such as rolling. For instance, exploratory activity by the robot resulted in the displacement of a solid object against a static (*a priori* unknown) background, generating a correlated motion field that closely corresponded to the shape of the object (Figure 7.3). These motion signals represent structured information that was absent before the robot's exploratory actions. Note that active exploration can be applied even if the background changes, and that allows extracting information concerning the affordances of the segmented objects—a "passive" strategy is not sufficient. This example demonstrates the pivotal role of self-generated embodied interaction in inducing statistical structure. In other words, information structure emerges while the interaction is taking place.

Lungarella and Sporns attempted to quantify information on self-structuring in a simple robotic active vision platform equipped with a saliency-driven visual attention mechanism (Lungarella and Sporns, 2005; Lungarella and Sporns, 2006). Information in visual inputs sampled by

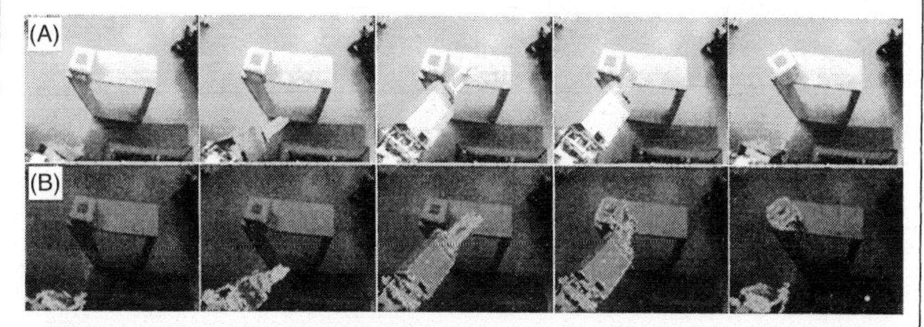

FIGURE 7.3 Disambiguation and segmentation of visual scene through embodied interaction. (A) Arm extending into a workspace, poking an object, and retracting. (B) Shape of the object is identified from the tap using simple image differencing (Metta & Fitzpatrick, 2003). Segmentation in this case is not a trivial task—the edges of the table and cube are aligned, the colors of the cube and the table are not well separated, and there are shadows that may change. (See color plate)

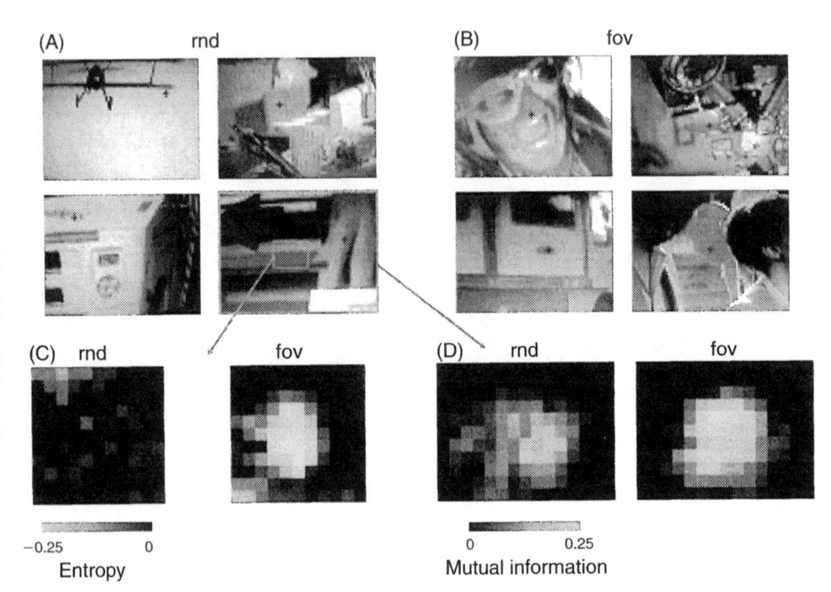

FIGURE 7.4 Information structure in the visual field as a function of embodiment. Images show sample video frames obtained from a disrupted (A; "rnd," "low embodiment"—no sensory–motor coupling) and normally tracking (B; "fov," "high embodiment"—high sensory–motor coupling), and active vision system. Plots at the bottom show spatial maps of entropy and mutual information, expressed as differences relative to the background. There is a significant decrease in entropy (C) and an increase in mutual information (D) in the center of the visual field for the "fov" condition, compared to little change in the "rnd" condition. Data replotted from Lungarella et al., 2005. (See color plate)

the active vision system, as well as between sensory and motor variables, was measured using a broad spectrum of tools from statistical information theory. To allow for quantification of the amount of information structure gained by maintaining a high degree of sensorimotor coupling, sensory and motor data were collected from a system actively tracking specific salient objects ("fov," Figure 7.4), as well as from a system in which the coupling between sensory inputs and motor activity had been disrupted ("rnd," Figure 7.4). Significant differences in the amount of information structure, including entropy, mutual information, integration, and complexity were found. With this approach the specific effects of embodiment (here, sensorimotor coupling) on inducing statistical relationships between different sensory (visual) channels can be quantitatively measured. Information measures can also be used as objective functions to guide evolving robots toward optimal embodiment (Sporns & Lungarella, 2006). Initial results indicate that robots capable of generating optimal amounts of information structure show high degrees of sensorimotor coordination.

The theoretical concepts outlined above receive support from experiments with human subjects showing that most of our sensory experiences involve active (i.e., sensory–motor) exploration of the world (e.g., through manipulation or visual inspection) (Noe, 2004). Such exploration promotes not only object recognition (Harman et al., 1999; Woods & Newell, 2004), but also, for instance, the learning of the 3D structure of objects (James et al., 2001), and depth perception (Wexler & van Boxtel, 2005).

LEARNING AND DEVELOPMENT

There is an interesting implication of information self-structuring for learning. Information structure does not exist before the interaction occurs, but emerges only while the interaction is taking place. However, once it has been induced, learning can pick up on it such that next time around, the responsible sensory–motor information structure is more easily reactivated. It follows that embodied interaction lies at the root of a powerful learning mechanism as it enables the creation of time-locked correlations and the discovery of higher order regularities that transcend the individual sensory modalities.

These ideas also extend to development. It is generally recognized that structured information and statistical regularities are crucial for perception, action, and cognition—and their development (Clark, 1999; Gibson & Pick, 2000; Barlow, 2001). At a very young age, babies frequently use several sensory modalities for the purpose of categorization: they look at objects, grasp them, stick them into their mouths, throw them on the floor, and so on. The resulting intra- and intermodal sensory stimulation appears to be essential for concept formation (Clark, 1999; Bahrick et al., 2004; Gallese & Lakoff, 2005). As they grow older, infants can perform categorization based on visual cues alone which implies that they must have learned something about how to predict sensory stimulation in one modality using the information available through another modality, for instance, the haptic from the visual one. By virtue of its continuous influence on the development of specialized neurons and their connections that incorporate consistent statistical patterns in their inputs, information structure plays a critical role in development. It is easier for neural circuits to exploit and learn sensory–motor patterns containing regularities and recurrent statistical features.

CASE STUDY 1: EMBODIED CATEGORIZATION

For an autonomous embodied agent acting in the real world (e.g., an animal, a human, or a robot), perceptual categorization—the ability to make distinctions—is a hard problem (Harnad, 2005). First, based on the stimulation impinging on its sensory arrays (sensation) the agent has to rapidly determine and attend to what needs to be categorized. Second, the appearance and properties of objects or events in the environment being classified fluctuate continuously, for example owing to occlusions, or changes of distances and orientations with respect to the agent. And third, the environmental conditions (e.g., illumination, viewpoint, and background noise) vary considerably. There is much relevant work in computer vision that has been devoted to extracting scale- and translation-invariant low-level visual features and high-level multidimensional representations for the purpose of robust perceptual categorization (Riesenhuber & Poggio, 2002). Following this approach, however, categorization often turns out to be a very difficult if not an impossible computational feat, especially when sufficiently detailed information is lacking.

A solution that can only be pursued by embodied agents—but is not available when using a purely disembodied (i.e., computational) approach—is that through their interaction with the environment, agents generate the sensory stimulation required to perform the proper categorization and thus drastically simplify the problem of mapping sensory stimulation onto perceptual categories. The most typical and effective way is through a process of sensory–motor coordination. One demonstration of how sensory–motor coordination influences category formation can be found in the experiments by Pfeifer & Scheier (1997). These experiments show that mobile robots can reliably categorize big and small wooden cylinders only if their behavior is sensory–motor coordinated. A similar point is illustrated by the artificial evolution experiments of Nolfi (2002) and Beer (2003): the fittest agents, that is, those that most reliably categorized different kind of objects belonged, were those engaging in sensory–motor coordinated behavior. Intuitively, in these examples, the interaction with the environment (a physical process) creates additional (i.e., previously absent) sensory stimulation, which is highly structured, thus facilitating subsequent information processing. Computational economy and temporal efficiency are purchased at the cost of behavioral interaction, so to speak.

CASE STUDY 2: APPLICATION OF EMBODIED COGNITION TO PROSTHETICS

The synthetic methodology and the notions of morphology and information self-structuring benefit working systems and can lead to spin-offs such as assistive prosthetic devices. In particular, they give indications as to how to augment the sensory–motor "intelligence" of the coupled man–machine system. For instance, in the case of a lower arm robotic prosthesis driven by electromyographic (EMG) signals taken from the amputee's upper arm, because of its particular construction—anthropomorphic shape, elastic, deformable materials (tendons, fingertips), and its decentralized control—when a "close" command is issued, the hand will close its fingers until a particular threshold in the pressure sensors is reached and will then stop. Without having to know about the shapes, the hand will adapt to a large variety of object shapes: the process is self-regulating (an instance of self-organization; Box 7.1).

For information self-structuring (Box 7.4), sensory feedback not only from the visual system, but also from the haptic system on the fingers and in the hand (touch and temperature sensors) must be generated (e.g., when a bottle is picked up). For prosthetics, therefore, if grasping with the artificial hand is to "feel" natural, information structure must be induced which implies that there must be rich sensory feedback (which is not available in many of the current prosthetic devices). Experiments with fMRI show that patients that are (in addition to the visual feedback) provided with even minimal, but correlated, sensory feedback (such as electrical stimulation to the skin or mechanical vibration), integrate their prosthesis into their body schema much faster (Hernandez-Arieta et al., in press). An embodied perspective on prosthetics also tells us that whenever an embodied agent performs an activity such as grasping, the entire agent will be involved, that is, proper hand control must take whole-body dynamics into account (Hernandez-Arieta et al., in press). In other words, the control of the hand, which is normally considered in isolation, will have to be altered depending on what the rest of the agent is doing, for example, sitting in a table, walking, or lying in the bed.

DISCUSSION: THE INTERACTION OF PHYSICAL AND INFORMATION PROCESSES

The importance of the interaction between physical and information processes can hardly be over-estimated. The complexity of perceptual categorization in the real world, for instance, cannot be managed by computational means only. We have therefore stressed the significance of sensory–motor coordination. The principle of information self-structuring illustrates how physical interaction with the real world induces structured sensory stimulation, which, given the proper morphology, substantially facilitates neural processing, and hence sets the foundations for learning and development of perception and cognition in general.

By looking at the dynamics of embodied systems, we can take the idea of interaction of physical and information processes a step further. Because of the constraints provided by their embodiment, the movements of embodied systems follow certain preferred trajectories. It turns out that in biological agents such dynamics typically leads to rich and structured sensory stimulation. For example, as grasping is much easier than bending the fingers of the hand backwards, grasping is more likely to occur, and owing to the morphology (e.g. the high density of touch sensors on the fingertips), the intended sensory stimulation is induced. The natural movements of the arm and hand are—as a result of their intrinsic dynamics—directed toward the front center of the body. This in turn implies that normally a grasped object is moved toward the center of the visual field thereby inducing correlations in the visual and haptic channels, which, as we pointed out earlier, simplify learning. So we see that an interesting relationship exists between morphology, intrinsic body dynamics, generation of information structure, and learning (Figure 7.1).

The ideas of action and cognition constrained by embodiment as sketched in this chapter are compatible with theories of grounded cognition (Barsalou et al., 2007; Barsalou, 2008) that stress the integrated nature of perceptual, emotional, and cognitive processes, as well as with the central role of developmental processes for the emergence of intelligence (Smith & Breazeal, 2007). How can these ideas be applied within a developmental framework? For instance, it is possible to explain the infant's immaturity and initial limitations in morphology (e.g., wide spacing of photoreceptors in the retina), as unique adaptations to the environmental constraints of the ecological niche (Bjorklund & Green, 1992). The specific effect of this adaptation is to filter out high spatial frequency information, and to make close objects most salient to the infant and hence reduce the complexity of the required information processing. Such complexity reduction may, for instance, facilitate learning about size constancy (Bjorklund, 1997). That is, the developmental immaturity of sensory, motor, and neural systems, which at first sight appears to be an inadequacy, may in fact be an advantage, because it reduces the "information overload" that otherwise might overwhelm the infant's developing cognitive architecture (Turkewitz & Kenny, 1982; Bahrick et al., 2004). A similar phenomenon occurs at the level of the motor system where musculo-skeletal constraints limit the range of executable movements and hence implicitly reduce the number of control variables. The neural system exploits such constraints and control is simplified by combining a rather small set of primitives, for example synergies (Bernstein, 1969) or force fields (Mussa-Ivaldi & Bizzi, 2000), in different proportions rather than individually controlling each muscle.

In this chapter, we outlined a view of sensory–motor coordination and natural dynamics as crucial causal elements for neural information processing because they generate information structure. Our argument has revolved mainly around brain areas directly connected to sensory and motor systems. It is likely, however, that embodied systems operating in a highly coordinated manner generate information structure and statistical regularities at all hierarchical levels within their neural architectures, including effects on neural activity patterns far removed from the sensory periphery. This hypothesis leads to several predictions, testable in animal or robot experiments. For example, activations or statistical relationships between neurons in cortical areas engaged in sensorimotor processing should exhibit specific changes across different states of sensorimotor coordination or coupling. Increased structuring of information through embodiment would be associated with increased multimodal synchronization and binding, or more efficient neural coding.

CONCLUSION

The conceptual view of perception as an active process has gained much support in recent years (Pfeifer & Scheier, 1997; Nolfi, 2002;Beer, 2003; Lungarella &

Sporns, 2005). In this chapter, we showed why perception cannot be treated as a purely computational problem that unfolds entirely *within* a given information processing architecture. Instead, we presented an emerging link between embodiment and information which views perception as naturally embedded within a physically embodied system, interacting with the real world. Thus, it is the *interplay* between physical and information processes that gives rise to perception. We identified specific contributions of embodiment to perceptual processing through the active generation of structure in sensory stimulation, which may pave the way toward a formal and quantitative analysis. The idea of inducing information structure through physical interaction with the real world has important consequences for understanding and building intelligent systems, by highlighting the fundamental importance of morphology, materials, and dynamics.

BIBLIOGRAPHY

Anderson, M. L. (2003). Embodied cognition: A field guide. *Artificial Intelligence, 149,* 91–130.

Bajcsy, R. (1988). Active perception. *Proceedings of the IEEE, 76*(8), 996–1005.

Ballard, D. (1991). Animate vision. *Artificial Intelligence, 48,* 57–86.

Bahrick, L. E., Lickliter, R., & Flom, R. (2004). Intersensory redundancy guides the development of selective attention, perception, and cognition in infancy. *Current Directions in Psychological Science, 13*(3), 99–102.

Barlow, H. B. (2001). The exploitation of regularities in the environment by the brain. *Behavioral and Brain Sciences, 24,* 602–607.

Barsalou, L. W. (2008). Grounded cognition. *The Annual Review of Psychology, 59,* 617–645.

Barsalou, L. W., Breazeal, C., & Smith, L. B. (2007). Cognition as coordinated non-cognition. *Cognitive Processing, 8,* 79–91.

Beer, R. D. (2003). The dynamics of active categorical perception in an evolved model agent. *Adaptive Behavior, 11*(4), 209–243.

Bernstein, N. (1969). *The Co-ordination and Regulation of Movements.* Oxford: Pergamon.

Bjorklund, E. (1997). The role of immaturity in human development. *Psychological Bulletin, 122*(2), 153–169.

Bjorklund, E. & Green, B. (1992). The adaptive nature of cognitive immaturity. *American Psychologist, 47*(1), 46–54.

Brooks, R. A. (1991). New approaches to robotics. *Science, 253,* 1227–1232.

Chiel, H. & Beer, R. (1997). The brain has a body: Adaptive behavior emerges from interactions of nervous system, body, and environment. *Trends in Neurosciences, 20,* 553–557.

Churchland, P. S., Ramachandran, V., & Sejnowski, T. (1994). A critique of pure vision. In C. Koch & J. Davis (Eds.), *Large-scale Neuronal Theories of the Brain.* Cambridge, MA: MIT Press.

Clark, A. (1999). An embodied cognitive science? *Trends in Cognitive Sciences, 3*(9), 345–351.

Collins, S., Ruina, A., Tedrake, R., & Wisse, M. (2005). Efficient bipedal robots based on passive-dynamic walkers. *Science, 307,* 1082–1085.

Dewey, J. (1896). The reflex arc concept in psychology. *Psychological Review, 3,* 357–370.

Edelman, G. M. (1987). *Neural Darwinism.* New York: Basic Books.

Edelman, S. (1999). *Representation and Recognition in Vision.* Cambridge, MA: MIT Press.

Gallagher, S. (2005). *How the Body Shapes the Mind.* Oxford: Clarendon Press.

Gallese, V. & Lakoff, G. (2005). The brain's concepts: The role of the sensory-motor system in conceptual knowledge. *Cognitive Neuropsychology, 22*(3/4), 455–479.

Gibson, J. J. (1979). *The Ecological Approach to Visual Perception.* Boston, MA: Houghton-Mifflin.

Gibson, E. J. & Pick, A. D. (2000). *An Ecological Approach to Perceptual Learning and Development.* Oxford: Oxford University Press.

Glenberg, A. M. & Kaschak, M. P. (2002). Grounding language in action. *Psychonomic Bulletin and Review, 9*(3), 558–565.

Harman, K. L., Humphrey, G. K., & Goodale, M. A. (1999). Active manual control of object views facilitates visual recognition. *Current Biology, 9*(22), 1315–1318.

Harnad, S. (2005). Cognition is categorization. In H. Cohen & C. Lefebvre (Eds.), *Handbook of Categorization in Cognitive Science.* Elsevier. http://eprints.ecs.soton.ac.uk/11725/

Hernandez-Arieta, A., Dermitzakis, K., Damian, D., Lungarella, M., & Pfeifer, R. (in press). Sensory-motor coupling in rehabilitation robotics. Service Robotics.

Iida, F., Pfeifer, R., Steels, L., & Kuniyoshi, Y. (Eds.) (2004). *Embodied Artificial Intelligence.* Berlin, Heidelberg: Springer-Verlag.

Iida, F., Rummel, J., & Seyfarth, A. (2008). Bipedal walking and running with spring-like biarticular muscles. *Journal of Biomechanics, 41*(3), 656–667.

James, K. H., Humphrey, G. H., & Goodale, M. A. (2001). Manipulating and recognizing virtual objects: Where the action is. *Canadian Journal of Experimental Psychology, 55*(2), 113–122.

Lakoff, G. & Johnson, M. (1999). *Philosophy in the Flesh: The Embodied Mind and its Challenge to Western Thought.* New York: Basic Books.

Lewkowicz, D. J. (2004). Perception of serial order in infants. *Developmental Science, 7*(2), 175–184.

Lichtensteiger, L. & Pfeifer, R. (2002). An optimal sensory morphology improves adaptability of neural network controllers. *Proceedings of International Conference on Artificial Neural Networks* (LNCS 2451, pp. 850–855). Heidelberg: Springer-Verlag.

Lungarella, M. & Berthouze, L. (2002). On the interplay between morphological, neural, and environmental dynamics: A robotic case-study. *Adaptive Behavior, 10*(3/4), 223–241.

Lungarella, M. & Sporns, O. (2005). Information self-structuring: Key principle for learning and development. *Proceedings of 4th International Conference on Development and Learning* (pp. 25–30).

Lungarella, M. & Sporns, O. (2006). Mapping information flow in sensorimotor networks. *PLoS Computational Biology, 2*(10), e144.

Lungarella, M., Metta, G., Pfeifer, R., & Sandini, G. (2003). Developmental robotics: A survey. *Connection Science, 15*(4), 151–190.

Lungarella, M., Pegors, T., Bulwinkle, D., & Sporns, O. (2005). Methods for quantifying the informational structure of sensory and motor data. *Neuroinformatics, 3*(3), 243–262.

Metta, G. & Fitzpatrick, P. (2003). Early integration of vision and manipulation. *Adaptive Behavior, 11*(2), 109–128.

Mussa-Ivaldi, F. A. & Bizzi, E. (2000). Motor learning through the combination of primitives. *Phil. Trans. of Roy. Soc. London B, 355,* 1755–1769.

Nehaniv, C., Mirza, N.-A., & Olsson, L. (2008). Development via information self-structuring of sensorimotor experiences and interaction. In M. Lungarella et al. (Eds.), 50th *Anniversary of Artificial Intelligence. Lecture Notes in Computer Science* (Vol. 4850). Berlin: Springer-Verlag.

Noe, A. (2004). *Action in Perception.* Cambridge, MA: MIT Press.

Nolfi, S. (2002). Power and limit of reactive agents. *Neurocomputing, 49,* 119–145.

Olshausen, B. & Field, D. J. (2004). Sparse coding of sensory inputs. *Current Opinion in Neurobiology, 14,* 481–487.

Palmeri, T. J. & Gauthier, I. (2004). Visual object understanding. *Nature Reviews Neuroscience, 5,* 291–304.

Piaget, J. (1953). *The Origins of Intelligence.* New York: Routledge.

Pfeifer, R. & Bongard, J. C. (2007). *How the Body Shapes the Way We Think – A New View of Intelligence.* Cambridge, MA: MIT Press.

Pfeifer, R. & Scheier, C. (1997). Sensory-motor coordination: The metaphor and beyond. *Robotics and Autonomous Systems, 20,* 157–178.

Pfeifer, R. & Scheier, C. (1999). *Understanding Intelligence.* Cambridge, MA: MIT Press.

Pfeifer, R., Iida, F., & Bongard, J. C. (2005). New robotics: Design principles for intelligent systems. *Artificial Life, 11*(1/2), 99–120.

Pfeifer, R., Lungarella, M., & Iida, F. (2007). Self-organization, embodiment, and biologically inspired robotics. *Science, 318,* 1088–1093.

Pitti, A., Lungarella, M., & Kuniyoshi, Y. (2005). Quantification of emergent behaviors induced by feedback resonance chaos. *Recent Advances in Artificial Life: Advances in Neural Computation, 3*(5), 199–213.

Raibert, M. H. (1986). *Legged Robots that Balance.* Cambridge, MA: MIT Press.

Riesenhuber, M. & Poggio, T. (2002). Neural mechanisms of object recognition. *Current Opinion in Neurobiology, 22,* 162–168.

Simon, H. (1976). *The Sciences of the Artificial.* Cambridge, MA: MIT Press.

Simoncelli, E. & Olshausen, B. (2001). Natural image statistics and neural representation. *Annual Review of Neuroscience, 24,* 1193–1216.

Smith, L. B. & Breazeal, C. (2007). The dynamic lift of developmental process. *Developmental Science, 10,* 61–68.

Sporns, O. (2003). Embodied cognition. In M. Arbib (Ed.), *Handbook of Brain Theory and Neural Networks.* Cambridge, MA: MIT Press.

Sporns, O. (2005). What neuro-robotic models can tell us about neural and cognitive development. In D. Mareschal (Ed.), *Neuroconstructivism: Perspectives and Prospects* (Vol. 2). Oxford, UK: Oxford University Press.

Sporns, O. & Lungarella, M. (2006). Evolving coordinated behavior by maximizing information structure. In L. Rocha et al. (Eds.), *Artificial Life X.* Cambridge, MA: MIT Press.

Tedrake, R., Zhang, T. W., & Seung, H. S. (2004). Stochastic policy gradient reinforcement learning on a simple 3D biped. *Proceedings of the 10th International Conference on Intelligent Robots and Systems* (pp. 3333–3338).

Thompson, E. & Varela, F. J. (2001). Radical embodiment: Neural dynamics and consciousness. *Trends in Cognitive Sciences, 5*(10), 418–425.

Turkewitz, G. & Kenny, P. (1982). Limitation on input as basis for neural organization and perceptual development: a preliminary theoretical statement. *Developmental Psychology, 15,* 357–368.

Webb, B. (2001). Can robots make good models of biological behavior? *Behavioral and Brain Sciences, 24,* 1033–1050.

Wexler, M. & van Boxtel, J. J. A. (2005). Depth perception by the active observer. *Trends in Cognitive Sciences, 9*(9), 431–438.

Wilson, M. (2002). Six views of embodied cognition. *Psychonomic Bulletin and Review, 9*(4), 625–636.

Woods, A. T. & Newell, F. (2004). Visual, haptic, and cross-modal recognition of objects and scenes. *Journal of Physiology – Paris, 98*(1–3), 147–159.

Yokoi, H., Hernández-Arieta, A., Katoh, R., Yu, W., Watanabe, I., & Maruishi, M. (2004). Mutual adaptation in a prosthetics application. In F. Iida, R. Pfeifer, L. Steels, & Y. Kuniyoshi (Eds.), *Embodied Artificial Intelligence, Lecture Notes on Computer Science* (pp. 146–159). Berlin: Springer-Verlag.

8

ANIMATE VISION, VIRTUAL ENVIRONMENTS, AND NEURAL CODES

DANA BALLARD

Department of Computer Science, University of Texas, Austin, TX, USA

EMBODIED INTELLIGENCE

The idea of robust intelligence comes easily to mind. When human beings get stuck while problem solving, they can readily come up with work-arounds, and can recognize a hopeless situation when everything else fails. They then move on to a different agenda. But how can we model this and build systems that have similar properties? Embodied intelligence explores this question via the use of realistic humanoid models that realize real-world constraints between themselves and their surroundings.

Computational models of cognition based on embodied intelligence are a relatively new idea. The original Artificial Intelligence systems were based on the twin ideas of formal logic and search (Fikes & Nilsson, 1971; Newell & Simon, 1995). The beauty of logic is that by binding its inference rules flexibly to the real world one can tackle a huge range of problems. However the downside occurs in the binding process itself: how does one attach the symbols in logical formulas to the real world? This turns out to be a substantial search problem, compounded by the intricacies of the real world that can easily produce inconsistent predicates owning to its unpredictability. So one step forward is simply to constrain the bindings based on prior knowledge that comes from experts. This has been the approach of expert systems models such as ACT-R, Icarus, CLARION, and SOAR, to name a few of the most prominent ones (Anderson, 1983; Laird et al., 1987; Langley & Choi, 2006; Sun, 2006). These systems use rules with variables and bind them by pattern matching. The broad intent is to search for a sequence of rules that will solve a problem. But since the formalisms have left the bounds of formal logic, the price is that there are no guarantees of solutions. This has been addressed recently, by the introduction of reinforcement learning into the rule chaining process (Fu & Anderson, 2006). Provided the designer can successfully avoid the perceptual aliasing problem where different pattern match instances produce contradictory rewards, one has at least formal convergence results.

Despite the enormous difficulties involved, expert systems have achieved many notable successes, particularly in intellectual problems where the symbol bindings can be modeled, such as in algebraic problem solving (Ritter et al., 1998). However, an area that these systems have tackled somewhat superficially is that of perception and action. To use ACT-R as an example, vision is appended as a module, with the ability to search for parts of the image by coordinates or features. The rules are based on early ideas from Treisman as to how images are stored (Treisman, 1980), and by Pylyshyn as to how image features are temporally accessed (Trick & Pylyshyn, 1994). They do not address the difficulty of image segmentation or search (cf. [Palmer, 1995; Caspi et al., 2004] for modern approaches based on signal detection theory). But more strikingly the general integration of image information reflects the traditional vision–cognition–action paradigm.

Although the issues regarding the relationship between vision and action with cognition have been added as a sideshow to the symbolic processing, *embodied cognition* takes these issues as central and regards cognition as the evolutionary add-on. The most important point made by embodied cognition is that the body is a kind of computer. It is rather unusual to think along these lines, so to orient us, we will use an example originally due to (Fischler & Firschein, 1987). Consider finding abstract problem of finding the shortest path in a graph. One could write an algorithm that works for an arbitrary graph, but if we had a particular graph in mind, we could construct it out of physical materials such as string for the arcs and wooden spheres for the nodes. Then to find the shortest path between any two nodes, we can simply grab them and pull them apart and presto, the intervening taut string defines the shortest path. Figure 8.1 shows these basic operations.

It is much harder to think of the body as doing computation but the fundamental premise of embodied cognition is that both sensory and motor systems do computation. In short, the eyes have special orienting circuitry to allow them to focus on desired points in the world. Think of reading. Provided we have normal vision, our eyes focus on almost every word being read, skipping some small function words, yet we are never conscious of this process. In the same way the motor systems of the body and its elaborate spinal cord have ingrown computational rhythms for walking and maneuvering obstacles that too we are never aware of. The point of this computation is that since it does not have to be repeated in terms of establishing elaborate cognitive representations in the

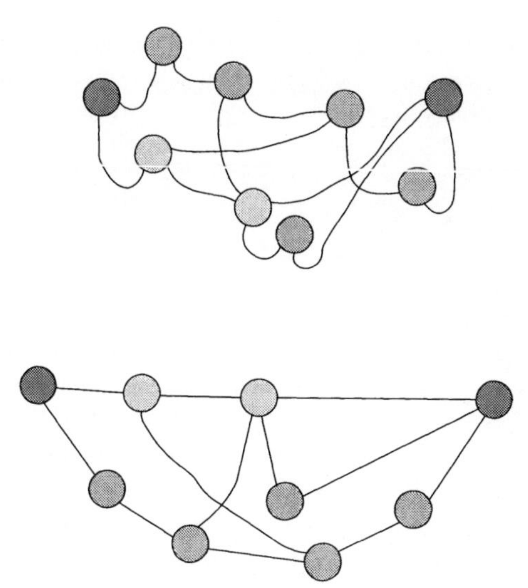

FIGURE 8.1 A physical device that does computation. Computing the shortest path through a graph can be accomplished by building the graph out of wooden spheres and string and grabbing the two end nodes and pulling them apart.

forebrain, the latter can take for granted the work of the body and communicate with it in a much more fundamentally compact way than would be possible if the body did not do this computation.

The original steps toward embodied cognition were taken by Badler (Phillips et al., 1990; Badler, 2001). His well-known avatar "Jack" demonstrated how to solve tough problems such as: "Can I reach my hand around that lamp and grab that cup" by trying to *do* this action. The success or failure of the operation was captured by the success or failure of the avatar. Badler was driven to this line of investigation as the classical way of trying to do this by describing the world in logic proved hopelessly impractical. Thanks to this pioneering work, we understand the enterprise, but another by-product is that it has spurred the development of several other avatars, several of which have impressive real-time performance.

Avatars have spurred another related development, that is advances in real-time control systems that tackle the complexities of behaving in natural worlds. Such systems attempt to model human constraints in realistic ways, eschewing algorithms based on logical formalisms that are typically too expensive to run in real time. A pioneer in this research has been Terzopoulos (Faloutsos et al., 2001). He has explored the design of operating systems needed to control avatars interacting with other objects in complex virtual environments. Such environments were initially greeted with skepticism since they cannot capture all real-world complications, but their hallmark has been faster than the real time speed of simulation that has allowed parametric studies of different design that were initially not possible in real-world robotics systems (Roy & Pentland, 2002; Roy, 2005; Gray, 2007; Sprague et al., 2007). In addition, new learning algorithms have been possible that interact with the real world in various ways. One such algorithm is to use real world data as an initial condition and then learn much more complex scenarios in simulation, an example being helicopter flight control (Abeel et al., 2006).

In embodied cognition, the central focus is on exactly how the human body uses its sensory–motor system to create variable bindings. The idea is that understanding the details of the body provides insight into how the body avoids costly searches and achieves very rapidly situated variable bindings in the course of natural behavior. The second focus is on how behaviors get coded in humans. Since humans repeat everyday tasks thousands of times—think of washing dishes—the learning of those tasks and, importantly for robust intelligence, how to handle certain steps that can go wrong, can be amortized over one's lifetime. The result is that the programs to perform hundreds of highly learned tasks could be coded in terms of look-up tables that tell cognition how to guide the body through the various steps. The macaque monkey cortex gives a hint as to how extensive these tables can be as about 30% is devoted to vision and an equivalent amount for sensory–motor functions. When other areas such as audition are accounted for, the frontal cortex, implicated as the "symbolic" portion of the cortical memory, takes up a mere 25% of the total. Another hint that the body plays a central role in the creation of symbolic tokens comes from our models of the role of the body in infant language learning. Eye and hand signals from a caregiver in the course of conversation with an infant may play a crucial role in learning basic nouns and verbs (Yu & Ballard, 2004; Yu & Ballard, 2007; Yu et al., 2005). All these are not to imply that symbolic processing is not important, but instead to emphasize the virtues of the strategy of pre-storing huge amounts of low level details about behavior based on learned experience.

How are human behaviors' variable bindings pre-computed and stored? The second focus of embodied cognition is to use actual human behavior to solve this problem. We were the first group to place an eye tracker in a head mounted display (HMD), and combine its measurements with head and hand movements used in the course of behavior. By recording these movements made by several different subjects, we can develop graphical models that code the most prominent and different ways of solving everyday behavioral problems (Yu et al., 2005; Yi & Ballard, 2006; Sprague et al., 2007). Thus the statistics of actual behavior can replace the guesswork used in formal pattern matching.

Once complete behaviors are stored as graphical models, complete with instructions as to how to direct the body, the next question is: how are they used? Here the central focus is on understanding the way a robustly intelligent agent can manage multiple behaviors in real time. One can take a cue from the most robustly intelligent agent that we have—the humans—to extract and model basic design commitments. The most important of these are that it is very difficult for the human body to do many things at once, and therefore the body is usually only committed to one or two behaviors at a time. However, our hypothesis has been that, to explain how humans juggle more than one behavior, they have to have at least a few behaviors active at any one instant. Thus, we take the psychological concept of working memory to be about "threads," or the number of independent behavioral processes that can be simultaneously active. How are these chosen? One hypothesis is that these can be chosen on the basis of their expected reward. An important first result that shows how to share reward amongst multiple active behaviors (Rothkopf & Ballard, 2007).

A crucial problem to solve in a real-time model of behavior is that of interrupts: How does one know when to change the current agenda? One idea is that of completely predicting aspects of the environment. If an agent could predict its perceptions in a literal way, then deviations from these predictions could be used to change focus. The huge complicating issue is that perceptions are hugely vary from instance to instance owing in part to view variations, and therefore, perhaps arguably impossible to predict. However, this problem could be tackled with time-varying histograms and linear prediction methods (Swain & Ballard, 1991; Rao, 1999).

In order to understand the role of the constraints of the human body in cognition, one must have a working human body. Standard humanoid graphics model have a very restricted repertoire of movements. One needs accurate physical models of reaching and grasping in order to simulate how long they take and the errors they make. Understanding such movements is a big problem, however we have chosen a lightweight approach that generates movements using equilibrium points (EP) and Jacobian approximations. With these we have generated a small number of parameter movements that are both realistic enough to be useful and compact enough to be accessible to our expected reward methods (Gu & Ballard, 2006).

AN AVATAR CONTROL SYSTEM DESIGN

Our work is a direct descendent of earlier works on cognitive modeling with avatars (Phillips et al., 1990; Faloutsos et al., 2001). It focuses on defining a human cognitive architecture that can integrate and organize a very large number of experimental findings into a hierarchical system for managing visual tasks (Sprague et al., 2007). We have developed a functional human model that has a binocular vision system together with head, hand, and postural constraints. We then designed an interface to the control architecture that allocates those resources in response to task demands. The result is a model of visually guided human behavior in temporally extended tasks that may be tested against human performance.

The virtual human vision model has physical extent and programmable kinematic degrees of freedom that closely mimic those of real humans. Its graphical representation and kinematics are provided by the DI-guy™ package developed by Boston Dynamics. This software is augmented by the Vortex™ package

FIGURE 8.2 A frame from the human embedded vision system simulation. The human visually guided behaviors simulation. The main panel shows a single video frame from the real-time simulation that has the model negotiating a sidewalk strewn with purple litter and blue obstacles, each of which must be dealt with. The insets show the use of vision to guide the humanoid through a complex environment. The upper inset shows the particular visual routine that is running at any instant. This instant shows the detection of the edges of the sidewalk that are used in navigation. The lower insert shows the visual field in a head-centered viewing frame.

developed by CMLabs for modeling the physics of collisions. This software base has been augmented with our control architecture for managing behaviors. We use "behavior" in a very specialized sense. Each behavior has a very specific goal and contains all the structures for the extraction of information from visual input that is in turn mapped onto a library of motor commands.

The model is illustrated on a sidewalk navigation task that requires the virtual human to walk down a sidewalk and cross a street while avoiding obstacles and collecting litter. The movie frame in Figure 8.2 shows the humanoid model in the act of negotiating the sidewalk, which is strewn with obstacles (blue objects) and litter (purple objects) on the way to crossing a street. The essential feature of its cognitive architecture is that behaviors are compositional, that is, they run independently and also can be combined in different subsets. The simple compositional setting wherein many such behaviors can be combined has allowed us to address an extraordinarily rich set of questions about the human cognitive architecture.

THE OPERATING SYSTEM MODEL

During the course of normal behavior humans engage in a wide variety of tasks, but at any moment the number of separate tasks that can be addressed is small. Thus the job of the "operating system" in Figure 8.3 is to choose the right suite of behaviors for the current task demands. In addition, once activated, each behavior requires different visual and motor resources. Thus there must be mechanisms that

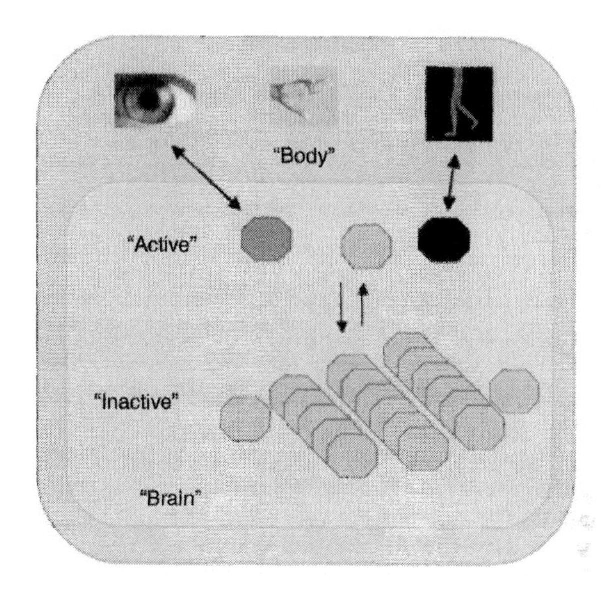

FIGURE 8.3 The main features of the operating system model. Following work in psychology and robotics (Bonasso et al., 1997), we developed a tripartite abstract cognitive architecture for realizing ordinary behaviors that has three levels: *central executive, arbitration*, and *behavior*. The central executive level of the hierarchy maintains an appropriate set of active behaviors from a much larger library of possible behaviors, given the agents current goals and environmental conditions. The composition of this set is evaluated at every simulation interval, which we took to be 300 ms. The arbitration level addresses the issue of managing competing active behaviors. Thus an intermediate task is the mapping action recommendations onto the body's resources. Since the set of active behaviors must share perceptual and motor resources, there must be some mechanism to arbitrate their needs when they make conflicting demands. The behaviors themselves, when they are active, each have their own distinct jobs to do, such interrogating the image array with the objective of computing the current state of the process.

allocate these resources to tasks. Understanding this resource allocation requires an understanding of the ongoing demands of behavior, as well as the nature of the resources available to the human sensorimotor system.

To test the architecture, we chose a scenario that involved walking along a sidewalk with tasks of picking up litter, staying on a sidewalk and avoiding obstructions and then crossing a street at a light, and finally continuing on a sidewalk on the opposite side of the street. To focus the management of behaviors, we deliberately chose *simple* behaviors that we could model as not needing any history, for example, avoiding an obstacle is assumed to be independent of what has been done previously (Rothkopf et al., in press). Figure 8.4 shows the total set of nine behaviors used in the scenario (the "Behavior List"). In this case, the central executive is a simple state machine that bundles the behaviors depending on the local venue. Behaviors that detect changes in venues trigger changes in the executive's state, and as a consequence, different sets of active behaviors are

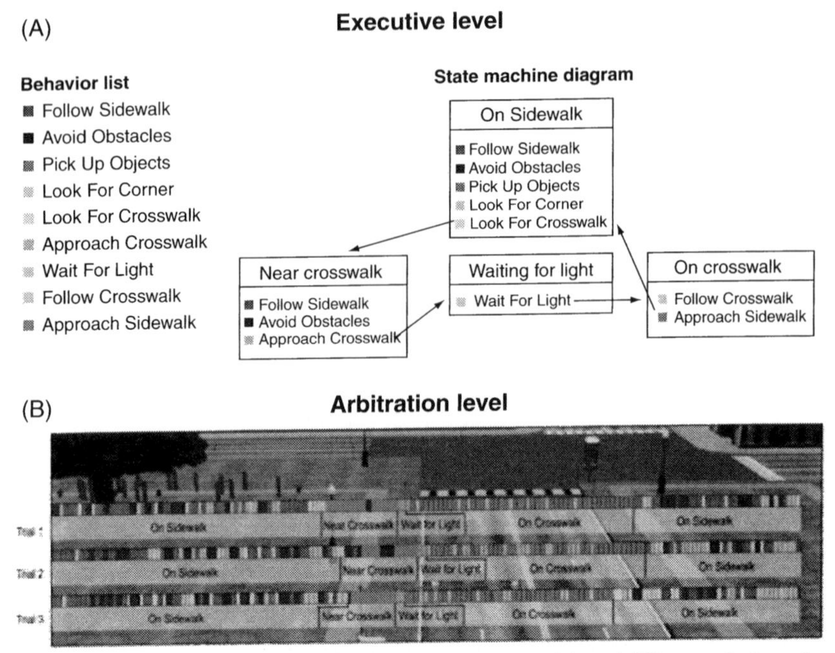

FIGURE 8.4 The humanoid model operating system's servicing of different tasks from three separate trials. The cognitive architecture has computations at three distinct levels, the two most abstract of which are shown here. (A) At the most abstract level the composition of behaviors must be continually adjusted by the Central Executive's state machine. (B) At the Arbitration level, behaviors compete with each other for the body's resources. The colored bars show the result of competition for eye fixations. Each bar denotes a fixation and its color indicates the behavior that initiated it. At the most basic level (not shown) visual routines extract specific information from each fixation to define the state of a behavior. (See color plate)

triggered. The figure also uses colors for each eye fixation to denote the individual behavior whose needs were serviced during three runs of the simulation. These colors illustrate our central tenet; and because we know the behavior that is directing gaze, we can interpret the goal of the fixation in terms of information needed for the task. Note that this is very different from models of fixation that direct the fixations to groups of salient image features (Itti & Koch, 2001) or to surprising distributions of such features (Itti & Baldi, 2005).

The state machine diagram illustrates our main hypothesis that was tested in the simulation. These everyday behaviors can be described in terms of sets of more primitive behaviors that can be composed (the groups of colors). In the model, for example, while on the sidewalk there are five things to do but while waiting for the light, there is only one. Reconciliation between a human's goals and the surrounding context determines the best set of primitive behaviors to use.

These simulations have yielded a number of specific results: (1) only a small amount of relevant state is typically needed for each behavior (Sprague & Ballard, 2003a; Yu & Ballard, 2004; Yu et al., 2005; Yi & Ballard, 2006), (2) these behaviors can be learned via reinforcement (Sprague & Ballard, 2003a; Sprague et al., 2007), (3) behaviors learned individually can be combined for more complex behaviors (Rothkopf et al., in press; Sprague & Ballard, 2007), and (4) their relevant rewards can still be calculated under the combined circumstances (Rothkopf & Ballard, 2007). However, the most important result is a model for the use of gaze control based on competition between multiple ongoing behaviors. We have modeled this competition based on reinforcement (Sprague & Ballard, 2003a) and have shown that it outperforms standard methods (see section "Modeling task-directed eye movements"). The central idea of the model is that fixation is used to reduce uncertainty in the state of the behavior that requested it.

LEARNING IN SIMPLE CONCURRENT BEHAVIORS

The actions in behaviors were programmed using temporal difference (TD) learning (Barto & Sutton, 1998). In TD learning, the error between the current estimated values of states and the observed reward is used to drive learning. In its most general form the quality value $Q(s_t, a_t)$ of a state-action pair (s_t, a_t) is adjusted by this error $\alpha\delta_Q$ (using a learning rate α). Our calculations use an on-line error correction rule,[1] where the update is based on the action actually taken:

$$\delta_Q = r_t + \gamma Q(s_{t+1}, a_{t+1}) - Q(s_t, a_t) \tag{8.1}$$

A vital aspect of the compositional approach is that an individual active behavior needs to be able to calculate its self-worth even when other behaviors are also active. This is known as the *credit assignment problem*: if reward is handed out for the group of active behaviors how should it be apportioned amongst them? It has a particularly distinct motivation in the context of reinforcement learning as this problem must be solved at each step. In the walking environment, this composition was pre-programmed, but more realistically, the right combination of behaviors for any context must be learned and that will be a central focus of our research plan. The problem can be seen in Equation (8.1) where r_t is the reward that is given out at each step. Although this value is needed to make the learning rule work, it would not be available if there were other active procedures. It is only reasonable to assume that the total reward of all the active behaviors is known. Recently we have discovered an elegant formulation to compute local reward from the global estimate that holds great promise (Rothkopf & Ballard, 2007). So far it has only been tested in very simple artificial situations, but in these it has proved to be very robust. Our algorithm that has only been tested on certain standard toy problems compares very favorably

[1] SARSA (Rummery & Niranjan, 1994).

to Kaelbling et al.'s (Chang et al., 2003). Our goal is to show that it will extend to the more complicated avatar environments.

The crux of the algorithm is its concept of *consumable rewards.* This is shorthand for the assumption that running behaviors can know which other behaviors are running, together with their corresponding reward estimates. Thus when global reward $G(t)$ is handed out, the reward estimate of the k-th behavior, $\hat{r}_k$ can be adjusted by:

$$\hat{r}_k(t+1) = \alpha\hat{r}_k(t) - (1-\alpha)\left(G(t) - \sum_{i \neq k}\hat{r}_i(t)\right) \tag{8.2}$$

In other words, the estimates of the co-active behaviors can be used to refine a behavior's own estimate. equation (8.1) only describes the updating of the reward estimate. In reinforcement learning, this estimate has to be folded into the microbehavior's Q-table, but this is readily done. Our preliminary tests of this formula in small problems have shown that it computes accurate reward estimates very quickly.

MODELING TASK-DIRECTED EYE MOVEMENTS

The previous section highlighted one fundamental problem, that of credit assignment, in utilizing reward in a multi-task environment, but another very important problem surfaces for task-directed vision and this concerns the direction of the gaze itself. The small fovea makes its use to obtain accurate measurement a premium. So the question is that in the case of multiple active behaviors, which of them should get the gaze vector at any instant? Sprague (Sprague & Ballard, 2003b) found an accurate solution to this problem. His RL formula calculates the amount each behavior stands to gain by updating its state. Where $Q(s_i, a)$ is the discounted value of behavior i choosing an action a in state s_i, an agent that chooses an action that is sub-optimal for the true state of the environment can expect to lose some return estimated as follows:

$$\text{loss}_b = E\left[\max_a\left(Q_b(s_b, a) + \sum_{i \in B, i \neq b} Q_i^E(s_i, a)\right)\right] - \sum_i Q_i^E(s_i, a_E) \tag{8.3}$$

The expectation in the left braces is the expected return if s_b, which were known, but the other state variables were not. The value on the right is the expected return if none of the state variables are known. The difference is interpreted as the cost of the uncertainty associated with s_b. This calculation is done for all the active behaviors and the one that has the most to lose gets the vector. Figure 8.5 shows this happening for a walking segment.

As Figure 8.5 shows, the improvement is small—5.03 versus 4.99—but nonetheless highly significant. The narrow margin highlights the difficulty of reward-based hypotheses about eye fixations that would attempt to have the results of

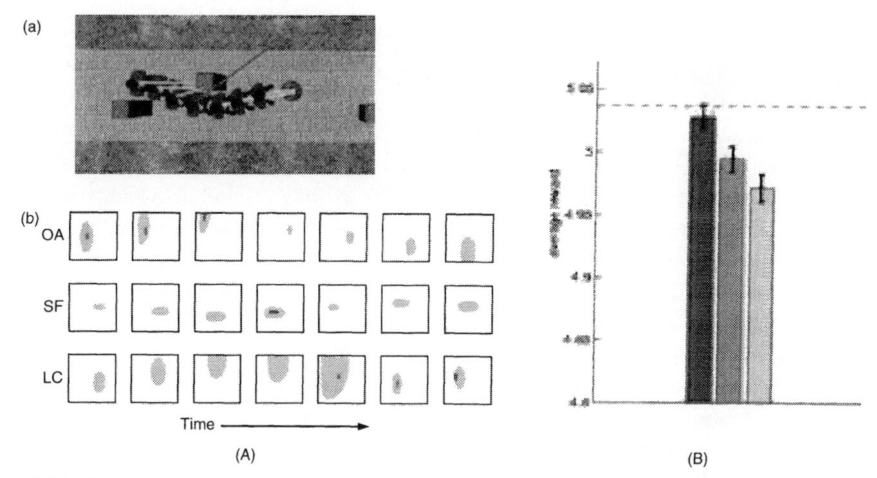

FIGURE 8.5 Behaviors compete for gaze in order to update their measurements. (A) The top panel shows seven time steps in walking and the associated gaze vector color coded for obstacle avoidance (OA – red), sidewalk finding (SF – blue), and litter pickup (LC – green). The corresponding boxes below show the state spaces where the a priori uncertainty is indicated in beige and the a posteriori uncertainty is indicated in the appropriate color. Uncertainty grows because the internal model has noise that adds to uncertainty in the absence of measurements. Making a measurement with a visual routine that uses gaze reduces the uncertainty. For example for litter collection (LC) panel five shows a large amount of uncertainty has built up that is greatly reduced by a visual measurement. Overall, OA wins the first three competitions, in addition to sidewalk finding, and finally LC wins the last three. (B) Tests of the Sprague algorithm (dark green) against the robotics standard round robin algorithm (light green) and random gaze allocation (yellow) show a significant advantage over both. (See color plate)

the fixation directly alter the Q-table. The relative value of any given fixation is small enough so as to be practically undetectable by the learning process. Our early trial simulations using Equation (8.1) were unable to detect any systematicity in these variations, and led us to propose the most-to-gain model which uses the state table to calculate the value of an eye fixation, but does not attempt to adjust the Q-table otherwise.

COMPLEX TASKS WITH SEQUENTIAL STEPS

Complex behaviors need more complex state representations. In the walking examples each behavior has very minimal state. Staying on the sidewalk just requires measuring the sidewalk edge. The history of the traverse is not needed. More complicated behaviors, such as making a sandwich, require much more state. If you want to put peanut butter on a slice of bread, you must be holding the knife and you must take the lid off the peanut butter jar. Modeling this state is not straightforward owing to a number of factors. Consider the problem of watching someone make a sandwich and describing what has transpired. The basic actions must be measured and recognized. However, all the steps in the process are noisy and hence the description must necessarily be probabilistic. Now consider describing the order of steps making a sandwich. Since there are over 1000 distinct ways of making it that differ in the order of the steps, any particular sequence of steps is best described probabilistically. A central way of handling probabilistic information goes under the name *graphical models* with their sub-case of Dynamic Bayes Nets (DBN) (Rimey & Brown, 1993; Murphy, 1994). These are particularly valuable when the basic dependencies are in the form of conditional probabilities, as in the sandwich making case. DBN have proven extraordinarily useful in the modeling of all kinds of sequential processing and there is a substantial literature on them. Their use here should only be regarded as illustrative of the methodology.

While some care has to be taken in developing a graphical model in any particular case, we were able to construct one to generate a simulation of a sandwich making sequences by a model human, complete with eye fixations, and the result is described in Yi and Ballard (2006).

To make this model, we collected data on four human subjects making peanut butter and jelly sandwiches. In the experimental setting objects are placed in the same positions as in the virtual reality. We used the ASL eye tracker to record gaze movements. Human eye movements are far more noisy and unpredictable than camera movement in a computer vision system, but the task setting makes the problem of determining gaze points tractable. We processed human data by clustering fixation points and associating them to objects in the scene. Fortunately this association is easy. Almost every fixation point is close to a certain object which is relevant to the task. We analyzed the transitions of fixation points and played them back using the same virtual environment as the model.

Next we were interested in the kinds of sub-task sequences used by the subjects. Table 8.1 summarizes the scheduling of 10 sub-tasks in making a peanut butter sandwich by 3 human subjects.[2] Despite that some chronological constraints, for example, BT, PLF and KH must precede POB and JOB, have ruled out most of the 10! orders, the number of possible orders remaining is still large.[3] However, experiments with much more subjects show that the orders picked by human subjects display some common features. For instance, BT is always the starting point and KH is always immediately followed by POB or JOB. These special constraints and the mechanism of scheduling these sub-tasks can be studied in the routine-based framework of visuomotor interaction between the agent and its surroundings.

In addition to the job of scheduling behaviors for the construction of a sandwich, we studied the complementary job of recognition of sandwich-making. This is a very demanding task, since the model must take head, hand, and eye data

[2] We make some assumptions such as the knife is picked up only once and is not put down until spreading finishes.

[3] If we divide the 10 sub-tasks into three stages: {BT, PLF, JLF, KH}, {POB, JOB}, and {PLO, JLO, KT, FB}, we have 4! x 2! x 4! = 1152 different orders.

TABLE 8.1 Scheduling of sub-tasks by human subjects.

	1	2	3	4	5	6	7	8	9	10
BT	abc									
PLF		a	c		b					
JLF		bc				a				
KH			ab	c						
POB				a	c	b				
JOB				b		c	a			
PLO					a				b	c
JLO									c	ab
KT							c	ab		
FB							b	c	a	

The task is decomposed into 10 sub-tasks including BT (putting bread on table), PLF (taking peanut butter lid off), JLF (taking jelly lid off), KH (grabbing knife in hand), POB (spreading peanut butter on bread), JOB (spreading jelly on bread), PLO (putting peanut butter lid on), JLO (putting jelly lid on), KT (putting knife on table), and FB (flipping bread to make an sandwich). Letters a, b, and c denote the orders of sub-tasks taken by three subjects, for example, in the first two steps subject c put bread on the table and took jelly lid off. Note that each of the three subjects chose a slightly different order.

from the subjects and at any given time, recognize what stage in the sandwich making is occurring. For this task we again used a graphical model in the form of a Bayes Net. Such a network is a suitable tool for this class of problems because it uses easily observable evidence to update or infer the probabilistic distribution of the underlying random variables. A Bayesian net represents the causalities with a directed acyclic graph, its nodes denoting variables and edges denoting causal relations. Since the state of the agent is dynamically changing, and the observations are being updated throughout the task execution process, we need to specify the temporal evolution of the network. Figure 8.6 illustrates two temporal slices of a DBN. Shaded nodes are observed; the others are hidden. Probability distribution matrices determine causalities, represented by straight arrows. The state of the lowest hidden node is determined by its prior distribution in the first time/slice, and thereafter jointly determined by its previous state, the observed data, and the transition matrix, as denoted by the curved arrow.

The two-slice representation shown in Figure 8.6 can be easily unrolled to address behaviors with arbitrary numbers of slices. At each moment, the observed sensory data (gray nodes), along with its history, are used to compute the probability of the hidden nodes being in certain states:

$$P(S^t \mid O^{[1,t]}) = P(S^1)P(O^1 \mid S^1)\prod_{t=2}^{t} P(S^t \mid S^{t-1})P(O^t \mid S^t)$$

where S^t is the set of states of hidden nodes at time t, and $O^{[1,t]}$ is the observations over time span $[1,t]$. Behavior recognition computes the states of each

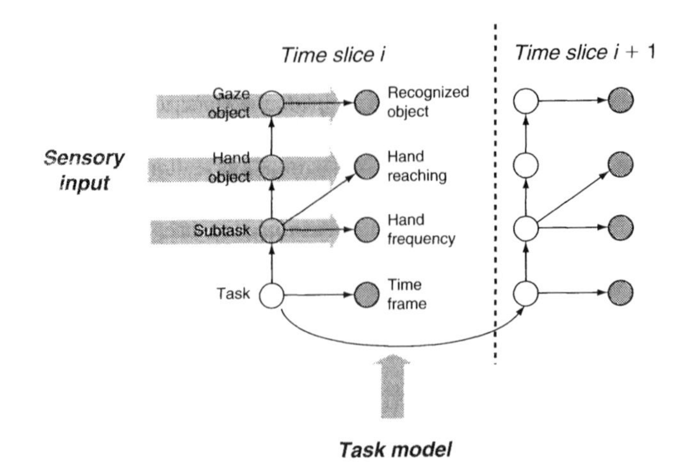

FIGURE 8.6 The basic structure of the DBN used to model sandwich making. Visual and hand measurements provide input to the shaded nodes, the set of which at any time t comprise the measurement vector O^t. The rest of the nodes comprise the set S^t whose probabilities must be estimated. The sequencing probabilities between sub-tasks are provided from a task model that in turn is based on human subject data.

hidden node S^t at time t that maximize the probability of observing the given sensory data:

$$S^t = \arg\max_S P(S^t) = S \mid O^{[1,t]}$$

Although the overall form of the network that records the probabilistic dependencies, shown in Figure 8.6 was designed by hand, the values of the probabilities that it contains were obtained directly from our experiments with the human sandwich makers. As the data in Figure 8.7 shows, the recognition algorithm exhibited very good performance, being 100% successful when allowed to use all of the data.

A COMPACT LANGUAGE FOR MOTOR COMMANDS

In order to realize the goal of having a robustly intelligent avatar, we need to extend the avatar's physical capabilities. Our previous work did this for vision, by adding image capture capability to the DI-Guy humanoid in a way that modeled human vision. However, the standard package has no fine motor coordination primitives and also has a very limited movement repertoire. The general problem of understanding and programming human movements is very much open at this time, but one can still add a limited additional capability to the DI-Guy (Gu & Ballard, 2006). This research shows that reaching, sitting, and grasping cylinders could be handled by modifying an idea by Torres and Zipser (2002) to use EP control (Feldman, 1966). The idea is that movements are planned in segments. Target points are found by gradient descent and the terminal state used to set a

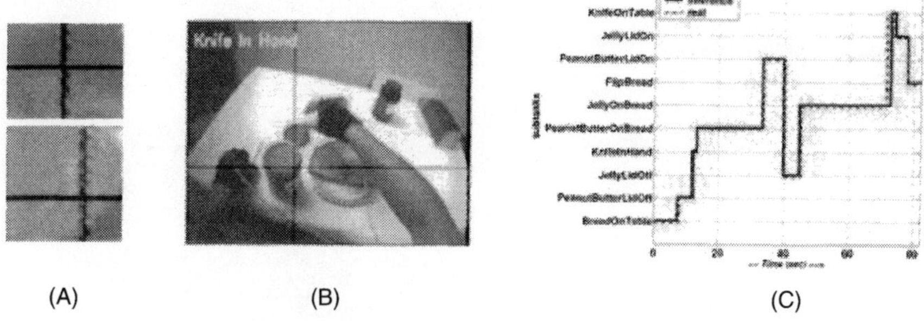

(A) (B) (C)

FIGURE 8.7 (A) Two fixations from different points in the task—(top) bread with peanut butter (bottom) peanut butter jar appear very similar, but do not confuse the DBN which uses task information. (B) A frame in the video of a human subject in the process of making a sandwich showing that the DBN has correctly identified the sub-task as "knife-in-hand." (C) A trace of the entire sandwich making process showing perfect sub-task recognition by the DBN. This trace uses the entire data set in an off-line mode. In the on-line mode where the classifications have to be made on the partial data sets, small errors can be made. (See color plate)

spring-damping model of muscle sets. Thus movements proceed by virtue of a set of planned segments. Before the movement is initiated, the end point is calculated using the motor simulation model elaborated below, and then it is used to set the damped spring natural lengths in movement execution. During the movement, as far as the spring actual length is deviated from the natural length, forces are generated to pull the limb to the end point. This control model proposes that human movement control can be pipelined with segments of planning and execution. That is, the brain first plans the endpoint for tasks at hand, and then sends the motor commands to the musculoskeletal system for execution. While the first movement is under operation, the second task can be simulated in the brain for planning. The distinction between planning and execution makes this approach different from both (Torres & Zipser, 2002) and (Sentis & Khatib, 2006).

Although this model is hardly the whole answer to the problem of understanding movements, it can give one enough realism to incorporate the temporal constraints that movements pose and also to represent, in the physical simulation, the likelihood of making mistakes, such as a grasp that slips.

Planning a Motion Segment

Given a task in Cartesian space, there usually exist an infinite number of solutions in joint space, because human have more degrees of freedom in joint space than the six coordination constraints in Cartesian space. In addition, the mapping from the task space to the joint space is non-linear, which makes the inverse kinematics problem more difficult. Torres and Zipser (2002) steered the end effector to the destination using gradient descent of an objective function that expresses the error between the current joint coordinates and the destination. Although we were able to replicate their results, our experience shows that this method is delicate and very sensitive to its various parameter settings. Additionally, pure gradient descent

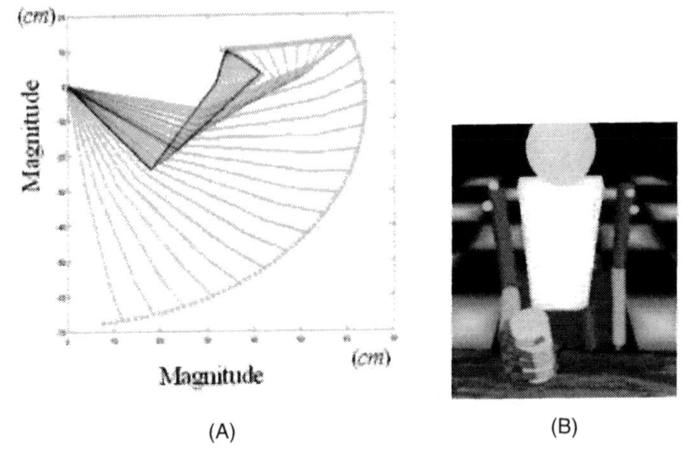

(A) (B)

FIGURE 8.8 First trials with the EP segment movement generator. (A) End posture planning. Blue curves: stages in the gradient descent used to find an initial joint coordinate solution for a target. The bold blue curve is the joint configuration generated by this stage. Note that the simulation trajectory is very non-biological. Black curves: An energy function is added to the simulation to adjust the end configuration. The bold black curve gives the optimal end posture for this stage. Note that the final configuration is very different than the previous end posture. (B) Preliminary trials with a cylindrical grip using the vortex physical simulator. (See color plate)

method can have non-biological trajectories as shown in Figure 8.8. However, it serves as the starting point for our method which uses the gradient method, not to generate the actual trajectory, but instead in simulation to generate end configuration in joint coordinates that are then used by subsequent processing stages.

Gradient descent requires an objective function which decreases as the posture approaches the destination. We define the objective function as the distance between the current hand configuration x and the target x^d, as $\|X - X^d\|^2$ Consider a three joint arm moving in a vertical plane. Let X, the task space, be the set of points that can be reached by the arm in the plane. Let Θ, the configuration space or joint space, be a subset of R^3, which specifies all the possible postures. There exists a vector function f, which maps Θ onto X, that is, every hand location $x \in X$ can be written as $f(\theta)$ for at least one $\theta \in \Theta$, and every θ maps to a certain $x \in X$. The objective function is defined as the distance from end effector $f(\theta)$ for the current posture to the destination x^d.

The normalized gradient of the distance function is given by chain rule as:

$$d\theta = -\beta \frac{(X - f(\theta)) * J(f(\theta))}{\sqrt{\sum_{i=1}^{2} (x_i^d - f_i(\theta))^2}} \tag{8.4}$$

where the β is the parameter to adjust the movements of each step, and J is the Jacobian of the arm system. Equation (8.4) is a vector equation, and each of the components specifies how much to change each joint value to bring the hand closer to the target. The simulation consists of repeatedly computing the gradient of the objective function and changing the posture a small amount for each step until the hand is at the target. This method is illustrated in Figure 8.8 with successive configurations in blue. The green curve shows the trajectory generated which is non-biological in two ways: (1) the trajectory goes through a big arc and subsequently traces a line going back to the destination, which is not the case for human movements; and (2) the end posture presented in bold blue lines also does not match the human posture in the same task. Torres and Zipser (2002) add a mathematical transformation to correct the curved trajectory to a straight line, and additional constraints to adjust the final posture problem. However, their methods are somewhat *ad hoc*.

In our formalism, the trajectory is unimportant since it is merely a simulation to find a feasible end joint configuration and does not represent the actual movement. Thus we only have to address the problem that the final posture needs correction. To solve this problem, we used the idea of energy minimization as has been studied by Alexander (1997), and Kashima and Isurugi (1998). Different from their study of energy during movement, the energy optimization we propose here is before the initiation of movements. We try to minimize the gravity potential energy during the planning, instead of using the kinetic energy which is not available until the movements begin. We argue that potential energy is at least proportional to the energy required for the movements, even if it cannot represent the exact energy consumption for the movements. We define another cost function given in Equation (8.5) using the Lagrange multiplier method for a constraint optimization problem. The goal is to minimize the first part, which is the energy, while keeping the second part, which constraints the end effector to the target, equal to zero. The black configurations in Figure 8.8 illustrate this approach, showing how the joint configurations are gradually changing from the blue bold one to the black bold one, the final biological posture.

$$E(\theta) = Mgh + \lambda(f(\theta) - \mathrm{x}^d) \tag{8.5}$$

where, M is the center of mass (c.g.) of the limb and h is the height of the c.g. The term Mgh gives the gravity potential energy of the arm system. $f(\theta)$ is the current end effector position with the current posture, and x^d is the destination. λ is the parameter for the constraint function, for which a detailed explanation can be referred in Ber (1982).

A further advantage of our method is that it can plan multi-goal movements, since the Jacobians for the different trajectories can be summed, that is, for two different goals $d\theta = d\theta_1 + d\theta_2$. This is another difference from Sentis and Khatib (2006) who use null spaces for different goals. Figure 8.8A shows the two simulation steps. The green trajectory finds the endpoint and then, with the endpoint pinned, the black trajectory adjusts the potential energy. These two steps are used to set the EP which then directs the movement (not shown). Details are contained in Gu & Ballard (2006).

SUMMARY: THE ADVANTAGES OF EMBODIED COGNITION

A human cognitive model that can robustly execute basic tasks would make a valuable contribution to in at least five interrelated areas of endeavor.

1. *Robust Agent Simulations*: Human avatars are a valuable tool in simulation in a variety of different training instances, for example, hostage rescue planning, emergency road condition scenarios, their hospital routine training, that require the ability to predict and react to specific decisions taken by the participants and their behaviors taken by participants. The ability to conduct simulations that include an accurate reflection of peoples' behavior would lead to improved performance in all these areas.

2. *A New Generation of Computer Interfaces*: Avatars in human–computer interfaces will become increasingly realistic. People are comfortable with human-like behavior as long as it spans the "uncanny valley." Psychologically, as avatars become more human-like, people become more and more demanding as to ward realism. The valley refers to human-like avatars that are very good, but are treated as unfamiliar since they do not meet the increasing expectations. Our work promises to make a significant contribution toward avatars that have realistic behaviors reflected in detailed and appropriate body movements.

3. *A Model of Human Cognition*: What we ultimately mean by robust cognition is the thing that human's exhibit. Thus if we can use the humans as a laboratory tool to produce usable behavioral data, it would be a valuable route toward ensuring that the ultimate model has human-like properties, at least in the area of basic behaviors.

4. *A Complementary Adjunct to Symbolic Methods*: A huge amount of research has been expended toward symbolic models of cognition. In much of this effort, the detailed study of human sensory–motor behaviors have been neglected, partly because there has been no systematic way of studying them, and partly because of their own intrinsic difficulty. Our work promotes the use of new tracking instrumentation for the codification of actual human performance. It also attempts to build systematic ways of achieving real-time performance of what has been characterized as peripheral, but, since it tackles symbol binding in a useful and direct way, is actually a central to understanding robust intelligence.

5. *Improved Medical Diagnosis*: A large variety of diseases can be diagnosed by characteristic unusual eye movements, for example, Tourettes, Huntingdons, Schizophernia, and ADHD (Crawford et al., 1989; 1998; Ross et al., 1994; Sweeney et al., 1996; Farber et al., 1999; Munoz et al., 2003; Blekher et al., 2006). Others, notably Parkinson's, can be diagnosed via characteristically unusual body movement patterns.

REFERENCES

Abeel, P., Quigley, M., & Andrew, Y., (2006). Using inaccurate models in reinforcement learning. *International Conference on Machine Learning*.

Alexander, R. McN. (1997). A minimum energy cost hypothesis for human arm trajectories. *Biological Cybernetics, 76*(2), 97–105.

Anderson, J. (1983). *The Architecture of Cognition.* Harvard Univeristy Press, Cambridge, MA.

Badler, N. (2001). Virtual beings. *Communications of the ACM, 44*, 33–35.

Barto, A. G. & Sutton, R. S. (1998). *Introduction to Reinforcement Learning.* MIT Press, Cambridge, MA.

Bertsekas, D. P. (1982). *Constrainted Optimization and Lagrange Multiplier Methods.* Academic Press.

Blekher, T., Johnson, S. A., Marshall, J., White, K., Hui, S., Weaver, M., Hui, S., Weaver, M., Gray, J., Yee, R., Stout, J. C., Beristain, X., Wojcieszek, J., & Foroud, T. (2006). Saccades in presymptomatic and early stages of Huntington disease. *Neurology, 67*, 394–399.

Bonasso, R. P., Firby, R. J., gat, E., Kortenkamp, D., Miller, D. P., & Slack, M. G. (1997). Experiences with an architecture for intelligent reactive agents. *Journal of Experimental and Theoretical Artificial Intelligence, 9*, 237–256.

Caspi, A., Beutter, B. R., & Eckstein, M. P. (2004). The time course of visual information accrual guiding eye movement decisions. *Proceedings of the National Academy of Science, 101*, 13086–13090.

Crawford, T. J., Henderson, L., & Kennard, C. (1989). Abnormalities of nonvisually-guided eye movements in Parkinson's disease. *Brain, 112*, 1573–1586.

Crawford, T. J., Sharma, T., Puri, B. K., Murray, R. M., Berridge, D. M., & Lewis, S. W. (1998). Saccadic eye movements in families multiply affected with schizophrenia: The maudsley family study. *American Journal of Psychiatry, 155*, 1703–1710.

Faloutsos, P., van de Panne, M., & Terzopoulos, D. (2001). The virtual stuntman: Dynamic characters with a repertoire of motor skills. *Computers and Graphics, 25*, 933–953.

Farber, R. H., Swerdlowand, N. R., & Clementz, B. A. (1999). Saccadic performance characteristics and the behavioural neurology of Tourette's syndrome. *Journal of Neurology, Neurosurgery and Psychiatry, 66*, 305–312.

Feldman, A. (1966). Functional tuning of the nervous system with control of movement or maintenance of a steady posture. II. controllable parameters of the muscle. *Biophysics, 11*(3), 498–508.

Fikes, R. E. & Nilsson, N. J. (1971). Strips: A new approach to the application of .theorem proving to problem solving. *Artificial Intelligence, 2*, 189–208.

Fischler, M. A. & Firschein, O. (1987). *Intelligence: The Eye, the Brain, and the Computer.* Addison-Wesley Longman.

Fu, W.-T. & Anderson, J. R. (2006). From recurrent choice to skill learning: A reinforcement-learning model. *Journal of Experimental Psychology: General, 135*, 184–206.

Gray, W. D. (Ed.) (2007). *Integrated Models of Cognitive Systems.* Oxford University Press, Oxford, UK.

Gu, X. & Ballard, D. (2006). An equilibrium point based model unifying movement control in humanoids. *Robotics: Science and Systems.*

Itti, L. & Baldi, P. (2005). A principled approach to detecting surprising events in video. *IEEE International Conference on Computer Vision and Pattern Recognition*, (Vol. 1, pp. 631–637).

Itti, L. & Koch, C. (2001). Computational modeling of visual attention. *Nature Reviews Neuroscience, 2*(3), 194–203.

Kashima, T. & Isurugi, Y. (1998). Trajectory formation based on physiological characteristics of skeletal muscles. *Biological Cybernetics, 78*(6), 413–422.

Laird, J. E., Newell, A., & Rosenblum, P. S. (1987). Soar: An architecture for general intelligence. *Artificial Intelligence, 33*, 1–64.

Munoz, D. P., Armstrong, I. T., Hampton, K. A., & Moore, K. D. (2003). Altered control of visual fixation and saccadic eye movements in attention deficit hyperactivity disorder. *Journal of Neurophysiology, 90*, 503–514.

Murphy, K. (1994). *Dynamic Bayesian Networks: Representation, Inference and Learning.* Ph.D. Thesis, University of California, Berkeley.

Newell, A. & Simon, H. A. (1995). *Computers and Thought.* MIT Press, Cambridge, MA.

Palmer, J. (1995). Attention in visual search: Distinguishing four causes of a set-size effect. *Current Directions in Psychological Science, 4*, 118–123.

Pat. Langley, & Dongkyu. Choi, (2006). Learning recursive control programs from problem solving. *Journal of Machine Learning Research, 7*, 493–518.

Phillips, C., Zhao, J., & Badler, N. (1990). Interactive real-time articulated figure manipulation using multiple kinematic constraints. *Computer Graphics, 24*, 245–250.

Rao, R. P. N. (1999). An optimal estimation approach to visual perception and learning. *Vision Research, 39*, 1963–1989.

Rimey, R. D. & Brown, C. M. (1993). Control of selective perception using Bayes Nets and decision theory. *International Journal of Computer Vision, 12*, 173–207.

Ritter, S., Anderson, J. R., Cytrynowicz, M., & Medvedeva, O. (1998). Authoring content in the pat algebra tutor. *Journal of Interactive Media in Education, 98*(9).

Ross, R. G., Hommer, D., Breiger, D., Varley, C., & Radant, A. (1994). Eye movement task related to frontal lobe functioning in children with attention deficit disorder. *Journal of the American Academy of Child and Adolescent Psychiatry, 33*, 869–874.

Rothkopf, C.A. & Ballard, D.H. (2007). Credit assignment with bayesian reward estimation. *Computational and Systems Neuroscience Conference.*

Rothkopf et al. (in press) Task and context determine where you look. *Journal of Vision, 7*(14), 1–20.

Roy, D. (2005). Semiotic schemas: A framework for grounding language in action and perception. *Artificial Intelligence, 167*, 170–205.

Roy, D. K. & Pentland, A. P. (2002). Learning words from sights and sounds: A computational model. *Cognitive Science, 26*, 113–146.

Rummery, G.A. & Niranjan, M. (1994). On-line q-learning using connexionist systems. *Technical Report*, CUED/F-INFENG/TR 156. Cambridge, England: Cambridge University Engineering Department.

Sentis, L. & Khatib, O. (2006). A whole-body control framework for hu-manoids operating in human environments. *Proceedings of the IEEE International Conference in Robotics and Automation.*

Sprague, N. & Ballard, D. (2003a). Multiple-goal reinforcement learning with modular sarsa(0). *International Joint Conference on Artificial Intelligence*, Morgan Kaufmann.

Sprague, N. & Ballard, D. (2003b). Multiple-goal reinforcement learning with modular sarsa(0). *Technical Report*, 798, University of Rochester: Computer Science Department.

Sprague, N. & Ballard, D. (2007). Modeling embodied visual behaviors. *ACM Transactions on Applied Perception, 4*(2), 11.

Sprague, N., Ballard, D., & Robinson, Al. (2007). Modeling embodied visual behaviors. *ACM Transactions on Applied Perception, 4.*

Sun, R. (2006). *Cognition and Multi-Agent Interaction.* Cambridge Univeristy Press, Cambridge, Chapter 4, pp. 79–99.

Swain, M. J. & Ballard, D. H. (1991). Color indexing. *International Journal of Computer Vision, 7*, 11–32.

Sweeney, J. A., Mintun, M. A., Kwee, S., Wiseman, M. B., Brown, D. L., Rosenberg, D. R., & Carl, J. R. (1996). Positron emission tomography study of voluntary saccadic eye movements and spatial working memory. *Journal of Neurophysiology, 1*, 454–468.

Torres, E. & Zipser, D. (2002). Reaching to grasp with a multi-jointed arm. I. Computational model. *Journal of Neurophysiology, 88*, 2355–2367.

Treisman, A. M. (1980). A feature-integration theory of attention. *Cognitive Psychology, 12*, 97–136.

Trick, L. M. & Pylyshyn, Z. W. (1994). Why are small and large numbers enumerated differently? A limited-capacity preattentive stage in vision. *Psychological Review, 101*, 80–102.

Yi, W. & Ballard, D. (2006). Behavior recognition in human object interactions with a task model. *IEEE International Conference on Advanced Video and Signal Based Survellance*, November 2006.

Yu, C. & Ballard, D. (2004). A multimodal learning interface for grounding spoken language in sensory perceptions. *ACM Transactions on Applied Perception, 1*, 57–80.

Yu, C. & Ballard, D. (2007). A unified model of early word learning: Integrating statistical and social cues. *Neurocomputing, 70*, 2149–2165.

Yu, C., Ballard, D., & Aslin, R. (2005). The role of embodied intention in early lexical acquisition. *Cognitive Science, 29*, 961–1005.

Yu-Han Chang, Tracey Ho, & Leslie Pack Kaelbling (2003). All learning is local: Multi-agent learning in global reward games. *Proceedings of Neural Information Processing Systems*, MIT Press.

SECTION III

PERCEIVING AND ACTING

9

Ecological Psychology: Six Principles for an Embodied–Embedded Approach to Behavior

Michael J. Richardson[1], Kevin Shockley[2], Brett R. Fajen[3], Michael A. Riley[2] and Michael T. Turvey[4]

[1]*Department of Psychology, Colby College, Wateville, ME, USA*
[2]*Department of Psychology, University of Cincinnati, Cincinnati, OH, USA*
[3]*Department of Cognitive Science, Rensselaer Polytechnic Institute, Troy, NY, USA*
[4]*Department of Psychology, University of Connecticut, Storrs, CT, USA*

Traditionally, cognitive science has taken the perspective that the causal system that underlies behavior is a representation-based information processing system. The appeal of this approach is that the regularity of behavior can be attributed to centralized computational processes, whereby an understanding of behavior requires understanding how a system that receives, stores, manipulates, computes, and outputs information by means of symbolic structures could account for such behavior (von Eckardt, 1993; Lakoff & Johnson, 1999). Additional motivation for this approach is that the cognitive capabilities of mind are *disembodied* and that any material substrate that allows for symbolic computation can provide an effective framework for studying and understanding behavior. Such a disembodied approach to cognition encourages the study of cognitive phenomena that are trivially dependent upon the environment. This makes it possible for computational processes and representational structures to be lifted away from the organism–environment system and be studied on their own, permitting cognitive scientists to proceed whereas other specialists work to understand the body and environment of the knower.

Although pragmatically attractive (Kirsch, 1991), debilitating issues for the traditional approach are the origin and grounding of the representational structures that are fundamental to its realization (Turvey & Shaw, 1979; Searle, 1981; Brooks; 1991; Haugeland, 1998; Fodor, 2000; Shaw, 2003) and the implicit recourse to an internal *executive* or *homunculus* that the reliance on such representational structures requires (Turvey et al., 1981; Turvey et al., 1982). Figure 9.1 illustrates this recourse with respect to perception and action, respectively. In both cases the executive—*the ghost in the machine*—plays a centralized role and frequently intervenes. Moreover, the reliance on memorized representational and computational processes endows the internal executive with knowledge about the meaning of objects, surfaces, and events in the world, as well as how to appropriately select and order actions in response to the perceived objects, surfaces, and events in that world.

To endow the executive with knowledge, however, is to take out one or more *loans of intelligence* (Dennett, 1978; Kugler & Turvey, 1987). These loans ensure the competence of representational inference engines and the means by which they can account for the subsequent regularity of behavior, yet it is never clear as to how these loans of intelligence are to be repaid (Kugler & Turvey, 1987; Turvey & Fonseca, 2008). In truth, many researchers pay little attention to this issue. A deeply rooted acceptance that behavior's organization reflects entirely internal, locally defined, representational processes has made

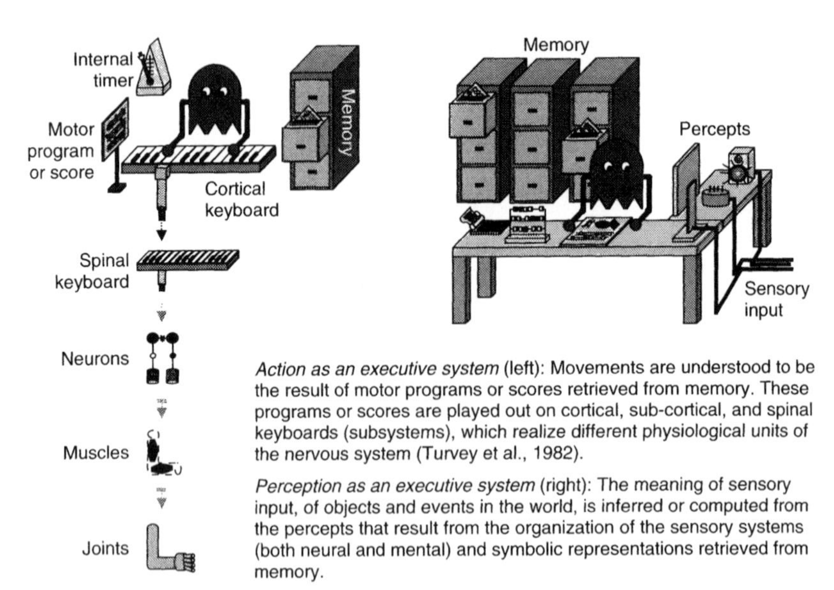

Internal timer
Motor program or score
Memory
Cortical keyboard
Spinal keyboard
Neurons
Muscles
Joints
Memory
Percepts
Sensory input
Action as an executive system (left): Movements are understood to be the result of motor programs or scores retrieved from memory. These programs or scores are played out on cortical, sub-cortical, and spinal keyboards (subsystems), which realize different physiological units of the nervous system (Turvey et al., 1982).
Perception as an executive system (right): The meaning of sensory input, of objects and events in the world, is inferred or computed from the percepts that result from the organization of the sensory systems (both neural and mental) and symbolic representations retrieved from memory.

FIGURE 9.1 Action and perception as executive systems.

the dependence of standard theory on executive function so pervasive as to be almost invisible (Haugeland, 1998). Others simply assume that the role of the executive and the required loans of intelligence will eventually be repaid by a more ardent and thoroughgoing appeal to epistemic mediators, including a more detailed analysis of how symbolic computational processes might be realized by the anatomical and neural substrates of the brain and the nervous system. There is a growing awareness, however, that traditional theory, however embellished, is unable to repay the intelligence loan (Turvey et al., 1981; Turvey, 1990, 2004; Brooks, 1991; Beer, 1995; Turvey & Shaw; 1995; Clark, 1997; Fodor, 2000). A radically different kind of theory is implied, one that attempts to understand and explain behavior, knowledge, and meaning in a predominantly noncentralized, nonrepresentational, way.

The inspiration for the present volume is the mundane observation that each and every animal *qua* cognitive agent has a body, inhabits an environment, and lives by the constraints of both (Warren, 2006). The present volume reflects an increasing tendency to regard this observation as the appropriate starting point for an understanding of cognition as neither centralized nor representational (Brooks, 1991; Thelen & Smith, 1994; Beer, 1995; Hutchins, 1995; Clark, 1997; Pfeifer & Lida, 2005; Gibbs, 2006; Pfeifer & Bongard, 2006). For this understanding to take root the key notions of "embodied" and "embedded" must be used and interpreted clearly and consistently. Weakly constrained uses and interpretations that cling to the orthodox explanatory language of the sciences of cognition hinder rather than promote development. Worse, perhaps, they invite debate and skepticism about the authenticity and uniqueness of the embodied–embedded approach (Wilson, 2002) resulting in the risk of the approach losing the considerable traction it has gained recently.

Although not often acknowledged, this debate is centered on the degree to which cognitive researchers are willing to let go of the various forms of dualism that have shaped the history of traditional cognitive and psychological science—at the forefront: mind–body dualism, but in the background: semantics–syntax, perception–action, and, most importantly, organism–environment dualisms. It is our view that addressing matters of knowing as embodied and embedded requires flatly and completely rejecting all of these classical dualisms. Only then can an embodied–embedded approach cleanly break away from the traps of the traditional disembodied approach. Until that break occurs, the full promise of an embodied–embedded approach cannot be achieved, and such approaches will amount to little more than incremental revisions of cognitive science that leave the core beliefs intact, making the classical approach intractable.

What would be required for embodied–embedded approaches to accomplish a real revolution in cognitive science? We argue that the embodied–embedded approach should draw its foundation from the ecological approach to perception–action as originally conceived by J. J. Gibson (1966, 1979/1986). Gibson sought to work out an approach that would not require recourse to central executives or representations. Thus to advance an embodied–embedded approach we present six principles central to the ecological perspective that reduce the need for representational–computational explanation and the implicit reliance on executive cause that traditional explanations require. These principles do not define the complete scope of ecological psychology, but are illustrative of a way of thinking about perception, action, and cognition, that does not require symbolic representations and constructive computations.

ECOLOGICAL PRINCIPLE I: ORGANISM–ENVIRONMENT SYSTEMS ARE THE PROPER UNITS OF ANALYSIS

As noted, psychologists and neuroscientists have tended to endorse, implicitly and explicitly, a number of dualisms or polarities of which mind–body is the most common. Arguably, as a group, these multiple dualisms are reflections of an overarching dualism, that of organism–environment—the orthodox historical position that *organism* and *environment* are logically distinct, functionally separate systems (Turvey & Shaw, 1995; Järvilehto, 1998a). Such a position seems unquestionable from a common sense point of view. One might say that it is "self-evident." Casual everyday observation is of animals (mainly humans) as one kind of thing acting in the surround, in the environment, which is another kind of thing. From one's personal perspective, "[t]he vista that results from the positioning of the eyes, the resonating tones and muscle activation that spoken language creates in the head, the physical distance between the 'me' and the 'you' (Richardson et al., in press, p. 4)" localize mental activity here, in one's mind and brain, and not there, in the surrounding. In an earlier time, we might have commented that the separation between an organism and its environment is as self-evident as the fact that the sun rotates around the earth.

The allusion to the "fact" of geocentricism as falsified by Copernicus and Kepler is of considerable relevance to the enterprise of embodied–embedded cognition. Presuming a self-evident separation between animal and environment (knower and known, inner and outer; see Bentley, 1941) motivates explanations of cognitive activity centered at the organism. Figure 9.2 compares the earth-centered and organism-centered explanations of their respective discourses. With respect to the former, many of the earthly behaviors that appeared mysterious from a geocentric perspective (and apparently the work of some force beyond the purview of scientific reasoning), such as the changing seasons, tides, and weather, suddenly appear lawful, mandatory, and coherent once the heliocentric view of the universe is accepted (Humphrey, 1933; Turvey & Shaw, 1979). Indeed, attempts to understand such earthly phenomena without acknowledging the earth's noncentrality is what require recourse to other, nonobservable, and often heavenly, causes (Richardson et al., in press).

Elaborating on the work of Ashby (1952, 1963), Turvey and Shaw (1979) have exemplified the latter analogy with respect to memory. This example is

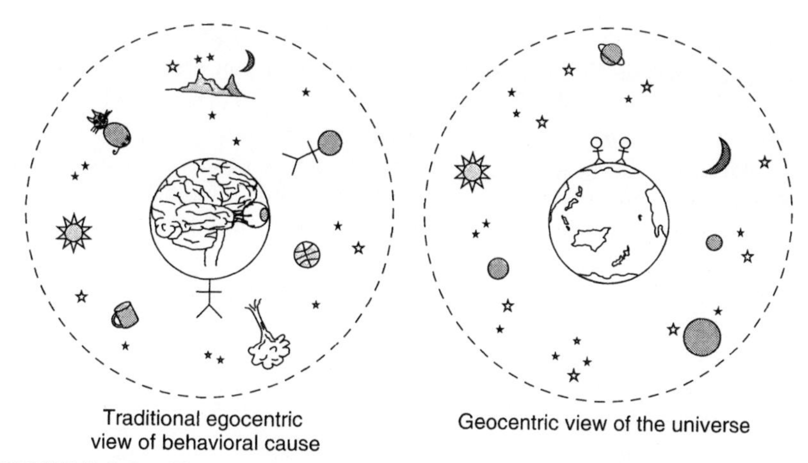

FIGURE 9.2 The geocentric view of the universe and the analogous egocentric view of human behavior. Adapted from Richardson et al. (in press).

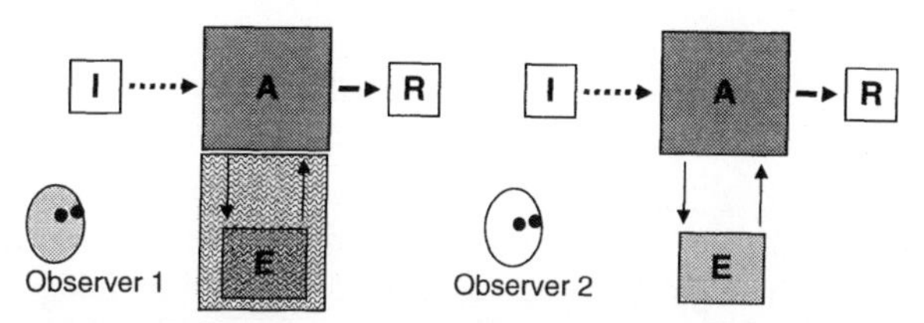

FIGURE 9.3 Two observers attempt to explain *A*'s behavior *R*. When *E* is not observed, *A* and *I* do not predict *R*. Observer 1 discerns that *R* results when *I* passes successively through states and concludes that $R = I + A +$ (*A*'s memory of *I*). When *E* is observed Observer 2 discovers that *R* occurs when $E = z$ and $I = y$ and concludes that *R* is a result of the total system, $R = I + A + E$. See text for more details.

illustrated in Figure 9.3 and can be described as follows. Imagine an organism *A*, whose behavior *R* is the function of *I*, *A* and *E*. For instance, *A* shows behavior *R* when *A* detects that $E = z$ and $I = y$. Note, however, that because of the mutual and reciprocal union of *A* and *E* (denoted by the solid lined arrows), *E* is also influenced by *I*, such that $E = z$ is only subsequent to $I = w$. Now imagine that there are two observers (scientists) of *A*, both attempting to understand the cause of behavior *R*. For the first observer, *E* is not observed or assumed to be of little consequence. Thus, to the confusion of Observer 1, *A* and *I* do not predict *R* directly. *I* is sometimes *y* and sometimes some other state (*z*, *w*, *v*, etc.). As a result, Observer 1 discerns (after a while) that *R* results when *I* passes successively through states *w* and *y* and hypothesizes that $R = I + A +$ (*A*'s *memory* of *I*). In other words, Observer 1 endows *A* with other causal structure. In contrast, Observer 2 does observe *E* in addition to *A* and *I*. Thus, Observer 2 discovers (after only a short period of time) that R occurs when $E = z$ and $I = y$. As a result, Observer 2 concludes that R is a direct result of the total system, $R = I + A + E$. That is, Observer 2 makes no hypothesis about "other" cause, internal to A, as such cause is not required. This example, though somewhat obvious in its simplicity, is by no means trivial, nor is its facetious criticism of traditional theory unjustified. To be blunt, when organism is considered separate from environment, and the partial system (organism) deputizes for the whole system (organism and environment), there is a tendency to fashion explanation through variables that are beyond immediate observation. Gratuitous appeals to internal states as explanations of everyday behaviors exemplify this tendency (Ashby, 1963; Turvey & Shaw, 1979; Clancey, 1997).

As might be expected from its name, the ecological approach opposes the separation of organism and environment. In Gibson's (1979/1986, p. 8) words, "animal and environment make an inseparable pair. Each term implies the other. No animal could exist without an environment surrounding it. Equally, although not so obvious, an environment implies an animal (or at least an organism) to be surrounded." The animal and environment are therefore *mutual* and *reciprocal*, in that the existence and influence of animal on environment and the existence and influence of environment on animal are both equivalent and complementary (Gibson, 1979/1986; Michaels & Carello, 1981; Shaw & Turvey, 1981; Turvey et al., 1981). More than just mutual and reciprocal, however, organism and environment are a combined whole, a synergy or coalition (Turvey et al., 1978; Shaw & Turvey, 1981), such that the organism-in-its-environment—the organism (O)–environment (E) system—should be taken as the proper unit of analysis for studying and understanding behavior (Chemero & Turvey, 2007). Returning to the allegory of earth-as-center versus earth as an integral part of a system, one could expect psychological explanation from an O-separate-from-E perspective and an O–E system perspective to differ in fundamental ways.

Järvilehto (1998a, b, 1999, 2000) suggests that, in regard to theory and understanding, the implications of O–E as the unit of analysis are radical and potentially profound. From the single system perspective (a) behavior is a reorganization of O–E, not an interaction of O and E and (b) mental activities are different aspects of the organization and dynamics of O–E, not local processes of O. In respect to the analysis of O–E systems, the key is as such not behavior nor mental activity, but rather the system's outcomes. An O–E organization stands in correspondence to its outcome, and the dynamics of an O–E organization is understandable only in light of the events that established the necessary conditions for achieving the outcome (Järvilehto, 1998a).

The one-system perspective with emphasis upon the system's outcomes connects with the themes of developmental systems theory (Oyama et al., 2001). Among this theory's aims is the dissolution of the dichotomy of development and evolution, as the following quotations suggest. "Fundamentally, the unit of both development and evolution is the developmental system, the entire matrix of interactants involved in a life cycle (Griffiths & Gray, 2001, p. 206)." "Selection acts

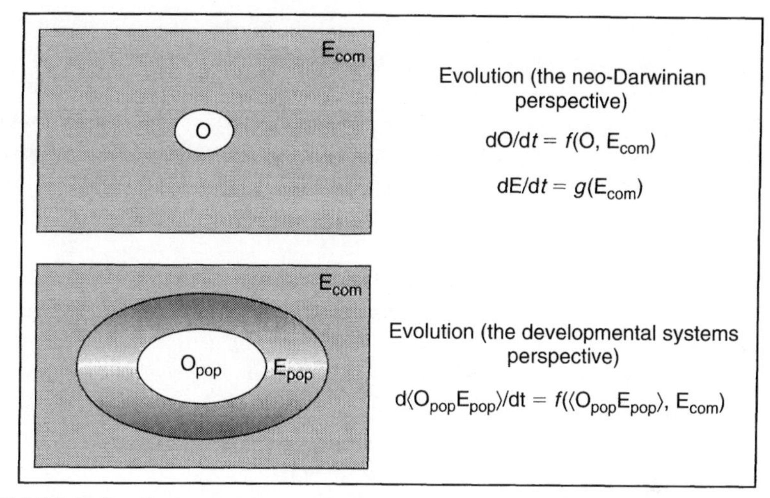

FIGURE 9.4 Two perspectives on the dynamics of evolution. O is organism, O_{pop} is a population of organisms, E_{com} is the physical environment described without reference to any organisms and common to all organisms, and E_{pop} is the environment defined with reference to O_{pop}. (a) Evolutionary change in O is a function of states of O and E_{com} at each previous instant. E_{com} selects from the pool of variation the best fitting Os. E_{com} also changes, but not dependently on O. (b) The bracket $\langle\ \rangle$ signifies a unitary system and $\langle O_{pop}E_{pop}\rangle$ signifies a population of unitary developmental systems. Evolutionary change occurs in the nature of $\langle O_{pop}E_{pop}\rangle$. That is, developmental systems evolve as a function of themselves, how they modify the resources for future generations, and of E_{com}, how it modifies the (same) resources for future generations. Effect of changes in E_{com} can be understood only in terms of how they induce changes in E_{pop}. Equations based on those identified in table 16.1, Griffiths and Gray (2001).

not only on the developmental outcome but also on the entire developmental process leading up to that outcome ... including the context or contexts in which development takes place and those in which the outcomes are expressed (Miller, 1997, p. 495)." Evolution from a developmental systems' perspective is sharply distinguished from evolution in the standard theory, as Figure 9.4 depicts.

The capstone notion of a unitary O–E system will figure prominently in the exposition of *Ecological Principles (EP) II–VI*. As will be shown, recognition of the environment as organism referential and recognition of behavior as dynamical reorganization at the level of the O–E system promote hypotheses and research that refine the interpretation of embodied–embedded cognition and discourage gratuitous uses of representation and inference.

ECOLOGICAL PRINCIPLE II: ENVIRONMENTAL REALITIES SHOULD BE DEFINED AT THE ECOLOGICAL SCALE

In accordance with EP I, the ecological approach challenges traditional notions of behavior by recognizing that to understand perception, action, and cognition one must identify the organism-relevant properties of the environment that define what is perceived, acted upon, and known. It recognizes and respects the need for an ontological theory of E_{pop} (Figure 9.4), a theory of the environments in which organisms live and move (Turvey & Shaw, 1999; Smith, 2001). For those who wish to pursue embodied–embedded cognition, the ontological theory in question is the theory of what embeds.

Organism-relevant descriptions of E_{com} begin with *substances*, *surfaces*, *places*, *objects*, and *events* (Gibson, 1979/1986). These are the realities that ground E_{pop}. They are realities at the ecological scale (the scale of nature at which O–E systems are defined) and are environmental facts of direct pertinence to adaptive behavior. For example, substances vary in hardness, viscosity, density, cohesiveness, elasticity, and plasticity—variations that have implications for the organizations and dynamics of O–E systems. The realities of the mesoscopic ecological scale are to be contrasted (Table 9.1) with the realities at nature's more microscopic and macroscopic scales, as detailed in physics textbooks. The latter realities have held sway over most past philosophical and psychological treatises on the knower, knowing, and the known.

Surfaces, substances, places, objects, and events are opportunities or possibilities for action. Referred to by Gibson (1979/1986) as affordances (see EP VI), these action possibilities are defined by the complementary relations that exist between the properties of ecological realities and the properties of the organism under consideration. A surface that supports human locomotion by being sufficiently hard and flat affords walking and/or running and is perceived as such. Similarly, a detached object (an outward facing layout of surfaces completely surrounded by the medium) that is sufficiently small and can be grasped in an individual's hand is perceived to afford throwing and when such an object is thrown with sufficient force and within sufficient range of another individual, it is perceived by that other individual to afford catching.

The implication is that for an organism to perceive what an environmental surface, substance, place, object, or event affords is for that organism to perceive what an environmental surface, substance, place, object, or event *means* (Gibson, 1979/1986; Turvey, 1992; Michaels, 2003). In other words, what a substance, surface, etc., *is* and what a substance, surface, etc., *means* are one and the same thing (Reed & Jones, 1982; Reed, 1988). As such, meaning is not a subjective or phenomenal property of mind, nor does it need to be imposed, constructed, or computed by mental or executive processes. Rather, meaning can be understood and studied as an objective and real property of an O–E system (see, additionally, Dewey & Bentley, 1949).

Formal development of the realities that embed behaving organisms—achieving the desired ontological theory referred to above—is challenging on several fronts. Consider the apparently simple notion of *place*. We can readily intuit that organisms can orient to places—for mammals, the surface and substance layouts to go to in order to sleep, hide, eat, drink, and so on. Further, by learning the places reachable by locomotion (say, from a place called home) they can become

TABLE 9.1 Environmental realties defined at the ecological scale.

Substances: Aspects of the environment (e.g., rock, soil, wood, plant tissue) that are (with respect to the physical properties of an organism) rigid, nondeformable, impenetrable and unyielding in shape. They differ in hardness, viscosity, density, and elasticity, as well as in solubility and stability. They can persist over some transformations, but not over others (e.g., for an animal, plant tissue cases to exist once eaten).	Not to be confused with the physical notion of matter (e.g., atoms or molecules), which always persist and never go out of existence.
Surfaces: The interfaces between substances and the medium (e.g., air or water) that surrounds an organism. They are the one sided, visible aspect, of a substance. At the scale of living systems they are indefinitely nested within other surfaces. Surfaces can persist or change, such as their layout, texture, or state of illumination (shaded or unshaded). They structure light, transform chemicals, transmit substance vibrations, and make contact with limbs or bodies.	Not to be confused with the geometrical notion of a plane that is completely level, textureless, and two sided.
Places: Extended surfaces of the environment. A place can be a "point of observation," yet places do not have an absolute boundary. Places are nested and are thus located by their inclusion in other places.	Not to be confused with the geometrical notion of point. Places are not singularities and cannot be located using coordinates.
Objects: The attached or detached substances and, respectively, can be either completely or partially surrounded by medium (e.g., water or air). The surface layout of detached objects is topographically closed. The surface layout of attached objects is continuous with surface layout of other substances (or objects)	Not to be confused with notion of body or particle portrayed by physics. Nor with the philosophical dichotomy between subject–object.
Events: Changes in the layout, texture (and or color), and existence of environmental surfaces. Events are reversible in some instances, but not in others. Examples include: the movement of an organism or object from one place to another; the ripening of fruit; and the melting of ice. Like the other ecological realties, events are nested, and thus are defined by their inclusions in other events.	Not to be confused with "clock" time or time defied by the second law of thermodynamics, both of which are irreversible. Nor are events restricted to the translations and rotations of classic mechanics.

oriented to, and can be said to know, their habitat. One might contend that *place*, like *point*, can be put into correspondence with *coordinates*. That is, a place can be located within a coordinate system. Alternatively, a place can be located by the ecological reality of *inclusion*, by how it is nested in other, larger places (Gibson, 1979/1986; Meng & Sedgwick, 2001). Inclusion motivates a geometric system different in kind from the Euclidean system (Huntington, 1913). It also motivates new hypotheses about what it means to be oriented to a place. In the absence of a change in a place's coordinates, a modification in how it is included in other places should alter (and does alter) a perceiver's orientation to it (Harrison, 2007).

For many species, but most especially humans, cognitive activity is embedded in social settings (e.g., a courtship ritual, a conversation, lunch with friends, a lecture, a football game), raising the question of whether the ontological theory of E_{pop} should include properties marked by inter-organism or extra-individual dimensions, and if so how (see, additionally, Schmidt, 2007). Members of this potential class of ecological properties have been referred to, alternatively, as *physical-behavioral units* and *behavior settings* (Barker, 1968; Schoggen, 1989). As currently interpreted, each is an approximately invariant array of physical objects and physical infrastructure coordinate with an approximately invariant pattern of individual participant and inter-participant behaviors (Smith, 2001). Although the boundary of each such unit/setting cannot be simple, and may be context sensitive, it must nonetheless be perceptible by both participants of the social unit/setting and by other individuals outside the social unit/setting. That is, the boundary of the social unit/setting must be an ecological reality that grounds the separation of any one unit/setting from the multitude of others, those that encompass it and those that it encompasses. What may compel consideration of these socially marked properties of E_{pop} is the fact that to the organism (here, human) "[t]hey are as objective as rivers and forests—they are parts of the objective environment that are experienced directly as rain and sandy beaches are experienced (Barker, 1968, p. 11)."

ECOLOGICAL PRINCIPLE III: BEHAVIOR IS EMERGENT AND SELF-ORGANIZED

The vestiges of organism–environment dualism provoke the traditional assumption that behavior is reducible to components that interact mechanistically and locally. Coupled with a tendency to define components in terms of context-independent anatomical mechanisms, this assumption posits that the identification of such mechanisms counts as the proper explanation of a given behavior. By this view, a behavioral system as a whole does not exhibit any properties that are not, to some extent, identifiable in its fundamental component structures. The paradox, however, is that organisms exhibit emergent properties that are not found in any component structure. When an animal produces a coordinated action, for instance, the coordination among the components cannot be tied to any specific componential property of the action. The source of the coordination is not to be found in any individual muscle, neuromotor unit, joint, or any other component structure. Yet, for the orthodox, mechanistic view of behavioral systems, the coordination must originate from somewhere within the system. Thus, as illustrated in Figure 9.1, the recourse of orthodox cognitive science to some other entity or process (e.g., an internal motor program, forward model, or schema) as the source of coordinated action is inevitable.

By eschewing centralized executive function and employing a broadened notion of mechanism that permits emergence and context-dependent (functional)

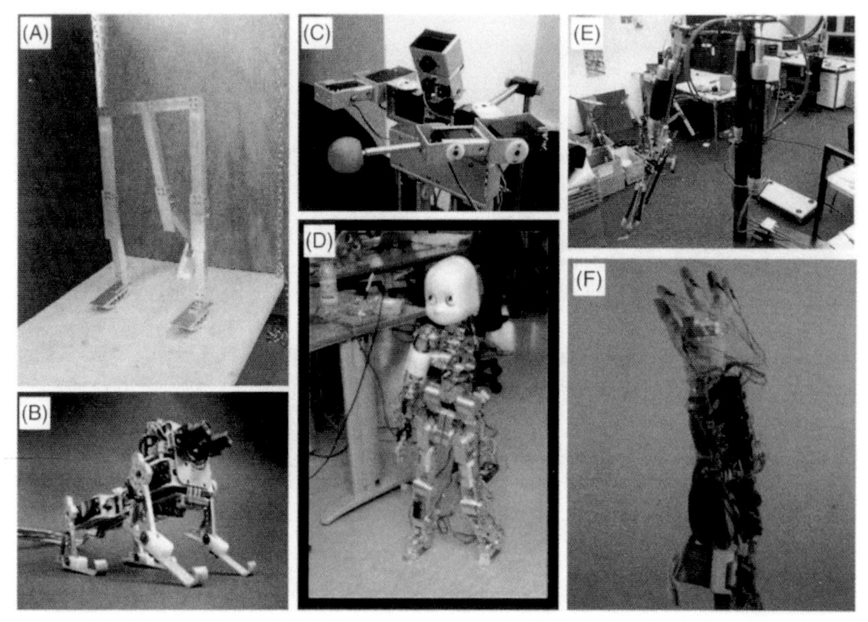

PLATE 1 Collection of robots used in the synthetic approach to embodied intelligence. (A) Passive dynamics walker. (B) Quadruped used for experiments on the influence of morphology and control on behavior. (C) Humanoid used for experiments on information self-structuring. (D) Developmental humanoids iCub. (E). Anthropomorphic arm with pneumatic actuators mimicking, among other things, the loosely swinging arm described in the text. (F) Prosthetic hand–arm complex (see Figure 7.2).

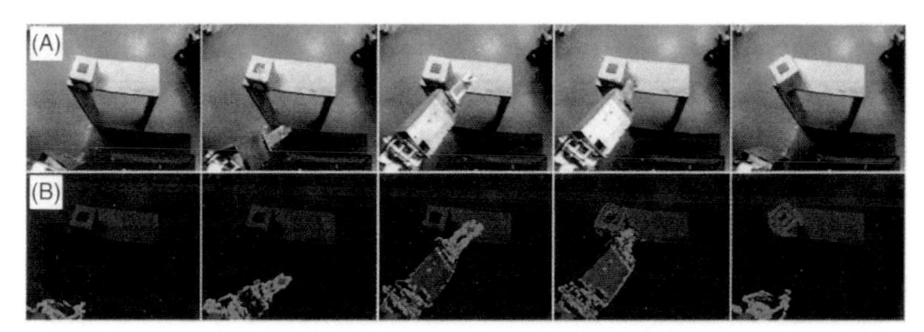

PLATE 2 Disambiguation and segmentation of visual scene through embodied interaction. (A) Arm extending into a workspace, poking an object, and retracting. (B) Shape of the object is identified from the tap using simple image differencing (Metta & Fitzpatrick, 2003). Segmentation in this case is not a trivial task—the edges of the table and cube are aligned, the colors of the cube and the table are not well separated, and there are shadows that may change (see Figure 7.3).

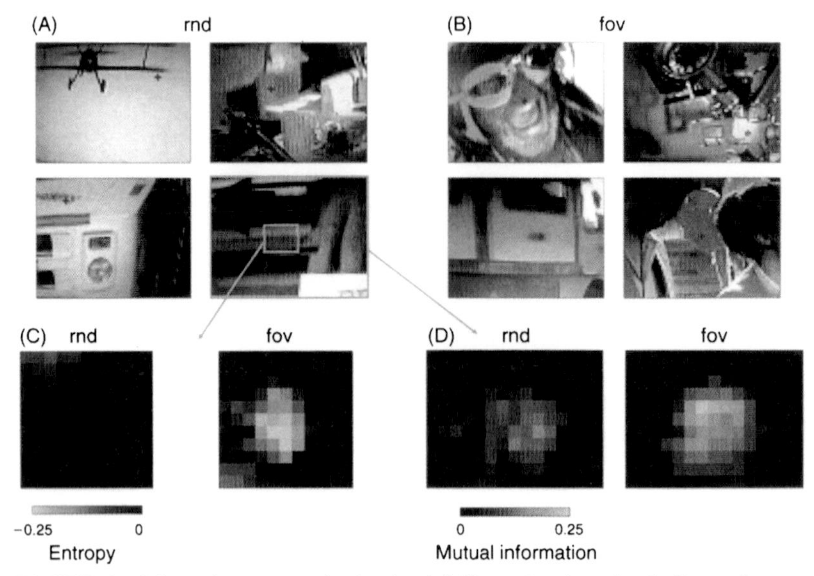

PLATE 3 Information structure in the visual field as a function of embodiment. Images show sample video frames obtained from a disrupted (A; "rnd," "low embodiment"—no sensory–motor coupling) and normally tracking (B; "fov," "high embodiment" —high sensory–motor coupling), and active vision system. Plots at the bottom show spatial maps of entropy and mutual information, expressed as differences relative to the background. There is a significant decrease in entropy (C) and an increase in mutual information (D) in the center of the visual field for the "fov" condition, compared to little change in the "rnd" condition. Data replotted from Lungarella et al. (2005) (see Figure 7.4).

PLATE 4 A frame from the human embedded vision system simulation. The human visually guided behaviors simulation. The main panel shows a single video frame from the real-time simulation that has the model negotiating a sidewalk strewn with purple litter and blue obstacles, each of which must be dealt with. The insets show the use of vision to guide the humanoid through a complex environment. The upper inset shows the particular visual routine that is running at any instant. This instant shows the detection of the edges of the sidewalk that are used in navigation. The lower insert shows the visual field in a head-centered viewing frame (see Figure 8.2).

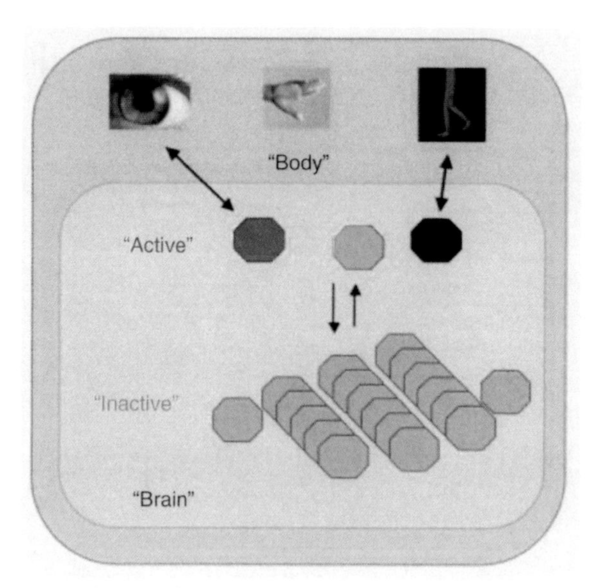

PLATE 5 The main features of the operating system model. Following work in psychology and robotics (Bonasso et al., 1997), we developed a tripartite abstract cognitive architecture for realizing ordinary behaviors that has three levels: *central executive, arbitration*, and *behavior*. The central executive level of the hierarchy maintains an appropriate set of active behaviors from a much larger library of possible behaviors, given the agents current goals and environmental conditions. The composition of this set is evaluated at every simulation interval, which we took to be 300 ms. The arbitration level addresses the issue of managing competing active behaviors. Thus an intermediate task is the mapping action recommendations onto the body's resources. Since the set of active behaviors must share perceptual and motor resources, there must be some mechanism to arbitrate their needs when they make conflicting demands. The behaviors themselves, when they are active, each have their own distinct jobs to do, such interrogating the image array with the objective of computing the current state of the process (see Figure 8.3).

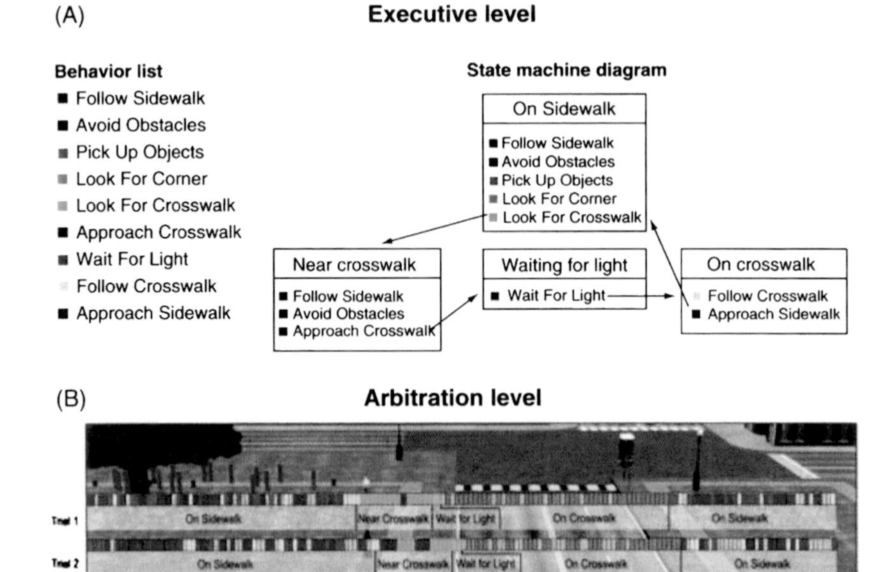

PLATE 6 The humanoid model operating system's servicing of different tasks from three separate trials. The cognitive architecture has computations at three distinct levels, the two most abstract of which are shown here. (A) At the most abstract level the composition of behaviors must be continually adjusted by the Central Executive's state machine. (B) At the Arbitration level, behaviors compete with each other for the body's resources. The colored bars show the result of competition for eye fixations. Each bar denotes a fixation and its color indicates the behavior that initiated it. At the most basic level (not shown) visual routines extract specific information from each fixation to define the state of a behavior (see Figure 8.4).

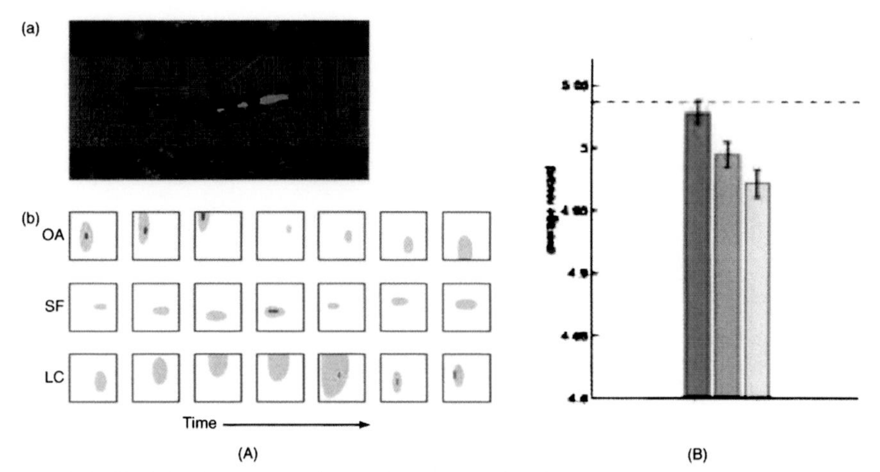

PLATE 7 Behaviors compete for gaze in order to update their measurements. (A) The top panel shows seven time steps in walking and the associated gaze vector color coded for obstacle avoidance (OA – red), sidewalk finding (SF – blue), and litter pickup (LC – green). The corresponding boxes below show the state spaces where the a priori uncertainty is indicated in beige and the a posteriori uncertainty is indicated in the appropriate color. Uncertainty grows because the internal model has noise that adds to uncertainty in the absence of measurements. Making a measurement with a visual routine that uses gaze reduces the uncertainty. For example for litter collection (LC) panel five shows a large amount of uncertainty has built up that is greatly reduced by a visual measurement. Overall, OA wins the first three competitions, in addition to sidewalk finding, and finally LC wins the last three. (B) Tests of the Sprague algorithm (dark green) against the robotics standard round robin algorithm (light green) and random gaze allocation (yellow) show a significant advantage over both (see Figure 8.5).

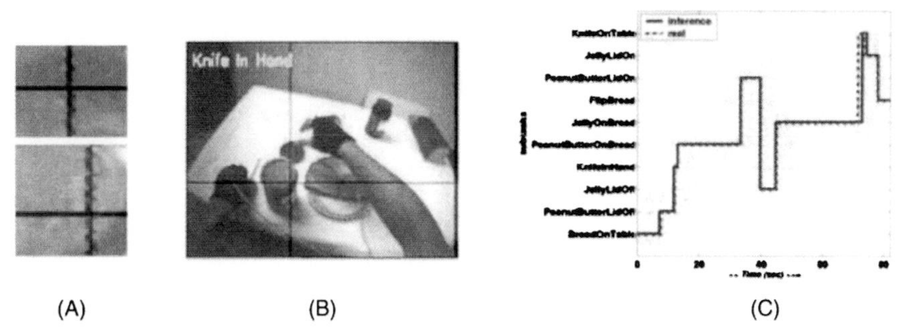

PLATE 8 (A) Two fixations from different points in the task-(top) bread with peanut butter (bottom) peanut butter jar appear very similar, but do not confuse the DBN which uses task information. (B) A frame in the video of a human subject in the process of making a sandwich showing that the DBN has correctly identified the sub-task as "knife-in-hand." (C) A trace of the entire sandwich making process showing perfect sub-task recognition by the DBN. This trace uses the entire data set in an off-line mode. In the on-line mode where the classifications have to be made on the partial data sets, small errors can be made (see Figure 8.7).

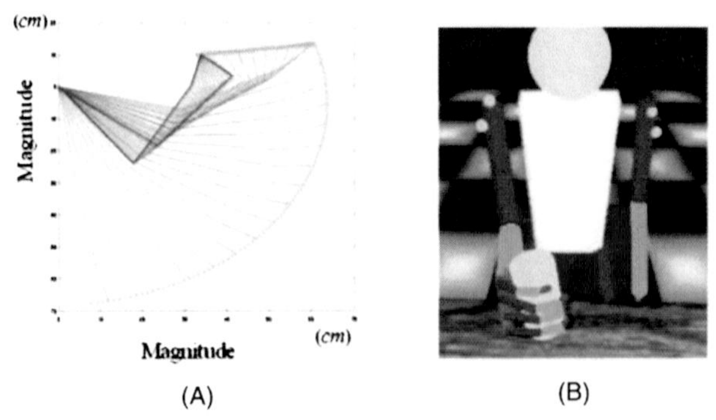

PLATE 9 First trials with the EP segment movement generator. (A) End posture planning. Blue curves: stages in the gradient descent used to find an initial joint coordinate solution for a target. The bold blue curve is the joint configuration generated by this stage. Note that the simulation trajectory is very non-biological. Black curves: An energy function is added to the simulation to adjust the end configuration. The bold black curve gives the optimal end posture for this stage. Note that the final configuration is very different than the previous end posture. (B) Preliminary trials with a cylindrical grip using the vortex physical simulator (see Figure 8.8).

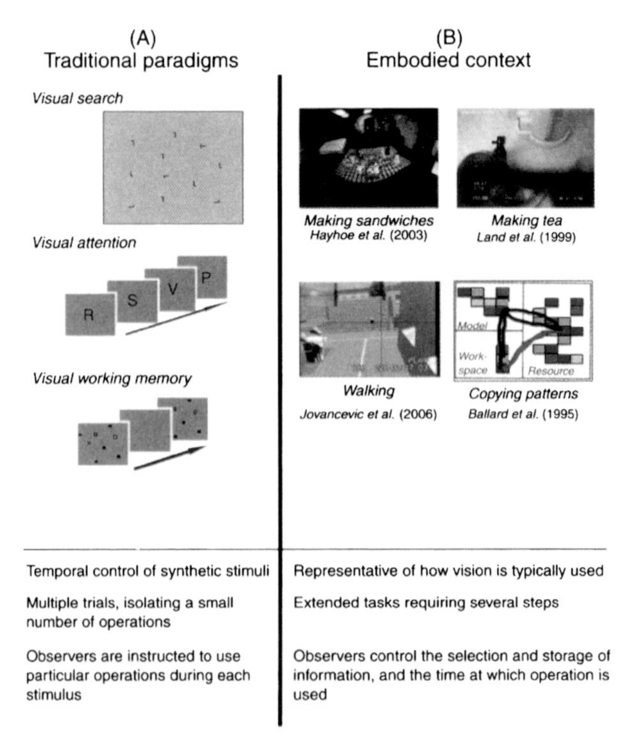

PLATE 10 Vision is either studied using (A) traditional paradigms with few behavioral requirements and the computations are controlled by the trial structure, or (B) observed in the context of embodied behavior, where observers control the timing of when visual information is acquired and what actions are performed (see Figure 10.1).

(A)

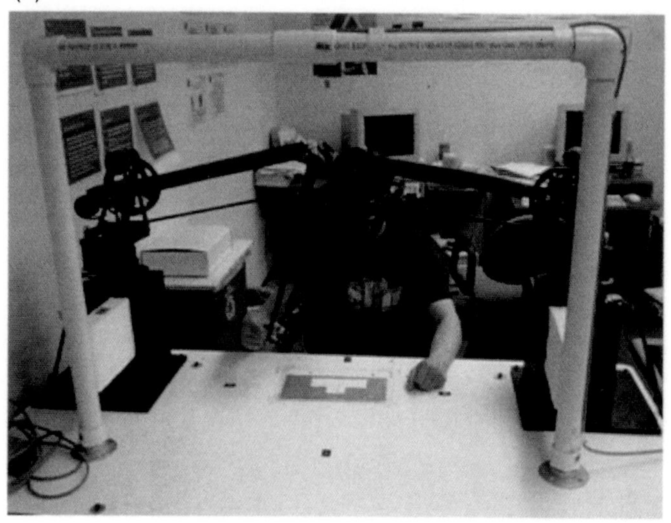

(B)

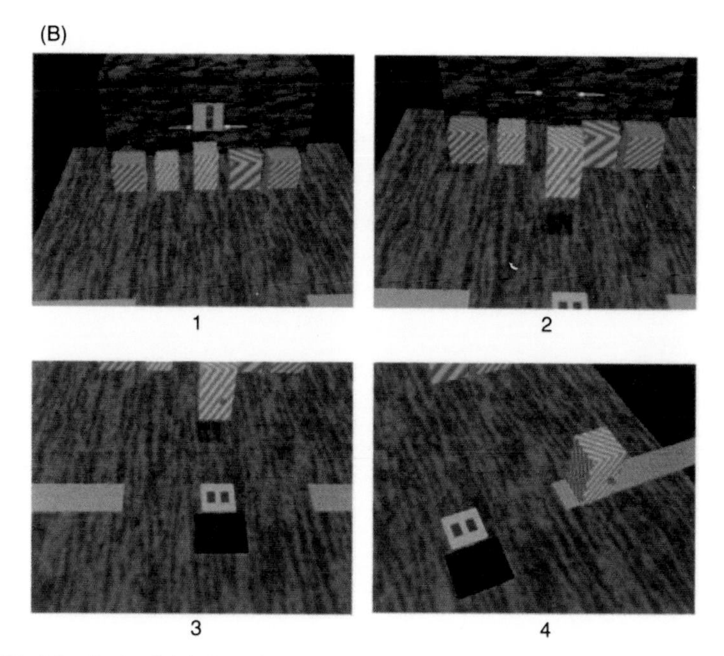

PLATE 11 Sorting bricks in a virtual environment. (A) Subjects wore a head-mounted display and received haptic feedback through mechanical arms attached to index finger and thumb. (B) Scene during a *One-Feature* trial when brick color was task relevant. Fingertips are represented as small red spheres. In a single trial, a subject (1) selects a brick based on the pick-up cue, (2) lifts the brick, (3) brings it towards themselves, (4) decides on which conveyor belt the brick belongs based on a put-down cue (see Figure 10.2).

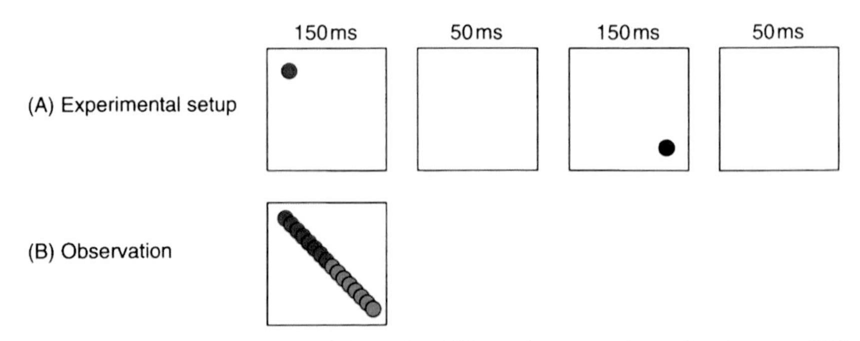

PLATE 12 An example setup of the "color phi" experiment as referenced by Dennett (1991, 1992) and conducted in a similar way by Kolers & von Grünau (1976). Top: A red spot is presented for 150 ms followed by a short delay of 50 ms, followed by the green spot for 150 ms, followed by a second delay of 50 ms. This "flickering" of the spots is repeated continuously. Bottom: The observers of this experimental setup report that they see a moving spot changing the color roughly in the middle of the way (see Figure 12.1).

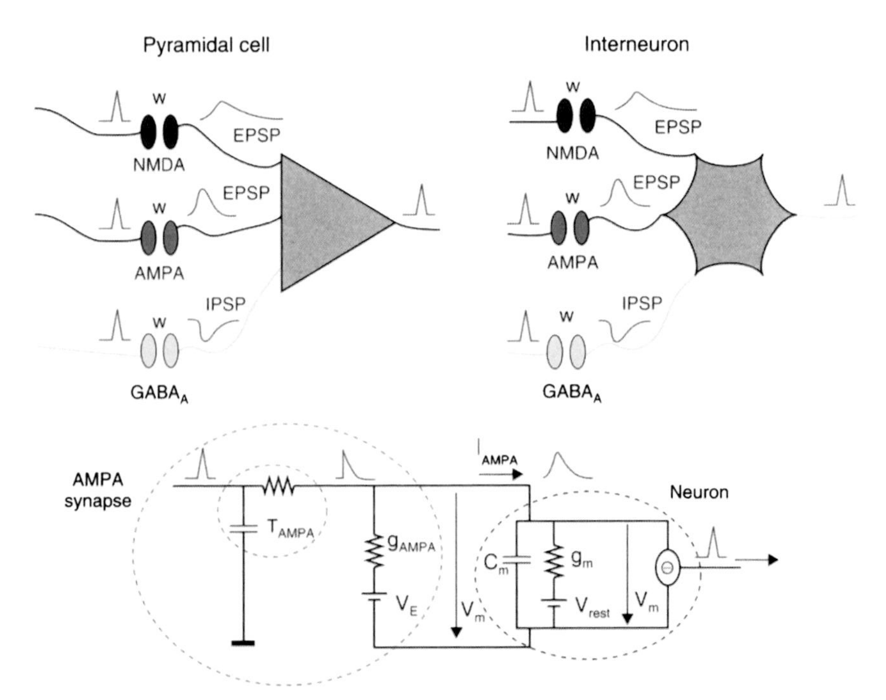

PLATE 13 The "Integrate-and-Fire" model describes the development of the subthreshold membrane potential for neurons by an equivalent electrical circuit. Different neuronal and synaptic connection types are possible. EPSP—excitatory postsynaptic potential. IPSP—inhibitory postsynaptic potential. Top left: schematic diagram of a pyramidal cell and different connections types. Top right: schematic diagram of an interneuron and connections. Bottom: The equivalent circuit used to calculate the subthreshold development of the membrane potential for either of the neurons with an exemplary AMPA synapse connected (compare also well Stemme, 2007, p. 73, Figure 6.1) (see Figure 12.2).

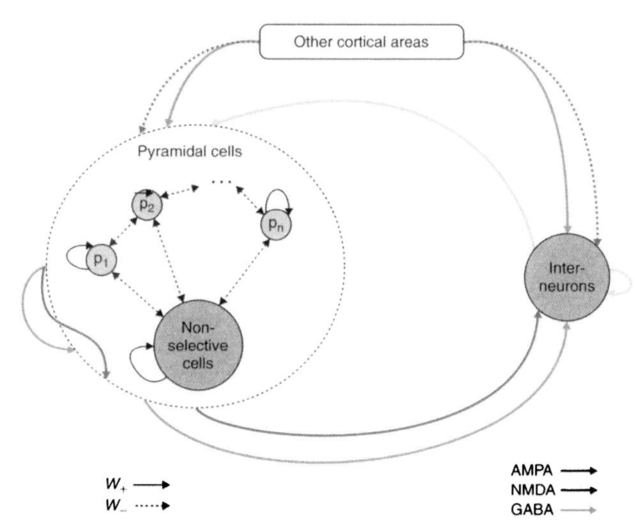

PLATE 14 Overview of the cortical module. Repainted according to Brunel & Wang (2001, Figure 2, compare also Stemme, 2007, Figure 6.2). Interneurons send inhibitory connections to the pyramidal cells and to the other interneurons within the inhibitory pool. Pyramidal cells in turn send AMPA and NMDA connections to the interneurons and are also interconnected with the relatively strong strength w_+ within a selective pool or a comparatively weaker weight w_- between different selective pools, respectively. External glutamatergic input from other cortical areas can be mediated by AMPA and NMDA receptors (see Figure 12.3).

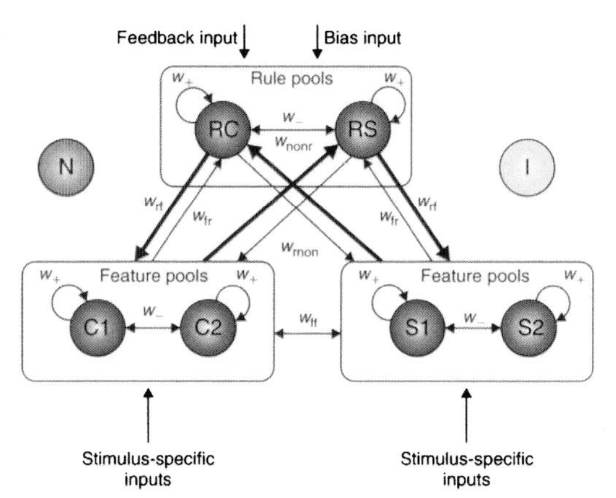

PLATE 15 Neurodynamical model for the set shifting task (compare also Stemme, 2007, p. 123, Figure 8.5). The model comprises two pools responsible for different rules; RC color rule; RS shape rule. Furthermore, there are two pools representing two different colors (C1 and C2) and two pools representing different shapes (S1 and S2). Every rule pool supports "its" feature pools via the weight "w_{rf}." In the opposite direction, the weight "w_{nonr}" is comparatively stronger than the weight "w_{rf}," a configuration that enables the rule change. The neuronal pool named "N" comprises a pool of non-selective neurons. This neuronal pool accounts for the circumstance that not all neurons within a given cortical area are engaged in a specific task. The pool named "I" comprises the inhibitory neurons. Altogether, we chose to model $N_E = 1600$ excitatory neurons and $N_I = 400$ inhibitory neurons. The network of neurons is fully connected with different connection strengths as indicated (see Figure 12.4).

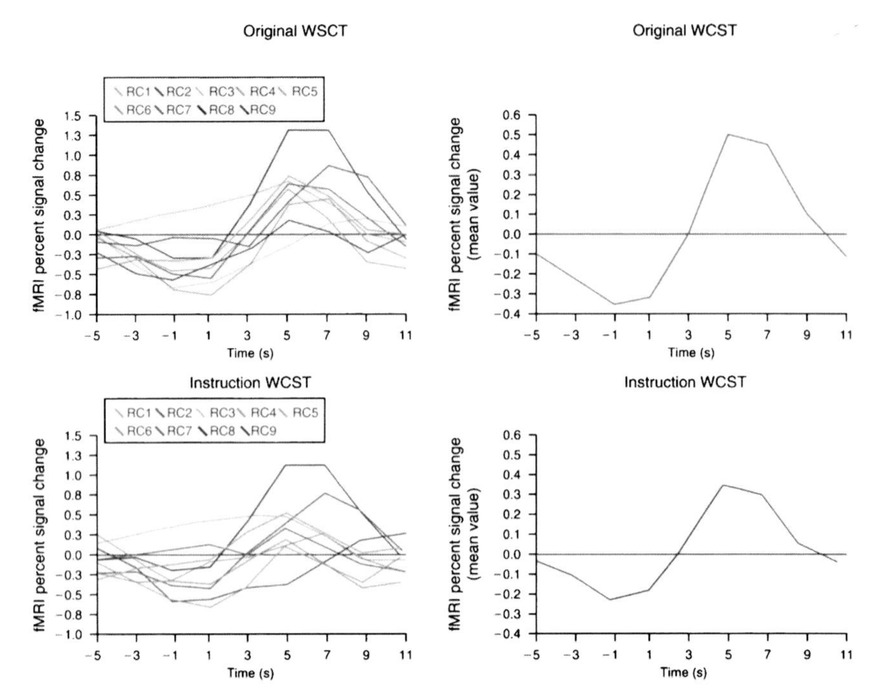

PLATE 16 Resulting fMRI signal for the neurodynamical simulations. The top left diagram shows the resulting fMRI signal for nine single rule changes for the original WCST condition. The right diagram shows the corresponding mean value. The bottom diagrams show resulting fMRI signals for the instruction variant of the WCST. For all cases there is a close match with the experimentally determined values (compare also Stemme et al., 2005) (see Figure 12.5).

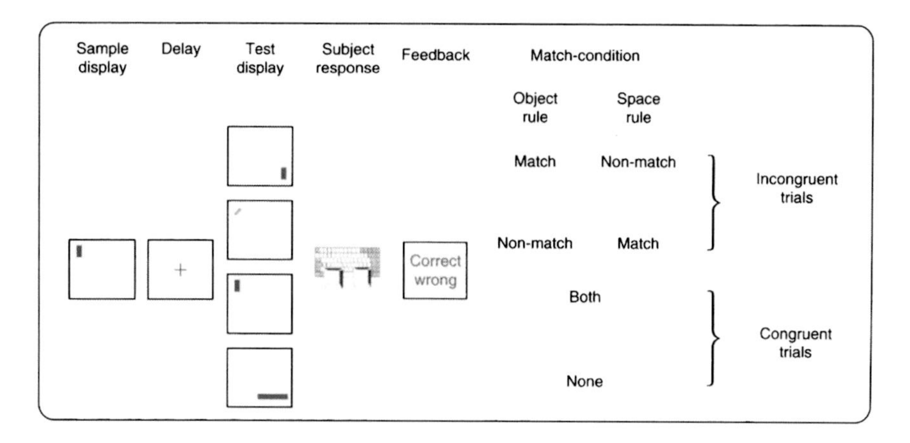

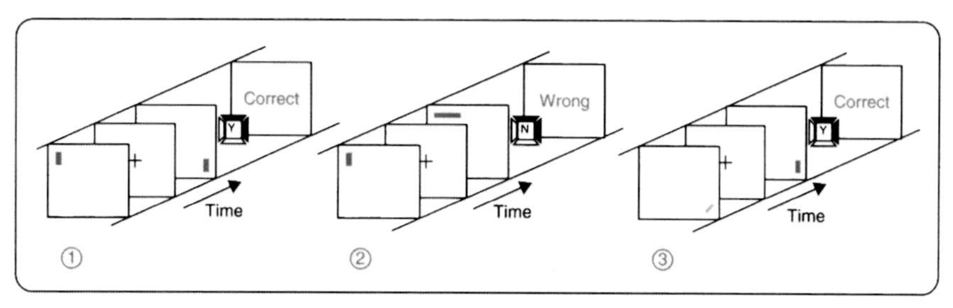

PLATE 17 Task setup for the WDMS experiments. Setup of the experiments to collect experimental response data (WDMS experiment—"Wisconsin Delayed Match to Sample" experiment). Top: Timing dynamics for the single trials and possible stimulus (match) conditions. Bottom: Example trial sequence including a rule change in the second trial (feedback message: "wrong") (see Figure 12.6).

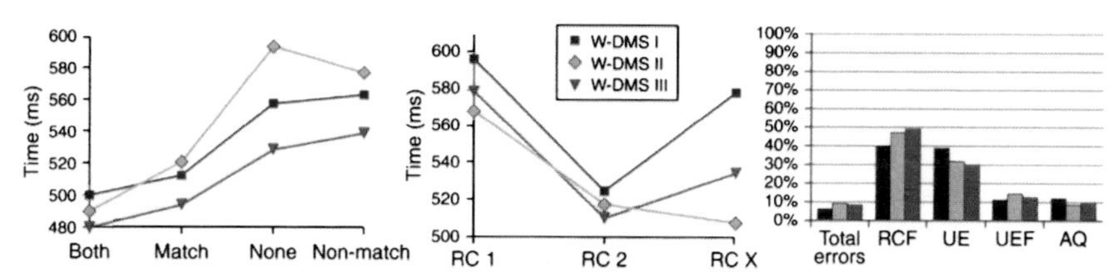

PLATE 18 Experimental results. The results were obtained in three variants of the WDMS experiments (WDMS I, II, III). For these variants the feedback times and hence the intertrial time was varied (1500, 1000, 500 ms), which did not produce any significant effect on the results. Left: response times for the different match conditions. Middle: response times relative to the rule change. RC1: first trial after a rule change; RC2: second trial after a rule change. RC X: all other trials. Right: average error rates and types for the experiments. RCF: errors in the context of a rule change (rule change follow-up). UE: unmotivated, that is, attentional errors. UEF: errors following a previously unmotivated or attentional error; AQ: rule acquisition errors, which occur at the beginning of an experiment or an experimental block to determine the first valid rule (see Figure 12.7).

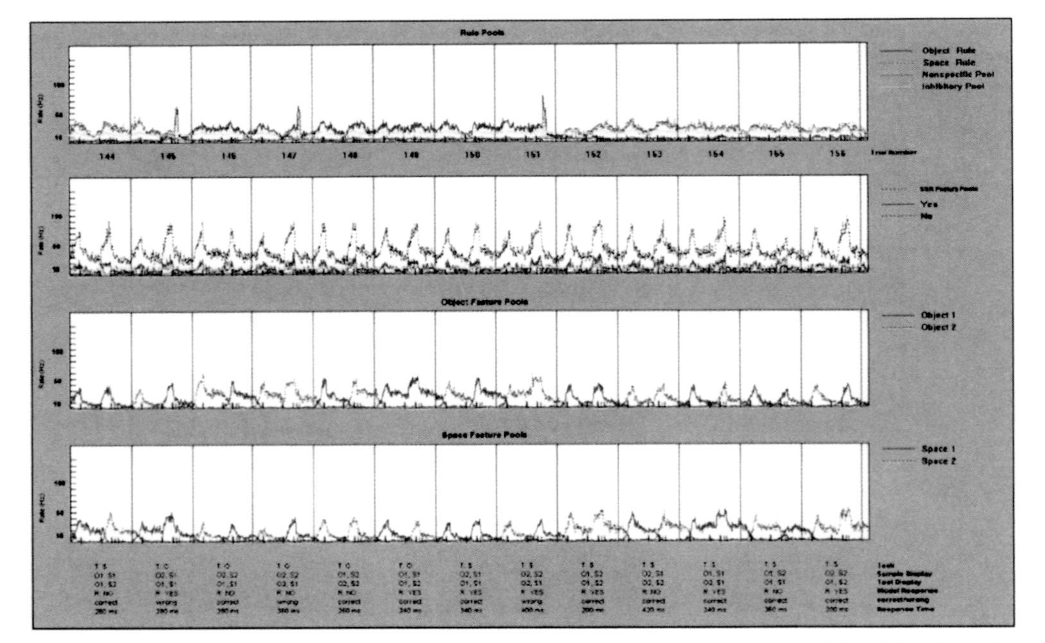

PLATE 19 Spiking dynamics during an example simulation. Excerpt from a simulation showing the spiking activity of the rule pools (top diagram), the object feature pools (third diagram), the space feature pools (bottom diagram), and the summed spiking rate (ssr) of all feature pools. In other words, the time course of the ssr as depicted in the second diagram was obtained by adding the spiking activity of O1 (in Figure 12.4 labeled "C1"), O2 (in Figure 12.4 labeled "C2"), S1 and S2 for any given time point. As response times depend on both stimulus dimensions, the ssr constitutes the input provided to optional response pools of the model. Thus, a "yes" or "no" response of the model has to be based necessarily on differences in this input. The resulting model responses as presented in Figure 9 were calculated on the basis of ssr where an answer of the model was considered to be "yes" if the ssr passed a certain threshold and "no" if the ssr failed the threshold for a certain period of time. The threshold values constituted important model parameters (compare also Stemme, 2007; Stemme et al., 2007a). The bottom lines in the diagram indicate the individual trial setup, the calculated model response, the response time, and the feedback the model received (i.e., "correct" or "wrong") (see Figure 12.8).

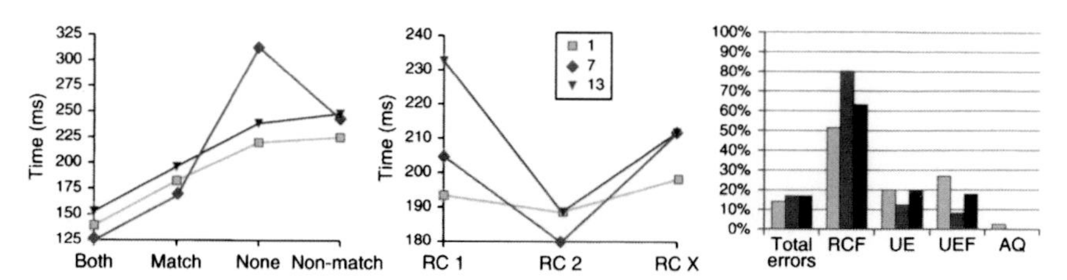

PLATE 20 Simulation results. Results of the WDMS simulations with respect to response times and error rates for three example model configurations using different parameter values. A constant factor accounts for the response time differences compared to the experimental results (see Figure12. 7), as an explicit motor response was not considered. Further investigations revealed the importance of considering individual participant results rather than averaged experimental data (see Figure 12.9).

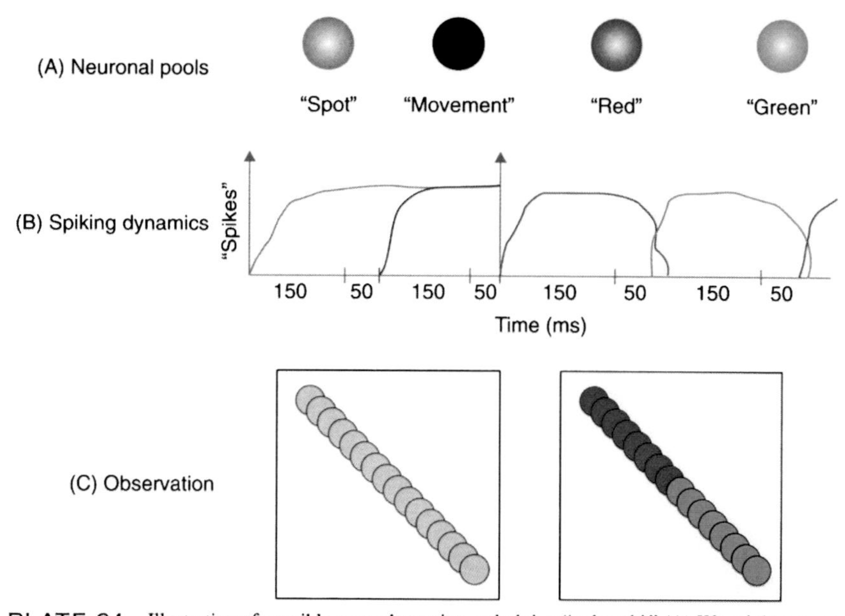

PLATE 21 Illustration of possible neurodynamics underlying "color phi." (A) We might assume for example four neuronal pools selective or "responsible," respectively, for a spot presented in the visual field, for a specific movement occurring within the visual field, and for the colors red and green, in a way very similar to that described for Figure 12.4. Neuronal selectivity for colors and forms follows the neurophysiological findings, as outlined for the design of the set shifting model. Similarly, it has been demonstrated that neurons respond to specific movements within the visual field. (B) Estimated spiking dynamics of the neuronal pools. It is assumed that neurons respond according to the visual stimuli presented, which implicates that the "movement" neurons start spiking with the presentation of the second stimulus in the first trial and continue spiking for the remainder of the experiment, as do the "spot" neurons (both pools enter a state of persistant activity). The "color" neurons respond in an alternating manner according to the presented color of the stimulus. (C) Under the assumption that the activity of a neuronal pool leads to a certain (subjective) perception, that is the activity of the neuronal pool responsible for "red" leads to a "red perception", we are able to provide a rather easy explanation for "color phi" in considering the spiking dynamics; these indicate why observers see a moving spot (activity of the corresponding pools) which abruptly changes the color (for 150 ms "red" is presented and for another 150 ms "green"), at least starting with the presentation of the second stimulus in the first trial and thereafter for the remainder of the continuous experiment (see Figure 12.10).

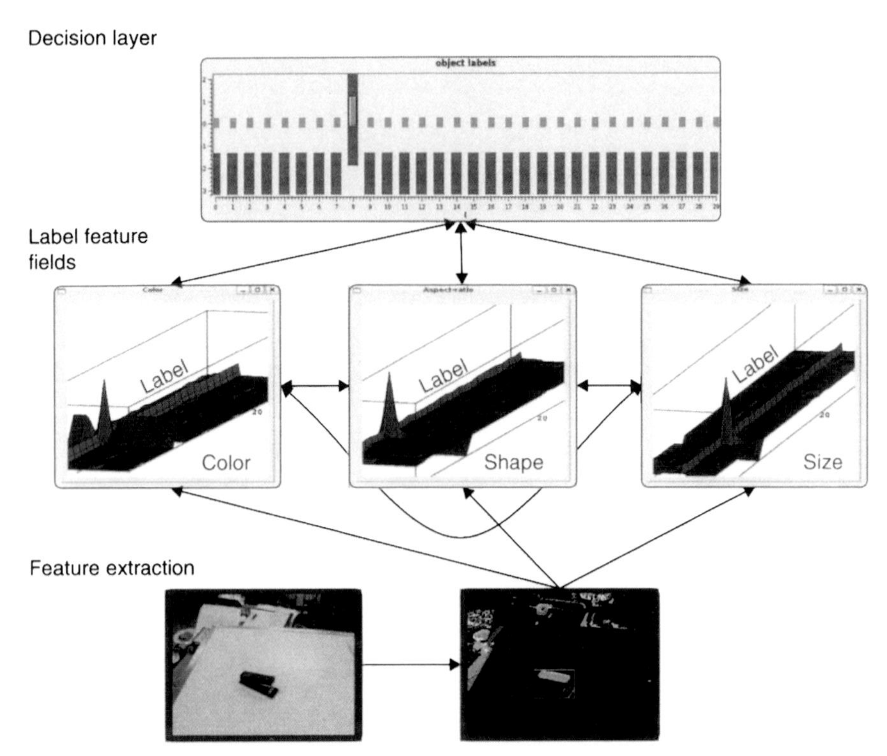

PLATE 22 For the object recognition procedure, three simple object features are extracted from an image and fed into a network of coupled dynamic fields. The interactions within and between these fields result in the selection of one label, which can be read out from the decision layer (see Figure 13.11).

PLATE 23 The speaker, at right, is referring to the Aymara expression *aka marat(a) mararu*, literally "from this year to next year." (A) When saying *aka marat(a)*, "from this year," he points with his right index finger downward and then (B) while saying mararu, "to next year," he points backwards over his left shoulder. (©2008 Rafael Núñez. Published by Elsevier Ltd. All rights reserved) (see Figure 17.3).

PLATE 24 The speaker, at left, is talking about the Aymara phrase *nayra timpu,* literally "front time," meaning "old times." When he translates that expression into Spanish, as he says *tiempo antiguo* he points straight in front of him with his right index finger. (©2008 Rafael Núñez. Published by Elsevier Ltd. All rights reserved) (see Figure 17.4).

Breakfast on the farm

Ben needs to feed the animals.
He pushes the hay down the hole.
The goat eats hay.
Ben gets eggs from the chicken.
He puts the eggs in the cart.
He gives the pumpkins to the pig.
All the animals are happy now.

PLATE 25 The farm toys and one text. The green traffic light signaled that a sentence was to be reread or used to direct manipulation (see Figure 18.1).

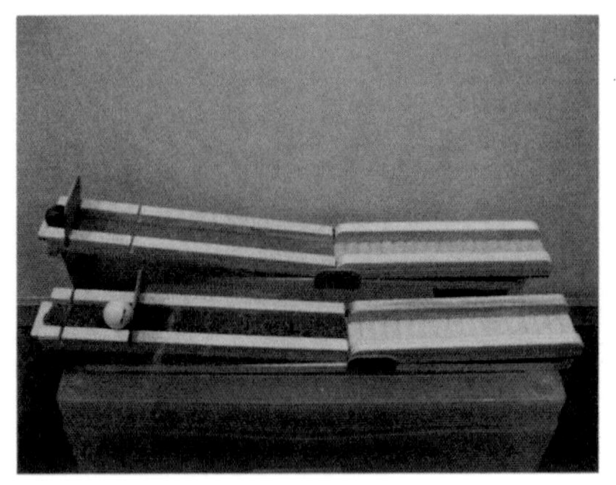

PLATE 26 Ramps context for CVS. A confounded design is illustrated because the ramps differ in ball starting location, type of ball, ramp angle, and ramp surface (see Figure 18.2).

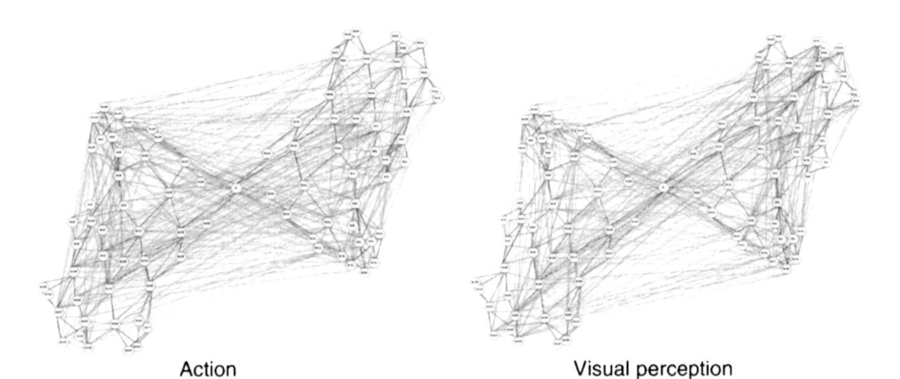

PLATE 27 Cortex represented as adjacency + co-activation graphs. Here the Brodmann areas are nodes, with black lines between adjacent areas and orange lines between areas showing significant co-activation. The graph on the left shows co-activations from 56 action tasks, and the graph on the right shows co-activations from 57 visual perception tasks. Graphs rendered with aiSee v. 2.2 (see Figure 21.1).

descriptions of behavior, the ecological approach motivates law-based accounts of the origins of order in behavior (Turvey, 2005; Turvey & Shaw, 1995). As such, an "other-organized" approach, in which order is prescribed by some homuncular entity, is contrasted sharply with an approach that is informed by an understanding of self-organizing systems. As lawful consequences of nonlinearity and complexity self-organizing systems exhibit macroscopic, novel (emergent) properties that cannot be reduced to properties of the components. Such systems are characterized by nonlocal interactions among components that play highly context-dependent (i.e., functionally defined) roles. From this perspective, behavior emerges from the interplay of mind, brain, body, information, and environment, functioning as a unitary complex system at the ecological scale.

The study of coordinated movement patterns has revealed numerous examples of emergent phenomena that are representative of this idea. One of the most well-known examples is the spontaneous transition between coordination patterns defined over a person's rhythmically moving left and right index fingers (see Kelso, 1995 for a detailed discussion). As movement frequency increases to a critical level, coordination patterns initially prepared in the anti-phase mode (fingers moving in opposite directions but at the same frequency) transition to the more stable in-phase mode (fingers moving in the same direction and at the same frequency). Movement frequency, the control parameter, does not "represent" or "code" the transition in the phase mode of the moving fingers. The stable relative phase modes, and the transition from anti-phase to in-phase, are not prescribed a priori by any executive function or entity, and are not identifiable in the properties of the components. Instead, the dynamics emerge from the nonlinear interplay of the two component oscillators, the nonlinear coupling between them (Haken et al., 1985), and extant constraints such as those imposed by the value of a control parameter like movement frequency. In this way, the stabilities and patterning of movement are understood to result lawfully from the physical and biomechanical constraints that naturally couple together the different limbs of the body (Kugler & Turvey, 1987; Kelso, 1995) and can be modeled as such (Figure 9.5). Importantly, this understanding provides researchers with a much deeper understanding of how the perceptual-motor system self-regulates and orders its many degrees of freedom than motor programming accounts, in that the rhythmic coordination of two limbs (and their many neurons, muscles, etc.) is conceived as a single synergetic system or coordinative structure[1] (Kugler & Turvey, 1987).

On its own, the research on within-person rhythmic coordination provides evidence that the complex patterns of coordinated action can arise without recourse

[1] The term coordinative structure is related to the notion of cooperativity from the field of thermodynamics and has been used in the human movement literature to refer to set of relatively independent units (e.g. muscles, limbs, animals, or substances) that are temporarily constrained, both at short and long time scales to act as a unitary functional unit (for more details, see Kugler et al., 1980; Kugler & Turvey, 1987).

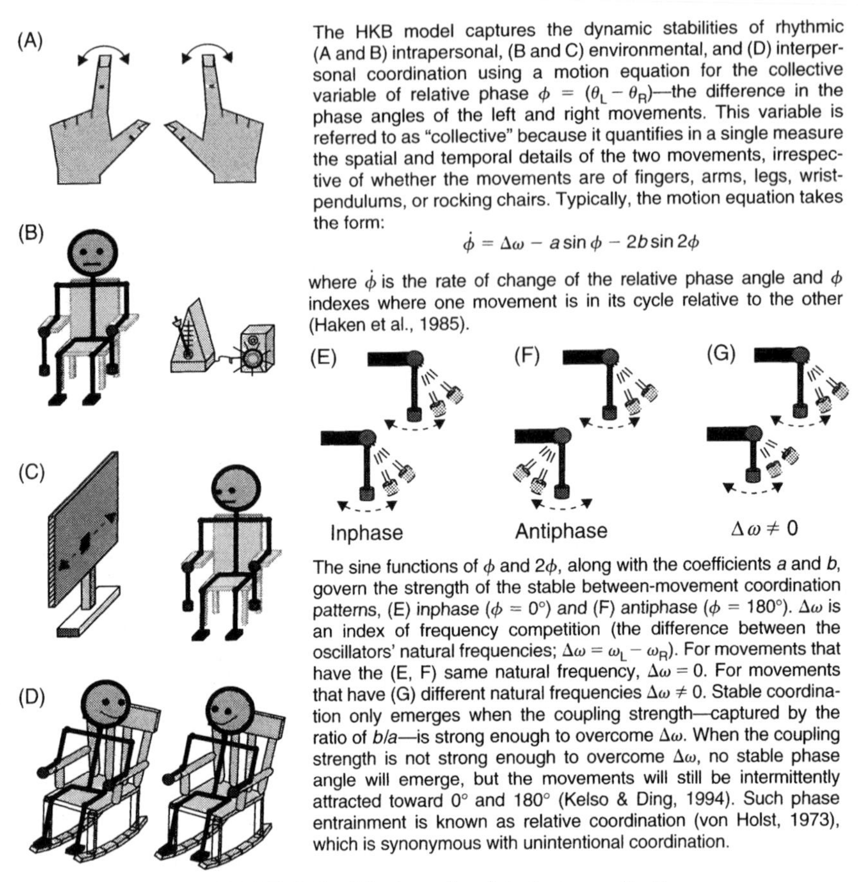

FIGURE 9.5 The dynamics of rhythmic coordination.

to internal or centrally defined mental causes or controllers. This evidence is compounded by the fact that the very same dynamics operate to constrain the rhythmic coordination that occurs between the rhythmic limb movements of an individual and a visual environmental rhythm (Bingham, 2004; Schmidt et al., 2007) and between the rhythmic limb movements of two interacting individuals (Schmidt et al., 1990). Moreover, such environmental or interpersonal coordination does not only occur *intentionally*, but also *unintentionally* (Schmidt & O'Brien, 1997; Richardson et al., 2007a). In each case, the emergent properties of the coordinated behavior result from the functional couplings among system components that arise and are dissolved spontaneously, depending on the values of control parameters and presence of certain constraints. In each case, the organized "system" as a whole is said to be a "soft-assembled" system, as opposed to a "hard-assembled" system with fixed components and fixed connections among

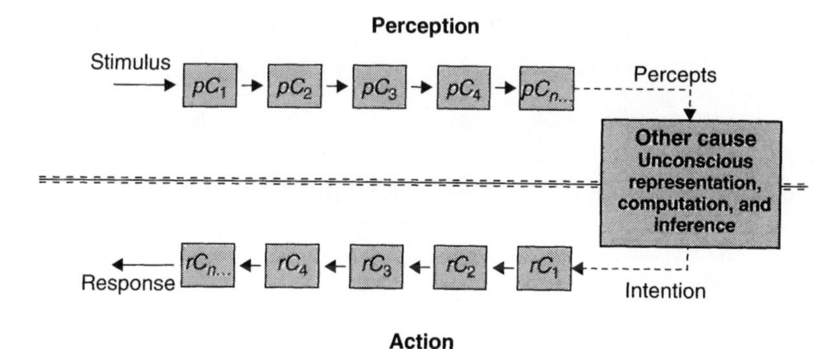

FIGURE 9.6 Traditional view of perception and action. Perception and action are envisaged as linear processes—from stimulus to percept and from intention to response via a linear sequence of causes (pC_1 to pC_n and rC_1 to rC_n) or mechanisms, respectively—separated and realized via unconscious representational processes hidden from view inside mind and brain. Figure adapted from Turvey (2004).

the components (Kugler & Turvey, 1987). Moreover, the causal system is not the brain, centralized mental or cognition structures, or even the animal (organism) itself, but a coordinative structure defined and distributed across an O–E system. Unnecessary recourse to motor programs or representations can thus give way to lawful equations of constraint that channel the dynamic unfolding of behavior (Turvey, 1990; Schmidt & Richardson, 2008).

ECOLOGICAL PRINCIPLE IV: PERCEPTION AND ACTION ARE CONTINUOUS AND CYCLIC

A more contemporary depiction of how traditional science approaches perception and action is illustrated in Figure 9.6. Equivalent to that portrayed in Figure 9.1, Figure 9.6 more directly highlights how the linear processes of mechanistic cause reify the "other" centralized causes argued to exist inside mind and brain. As a consequence of mind–body and organism–environment dualisms, Figure 9.6 also captures how the reliance on centralized representational–computational processing leads to the view that perception and action are distinct and separate processes. Perception, although important, is implicated as subservient to centralized representational–computational processing, with the environment, its objects, events, and surfaces being reduced to a system input or stimulus. Similarly, observable action is implicated as a subservient or secondary consequence of centralized representational–computational processing and is simply reduced to a system output or response (Turvey & Shaw, 1979; Hurley, 1998).

Gibson (1966) criticized the above conception by noting that the appropriate organs of sensitivity for perception are not passively stimulated receptors or nerves, but active perceptual systems. For instance, visual perception entails a pair of eyes, set apart, in a head that can turn and that is attached to a body that can move from place to place. Significantly, such systems are never passively stimulated, but are rather actively engaged in the detection of information (see EP V). The ecological approach is therefore adamantly opposed to any separation between perception and action, arguing instead that to study perception is to study action (and vice versa). To paraphrase Shaw and Kinsella-Shaw (1988, p. 159), perception and action are conjoint in that they serve a mutual aim—the satisfaction of a goal (see EP VI). Perception and action serve that aim in reciprocal ways—by detecting information that dynamically constrains action and by the control of action that dynamically constrains perception (see EP V). In a circular-causal manner, perceiving constrains action and action constrains perception.

Linking EP IV with EP I, EP II, and EP III, Figure 9.7A captures the cyclic nature of perception and action following the ideas of Kugler and Turvey (1987). In this case the perception–action cycle is a continuous relation between transformations or *flow* of the optic array, illustrated as a velocity vector field, and the *forces* that an animal produces to move from one point of observation to another (Turvey & Carello, 1986; Turvey, 2004). The cycle is that of forces resulting in flows and flows resulting in forces—of perception entailing action and of action entailing perception—whereby the time-evolution of behavior both generates and is constrained by the information revealed by the transformations of the optic array (e.g., direction of heading, time to contact; see EP V). As recently clarified by Warren (1998, 2006), this approach expresses how behavior is self-organized (Figure 9.7B), emerging from an O–E system via the detection of information (e.g., transformations of the optic array) and the modulation of action (e.g., the forces exerted in the environment by the organism, or by other objects, or by both).

On arguing that perception and action are cyclic, the ecological approach is not simply stating that perception and action *influence* or *interact* with each other (Figure 9.7C), but that perception and action are of the same logical kind, and are mutual, reciprocal, and symmetrically constraining (Shaw & Turvey, 1980). This distinction is not a trivial one. To argue that perception and action interact with each other is to support a distinction between perception and action and ultimately a disembodied account of behavior. The recent arguments for a common-coding theory of perception and action, which hold that the representational codes of perceived events are written in the same representational language as to-be-produced events (Prinz, 1997; Hommel et al., 2001), provide a good example. Such a theory maintains that knowing and acting are largely separate, linked only indirectly via representational processes. As a result, it reinforces the very thing it strives to undermine—the irrelevance of body and environment to cognition.[2]

[2]Research aimed at demonstrating the interaction of sensory motor states on traditionally defined cognitive processes (i.e., memory, affective evaluations, and emotions) suffer from a similar plight, in that they reinforce the classic dualisms by theoretically pre-supposing that such processes exist as centrally defined, trait- or state-like corporeal processes (Richardson et al., in press).

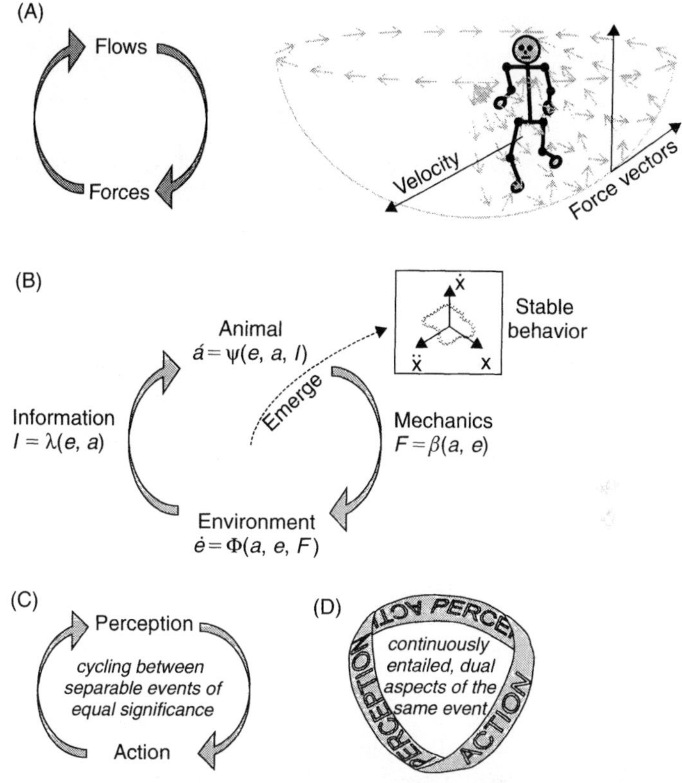

FIGURE 9.7 The perception–action cycle. (A) The cyclic nature of perception and action following the ideas of Kugler and Turvey (1987). (B) Nonlinear dynamical perspective on the perception–action cycle as outlined by Warren (1998, 2006). (C) Interactionist view of the perception–action cycle. (D) Perception–action Möbius band (Turvey, 2004).

In contrast to such "interactionist" notions, Figure 9.7D depicts a perception–action Möbius band, a depiction that realizes perception and action as continuously unified, dual aspects of an ongoing organism–environment event (Turvey, 2004). A comparison between the leftward plane (which shows action without the perception of environmental referents) and the rightward panel (which shows the perception–action event of changing a tire) of Figure 9.8 gives expression to this ecological principle and the implication that behavior is not the result of executive functions that reside inside the organism (here, human), but is a dynamic process distributed over the O–E system (Gibson, 1979/1986; Turvey & Fonseca, 2008). To anticipate the implications of EP V and EP VI below, Figure 9.8 also reveals how behavior is intrinsically functional rather than intrinsically mechanical and only extrinsically (secondarily) functional. In other words, the regularity of behavior emerges to realize functionally specific acts based on the direct perception of affordances (Turvey, 1992; Reed, 1996; Turvey & Fonseca, 2008).

FIGURE 9.8 Behavior is not defined with respect to mechanically specific postures and movements of the body, but to functionally specific descriptions of an ongoing organism–environment event. Adapted with permission from Turvey and Fonseca (2008).

ECOLOGICAL PRINCIPLE V: INFORMATION IS SPECIFICATIONAL

In seeking to provide an account of the tight coupling of perception and action, Gibson recognized that direct epistemic contact with the environment must be possible. His theory of *direct perception* (Gibson, 1979/1986; Michaels & Carello, 1981) can be sharply contrasted with most conventional theories, according to which perception of the world is mediated by inferential mechanisms and mental representations. Lying behind these contrasting views are basic assumptions about the nature of the stimulus information upon which perception is based.

According to classic views of perception (Figures 9.1 and 9.6), inference-like executive processes are needed because the informational support for perception is inherently ambiguous. Although this idea can be traced back to Müller (1826) and Helmholtz (1867/1925), it continues to influence modern theories of perception. To illustrate this point, proponents often point to the inverse projection problem, whereby each proximal stimulus defines an infinite family of equivalent configurations, or to illusions, such as the well-known Ames Room (Ittelson, 1968). Thus, an animal's perception of the world is viewed as a guess, based on past experience together with cues provided by the senses. An animal can only perceive the world indirectly, mediated by an inference or interpretation.

Following Gibson (1979/1986), the ecological response to the classic puzzles of perception is to rethink deeply rooted assumptions about both the properties of the world that are perceived (see EP V1) and the nature of the stimulus for perception, leading to a rejection of Müller's doctrine of specific nerve energies and its implications for how stimuli relate to the environment. Although each proximal stimulus is indeed consistent with an infinite number of configurations, there are many instances in which all but one configuration constitutes a serious violation of ecological constraints. In other words, ecological constraints render certain patterns found in ambient energy arrays as unambiguous with respect to certain properties of the world (Runeson, 1988). These constraints need not be internalized as representational structure or executive assumptions because perceptual systems need not function in every imaginable situation. Of primary concern then is a general theory of *specificity*, not a general theory of representation, which presupposes specification. Indeed, resolving the so-called grounding problem requires a theory of specification.

The term *specification* is used to characterize the relation between certain patterns in the distributions of energy surrounding an organism and those properties to which they bear a 1:1 correspondence. Likewise, the term *information* is reserved for those patterns that uniquely specify properties of the world. Accordingly, much of the research agenda for ecological psychology is aimed at identifying sources of information, which often requires a careful mathematical analysis of patterns found in ambient energy arrays as well as consideration of ecological constraints. The research on optic flow fields and their role in the guidance of locomotion is well known (Warren, 1998). In particular, information that specifies one's direction of heading (Warren, 2004) and time-to-contact with approached surfaces (Lee, 1976; Hecht & Savelsbergh, 2004) has been identified, and the role of optic flow in guiding locomotion has been verified (Warren et al., 2001). Other sources of information have been identified for such tasks as steering toward a goal (Wilkie & Wann, 2003), braking to avoid a collision (Fajen, 2005a), running to catch a fly ball (McLeod et al., 2006), and intercepting moving targets (Chardenon et al., 2002; Fajen & Warren, 2004).

Dynamic touch perception further illustrates how information relevant to object properties and the control of action is available in the changing flux of stimulation. Dynamic touch refers to perceiving via deformations of muscle spindles and Golgi tendon organs involved in manipulating an object about a joint. This form of perceiving epitomizes both the perception–action cycle and sensitivity to quantities that conform to information as specified.

Although the patterns of muscular activation involved in manipulating a hand-held object are constantly changing, these patterns are not ambiguous with respect to the object. The physics of rotations (i.e., rotational dynamics) dictates that the pattern of muscular activation about a joint is related in a 1:1 fashion to an object's rotational motion by an invariant quantity that captures its resistance to rotational motion—the object's rotation inertia (Figure 9.9). The various moments of inertia (i.e., quantities that specify the muscular torque required to hold an object against gravity or to rotate an object) are *relational* quantities defined by the distribution of mass of an object *relative* to the location about which the object is held/rotated (e.g., the wrist joint). For example, a long, narrow rod grasped at its distal end has most of its mass distributed away from the rotation point (the wrist) yielding greater resistance to up/down rotation than a shorter, wide rod of equal mass. Moments of inertia have been implicated in a broad range of dynamic touch perceptual domains including object length

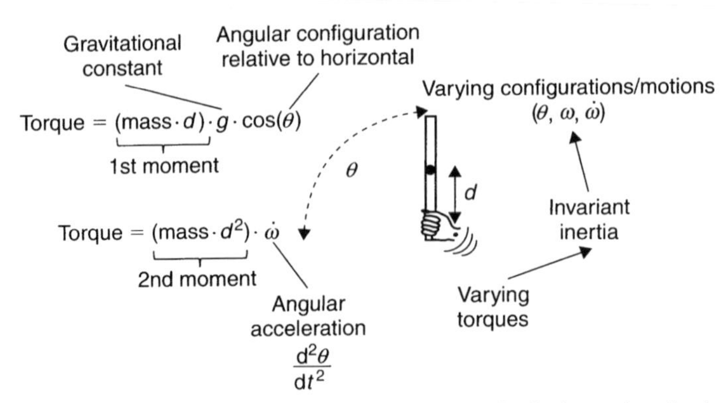

FIGURE 9.9 Relation between torque, moments of mass distribution, and motion in wielding a hand-held object.

(Solomon & Turvey, 1988; van de Langenberg et al., 2006), orientation (Pagano & Turvey, 1992, 1995), width (Wagman et al., 2001), and heaviness (Turvey et al., 1999; Shockley et al., 2001; Kingma et al., 2002), and also including perception of one's own limb orientation (Pagano et al., 1996; Riley & Turvey, 2001).

The ecological position that perception is sensitive to perceiver-scaled (i.e., relational) quantities such as rotational inertia reflects how ontological and epistemological assumptions drive empirical questions and interpretations (see EP VI below). If perception is assumed to involve inferences based on ambiguous proximal stimulation, we are led to a completely different conclusion about perceptual competence than if it is assumed that perception is directly sensitive to action-relevant, relational quantities. For example, a rod will feel differentially heavy depending on where it is grasped (e.g., at the distal end vs. the middle), an apparent illusion. However, this conclusion obtained only if the characterization of the object is perceiver-neutral (i.e., in terms of its mass as weighed on a scale). If, however, a perceiver-scaled quantity (rotational inertia) is the relevant information for perception, then the object *should* feel different depending upon where it is grasped because the mass distribution has changed with respect to the wrist across the two instances. Equally important is the related implication that although we may ask a perceiver to report on physical, perceiver-neutral primitives (e.g., weight), perceptual reports may nevertheless tacitly reflect the perceiver's sensitivity to action-relevant, relational properties (e.g., maneuverability; Turvey et al., 1999; Shockley et al., 2001, 2004).

ECOLOGICAL PRINCIPLE VI: PERCEPTION IS OF AFFORDANCES

It is the assumption that perception is unreliable, even fallible, that leads to a focus on mental representation and unconscious inference in explaining how animals can know their surroundings. Sanctioned by rationalism, knowing and knowledge are thus understood in terms of conceiving rather than perceiving (Turvey & Shaw, 1999). It should come as no surprise, however, that the ecological perspective views the classical distinction between *conception* and *perception* as misguided (Brooks, 1991; Kirsch, 1991; Turvey & Shaw, 1999). Ecologically, *knowing* is viewed as an epistemic relation between an animal, as a knowing agent, and the environment as what is to be known (Gibson, 1979/1986; Turvey & Shaw, 1979; Shaw, 2003). As noted in EP II and EP IV, affordances constitute this epistemic relation. Thus, for the ecological approach, to perceive, fundamentally, is to perceive affordances—opportunities for action.

Counteractive to the traditional view that the meanings that constrain behavior are represented in the mind or brain, affordances reveal meaning to be an objective property of an O–E system. That is, the use of an object or surface—what it affords and what it means for an animal—is a functional relation between animal and environment; affordances are not subjectively imposed by an animal, nor do they exist within the object in isolation from the animal. Consistent with EP V, affordances are perceived by detecting lawfully structured information (see EP V) that invariantly specifies features (capabilities) of a *particular* perceiving–acting agent in relation to features of a *particular* substance, surface, object, or event. A water surface with adequate tension can afford locomotion for an insect but not a human. Similarly, a Frisbee flying through the air affords catching for an animal with the appropriate limbs or mouth in which to catch it; an adult, child, or dog may perceive a successfully thrown Frisbee as catch-able, but an infant, snail, or beetle will not. Thus, animals do not perceive the environment in units of an absolute (perceiver-neutral) metric (e.g., meters), but rather in ecological units of action. The ontological assumption that affordances are the meaningful objects of perception which are specified and can, therefore, be perceived directly, mitigates the reliance on representational–computational structures and concepts (which ascribe meaning) by displacing the problem of meaning from epistemology—how one can know—to ontology—how the world is constituted (Gibson, 1979/1986; Turvey & Shaw, 1979).

There is much empirical support for this sixth ecological principle, with researchers having investigated the perception and informational specification of a wide variety of affordances, including step-across-ability (Cornus et al., 1999) and sit-on-ability of surfaces (Mark et al., 1990), reachability in the horizontal (Carello et al., 1989; Rochat & Wraga, 1997) and vertical planes (Pepping & Li, 1997), pass-through-ability (Warren & Whang, 1987) and pass-under-ability of apertures (White & Shockley, 2005), and stand-on-ability of slopes (Fitzpatrick et al., 1994). In the most well known of these investigations, Warren (1984) not only demonstrated that individuals accurately perceive the boundary between what is step-up-on-able or not, but also that the perception of this boundary is determined by information that specifies an invariant ratio of riser-height to leg-length (Figure 9.10). Subsequent work by Mark (1987) demonstrated that the optical information about step-up-on-ability (and sit-on-ability) of an object or

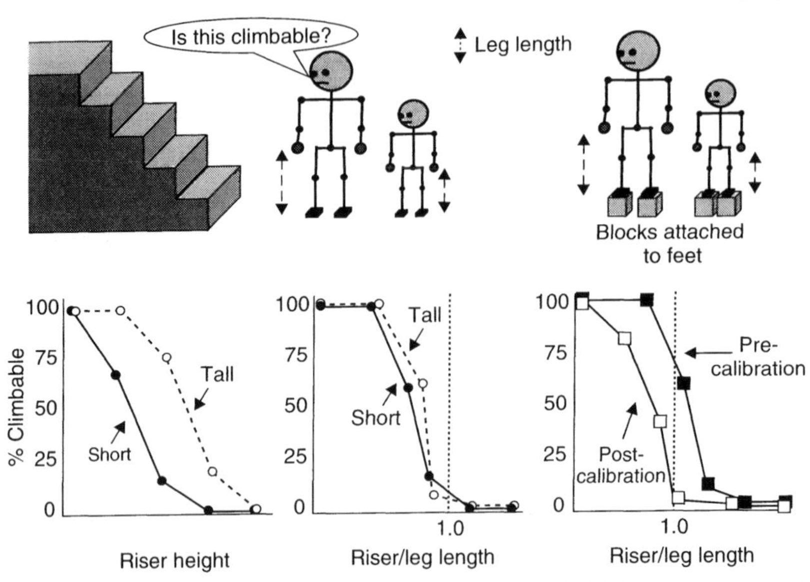

FIGURE 9.10 Perceived stair climb-ability. Short and tall individuals perceive the boundary between step-up-on-able and not step-up-on-able at different riser heights but at the same body-scaled ratio of riser height to leg length (Warren, 1984). Blocks attached to the feet of a participant raise the participant's eye-height and change the information for this affordance, such that step-up-on-ability is overestimated prior to the participant recalibrating to a new block + leg action system (Mark, 1987).

surface is related to a perceiver's effective eye height. Mark manipulated the effective eye height of perceivers by having them strap 10cm blocks to their feet. This manipulation changed the information for this affordance (the relation between object height and effective eye height) but did not change the actual height of the object that was step-up-on-able. As expected, perception corresponded to the optical information, such that participants overestimated the step-up-on-ability of risers prior to recalibrating to their new leg + block height (Figure 9.10).

Affordances, however, are not only a function of the geometric fit of the perceiver to the environment, but also of the action capabilities of the perceiver–actor (Fajen et al., in press). For example, a perceiver's performance on a braking task is a function of *both* the optical information about time to contact *and* the optical information relative to the perceiver's braking capabilities (Fajen, 2005b). Similarly, one's locomotional capabilities constrain the "catchability" of a moving target (Bastin et al., under review; Oudejans et al., 1996a), and the "crossability" of a busy intersection (Oudejans et al., 1996b). With respect to understanding the organization of behavior, the perception of affordances is thus crucial for selecting among different modes of action (Warren, 1988), allowing one to select only those modes for which the goal is afforded, and to abandon

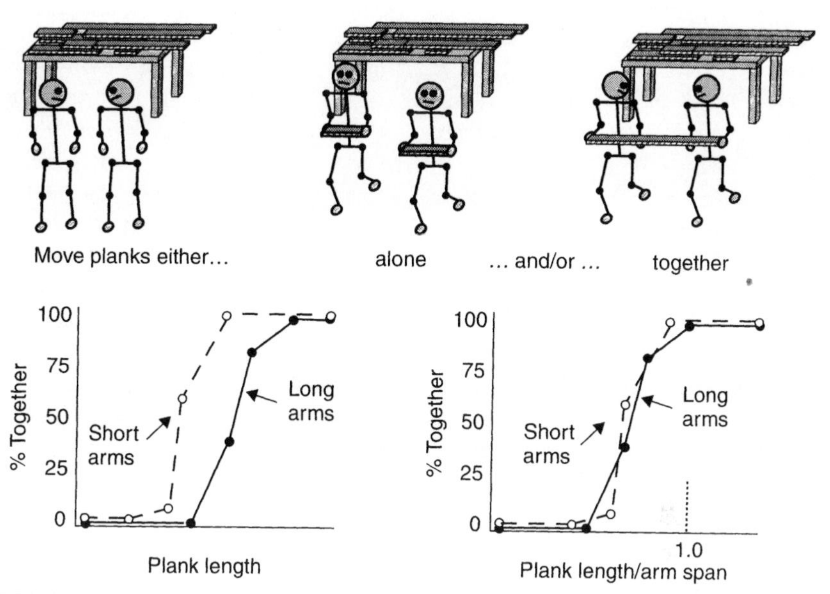

FIGURE 9.11 Interpersonal affordance and emergence of cooperative action. Compared to two people with shorter arms, two people with longer arms switch at a larger plank length from individually lifting planks alone to lifting planks as a pair. The switch is found to occur at a common value, however, when the threshold plank length of the "longer" and the "shorter" people is scaled to arm span (Isenhower et al., 2005; Richardson et al., 2007a). See text for details.

(before it is too late) modes that have no chance of success. Aligned with EP IV, affordance perception continuously guides and constrains action (Turvey, 1992; Stoffregen, 2000; Fajen, 2007), ensuring that a task can be completed within the limits of an animal's action capabilities.

It is worth noting that the perception of affordances is not restricted to the actions possible for oneself and that one can perceive the action possibilities of conspecifics (Rochat, 1995; Stoffregen et al., 1999; Ramenzoni et al., 2008; Ramenzoni et al., in press). Interestingly, however, although perceivers can differentiate among others' action capabilities (e.g., the maximum height a taller person can reach is proportionally higher than that a shorter person can reach), this perception appears to be scaled to the perceiver's own action capabilities. For example, Ramenzoni et al. (in press) demonstrated that when wearing ankle weights, one perceives the maximum reachable height by jumping to be lower for oneself *and* for others, compared to when not wearing ankle weights.

Affordances also exist and are perceived with respect to interpersonal or social action systems (Marsh et al., 2006). As illustrated in Figure 9.11, Richardson and colleagues have demonstrated how the affordances of an interpersonal plank moving task—planks movable alone or together—are determined by the size of a pair's arm span taken with respect to the length of the plank (Isenhower et al., 2005; Richardson et al., 2007b; Fowler et al., in press). Thus, the implicit commitment to act as a "plural subject" of action (Gilbert, 1996) is something that emerged without prior planning in response to a meaningful relation defined across an animal–(animal)–environment system (Richardson et al., in press). Understood in conjunction with the research that demonstrates that individuals can accurately perceive tool-based affordances in which the relevant action system is a functional synergy of body-and-tool (e.g., hit-able with a hammer, reach-able with a stick; grasp-able with an extendable claw; Hirose, 2002; Richardson et al., 2007b; Wagman & Carello, 2003), such research reveals how the boundary between what constitutes "animal" and what constitutes "environment" constantly shifts. Neither strictly "animal," nor strictly "environment," but both, the coordinative structures or perception–action synergies that actualize affordances are emergent properties of an O–E system, whereby that which is knowable, that which holds meaning, only does so in relation to the O–E system and cannot be reduced to any individual part (Marsh et al., 2006; Richardson et al., in press).

CONCLUSION

A truly embodied-embedded approach to behavior promises a radical change in how scientists conceptualize cognitive agents (both biological and non-biological) and how they proceed to understand the behavioral order of such agents, both empirically and theoretically. In our view, cashing in the promissory note requires that perceiving, acting and knowing be studied as emergent properties of an O-E system. The six principles described in the present chapter are proposed as an appropriate framework for that study. Our presumption is that the persistent application of the principles should enable cognitive and psychological science to repay the many loans of intelligence thus far accrued.

REFERENCES

Ashby, R. A. (1952). *Design for a Brain.* London: Chapman and Hall.

Ashby, R. A. (1963). *An Introduction to Cybernetics.* New York: Wiley.

Barker, R. G. (1968). *Ecological Psychology. Concepts and Methods for Studying the Environment of Human Behavior.* Stanford, CA: Stanford University Press.

Bastin, J., Fajen, B. R., & Montagne, G. (under review). Controlling speed and direction during interception: An affordance-based control model. *Journal of Experimental Psychology: Human Perception and Performance.*

Beer, R. D. (1995). A dynamical systems perspective on agent–environment interaction. *Artificial Intelligence, 72,* 173–215.

Bentley, A. F. (1941). The human skin: Philosophy's last line of defense. *Philosophy of Science, 8,* 1–19.

Bingham, G. P. (2004). A perceptually driven dynamical model of bimanual rhythmic movements (and phase entrainment). *Ecological Psychology, 16,* 45–53.

Brooks, R. A. (1991). New approaches to robotics. *Science, 253,* 1227–1232.

Carello, C., Grosofsky, A., Reichel, F. D., Solomon, H. Y., & Turvey, M. T. (1989). Visually perceiving what is reachable. *Ecological Psychology, 1,* 27–54.

Chardenon, A., Montagne, G., Buekers, M. J., & Laurent, M. (2002). The visual control of ball interception during human locomotion. *Neuroscience Letters, 334,* 13–16.
Chemero, A. & Turvey, M. T. (2007). Hypersets, complexity, and the ecological approach to perception–action. *Biological Theory, 2,* 23–36.
Clancey, W. J. (1997). *Situated Cognition: On Human Knowledge and Computer Representations.* New York: Cambridge University Press.
Clark, A. (1997). The dynamical challenge. *Cognitive Science, 21,* 461–481.
Cornus, S., Montagne, G., & Laurent, M. (1999). Perception of a stepping-across affordance. *Ecological Psychology, 11,* 249–267.
Dennett, D. I. (1978). *Brainstorms: Philosophical Essays on Mind and Psychology.* Montgomery, VT: Bradford Books.
Dewey, J. & Bentley, A. F. (1949). *Knowing and the Known.* Boston, MA: The Beacon Press.
Fajen, B. R. (2005a). Calibration, information, and control strategies for braking to avoid a collision. *Journal of Experimental Psychology: Human Perception and Performance, 31,* 480–501.
Fajen, B. R. (2005b). The scaling of information to action in visually guided braking. *Journal of Experimental Psychology: Human Perception and Performance, 31,* 1107–1123.
Fajen, B. R. (2007). Affordance-based control of visually guided action. *Ecological Psychology, 19,* 383–410.
Fajen, B. R. & Warren, W. H. (2004). Visual guidance of intercepting a moving target on foot. *Perception, 33,* 675–689.
Fajen, B. R., Riley, M. R., & Turvey, M. T. (in press). Information, affordances, and the control of action in sport. *International Journal of Sports Psychology.*
Fitzpatrick, P., Carello, C., Schmidt, R. C., & Corey, D. (1994). Haptic and visual perception of an affordance for upright posture. *Ecological Psychology, 6,* 265–287.
Fodor, J. (2000). *The Mind Doesn't Work That Way: The Scope and Limits of Computational Psychology.* Cambridge, MA: MIT Press.
Fowler, C. A., Richardson, M. J., Marsh, K. L., & Shockley, K. D. (2008). Language use, coordination, and the emergence of cooperative action. In A. Fuchs & V. Jirsa (Eds.), *Coordination: Neural, Behavioral and Social Dynamics* (pp. 261–280). Heidelberg: Springer-Verlag.
Gibbs, R. W. (2006). *Embodiment and Cognitive Science.* Cambridge: Cambridge University Press.
Gibson, J. J. (1966). *The Senses Considered as Perceptual Systems.* Boston, MA: Houghton Mifflin.
Gibson, J. J. (1979/1986). *The Ecological Approach to Visual Perception.* Boston, MA: Houghton Mifflin.
Gilbert, M. (1996). *Living together: Rationality, sociality, and obligation.* Lanham, MD: Rowman & Littlefield.
Griffiths, P. E. & Gray, R. D. (2001). Darwinism and developmental systems. In S. Oyama, P. E. Griffiths, & R. D. Gray (Eds.), *Cycles of Contingency: Developmental Systems and Evolution* (pp. 195–218). Cambridge, MA: MIT Press.
Haken, H., Kelso, J. A. S., & Bunz, H. (1985). A theoretical model of phase transitions in human hand movements. *Biological Cybernetics, 51,* 347–356.
Harrison, S. (2007). *Orienting to Place and Orienting to Distance Are Distinct Functions of Human Locomotion.* Unpublished Ph.D. Dissertation, University of Connecticut, Storrs, CT, USA.
Haugeland, J. (1998). *Having Thought.* Cambridge, MA: Harvard University Press.
Hecht, H., & Savelsbergh, G. (Eds.), (2004). *Time-to-contact.* Amsterdam: Elsevier.
Helmholtz, H.v. (1867/1925). *Treatise on Physiological Optics.* Rochester, NY: Optical Society of America.
Hirose, N. (2002). An ecological approach to embodiment and cognition. *Cognitive Systems Research, 3,* 289–299.
Hommel, B., Müsseler, J., Aschersleben, G., & Prinz, W. (2001). The theory of event coding (TEC): A framework for perception and action planning. *Behavioral and Brain Sciences, 24,* 849–937.
Humphrey, G. (1933). *The Nature of Learning.* New York: Harcourt Brace.

Huntington, E. V. (1913). A set of postulates for abstract geometry, expressed in terms of the simple relation of inclusion. *Mathematische Annalen, 73,* 522–559.

Hurley, S. L. (1998). *Consciousness in Action.* Cambridge, MA: Harvard University Press.

Hutchins, E. (1995). *Cognition in the Wild.* Cambridge, MA: MIT Press.

Isenhower, R. W., Marsh, K. L., Carello, C., Baron, R. M., & Richardson, M. J. (2005). The Specificity of Intrapersonal and Interpersonal Affordance Boundaries: Intrinsic Versus Absolute Metrics. In H. Heft & K. L. Marsh (Eds.), *Studies in perception and action VIII* (pp. 54–58). Mahwah, NJ: Lawrence Erlbaum Associates.

Ittelson, W. H. (1968). *The Ames Demonstrations in Perception.* New York: Hafner.

Järvilehto, T. (1998a). The theory of the organism–environment system: I Description of the theory. *Integrative Physiological and Behavioral Science, 33,* 321–334.

Järvilehto, T. (1998b). The theory of the organism–environment system: II. Significance of nervous system activity in the organism–environment system. *Integrative Physiological and Behavioral Science, 33,* 335–342.

Järvilehto, T. (1999). The theory of the organism–environment system: III. Role of efferent influences on receptors in the formation of knowledge. *Integrative Physiological and Behavioral Science, 34,* 90–100.

Järvilehto, T. (2000). The theory of the organism–environment system: IV. The problem of mental activity and consciousness. *Integrative Physiological and Behavioral Science, 35,* 35–57.

Kelso, J. A. S. (1995). *Dynamic patterns: The self-organization of brain and behavior.* Cambridge, MA: MIT Press.

Kingma, I., Beek, P., & van Dieen, J. H. (2002). The inertia tensor versus static moment and mass in perceiving length and heaviness of hand-wielded rods. *Journal of Experimental Psychology: Human Perception and Performance, 28,* 180–191.

Kirsch, D. (1991). Foundations of AI: The big issues. *Artificial Intelligence, 47,* 3–30.

Kugler, P. N., Kelso, J. A. S., & Turvey, M. T. (1980). On the concept of coordinative structures as dissipative structures: I. Theoretical lines of convergence. In G. E. Stelmach & J. Requin (Eds.), *Tutorials in Motor Behavior* (pp. 3–47). New York: North-Holland.

Kugler, P. N. & Turvey, M. T. (1987). *Information, Natural Law, and the Self-assembly of Rhythmic Movement.* Hillsdale, NJ: Lawrence Erlbaum Associates.

Lakoff, G. & Johnson, M. (1999). *Philosophy in the Flesh: The Embodied Mind and Its Challenge to Western Thought.* New York: Basic Books.

Lee, D. N. (1976). A theory of visual control of braking based on information about time-to-collision. *Perception, 5,* 437–459.

Mark, L. S. (1987). Eye-height scaled information about affordances: A study of sitting and stair climbing. *Journal of Experimental Psychology: Human Perception and Performance, 13,* 360–370.

Mark, L. S., Balliett, J. A., Craver, K. D., Douglas, S. D., & Fox, T. (1990). What an actor must do in order to perceive the affordance for sitting. *Ecological Psychology, 2,* 325–366.

Marsh, K. L., Richardson, M. J., Baron, R. M., & Schmidt, R. C. (2006). Contrasting approaches to perceiving and acting with others. *Ecological Psychology, 18,* 1–37.

McLeod, P., Reed, N., & Dienes, Z. (2006). The generalized optic acceleration cancellation theory of catching. *Journal of Experimental Psychology–Human Perception and Performance, 32,* 139–148.

Meng, J. C. & Sedgwick, H. A. (2001). Distance perception mediated through nested contact relations among surfaces. *Perception and Psychophysics, 63,* 1–15.

Michaels, C. F. (2003). Affordances: Four points of debate. *Ecological Psychology, 15,* 135–148.

Michaels, C. F. & Carello, C. (1981). *Direct Perception.* Englewood Cliffs, NJ: Prentice Hall.

Miller, D. B. (1997). The effects of nonobvious forms of experience on the development of instinctive behavior. In C. Dent-Read & P. Zukow-Goldring (Eds.), *Evolving Explanations of Development: Ecological Approaches to Organism-Environment Systems* (pp. 457–507). Washington, DC: American Psychological Association.

Müller, J. (1826). *Zur vergleichenden Physiologie des Gesichtssinnes des Menschen und der Thiere.* London: Cnobloch.

Oudejans, R. R. D., Michaels, C. F., Bakker, F. C., & Dolne, M. A. (1996a). The relevance of action in perceiving affordances: Perception of catchableness of fly balls. *Journal of Experimental Psychology-Human Perception and Performance, 22,* 879–891.

Oudejans, R. R. D., Michaels, C. F., van Dort, B., & Frissen, E. J. P. (1996b). To cross or not to cross: The effect of locomotion on street-crossing behavior. *Ecological Psychology, 8,* 259–267.

Oyama, S., Griffiths, P. E., & Gray, R. D. (2001). *Cycles of contingency: Developmental Systems and Evolution.* Cambridge, MA: MIT Press.

Pagano, C. C. & Turvey, M. T. (1992). Eigenvectors of the inertia tensor and perceiving the orientation of a hand-held object by dynamic touch. *Perception and Psychophysics, 52,* 617–624.

Pagano, C. C. & Turvey, M. T. (1995). The inertia tensor as a basis for the perception of limb orientation. *Journal of Experimental Psychology: Human Perception and Performance, 21,* 1070–1087.

Pagano, C. C., Garrett, S., & Turvey, M. T. (1996). Is limb proprioception a function of the limbs' inertial eigenvectors. *Ecological Psychology, 8,* 43–69.

Pepping, G. J. & Li, F. X. (1997). Perceiving action boundaries in the volleyball block. In S. M. A. Schmuckler & J. M. Kennedy (Eds.), *Studies in perception and action IV* (pp. 137–140). Mahwah, NJ: Lawrence Erlbaum Associates.

Pfeifer, R. & Bongard, J. (2006). *How the Body Shapes the Way We Think: A New View of Intelligence.* Boston, MA: MIT Press.

Pfeifer, R. & Lida, F. (2005). New Robotics: Design Principles for Intelligent Systems. *Artificial Life, 11,* 99–120.

Prinz, W. (1997). Perception and action planning. *European Journal of Cognitive Psychology, 9,* 129–154.

Ramenzoni, V. C., Riley, M. A., Shockley, K., & Davis, T. (2008). An information-based approach to action understanding. *Cognition, 106,* 1059–1070.

Ramenzoni, V. C., Riley, M. A., Davis, T., Shockley, K., & Armstrong, R. (in press). Carrying the height of the world on your ankles: Encumbering observers reduces estimates of how high another actor can jump. *Quarterly Journal of Experimental Psychology.*

Reed, E. S. (1988). *James J. Gibson and the Psychology of Perception.* New Haven, CT: Yale University Press.

Reed, E. S. (1996). *Encountering the World: Toward an Ecological Psychology.* New York: Oxford University Press.

Reed, E. S. & Jones, R. (1982). *Reasons for Realism: Selected Essays of James J. Gibson.* Hillsdale, NJ: Lawrence Erlbaum and Associates.

Richardson, M. J., Marsh, K. L., Isenhower, R., Goodman, J., & Schmidt, R. C. (2007a). Rocking together: Dynamics of intentional and unintentional interpersonal coordination. *Human Movement Science, 26,* 867–891.

Richardson, M. J., Marsh, K. L., & Baron, R. M. (2007b). Judging and actualizing intrapersonal and interpersonal affordances. *Journal of Experimental Psychology: Human Perception and Performance, 33,* 845–859.

Richardson, M. J., Marsh, K. L., & Schmidt, R. C. (in press). Challenging the egocentric view of coordinated perceiving, acting and knowing. In L. F. Barrett, B. Mesquita, & E. Smith (Eds.), *Mind in Context.* Guilford.

Riley, M. A. & Turvey, M. T. (2001). Inertial constraints on limb proprioception are independent of visual calibration. *Journal of Experimental Psychology: Human Perception and Performance, 27,* 438–455.

Rochat, P. (1995). Perceived reachability for self and for others by 3- to 5-year old children and adults. *Journal of Experimental Child Psychology, 59,* 317–333.

Rochat, P. & Wraga, M. (1997). An account of the systematic error in judging what is reachable. *Journal of Experimental Psychology: Human Perception and Performance, 23,* 199–212.

Runeson, S. (1988). The distorted room illusion, equivalent configurations, and the specificity of static optic arrays. *Journal of Experimental Psychology–Human Perception and Performance, 14,* 295–304.

Schmidt, R. C. (2007). Scaffolds for social meaning. *Ecological Psychology, 19,* 137–151.

Schmidt, R. C. & O'Brien, B. (1997). Evaluating the dynamics of unintended interpersonal coordination. *Ecological Psychology, 9*(3), 189–206.

Schmidt, R. C. & Richardson, M. J. (2008). Dynamics of interpersonal coordination. In A. Fuchs & V. Jirsa (Eds.), *Coordination: Neural, Behavioral and Social Dynamics* (pp. 281–308). Heidelberg: Springer-Verlag.

Schmidt, R. C., Carello, C., & Turvey, M. T. (1990). Phase transitions and critical fluctuations in the visual coordination of rhythmic movements between people. *Journal of Experimental Psychology: Human Perception and Performance, 16*(2), 227–247.

Schmidt, R. C., Richardson, M. J., Christine, A., & Galantucci, B. (2007). Visual tracking and entrainment to an environmental rhythm. *Journal of Experimental Psychology: Human Perception and Performance, 33,* 860–870.

Schoggen, P. (1989). *Behavior Settings. A Revision and Extension of Roger G. Barker's Ecological Psychology*. Stanford: Stanford University Press.

Searle, J. R. (1980). Minds, brains, and programs. *Behavioral and Brain Sciences, 3,* 417–457.

Shaw, R. E. (2003). The agent–environment interface: Simon's indirect or Gibson's direct coupling? *Ecological Psychology, 15,* 37–106.

Shaw, R. E. & Kinsella-Shaw, J. M. (1988). Ecological mechanics: A physical geometry for intentional constraints. *Human Movement Science, 7,* 155–200.

Shaw, R. E. & Turvey, M. T. (1980). Methodological realism. *The Behavioral and Brain Sciences, 3,* 94–96.

Shaw, R. E. & Turvey, M. T. (1981). Coalitions as models of ecosystems: A realist perspective on perceptual organization. In M. Kubovy & J. Pomeranz (Eds.), *Perceptual Organization* (pp. 343–415). Hillsdale, NJ: Erlbaum.

Shockley, K., Grocki, M., Carello, C., & Turvey, M. T. (2001). Somatosensory attunement to the rigid body laws. *Experimental Brain Research, 136,* 133–137.

Shockley, K., Carello, C., & Turvey, M. T. (2004). Metamers in the haptic perception of heaviness and moveableness. *Perception and Psychophysics, 66,* 731–742.

Smith, B. (2001). Objects and their environments: From Aristotle to ecological ontology. In A. Frank, J. Raper, & J.-P. Cheylan, (Eds.), *The Life and Motion of Socio-Economic Units* (pp. 79–97). London: Taylor & Francis.

Solomon, H. Y. & Turvey, M. T. (1988). Haptically perceiving the distances reachable with hand-held objects. *Journal of Experimental Psychology: Human Perception and Performance, 14,* 404–427.

Stoffregen, T. A. (2000). Affordances and events: Theory and research. *Ecological Psychology, 12,* 93–107.

Stoffregen, T. A., Gorday, K. M., Sheng, Y.-Y., & Flynn, S. B. (1999). Perceiving affordances for another person's actions. *Journal of Experimental Psychology: Human Perception and Performance, 25,* 120–136.

Thelen, E. & Smith, L. B. (1994). *A Dynamic Systems Approach to the Development of Cognition and Action*. Cambridge, MA: MIT Press.

Turvey, M. T. (1990). Coordination. *American Psychologist, 45,* 938–953.

Turvey, M. T. (1992). Affordances and prospective control: An outline of the ontology. *Ecological Psychology, 4,* 173–187.

Turvey, M. T. (2004). Impredicativity, dynamics, and the perception–action divide. In V. K. Jirsa & J. A. S. Kelso (Eds.), *Coordination Dynamics: Issues and Trends. Vol.1: Applied Complex Systems* (pp. 1–20). New York: Springer-Verlag.

Turvey, M. T. (2005). Theory of brain and behavior in the 21st century: No ghost, no machine. *Japanese Journal of Ecological Psychology, 2,* 69–79.

Turvey, M. T. & Carello, C. (1986). The ecological approach to perceiving–acting: A pictorial essay. *Acta Psychologica, 63,* 133–155.

Turvey, M. T. & Fonseca, S. (2008). Nature of motor control: Perspectives and issues. In D. Sternad (Ed.), *Progress in Motor Control: A Multidisciplinary Perspective* (pp. 93–123). New York: Springer-Verlag.

Turvey, M. T. & Shaw, R. E. (1979). The primacy of perceiving: An ecological reformulation of perception for understanding memory. In L. G. Nilsson (Ed.), *Perspectives on Memory Research: Essays in Honor of Uppsala University's 500th Anniversary* (pp. 167–222). Hillsdale, NJ: Lawrence Erlbaum Associates, Inc.

Turvey, M. T. & Shaw, R. E. (1995). Toward an ecological physics and a physical psychology. In R. L. Solso & D. W. Massaro (Eds.), *The science of the mind: 2001 and beyond.* New York, NY: Oxford.

Turvey, M. T. & Shaw, R. E. (1999). Ecological foundations of cognition: I. Symmetry and specificity of animal–environment systems. *Journal of Consciousness Studies, 6,* 95–110.

Turvey, M. T., Shaw, R., & Mace, W. M. (1978). Issues in the theory of action: Degrees of freedom, coordinative structures and coalitions. In J. Requin (Ed.), *Attention and Performance VII* (pp. 557–595). Hillsdale, NJ: Lawrence Erlbaum Associates, Inc.

Turvey, M. T., Shaw, R. E., Reed, E. S., & Mace, W. M. (1981). Ecological laws of perceiving and acting: In reply to Fodor and Pylyshyn. *Cognition, 9*(3), 237–304.

Turvey, M. T., Fitch, H. L., & Tuller, B. (1982). The Bernstein perspective: I. The problems of degrees of freedom and context-conditioned variability. In J. A. S. Kelso (Ed.), *Human Motor Behavior: An Introduction* (pp. 239–252). Hillsdale, NJ: Erlbaum.

Turvey, M. T., Shockley, K., & Carello, C. (1999). Affordance, proper function, and the physical basis of perceived heaviness. *Cognition, 73,* 1317–1326.

Van de Langenberg, R., Kingma, I., & Beek, P. (2006). Mechanical invariants are implicated in dynamic touch as a function of their salience in the stimulus flow. *Journal of Experimental Psychology: Human Perception and Performance, 32,* 1093–1106.

von Eckardt, B. (1993). *What is Cognitive Science?* Cambridge, MA: The MIT Press.

von Holst, E. (1973). The collected papers of Eric von Holst. Vol. 1: The behavioral physiology of animal and man. In R. Martin (Ed.), *The Collected Papers of Eric von Holst* (Vol. 1). Coral Gables, FL: University of Miami Press.

Wagman, J. B. & Carello, C. (2003). Haptically creating affordances: The user–tool interface. *Journal of Experimental Psychology: Applied, 9*(3), 175–186.

Wagman, J. B., Shockley, K., Riley, M. A., & Turvey, M. T. (2001). Attunement, calibration, and exploration in fast haptic perceptual learning. *Journal of Motor Behavior, 33,* 323–327.

Warren, W. H. (1984). Perceiving affordances: Visual guidance of stair climbing. *Journal of Experimental Psychology: Human Perception and Performance, 10,* 683–703.

Warren, W. H. (1988). Action modes and laws of control for the visual guidance of action. In O. G. Meijer & K. Roth (Eds.), *Movement Behavior: The Motor-Action Controversy.* Amsterdam: North Holland.

Warren, W. H. (1998). Visually controlled locomotion: 40 years later. *Ecological Psychology, 10,* 177–220.

Warren, W. H. (2004). Optic flow. In L. Chalupa & J. Werner (Eds.), *The Visual Neurosciences* (Vol. 2, pp. 315–358). Cambridge, MA: MIT Press.

Warren, W. (2006). The dynamics of perception and action. *Psychological Review, 113,* 358–389.

Warren, W. H. & Whang, S. (1987). Visual guidance of walking through apertures: Body-scaled information for affordances. *Journal of Experimental Psychology: Human Perception and Performance, 13,* 371–383.

Warren, W. H. Jr., Kay, B. A., Zosh, W. D., Duchon, A. P., & Sahuc, S. (2001). Optic flow is used to control human walking. *Nature Neuroscience, 4,* 213–216.

White, E. & Shockley, K. (2005). Perceiving pass-under-ability while walking and sitting. Paper presented at the *13th International Conference on Perception and Action,* Monterey, CA.

Wilkie, R. & Wann, J. (2003). Controlling steering and judging heading: Retinal flow, visual direction, and extraretinal information. *Journal of Experimental Psychology: Human Perception and Performance, 29,* 363–378.

Wilson, M. (2002). Six views of embodied cognition. *Psychonomic Bulletin and Review, 9,* 625–636.

10

Seeing What We Can Do: Insights into Vision and Action through Observations of Natural Behavior

Jason A. Droll[1] and Mary M. Hayhoe[2]

[1]*Department of Psychology, University of California, Santa Barbara, CA, USA*
[2]*Department of Psychology, University of Texas, Austin, TX, USA*

INTRODUCTION

While cognitive processes are often considered strictly internal operations, it is clear that these processes are reflected through behavior. Shifts in the direction of gaze indicate shifts in attention, a reach indicates the selection of an object of interest, and brief hesitations may reveal uncertainty as to what action to perform next. Yet the intuition that overt behavior reveals covert cognitive operations during natural behavior has not been extensively explored. The purpose of this chapter is to argue that observations of natural behavior allow us to make inferences about general principles of cognitive processes. We first describe the ways in which traditional laboratory experiments differ from everyday visually guided behavior, and then we move on to discuss recent experiments that analyze the sequence of cognitive operations involved in a simple sorting task. Finally, we offer suggestions on how to reconcile traditional approaches toward the study of vision to understand the principles of embodied cognition.

METHODS OF ASSESSING VISUAL PROCESSES IN ISOLATION AND IN CONCERT

Visual scenes in everyday environments are composed of a variety of objects of different shapes, sizes, and colors, and in each environment, an observer may have a variety of behavioral goals that can be executed through several possible action sequences. Although it is reasonable to approach this complexity by a separate analysis of the components of the image and the behavior, as has been done traditionally, there is also something to be gained by observing integrated behavioral sequences orchestrated by an active observer. In particular, it appears that cognitive mechanisms such as working memory and attention may function differently in the context of ongoing behavior than when isolated in a particular experimental context.

TRADITIONAL EXPERIMENTAL PARADIGMS VERSUS NATURAL BEHAVIOR

Traditional Paradigms

In a typical experiment studying eye movements or visual attention, subjects may be asked to report the presence or location of a target surrounded by a field of distractors presented on a computer monitor. The "objects" are usually artificial, composed of simple colors and shapes or letters, briefly presented in strict temporal order, under the control of the experimenter. Studies in working memory typically employ a similar trial structure, where an array of objects is briefly flashed, and subjects are asked to report if a subsequently presented array is the exact same as the first, or if it differs. Subjects are also told exactly what information is relevant at each stage of the task, and what actions or operations to execute at each step (e.g., perceptual encoding, storage of information during the inter-stimulus interval, and final button press). Reaction time, measures of percent correct, and the pattern of eye movements (if allowed) across hundreds of trials are used to make inferences about a particular cognitive process, or about perceptual thresholds. These measures usually result in estimates of processing or storage capacity, reflecting the limits of human performance.

Natural Behavior

Our everyday use of vision is markedly different from what is required in most laboratory experiments in several ways: the visual stimulus and task context is more complex, movement planning is required, and the cognitive subcomponents must be composed to generate a coordinated behavioral sequence.

Visual Scenes

Natural visual contexts differ from traditional experiments with respect to temporal structure, content, and scale. Rather than brief stimulus presentations, natural behavior allows observers to be situated within the same visual environment for extended time periods, where they are free to attend or fixate any of several complex objects. The continued presence of visual information allows for observers to accumulate information over time as dictated by the internal cognitive agenda. This may explain why measures of working memory capacity estimate storage of about four objects (Luck & Vogel, 1997; Vogel et al., 2001; Wheeler & Treisman, 2002), whereas in natural behavior, observers appear to store far less than this (Ballard et al., 1995, 1997). It may be more convenient to access this information by simply fixating the relevant information the moment it is needed for the task.

The content of visual scenes is also markedly different in ordinary behavior. Laboratory experiments often tailor their stimuli and present items in isolation, or with a small number of elements, each of which are simple geometric shapes, contrast increments, or gratings designed to elicit responses in a particular brain area (e.g., colored gratings for V4) or perceptual channel. Natural environments, of course, contain many complex objects, with a variety of shapes, sizes, and features. This complexity introduces challenges in perceptual encoding due to effects of crowding, object segmentation, and depth cues. Depth cues, in particular, introduce an additional level of spatial complexity and pose a greater challenge for the visuo-motor apparatus for maintaining coordination. Artificial displays also differ from normal scenes not just in their complexity but also in their spatial structure and scale. For example, the visual angle subtended by an image of a room in a typical experimental display is much smaller than being in a real room and it is not clear how such infidelities in spatial scale might affect observers' representations of the spatial structure of the scene. Perhaps, most importantly, in natural behavior the visual input varies contingent on the observer's actions. A complex image sequence is generated on the retina in response to the observer's movements and it is likely that this influences cognitive operations in profound ways.

Task Context

A second aspect of natural behavior that differs from conventional paradigms is the task context. Traditional experiments often require observers to use cognitive operations in ways that may not be optimal for the visual system. For example, measures of covert attention are often assessed in tasks requiring observers to maintain fixation in the center of a computer monitor while having to make decisions regarding stimuli in the periphery. Maintaining fixation, and withholding a saccade, can be difficult and requires practice; perhaps because this is a task that the visual system was not designed to perform. A more natural way of accomplishing this task is not only to use covert attention for peripheral detection alone, but also to allow gaze to be directed the region of interest. Some stimuli such as optic flow require global analysis across the entire retina, but spatially restricted stimuli are almost always fixated in natural behavior (Land, 2004; Hayhoe & Ballard, 2005).

Experiments in working memory require the comparison of sequentially presented images, where it is not clear to the observer which information is critical for selecting and storing internally, and all stimuli in the image is potentially relevant. During ordinary behavior, not only does the visual scene remain visible, allowing for re-inspection if necessary, but also the observer defines the long-term goal of the action sequence or task, and thus knows exactly what information is relevant and necessary to select and maintain in memory. Circumscribing the goals of the task, thus, reduces the load imposed on the observer. How observers choose to define what information is relevant is difficult to study in traditional experimental paradigms, because the timing of computations, and the relevance of the information, is controlled by the fixed trial structure of the experiment.

How can task structure be studied without explicitly defining which visual information is relevant at each stage of a task? If the experimenter establishes the long-term goal of the subject (e.g., to assemble a peanut-butter and jelly sandwich), then the subject's decision on how to structure the task as a sequence of behaviors and cognitive operations becomes clear when examining their actions. Although an incomplete measure, eye movements are an overt manifestation of the momentary deployment of attention in a scene. Through the sequence of eye movements, the task structure is evident. Across a variety of extended visuo-motor tasks. such as driving, walking, sports, playing a piano, hand-washing, and making tea or sandwiches, the central finding is that fixations are tightly linked to the performance of the task (Land et al., 1999; Land & Hayhoe, 2001; Pelz & Canosa, 2001; Pelz et al., 2001; Turano et al., 2001, 2003; Hayhoe et al., 2003; Turano et al., 2003; Land, 2004). Subjects exhibit regular, often quite stereotyped fixation sequences as they perform sequential operations. Very few irrelevant areas are fixated. Figure 10.1 shows an example of the clustering of fixations on task-specific regions when a subject makes a sandwich. A feature of the relationship of the fixations to the task is that they are tightly linked, in time, to the actions. This is hard to capture in a still image but can be clearly appreciated in video sequences such as those in Hayhoe et al. (2003).

This aspect of natural behavior, where observers acquire the specific information they need just at the point it is required in the task, is called a "just-in-time" strategy (Ballard et al., 1995, 1997). In the Ballard et al. (1995) experiment, subjects copied a pattern of colored blocks (the Model) using pieces in a Resource area, which they picked up and placed in the area where the copy was being made (Figure 10.1). When subjects copied a particular block, they typically fixated a block in the Model, then looked at a block of the same color in the Resource area while they picked it up, then looked back at the Model block, presumably to get information about location for placement, and then finally to the copy area where the block was placed in the appropriate location. Thus subjects appeared not to memorize the relatively simple model patterns, but simply to fixate individual blocks to get the information they need at that moment.

Note the critical need for establishing clearly defined goals when observing active behavior. For example, in experiments where subjects simply view images passively, the experimenter has no access to what the observer is doing. Observers may be engaged in object recognition, remembering object locations and identity, or performing some other visual operation. When viewing images of scenes, some regularities in fixation patterns can be explained by image properties such as contrast or chromatic salience. However these factors usually account for only a modest proportion of the variance (Itti & Koch, 2000, 2001; Parkhurst & Niebur, 2003, 2004). In natural behavior where the task is well defined, the demands of the task appear

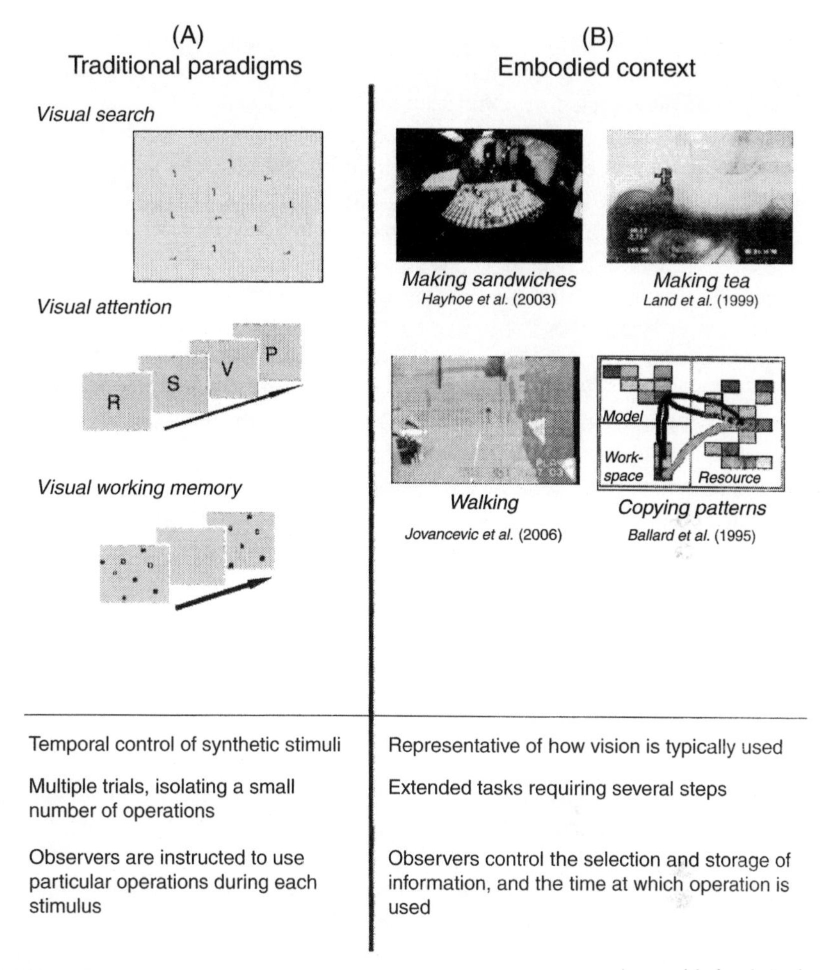

FIGURE 10.1 Vision is either studied using (A) traditional paradigms with few behavioral requirements and the computations are controlled by the trial structure, or (B) observed in the context of embodied behavior, where observers control the timing of when visual information is acquired and what actions are performed. (See color plate)

to be the overwhelming influence on gaze control. Only a small percentage of the observed fixations can be attributed to task-irrelevant locations. Thus, if the behavioral goals are clearly defined, even such as making a peanut-butter and jelly sandwich, then an observer's actions can be reasonably assumed to reflect the internal cognitive representation of the task.

Movement Planning

A third complexity introduced by ordinary behavior is the planning of movements required to obtain and manipulate relevant information. Although the continued presence of the visual scene may lessen the need for storing certain visual details, the need to plan eye and hand movements may increase the demand for maintaining information on spatial position to allow for accurate targeting. Motor planning must compete with visual operations for time and cognitive resources. Analysis of eye and reaching movements in sandwich-making revealed several features suggesting that planning for both eye and hand movements is frequently based on information acquired during a few fixations done previously (Hayhoe et al., 2003). In a more controlled task in a three-dimensional virtual environment, Aivar et al. (2005) found evidence that prior fixations facilitate saccadic target selection for locating the component pieces used in building a model. It is known that accurate saccades can be made on the basis of memory for stimulus location when the original stimulus is no longer present (Gnadt et al., 1991; Hayhoe et al., 1991; Colby, 1998), and the remembered target paradigm is frequently used in physiological investigations of movement planning. These studies show that visual memory information is also used in normal viewing, even when the target is often continuously present in the peripheral retina, and can be located on the basis of stimulus features. It is not immediately obvious that spatial memory would be useful in target selection in this case, but its usefulness becomes apparent when the need for motor planning is taken into account. In the case of reaching movements, the slower velocity of the arm and head relative to the eye makes early initiation of the movements particularly useful.

Composition of Behavioral Sequences

Fourth and most critical to embodied cognition, active agents in natural environments are not afforded the luxury of being told exactly what task to perform and when to execute each decision. This uncertainty is in contrast to traditional experiments that include a fixed sequence of stimuli, requiring a final response. This temporal structure is determined by the experimenter. In natural behavior, the onus of organization is on the observer, having to choose which cognitive operation to use at each moment. This includes deciding where to look, what to attend during each fixation, and what information to store across successive eye movements. How this organization is determined is hard to study in traditional paradigms, but can be more easily addressed in the context of natural behavior, where the sequence and timing of operations can be observed and indirectly manipulated. This then becomes an aspect of the data that we can attempt to account for. This is a particularly interesting question in dynamic environments (Jovancevic et al., 2006).

ISOLATING VISUAL PROCESSES WITHIN AN EMBODIED CONTEXT

Given the above discussion on the complexities of natural behavior in embodied contexts, how can researchers maintain experimental control? In this section, we discuss some recent experiments that demonstrate how naturalistic tasks with flexible structure can serve as controlled experiments and allow inferences about the operation of normal cognitive processes.

Building upon earlier experiments by Triesch et al. (2003), Droll et al. (2005), and Droll and Hayhoe (2007) performed a series of experiments in which subjects sorted bricks in a virtual environment that provided both a virtual visual environment delivered by a head mounted display and a virtual haptic environment, via a force-feedback device (Figure 10.2). The basic task was to select one brick from an array and to sort this brick onto one of two conveyor belts. The bricks were defined by several features (color, height, width, and texture), and a pick-up cue indicated which feature value was relevant for a particular trial. After picking up the brick, a put-down cue was displayed to guide the sorting decision. The brick was placed on the appropriate conveyor belt, removing the brick from the scene, and initiating a new trial with a new pick-up cue and array of bricks. Thus, because the put-down cue was presented after pick-up, the put-down decision was separated in time and space from pick-up, and the representations of the relevant object feature could be stored until the put-down decision was made. In the *One-Feature* task, subjects performed a task in which only one feature dimension was relevant for both pick-up and put-down (e.g., color). In another condition, the *Two-Feature* task, different features were used for pick-up and put-down (e.g., color for pick-up, height for put-down). The task sequence is illustrated in Figure 10.2.

This experiment also used the strategy of changing a feature of the brick that the subject was carrying on a small proportion of trials. Subjects indicated whether they saw a change by placing the brick in a "trash can" (the black square in Figure 10.2B). The purpose of this manipulation was to assess whether subjects were storing features of the brick internally after pick-up, or whether they would not store this information and thus be insensitive to the change. This manipulation also allowed a second more subtle analysis, determining the time at which visual information is acquired by comparing the sequence of eye movements to behavioral decisions requiring visual information either stored internally or present in the scene.

ANALYZING NATURAL BEHAVIOR DURING SORTING

Change Detection

Following a change to a feature in the brick being carried, subjects were unlikely to notice this change. Although the rates of change detection were quite low, they also depended on the relevance of the changed feature (approximately 47% across all task conditions). Subjects were about twice as likely to notice the feature change when that feature was relevant to the task as when it was irrelevant (either for pick-up, or put-down, or both). This supports the hypothesis that subjects preferentially represent the task-specific features of the objects. Objects are not necessarily stored in working memory as bound entities, as suggested by measures of capacity from traditional paradigms (Luck & Vogel, 1997). Thus, observers appear to use working memory much less than their capability.

This lack of reliance on working memory posed two questions. First, if observers are not relying on working memory, then how are they acquiring the

(A)

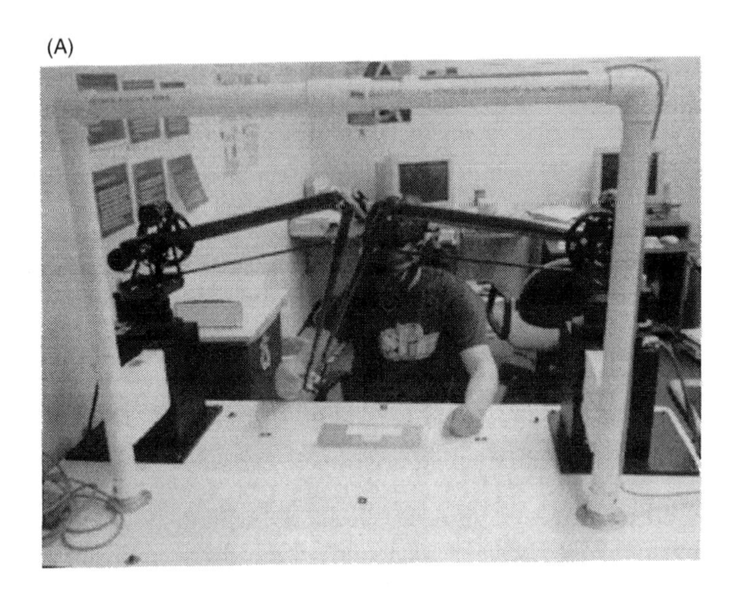

(B)

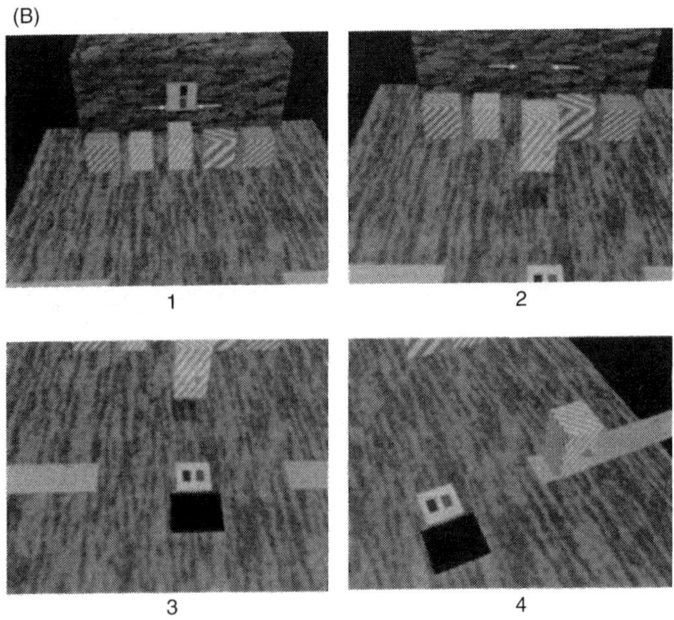

FIGURE 10.2 Sorting bricks in a virtual environment. (A) Subjects wore a head-mounted display and received haptic feedback through mechanical arms attached to index finger and thumb. (B) Scene during a *One-Feature* trial when brick color was task relevant. Fingertips are represented as small red spheres. In a single trial, a subject (1) selects a brick based on the pick-up cue, (2) lifts the brick, (3) brings it towards themselves, (4) decides on which conveyor belt the brick belongs based on a put-down cue. (See color plate)

different visual information used at different stages of the task? Secondly, under what conditions might subjects revert to a strategy in which memory is used?

Answers to these questions can be found in the subjects' performance across different task conditions, when more brick features were potentially relevant for sorting. Potential relevance for brick features was modulated not only by whether one or two features were used for pick-up and put-down, but also by whether subjects could predict which feature would be relevant for the final decision of put-down. When performing a series of trials in which a particular feature was either consistently relevant for put-down, or the relevance of each feature was randomized, anticipating which brick feature was relevant was either *Predictable* or *Unpredictable*. The different trial types are shown in Figure 10.3.

Fixation Patterns

Figure 10.4A plots the frequency of two eye movement sequences in each task condition. As subjects brought the brick toward the conveyor belts, they fixated the put-down cue before making the decision on which belt to place the brick. If subjects look directly to the appropriate belt immediately after the put-down cue (Figure 10.3A, top left), and begin to move the brick toward it, then

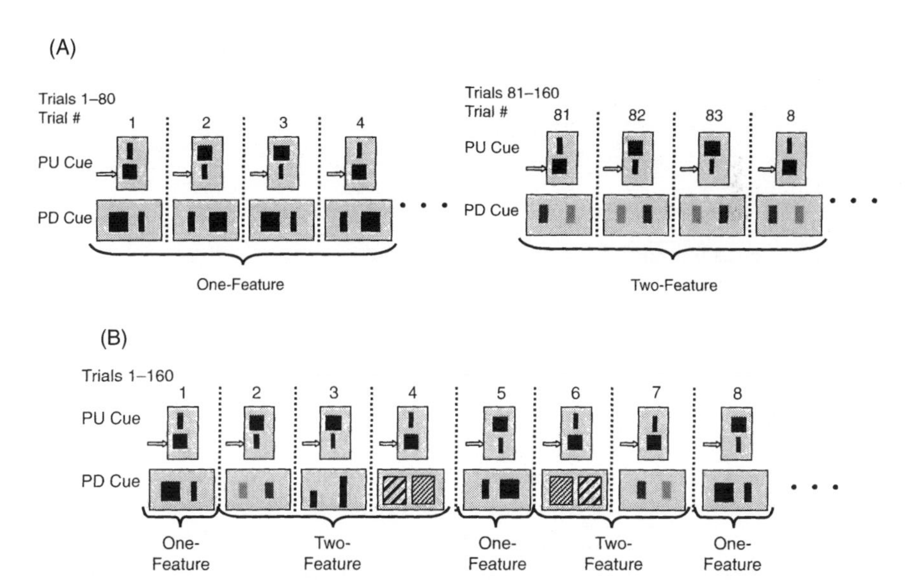

FIGURE 10.3 Example pick-up and put-down cues for each trial type in (A) Predictable and (B) Unpredictable conditions. In the Predictable condition, the same one or two features were relevant during all trials. In the Unpredictable condition, subjects always used the same feature for pick-up, but could not predict which brick feature would be relevant for put-down. One-Feature trials used the same feature for both pick-up and put-down; Two-Feature trials used separate features for pick-up and put-down. Adapted from Droll and Hayhoe (2007).

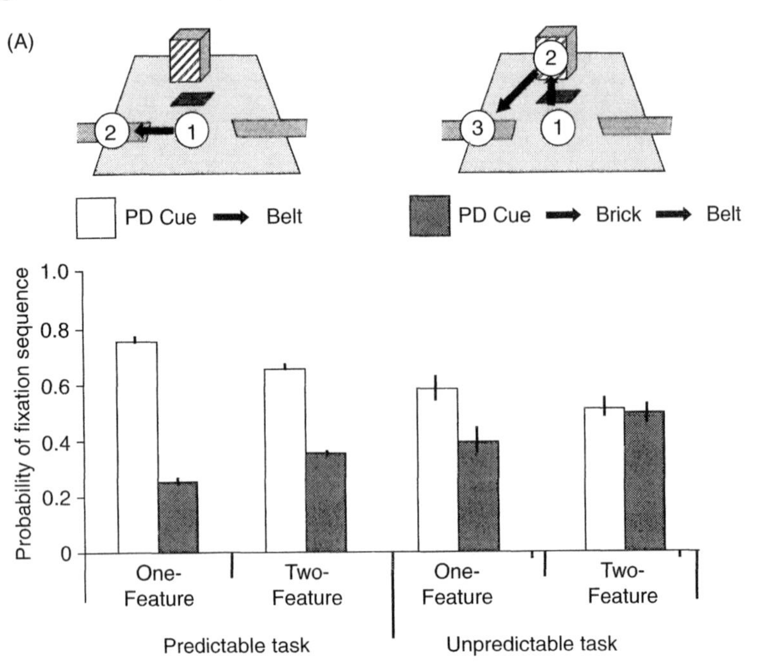

FIGURE 10.4 (A) The sequence of eye movements during the sorting tasks depended on memory load, in both the number of relevant features and the predictability of the which feature would be relevant. (B) During trials in which observes missed a feature change, the brick was more like to be sorted by the new, post-change feature, on trials where eye movements suggested that this information was not stored internally, before the change.

it is likely that the decision for brick placement was made on the basis of their working memory representation of the brick feature. If, however, the subject re-fixates the brick before looking at the belt to guide placement (Figure 10.4A, top right), then it is likely that the decision was made on the basis of feature information acquired during the re-fixation, using a just-in-time strategy.

The clear pattern is that subjects are increasingly likely to use a just-in-time strategy, re-fixating the brick in hand, with increasing unpredictability of what feature is relevant for put-down. While subjects were most likely to use working memory when using the same feature for pick-up and put-down could be anticipated, subjects were increasingly likely to re-fixate the brick when a different feature unexpectedly became relevant for put-down. In the *Predictable* condition, subjects may have used their behavioral strategy because they stored the task-relevant features in working memory as a consequence of the predictability with which features would be relevant in every trial, because of the blocked trial design. In comparison, in the *Unpredictable* condition, subjects were much more likely to re-fixate the brick in hand before fixating the conveyor belt to guide placement (Figure 10.4A). In this case subjects could not predict which of the four features would be used for put-down, and would have to store four brick features if they used a strategy that

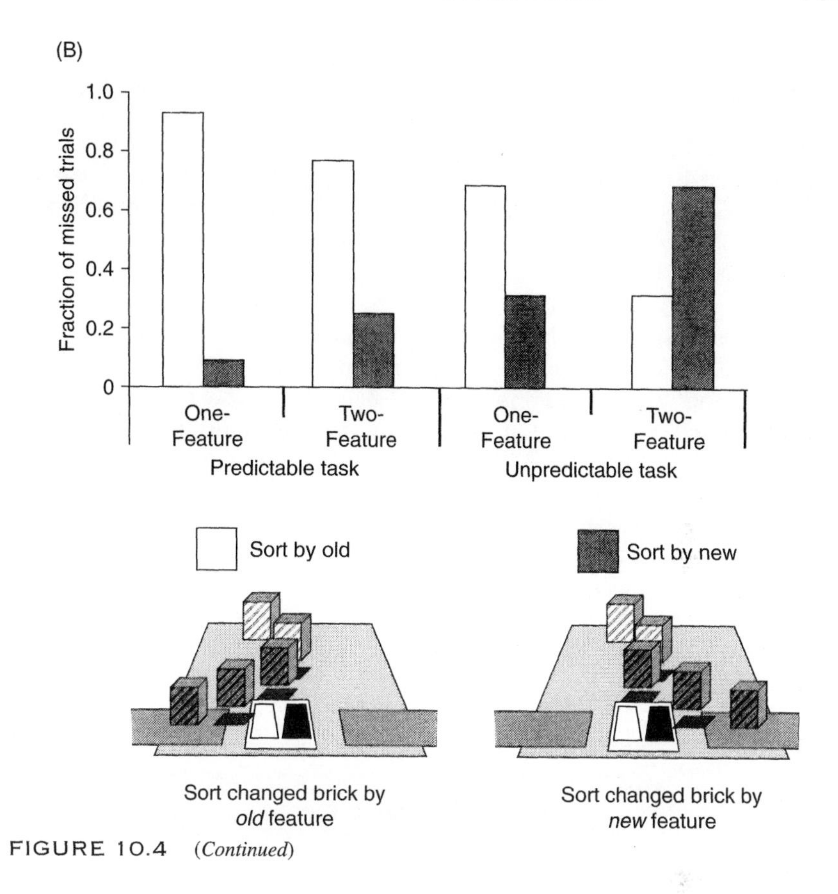

FIGURE 10.4 *(Continued)*

relied on working memory. Instead, subjects seemed to use a strategy that lessened their reliance on working memory. Consistent with this interpretation, the addition of a single extra feature in the *Two-Feature* trials increases the probability of a re-fixation, even in the *Predictable* condition.

Sorting Decisions

To more rigorously test the hypothesis that each of these two eye movement sequences reflected strategies of either using information stored in working memory for sorting, or information acquired from the scene, we analyzed subjects' sorting decisions on trials following missed changes. While changes relevant to sorting were more frequently detected than irrelevant changes, rates of detecting relevant changes were still quite low (from 38% to 57%, depending on conditions). On missed-change trials, subjects sorted the bricks onto one of the conveyor belts instead of placing it in the trash. How do the subjects sort the bricks? Figure 10.3B shows the two possibilities. Either the subject can treat the brick as if it retained its old feature, or else it can be sorted according to its current, new, feature.

The pattern of sorting decisions following missed changes parallels the pattern of eye movement sequences on non-change trials (Figure 10.4B). During trials in which the relevant information can be predicted, and eye movement sequences suggest a strategy of using working memory to store this information internally, subjects invariably sort the changed brick by its old, pre-change feature.[1] Conversely, during trials in which the relevant information was unpredictable, when eye movement sequences suggest a strategy of acquiring information from the scene just-in-time, subjects became increasingly likely to sort the changed brick by its new, post-change feature. In the task where the feature relevance was uncertain, using a memory-based strategy would require storage of all four features, acquired early in the task, at the time of pick-up. Thus, the use of one versus the other gaze strategy appears to depend on the memory load required to perform the task using information already in memory. Fixations to the brick were more frequent when the task required a high memory load (*Unpredictable* task), but subjects preferred to use working memory when they knew in advance which one or two features were needed (*Predictable* task).

Both these results suggest that the balance between eye movements and working memory reflects some kind of optimization or trade-off with respect to a set of constraints on the part of the observer. We next discuss possible factors that may influence the trade-off between using external behavior, or internal operations, to perform extended tasks.

TRADE-OFFS BETWEEN GAZE AND WORKING MEMORY USE

Other related work has also observed a trade-off between the use of working memory and eye movements. For example, in Ballard et al.'s (1995) block copying task, subjects often fixated a block twice while they were copying it, once before picking up a brick of that color, and then again, just before putting it down. In one condition, the experimenters increased the cost of the eye movements by placing the regions in the display further away from each other. Subjects had to make a larger head movement to fixate the model blocks, and this may have added to the cost, either in energy or in time, for re-fixating the blocks. Consequently, subjects reduced the frequency of fixations on the model bricks, and were more likely to remember the fixated information throughout the next few stages of the task. This suggests that as the cost of a change in gaze increased, observers offset this cost by increasing their reliance on working memory.

[1] Note that while these trials rarely included a re-fixation to the brick just before the sorting decision, these bricks were the target of fixation as the subject made the careful placement of the brick onto the conveyor belt. These placement fixations were of an average duration of 750 ms, after the feature change. This indicates that subjects are using their memory of the brick feature, rather than its actual current state, to make the sorting decision, and did not update this information after the change, during put-down.

What guides the decision to use one sequence of actions and cognitive processes instead of another? On contributing factor to the decision to use each of these strategies may be the cost of each policy. These costs may reflect the burden placed on each resource, either in metabolic cost for storing information (Haxby et al., 2000), re-directing attention (Yantis et al., 2002; Serences et al., 2004), or in perceptual or attentional load (Lavie et al., 2004). Thus, in the course of performing a task, observers may need to balance the use of each strategy to satisfy a set of such constraints (Gray et al., 2005). Throughout the course of a task, observers may find a balance point between using changes in gaze to acquire information in the scene, and storing information internally. Thus, opposite to the manipulation performed in Ballard et al., if a task increased the demand on working memory, then subjects in Droll et al. (2005) offset this cost by more frequently using changes in gaze to acquire the necessary visual information. A striking feature of the Droll et al. (2005) results is the ease with which the subjects shifted from using either gaze or working memory, in response to the increased memory load from one to four features. Such sensitivity suggests that trade-offs, between storing information internally and using eye movements to acquire information when needed, are an intrinsic aspect of natural behavior. The mechanisms by which observers settle on a particular strategy or balance between working memory use and just-in-time fixations are not clear. Note that subjects were instructed to perform the task at a pace they felt was comfortable and natural. The variability in performance between subjects is quite small, suggesting that this trade-off reflects a pervasive and stable aspect of natural behavior. Subjects also do not appear to be aware that they are making these trade-offs, or explicitly make decisions about the strategy. It is also unclear why subjects appear to opt for reduced memory strategies, well below traditional measures of working memory capacity, in the context of such natural tasks. One possibility is that there may be other demands on attention and working memory in natural tasks that are not obvious to the experimenter, such as controlling the grasp, or remembering the location of the conveyor belts to program the eye and hand movements for put-down. Prioritizing each of these many "micro-tasks" required for each trial may have also been influenced by the expected value or execution cost of performing each operation (Gray et al., 2005; Fu & Gray, 2006). Had we provided extra incentive, in the form of reward, or different task instruction, subjects' may have re-prioritized perceptual encoding or memory use, resulting in a different pattern of behavior. Another possibility is that organizing multiple sub-tasks requires the use of an executive control mechanism, demanding attentional resources that could otherwise be allocated toward working memory (Hester & Garavan, 2005).

The trade-offs between fixations and working memory use observed in the brick-sorting experiment serve as a caution against over-generalizing the performance from many standard experimental paradigms. Traditional visual paradigms are often designed with the intention of identifying the properties of particular processes, such as working memory (Vogel et al., 2001). However, such experiments may reflect the limits on performance in the context of the particular experiment, rather than the usage of these operations during natural tasks. In the present experiment, subjects were encouraged to perform in a manner in which they felt most comfortable. Thus, the decision of when to store information in memory or to use gaze may reflect the operation of sets of constraints, and the trade-offs between these constraints may vary from task to task. Assessing where this balance point lies, by monitoring trade-offs between gaze and working memory use, may be a more accurate way in which to characterize memory *usage*, rather than more traditional paradigms investigating working memory *capacity* (Vogel et al., 2001).

BRIDGING THE GAP BETWEEN LABORATORY EXPERIMENTS AND NATURAL BEHAVIOR

Subjects' use of eye movements, attention and working memory in the brick-sorting task would not have been predicted by traditional measures that attempt to measure properties of working memory and attention in isolation. Thus, it is worth considering how the different approaches relate to each other. We next review literature from fields of executive control, cortical mechanisms of learning, and other task paradigms to consider how to unite these superficially disparate fields of research.

EXECUTIVE CONTROL

Critical to the philosophy behind embodied cognition is the idea that vision is not simply the process by which the brain represents the visual environment. A central feature of vision is its versatility in performing complex tasks and assisting in a variety of decisions, even in a task as simple as picking up and putting down an object. Performance in the brick-sorting experiment suggests a model of vision in which visual operations (such as directing gaze, acquiring select visual information, and storage in working memory) are controlled by learnt procedures that somehow orchestrate the sequence of visual operations required to complete a task. For example, a single brick-sorting trial may be organized as a sequence of sub-tasks, such as the pick-up or put-down operation for each brick, each lasting 2–3 seconds. These sub-tasks will require yet smaller tasks, such as using an eye movement to acquire an individual brick feature (Ballard et al., 1997). Feature states of the brick that are not immediately relevant may simply not be evaluated, and thus their change will escape detection. This way of thinking about vision suggests that the purpose of vision, and visual attention, is not to construct internal representations for general-purpose use. Instead, visual attention is best understood in the context of what information must be selected, and evaluated, for the purpose of guiding behavior (Allport, 1989). For example, the visual operations performed at the time of pick-up may include not a single operation, in which all features are automatically acquired, but rather a process involving up to four discrete tasks; categorizing color, height, width, and texture into each of two possible feature states (e.g., red/blue, tall/short, etc.).

These individual feature judgments may be similar to processes required in traditional dual-task paradigms in which subjects are shown similar stimuli in all trials but are asked to make multiple decisions (e.g., parity and/or magnitude judgments when shown numbers). Rather than performing multiple operations simultaneously, in parallel, it is often more efficient to perform them in a serial manner due to limits in attentional capacity (Logan & Gordon, 2001). This conceptualization of performance is similar to some formal models of executive control in which high-level decision processes affect lower level sensory selection (Bundesen, 1990; Logan & Gordon, 2001; Logan, 2004). In these models, the task defines the sensory parameters to be used, and this constrains what information is acquired (see also Figure 10.1). Visual information that is not immediately relevant, or unlikely to be relevant, is simply not evaluated and thus never gains access to working memory. This general framework is also consistent with many theoretical ideas on how cognitive systems may be fundamentally organized (Ullman, 1984; Newell, 1990; Ballard et al., 1997) and can also be used to model complex behavior, such as navigating a cluttered sidewalk (Sprague et al., in press).

CONNECTING VISION AND ACTION THROUGH REWARD-SENSITIVE LEARNING MECHANISMS

The way in which tasks exert control over the acquisition of visual information must be clearly learnt (Hayhoe & Ballard, 2005). One contributing element to this learning may be the reward, or costs, that observers learn to associate with alternate courses of action. For example, recent work in the neurophysiological basis of eye movements has revealed that the saccadic eye movement circuitry is sensitive to the reward structure of the task (Platt & Glimcher, 1998; Hikosaka et al., 2000; Ikeda & Hikosaka, 2003; Sugrue et al., 2004). In these experiments, monkeys are often free to choose between visually guided alternative behaviors (direct gaze to the target on the left or to the right). The monkey's behavior can be generalized as seeking a strategy in which the direction of gaze is chosen by virtue of the reward they have learned to expect to receive following their behavior, and monkeys are quite sophisticated at learning how to maximize their overall gain (e.g., drops of juice). Hayhoe and Ballard (2005) have argued that this sensitivity to reward may serve as a substrate for mediating the tight linkage between fixations and task structure in natural behavior. There is also growing psychophysical evidence that observers use learned reward structure to guide reaching movements (Trommershauser et al., 2003a, b, 2005, 2006).

Similar to tasks in the reward literature, subjects in brick-sorting were allowed to make self-directed visually guided decisions. Whereas monkey subjects may be maximizing their expected reward, human subjects during brick-sorting may be making their decisions based on the expected cost of each alternative (e.g., store visual information or re-fixate this object later). Thus, the study of trade-offs need not use reward as an explicit variable, such as juice or monetary gain. Modulating the frequency of performing one action, or cognitive operation, at the expense of another, suggests that there is some sort of cost function being evaluated when executing tasks. Determining the nature of this cost function may be difficult, and perhaps reflect an amalgam of factors, such as metabolic cost, cognitive load, temporal urgency, and expectation of risk and task outcome.

It is also interesting to note that sensitivity to reward is often observed in cortical areas traditionally associated as modulating "attention" and that distinguishing between the two may be difficult (Maunsell, 2004). Thus, it seems reasonable to consider that this reward circuitry may serve as a substrate to mediate the effects of task structure, and visual attention, observed throughout embodied behavior.

FUTURE DIRECTIONS OF RESEARCH IN EMBODIED VISUAL COGNITION

Given the extensive research using non-embodied tasks, it is critical to understand how results from such paradigms may scale up to more natural contexts. Executing complex tasks extended in time requires coordination of cognitive processes and actions. This coordination often requires selecting a sequence of operations, both internal and external, that includes assessing trade-offs between alternate solution strategies. While capacity limits in each operation may rarely be reached in the course of natural behavior, it is possible that these limits contribute to the operational costs of executing each process. In the example of brick-sorting, eye movements were used rather than storing information internally. The use of each strategy will depend on the task, or other external circumstances, and experience of the observer. Although natural behavior during brick-sorting was relatively consistent across observers, what accounts for the differences? How might variation in capacity limits across observers translate into differences in how working memory and eye movements are used in natural behavior? Do observers with higher capacity measures have a lower cost of using these processes? Answers to an understanding of how observers select which course of action to perform during everyday behavior require studying tasks in which observers are required to make trade-offs between alternative cognitive operations and sequences of action. Exploration and adaptation to tasks with novel structure will reflect ways that the human cognitive system uses trade-offs to deal with attentional and memory limitations.

In conclusion, investigation of visual performance in natural tasks is now much more feasible, given the technical developments in monitoring eye, head, and hand movements in unconstrained observers as well as the development of complex virtual environments. This allows some degree of experimental control while allowing relatively natural behavior.

We have reviewed the ways in which natural tasks are different from standard experimental paradigms, and reviewed an experiment that was designed to reflect a class of simple, natural behaviors. The experiment revealed several aspects of visual cognition that are not easy to study in standard paradigms that try to isolate a particular cognitive mechanism. The experiment revealed an important aspect of natural vision, namely, that gaze is controlled by some kind of cost function that determines on a very fast timescale where and when to fixate, and the contents of working memory. We also observed that natural, visually guided behavior is surprisingly regular and responsive to manipulation of the experimental conditions. Thus there is little to lose, and much to gain by the study of cognition in its embodied context.

REFERENCES

Aivar, M. P., Hayhoe, M. M., Chizk, C. L., & Mruczek, R. E. B. (2005). Spatial memory and saccadic targeting in a natural task. *Journal of Vision, 4,* 1–3.

Allport, A. (1989). Visual attention. In M. Posner (Ed.), *Foundations of Cognitive Science* (pp. 683–726). Cambridge, MA: MIT Press.

Ballard, D. H., Hayhoe, M., & Pelz, J. B. (1995). Memory representations in natural tasks. *Journal of Cognitive Neuroscience, 7*(1), 66–80.

Ballard, D. H., Hayhoe, M. M., Pook, P. K., & Rao, R. P. (1997). Deictic codes for the embodiment of cognition. *Behavioral and Brain Sciences, 20*(4), 723–742, discussion 743–767.

Bundesen, C. (1990). A theory of visual attention. *Psychological Reviews, 97*(4), 523–547.

Colby, C. L. (1998). Action-oriented spatial reference frames in cortex. *Neuron, 20*(1), 15–24.

Droll, J. & Hayhoe, M. (2007). Trade-offs between working memory and gaze. *Journal of Experimental Psychology: Human Perception and Performance, 33*(6), 1352–1365.

Droll, J. A., Hayhoe, M. M., Triesch, J., & Sullivan, B. T. (2005). Task demands control acquisition and storage of visual information. *Journal of Experimental Psychology. Human Perception and Performance, 31*(6), 1416–1438.

Fu, W. & Gray, W. D. (2006). Suboptimal tradeoffs in information seeking. *Cognitive Psychology, 52,* 195–242.

Gnadt, J. W., Bracewell, R. M., & Andersen, R. A. (1991). Sensorimotor transformation during eye movements to remembered visual targets. *Vision Research, 31*(4), 693–715.

Gray, W. D., Schoelles, M. J., & Sims, C. R. (2005). Adapting to the task environment: Explorations in expected value. *Cognitive Systems Research, 6,* 27–40.

Haxby, J. V., Petit, L., Ungerleider, L. G., & Courtney, S. M. (2000). Distinguishing the functional roles of multiple regions in distributed neural systems for visual working memory. *Neuroimage, 11*(5 Pt 1), 380–391.

Hayhoe, M. & Ballard, D. (2005). Eye movements in natural behavior. *Trends in Cognitive Sciences, 9*(4), 188–194.

Hayhoe, M., Lachter, J., & Feldman, J. (1991). Integration of form across saccadic eye movements. *Perception, 20*(3), 393–402.

Hayhoe, M. M., Shrivastava, A., Mruczek, R., & Pelz, J. B. (2003). Visual memory and motor planning in a natural task. *Journal of Vision, 3*(1), 49–63.

Hester, R. & Garavan, H. (2005). Working memory and executive function: The influence of content and load on the control of attention. *Memory & Cognition, 33*(2), 221–233.

Hikosaka, O., Takikawa, Y., & Kawagoe, R. (2000). Role of the basal ganglia in the control of purposive saccadic eye movements. *Physiological Reviews, 80*(3), 953–978.

Ikeda, T. & Hikosaka, O. (2003). Reward-dependent gain and bias of visual responses in primate superior colliculus. *Neuron, 39*(4), 693–700.

Itti, L. & Koch, C. (2000). A saliency-based search mechanism for overt and covert shifts of visual attention. *Vision Research, 40*(10–12), 1489–1506.

Itti, L. & Koch, C. (2001). Computational modeling of visual attention. *Nature Reviews. Neuroscience, 2*(3), 194–203.

Jovancevic, J., Sullivan, B., & Hayhoe, M. (2006). Control of attention and gaze in complex environments. *Journal of Vision, 6*(12), 1431–1450.

Land, M., Mennie, N., & Rusted, J. (1999). The roles of vision and eye movements in the control of activities of daily living. *Perception, 28*(11), 1311–1328.

Land, M. F. (2004). The coordination of rotations of the eyes, head and trunk in saccadic turns produced in natural situations. *Experimental Brain Research, 159*(2), 151–160.

Land, M. F. & Hayhoe, M. (2001). In what ways do eye movements contribute to everyday activities? *Vision Research, 41*(25–26), 3559–3565.

Lavie, N., Hirst, A., de Fockert, J. W., & Viding, E. (2004). Load theory of selective attention and cognitive control. *Journal of Experimental Psychology: General, 133*(3), 339–354.

Logan, G. D. (2004). Cumulative progress in formal theories of attention. *Annual Review of Psychology, 55,* 207–234.

Logan, G. D. & Gordon, R. D. (2001). Executive control of visual attention in dual-task situations. *Psychological Reviews, 108*(2), 393–434.

Luck, S. J. & Vogel, E. K. (1997). The capacity of visual working memory for features and conjunctions. *Nature, 390*(6657), 279–281.

Maunsell, J. H. (2004). Neuronal representations of cognitive state: Reward or attention? *Trends in Cognitive Sciences, 8*(6), 261–265.

Newell, A. (1990). *Unified theories of cognition.* Cambridge, MA: Harvard University Press.

Parkhurst, D. J. & Niebur, E. (2003). Scene content selected by active vision. *Spatial Vision, 16*(2), 125–154.

Parkhurst, D. J. & Niebur, E. (2004). Texture contrast attracts overt visual attention in natural scenes. *European Journal Neuroscience, 19*(3), 783–789.

Pelz, J., Hayhoe, M., & Loeber, R. (2001). The coordination of eye, head, and hand movements in a natural task. *Experimental Brain Research, 139*(3), 266–277.

Pelz, J. B. & Canosa, R. (2001). Oculomotor behavior and perceptual strategies in complex tasks. *Vision Research, 41*(25–26), 3587–3596.

Platt, M. L. & Glimcher, P. W. (1998). Response fields of intraparietal neurons quantified with multiple saccadic targets. *Experimental Brain Research, 121*(1), 65–75.

Serences, J. T., Schwarzbach, J., Courtney, S. M., Golay, X., & Yantis, S. (2004). Control of object-based attention in human cortex. *Cerebral Cortex, 14*(12), 1346–1357.

Sprague, N., Ballard, D. & Robinson, A. (2007). Modelling embodied visual behaviors. *ACM Transactions on Applied Perception, 4*(2) Article 11. DOI = http://doi.acm.org/10.1145/1265957.1265960

Sugrue, L. P., Corrado, G. S., & Newsome, W. T. (2004). Matching behavior and the representation of value in the parietal cortex. *Science, 304*(5678), 1782–1787.

Triesch, J., Ballard, D. H., Hayhoe, M. M., & Sullivan, B. T. (2003). What you see is what you need. *Journal of Vision, 3*(1), 86–94.

Trommershauser, J., Maloney, L. T., & Landy, M. S. (2003a). Statistical decision theory and the selection of rapid, goal-directed movements. *Journal of the Optical Society of America. A, Optics, Image Science, and Vision, 20*(7), 1419–1433.

Trommershauser, J., Maloney, L. T., & Landy, M. S. (2003b). Statistical decision theory and trade-offs in the control of motor response. *Spatial Vision, 16*(3–4), 255–275.

Trommershauser, J., Gepshtein, S., Maloney, L. T., Landy, M. S., & Banks, M. S. (2005). Optimal compensation for changes in task-relevant movement variability. *Journal of Neuroscience, 25*(31), 7169–7178.

Trommershauser, J., Landy, M. S., & Maloney, L. T. (2006). Humans rapidly estimate expected gain in movement planning. *Psychological Sciences, 17*(11), 981–988.

Turano, K. A., Geruschat, D. R., Baker, F. H., Stahl, J. W., & Shapiro, M. D. (2001). Direction of gaze while walking a simple route: Persons with normal vision and persons with retinitis pigmentosa. *Optometry and Vision Science, 78*(9), 667–675.

Turano, K. A., Geruschat, D. R., & Baker, F. H. (2003). Oculomotor strategies for the direction of gaze tested with a real-world activity. *Vision Research, 43*(3), 333–346.

Ullman, S. (1984). Visual routines. *Cognition, 18*(1–3), 97–159.

Vogel, E. K., Woodman, G. F., & Luck, S. J. (2001). Storage of features, conjunctions and objects in visual working memory. *Journal of Experimental Psychology. Human Perception and Performance, 27*(1), 92–114.

Wheeler, M. E. & Treisman, A. (2002). Binding in short-term visual memory. *Journal of Experimental Psychology: General, 131*(1), 48–64.

Yantis, S., Schwarzbach, J., Serences, J. T., Carlson, R. L., Steinmetz, M. A., Pekar, J. J. et al. (2002). Transient neural activity in human parietal cortex during spatial attention shifts. *Nature Neuroscience, 5*(10), 995–1002.

11

Why We Don't Mind to Be Inconsistent

Jeroen B.J. Smeets and Eli Brenner

Research Institute MOVE, Faculty of Human Movement Sciences, VU University, Amsterdam, The Netherlands

Introduction

It is frequently assumed that perception involves the creation of a model of our environment. Our senses provide incomplete and noisy information about the objective world. One might think that what we perceive is the situation that best matches the incomplete and noisy information. However, this is not the case. Hermann von Helmholtz (1925) already noted that perception is unconscious inference about the situation that most likely caused the sensory state. This view has recently become very popular and has been formalized in terms of Bayesian inference (Kersten et al., 2004; Knill & Pouget, 2004; Körding & Wolpert, 2006). In this view, the likelihood that you perceive situation X depends on the likelihood that situation X is the cause of the present sensory state, multiplied by the a priori likelihood that situation X occurs. This description of perception as the formation of an internal model of the outside world, which represents the most likely cause of our sensory stimulation, is a very powerful approach to perception, that can explain various phenomena.

One clear prediction of the Bayesian/Helmholtzian approach is that you will never perceive a situation that is physically impossible. This seems a reasonable prediction, but M.C. Escher's drawing in Figure 11.1A shows a clear counter-example. You perceive a situation in which the figures can walk up or down the stairs infinitely while returning to their initial position after each turn. This situation is physically impossible, and thus has an a priori likelihood of exactly zero. Why can we see a situation with a zero a priori likelihood? One might argue that

(A)

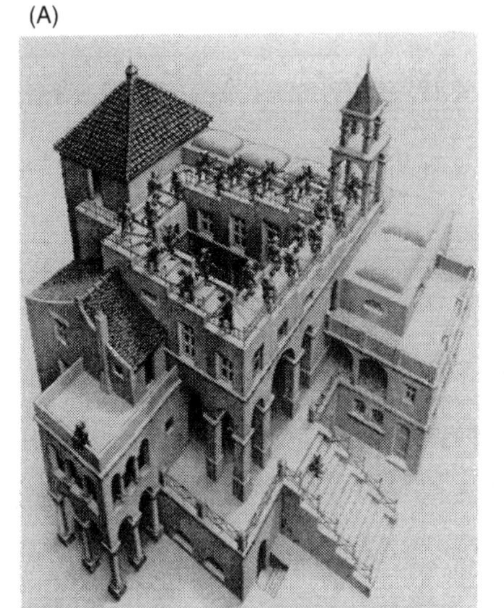

(B)

FIGURE 11.1 Perceiving "impossible" constructions. (A) M.C. Escher's "Ascending and Descending" © 2008 The M.C. Escher Company B.V.-Baarn — The Netherlands. All rights reserved. www.mcescher.com (B) A similar construction has been built in LEGO™. Despite the fact that there is a possible construction leading to this image, we perceive the impossible construction. Details of the construction can be found at the web site: http://www.andrewlipson.com/escher/ascending.html (© Andrew Lipson, reproduced with permission).

perceiving an impossible situation in a picture is not problematic because the a priori chance that something impossible is depicted is not zero at all. It is like looking at a photograph of yourself as a 6 months old baby; you perceive something that you know is not reality anymore, but history. In a similar way, Escher's drawing might be thought of depicting not reality, but fantasy. One might also argue that because there is no real 3D construction that gives this image, there is no a priori likelihood, so we cannot use Bayesian inference.

Unfortunately, the reasoning above is not correct, as there is a possible 3D construction that leads to an image as in Escher's drawing: Andrew Lipson built it as a *LEGO*™ construction and photographed it (Figure 11.1B). Everybody perceives this picture as depicting the same impossible 3D situation as Escher's original. However, as the a priori likelihood of an impossible object is zero, Bayesian inference would predict that this percept can never occur. Moreover, the a priori likelihood that the actual LEGO construction depicted here exists is definitely not zero, so Bayesian inference would predict that you perceive the actual construction. Many other "impossible" drawings (such as the Penrose triangle; Gregory, 1968) give the same retinal image as real objects. The question we will address in the remainder of this chapter is how we can understand the perception of the picture in Figure 11.1B within the Bayesian framework.

DETECTING ATTRIBUTES

We will start by sketching some textbook knowledge about the neural basis of visual perception. Textbooks start with the receptors at the level of the retina: the rods and three types of cones that respond to the incoming light. Subsequently, still at the level of the retina, information from various receptors is combined to improve the sensitivity for a certain aspect of information. For instance, lateral inhibition increases the sensitivity for local luminance differences at the expense of losing information about the absolute luminance level. This mechanism (very useful for detecting object edges) is commonly regarded as being on the basis of various illusions, such as Mach bands and the Cornsweet illusion. In these explanations, the luminance difference is caused by an erroneous integration of the luminance differences obtained by the edge detectors (Land & McCann, 1971; Arend, 1973). Why would the brain first take a spatial derivative and subsequently integrate it? It is an efficient way of coding, making it possible to transmit information about differences in reflectance that are orders of magnitude smaller than the variations in illumination. Despite the limited bandwidth of the optic nerve, we can recognize objects and other animals both in the sunlight and in the dark.

To explain the illusions using the image coding based on edge detectors, one implicitly assumes that the brain makes systematic errors in the differentiation and/or subsequent integration (the integration of a perfect derivative wouldn't yield any error). Thus the essence of the explanation is not the use of edge detectors but the systematic errors that are made. What is the reason for making such systematic errors? These illusions are the consequence of an ambiguity in images: Are differences in lightness due to differences in illumination or due to differences in the surface reflectance?

In most images, there are various other cues for the illuminant and surface properties. For instance, the illumination is likely to vary much more with position for curved surfaces than for flat surfaces (Figure 11.2). By varying the presence of such other cues, it has been shown that the perception of the equiluminant territories flanking the Cornsweet edge varies according to whether these regions are more likely to be equally illuminated surfaces having different material properties or unequally illuminated surfaces with the same properties (Knill & Kersten, 1991; Purves et al., 1999). The illusion is thus not the consequence of low-level processing errors, but a percept that is optimal from a Bayesian perspective. In a similar way, the presence of Mach bands can be explained in terms of the likelihood of photometric highlights near contrast edges (Lotto et al., 1999).

There is an interesting difference between the attributes luminance and local-luminance gradient. Luminance itself is very sensitive for naturally occurring slow variations of illumination over a smooth surface, whereas such variations are negligible at the scale of the edge detectors. So the prior information needed to reliably determine luminance itself is not useful for determining edges on the basis of local luminance gradients. Although the Bayesian approach yields optimal estimates for both attributes (luminance and luminance gradient), it has the

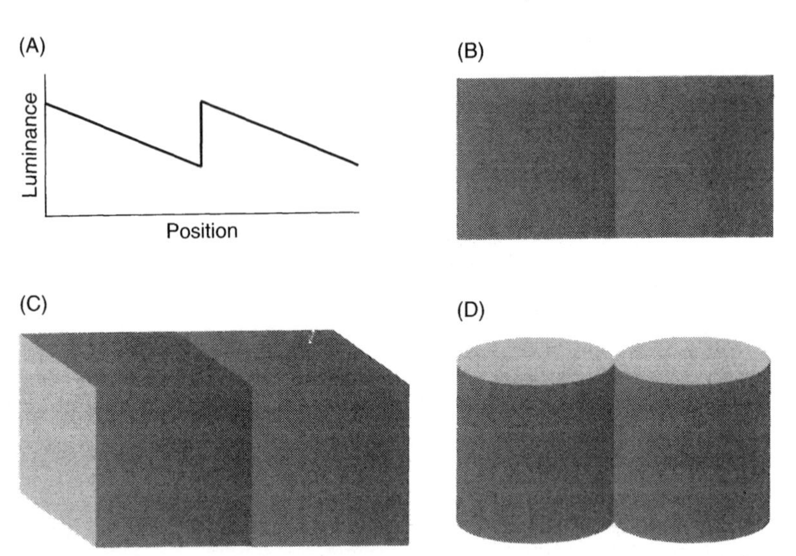

FIGURE 11.2 The Cornsweet illusion: if two surfaces with equal luminance gradients are presented next to each other, one perceives the two as having unequal brightness (A, B). This illusion depends on the interpretation of the scene: it is strong if the surfaces seem to be flat (rendered as part of cubes), (C) than if the surfaces that seem curved (rendered as part of cylinders), (D). For the cylinders, the luminance gradients and luminance step are assumed to be caused by differences in illumination of the surfaces (due to the varying orientation relative to the assumed light source left above), whereas for the cubes, a constant luminance gradient due to the illumination is assumed, with in addition a step in reflectance at the border of the two cubes.

side effect that the perceived luminance gradients might differ from the gradient of perceived luminance if the actual situation is not very likely. In other words, Bayesian perception might be inconsistent. Should this bother us?

SPATIAL PERCEPTION

This inconsistency between the perception of an attribute and the perception of related attributes, such as difference measures and derivatives, is also the basis of many illusions in the spatial domain. The absolute measure of interest here is position, and the related difference measures are distance, length, and velocity. We are notoriously imprecise in determining the absolute position of objects in space. The resolution is about 0.5° (van Beers et al., 1998), presumably a combination of a limited resolution of eye orientation of about 0.15° (Smeets & Brenner, 1994; Brenner & Smeets, 2000) and that of head orientation. The visual acuity of a person with normal vision is one minute of arc, which is about 10 times as precise as our perception of location. The reason is that visual acuity is only determined by the properties of the retina, and is therefore independent of the low resolution of information about the orientation of the eye.

Motion is determined by motion detectors that compare the activation of two areas in the visual field that are separated by a distance (span), with a characteristic time delay at which the activity of the two areas is compared. The smallest size ("span") of motion detectors is in the order of the retinal resolution; about two minutes of arc (van Doorn & Koenderink, 1982). This is much smaller than the resolution at which position can be determined. The reason for the high precision is that motion detectors do not differentiate egocentric position, but determine the motion relative to other retinal input (Smeets & Brenner, 1994). This way of calculating motion has the consequence that motion perception can be inconsistent: you can see an object moving without changing position. This can be perceived in the Duncker illusion (motion perception of a stationary object induced by motion of the background (Duncker, 1929)) or in the aftereffect after prolonged exposure to motion.

Binocular vision (the perception of spatial layout based on the difference in information between the two eyes) is regarded as normal when stereoacuity is better than one minute of arc. How is this achieved in a situation in which each of the eyes has a precision that is not better than 10 minutes of arc? The story is again in the information that is used: the threshold for stereoacuity is based on relative disparity, the differences between the images of the two eyes, irrespective of the location of these images on the retina. Perceiving spatial layout is based on a difference measure that is insensitive to the least precise information available for egocentric localization. This means that shapes and relative positions can be determined very accurately. However, when we need to localize an object in depth relative to ourselves (instead of relative to other visual items), we have to rely on information about eye orientation, so that the precision is limited by the 0.15° resolution of eye orientation (Brenner & Smeets, 2000).

A similar reasoning holds for the perception of the size of objects. If the perception of size would be based on the difference between the judged egocentric positions of the object's edges, than the precision would be rather limited. Fortunately, there is a much better solution. If one bases ones size judgment on the retinal size, scaled by an estimate of distance, precision can be enhanced, as long as one can get a reliable judgment of distance. This is possible only if one does not limit oneself to extra-retinal information, but uses all available visual cues like perspective, familiar size, and texture gradients in a Bayesian way (Gregory, 1968). The Müller-Lyer illusion has an interpretation in terms of depth perception (Figure 11.3A). As can be seen in the same figure, this interpretation affects the perceived length of the line, but not other aspects of space, such as the perceived orientation of the dashed lines connecting their endpoints. The Ebbinghaus illusion can also be interpreted in terms of perspective (Figure 11.3B), and again this affects the perceived size of the disks, but not other spatial aspect in the figure, such as the parallelity of the dashed lines connecting the edges of the black disks. Along the same lines, it has been shown that retinal and extra-retinal information used to judge an object's size, shape, and egocentric distance are combined in a way that yields the most likely value for each of these attributes independently, ignoring any resulting inconsistency between the attributes (Brenner & Van Damme, 1999).

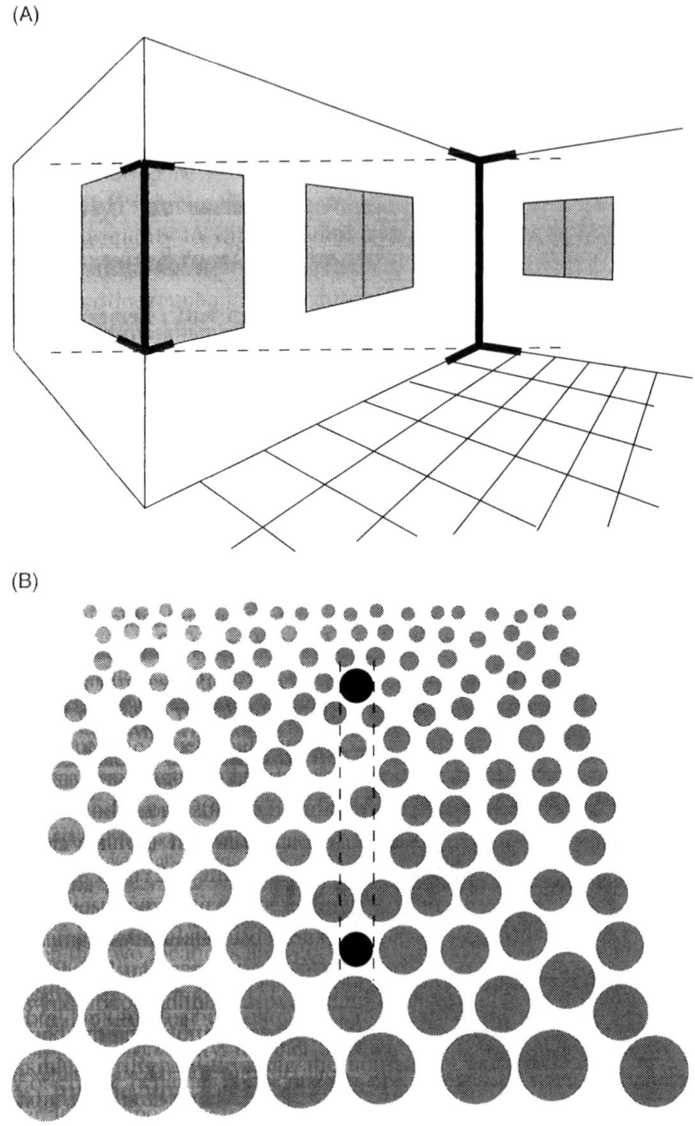

FIGURE 11.3 Examples of inconsistencies in illusions of depth. (A) The Müller-Lyer illusion and other perspective cues make the thick line on the left look smaller than the one on the right. At the same time, the horizontal dashed lines connecting the endpoints seem to be parallel (which they are). (B) The two black disks are equal in size, but the upper one seems to be larger due to the smaller surrounding gray disks. This version of the Ebbinghaus illusion only affects size, not other aspects of space. For instance, the dashed lines look parallel (which they are), which is inconsistent with the apparent difference in size of the two disks.

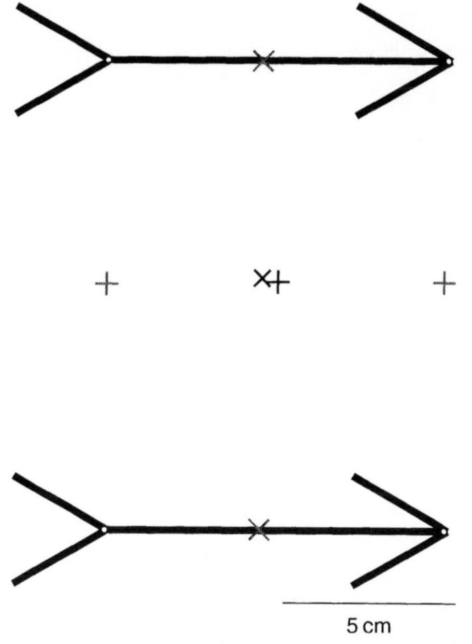

FIGURE 11.4 The center between the four white dots at the endpoints of the Judd-figures is determined in two ways. First, the pairs on each Judd-figure are bisected (gray x's), and then the midpoint between these two points (black x) is determined. Secondly, the midpoint between each vertical pair is found (gray +'s), and then the midpoint between these two points is determined (black +). Although the two ways should yield the same result (according to any affine geometry), the outcome is systematically different.

The inconsistency in the examples discussed earlier is between different attributes. The expected relationships between attributes are not present. Two line segments of different length are connected by parallel lines (Figure 11.3A), and the lines connecting sides of differently sized disks seem parallel (Figure 11.3B). This clearly defies Euclidian geometry. Is there another (non-linear) geometry that can describe human perception? A simple experiment shows that this is very unlikely. In a pencil-and-paper task, we asked subjects to judge the center of four white dots with two Judd-figures attached to them (Figure 11.4). We instructed them to perform the judgment in two ways. On the first sheet of paper, they were asked to bisect the horizontal distances (and thus the Judd-figures) first (gray "x"), and subsequently bisected the vertical distance between the resulting positions (black "x"). On the second sheet of paper, they started with two vertical bisections (gray "+"), and subsequently bisect the horizontal distance between the resulting positions (parallel to the Judd-figures, black "+"). The resulting center differed systematically between the two variants of the tasks. Such a result is not a simple consequence of a non-Euclidian (but nevertheless affine) space. A similar task has been performed to study the spatial deformation in 3D space (Todd et al., 2001). In that experiment, the center between four positions was systematically misjudged, but this misjudgment was independent of the order of bisections. Geometric illusions are thus essentially different from normal perceptual misjudgments.

INCONSISTENT ACTION

It might seem quite disturbing that there can be inconsistencies between attributes within perception. Is this inconsistency not far from optimal? It probably would if the purpose of perception would be to create an internal representation of the outside world. But the purpose of perception is to let an organism survive, for instance by allowing him to find food or to flee for a predator. One might think that organisms combine all information to make the best plan for a movement. For instance, if an animal wants to catch a running prey, it could combine all information about position and motion to predict the time and the position of interception. Is this how animals act? Experiments on human interception show that this is not the case. By using motion illusions, we showed that position, direction of motion and a priori estimate of speed are used to direct the hand, whereas speed information is used to time the action (Smeets & Brenner, 1995a, b; Brenner et al., 2002a).

Inconsistency in our actions can even be observed for the simple task of moving our arm from point A to point B. This inconsistency is caused by the fact that although knowing the target position is enough to move your hand to the target, this information is not enough to move along a straight line (which is what we normally tend to do). To move along a straight line to a target, we need to know in what direction to start our movement. This initial movement direction is often not correct (de Graaf et al., 1991; Brenner et al., 2002b), can be adapted independently of the location of the endpoint (Wolpert et al., 1995), and is easily influenced by illusions (Smeets & Brenner, 2004).

So, not only our conscious perception is inconsistent, but the use of spatial information in our actions is just as inconsistent. This is not surprising, as consistency is not important to find food or to flee for a predator. It is of utmost importance to have fast access to relevant information, such as the velocity of the predator and its position. Although it might be that further processing and combining can improve information, the animal would already be caught before the final percept was completed.

COMBINING INFORMATION

But even without the temporal constraints, it may not be useful to try to make all attributes consistent. As argued earlier, the brain uses all available information to make the best estimate for an attribute. By making attributes consistent, one has to change the values from these optimal values, which is—by definition—sub-optimal. This is similar to an issue in the cue-combination literature. It is well established that if two cues are in conflict, this conflict is not resolved: the cues remain in conflict, although this conflict might not be noted explicitly (Hillis et al., 2002; Muller et al., 2007).

The same holds for combination between senses. To know where our hand is, we have visual and proprioceptive information. We normally don't realize that we have these two sources, because we only have a single idea of where our hand is. But when closing our eyes, we realize that we still know where our hand is. The interesting aspect is that it is easy to induce conflicts between the senses, for instance by wearing wedge prisms. We don't perceive the discrepancy, only realize that we make errors, and adapt our behavior accordingly (van Beers et al., 1999).

But discrepancy between proprioception and vision is not restricted to experimental manipulations. If you put subjects in the dark, and let them make ample back-and-forth movements with their hands between visual targets without seeing their hand, they start making errors. These errors are not accidental: the same errors reoccur on repetition of the experiment the next day (Smeets et al., 2006). So our senses are not calibrated. The reason for this lack of calibration is similar to the reason why the inconsistencies are not resolved: if the combination of senses yields the most reliable estimate, recalibration can only reduce the reliability of the information. It is for instance not clear which of the modalities would need to be recalibrated. Is this lack of calibration problematic? Again, it is not. When controlling our hand movement, we don't use a single sense but use the optimal combination of all senses. And this does not only hold for our hand but also for any possible target we want to reach for with that hand. This means that the conflict might be present between attributes and between senses but that these conflicts do not interfere with our performance.

CONSCIOUS PERCEPTION

We made our argument in terms of the information needed for controlling our actions. We reached a radically different conclusion than for instance Goodale (2001), who claims that "accurate metrical information about an object" is needed to guide one's action. In our view, the same erroneous and inconsistent perceptual information can be used in both perception and controlling movements (Smeets et al., 2002). There is however one fundamental difference: whereas timely information is essential in the control of movements, our consciousness has ample time to reconsider information. Whereas control of action needs to rely on the fast feed forward processing of information, our cognition can wait until the information processing is recurrent (Lamme & Roelfsema, 2000). The consequence that our conscious percept is based on further processed information than used to control our actions, but this should not be taken as evidence for independent processing.

We started this chapter by discussing the limitations of the Bayesian approach, and argued that the inconsistent precepts are not very Bayesian. We continued by showing that inconsistency is very widespread in perception, and is a consequence of optimally determining information about each attribute of the world around us. These optimal estimates are enough to guide our actions. An exact calibration of perception is not needed for controlling ones actions. What we need are transformation rules between perceptual attributes and aspects of an action. That is what we learn very quickly while practicing an action.

REFERENCES

Arend, L. E. (1973). Spatial differential and integral operations in human vision: Implications of stabilized retinal image fading. *Psychological Review, 80,* 374–395.

Brenner, E. & Smeets, J. B. J. (2000). Comparing extra-retinal information about distance and direction. *Vision Research, 40,* 1649–1651.

Brenner, E. & Van Damme, W. J. M. (1999). Perceived distance, shape and size. *Vision Research, 39,* 975–986.

Brenner, E., De Lussanet, M. H. E., & Smeets, J. B. J. (2002a). Independent control of acceleration and direction of the hand when hitting moving targets. *Spatial Vision, 15,* 129–140.

Brenner, E., Smeets, J. B. J., & Remijnse-Tamerius, H. C. (2002b). Curvature in hand movements as a result of visual misjudgements of direction. *Spatial Vision, 15,* 393–414.

de Graaf, J. B., Sittig, A. C., & Denier van der Gon, J. J. (1991). Misdirections in slow goal-directed arm movements and pointer-setting tasks. *Experimental Brain Research, 84,* 434–438.

Duncker, K. (1929). Über induzierte Bewegung (Ein Beitrag zur Theorie optisch wahrgenommener Bewegung). *Psychologische Forschung, 12,* 180–259.

Goodale, M. A. (2001). Different spaces and different times for perception and action. *Progress in Brain Research, 134,* 313–331.

Gregory, R. L. (1968). Perceptual illusions and brain models. *Proceedings of the Royal Society London B, 171,* 279–296.

Hillis, J. M., Ernst, M. O., Banks, M. S., & Landy, M. S. (2002). Combining sensory information: Mandatory fusion within, but not between, senses. *Science, 298,* 1627–1630.

Kersten, D., Mamassian, P., & Yuille, A. (2004). Object perception as Bayesian inference. *Annual Review of Psychology, 55,* 271–304.

Knill, D. C. & Kersten, D. (1991). Apparent surface curvature affects lightness perception. *Nature, 351,* 228–230.

Knill, D. C. & Pouget, A. (2004). The Bayesian brain: The role of uncertainty in neural coding and computation. *Trends in Neurosciences, 27,* 712–719.

Körding, K. P. & Wolpert, D. M. (2006). Bayesian decision theory in sensorimotor control. *Trends in Cognitive Sciences, 10,* 319–326.

Lamme, V. A. F. & Roelfsema, P. R. (2000). The distinct modes of vision offered by feedforward and recurrent processing. *Trends in Neurosciences, 23,* 571–579.

Land, E. H. & McCann, J. J. (1971). Lightness and retinex theory. *Journal of the Optical Society of America, 61,* 1.

Lotto, R. B., Williams, S. M., & Purves, D. (1999). An empirical basis for Mach bands. *Proceedings of the National Academy of Sciences of the United States of America, 96,* 5239–5244.

Muller, C. M. P., Brenner, E., & Smeets, J. B. J. (2007). Living up to optimal expectations. *Journal of Vision, 7*(3), 1–10.

Purves, D., Shimpi, A., & Lotto, R. B. (1999). An empirical explanation of the Cornsweet effect. *Journal of Neuroscience, 19,* 8542–8551.

Smeets, J. B. J. & Brenner, E. (1994). The difference between the perception of absolute and relative motion: A reaction time study. *Vision Research, 34,* 191–195.

Smeets, J. B. J. & Brenner, E. (1995a). Perception and action are based on the same visual information: Distinction between position and velocity. *Journal of Experimental Psychology: Human Perception and Performance, 21,* 19–31.

Smeets, J. B. J. & Brenner, E. (1995b). Prediction of a moving target's position in fast goal-directed action. *Biological Cybernetics, 73,* 519–528.

Smeets, J. B. J. & Brenner, E. (2004). Curved movement paths and the Hering illusion: Positions or directions? *Visual Cognition, 11,* 255–274.

Smeets, J. B. J., Brenner, E., de Grave, D. D. J., & Cuijpers, R. H. (2002). Illusions in action: Consequences of inconsistent processing of spatial attributes. *Experimental Brain Research, 147,* 135–144.

Smeets, J. B. J., van den Dobbelsteen, J. J., de Grave D. D. J., van Beers, R. J., & Brenner, E. (2006). Sensory integration does not lead to sensory calibration, *Proceedings of the National Academy of Sciences of the United States of America, 103,* 18781–18786

Todd, J. T., Oomes, A. H. J., Koenderink, J. J., & Kappers, A. M. L. (2001). On the affine structure of perceptual space. *Psychological Science, 12,* 191–196.

van Beers, R. J., Sittig, A. C., & Denier van der Gon, J. J. (1998). The precision of propioceptive position sense. *Experimental Brain Research, 122,* 367–377.

van Beers, R. J., Sittig, A. C., & Denier van der Gon, J. J. (1999). Integration of proprioceptive and visual position-information: An experimentally supported model. *Journal of Neurophysiology, 81,* 1355–1364.

van Doorn, A. J. & Koenderink, J. J. (1982). Spatial properties of the visual detectability of moving spatial white noise. *Experimental Brain Research, 45,* 189–195.

von Helmholtz, H. (1925). Treatise on Physiological Optics, Vol. III, The Perceptions of Vision. Birmingam, Alabama: The Optical Society of America.

Wolpert, D. M., Ghahramani, Z., & Jordan, M. I. (1995). Are arm trajectories planned in kinematic or dynamic coordinates – an adaptation study. *Experimental Brain Research, 103,* 460–470.

SECTION IV

A Dynamic Brain

12

NEURONAL AND CORTICAL DYNAMICAL MECHANISMS UNDERLYING BRAIN FUNCTIONS

ANJA STEMME[1] AND GUSTAVO DECO[2]

[1]*Institute for Biophysics Computational Intelligence Group, University of Regensburg, Regenburg, Germany*
[2]*Theoretical and Computational Neuroscience, Universitat Pompeu Fabra, Barcelona, Spain*

INTRODUCTION

The "phi" phenomenon or the perception of "apparent motion" is closely related to the perception of motion pictures. It refers to the circumstance that observers report, under specific conditions, the perception of a moving object although only stationary stimuli had been presented. In 2000, Steinman et al. (2000) clarified the distinction between the perception of "phi" movements (i.e., "pure" movements) and "beta" movements (i.e., optimal movements) of two discrete stimuli as detected by Max Wertheimer in 1912, which launched the so called Gestalt revolution in perceptual psychology. Steinman et al. pointed out that the perception of pure movements ("phi") corresponds to the perception of a "background" movement without any specific form, as opposed to the perception of optimal movement ("beta") where two presented discrete stimuli appear as a single moving object.

In the "color phi" experiment (compare as well Figure 12.1), the participants are seated in front of a screen on which two spots in different colors at different positions are presented. Due to the spatial and timing conditions of the presented spots, the observers report that they see a moving spot (the classical "phi" phenomenon),

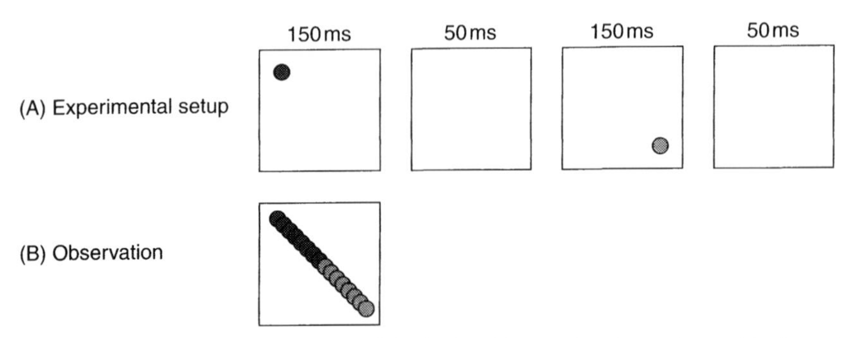

FIGURE 12.1 An example setup of the "color phi" experiment as referenced by Dennett (1991, 1992) and conducted in a similar way by Kolers & von Grünau (1976). Top: A red spot is presented for 150ms followed by a short delay of 50ms, followed by the green spot for 150ms, followed by a second delay of 50ms. This "flickering" of the spots is repeated continuously. Bottom: The observers of this experimental setup report that they see a moving spot changing the color roughly in the middle of the way. (See color plate)

changing its color midway between the spots actually presented (the "color phi" phenomenon). Following the explanations of Steinman et al. this "phi" movement should rather be called "beta" movement. The experimental question, however, which extended the "phi phenomenon" to the "color phi phenomenon" was raised initially by the philosopher Nelson Goodman and investigated by Kolers & von Grünau (1976, described in Dennett, 1992).

When Goodman found out about the "phi" phenomenon, that is, the illusory perception of motion following the presentation of two separate spots of the same color, he wondered whether the illusion of motion would disappear when the two spots have a different color or whether the color of the illusory moving spot would gradually change. Following the experimental investigation of Kolers and von Grünau, which revealed the abrupt change in color, Goodman concluded that the brain "fills in" the color of the moving spot with either of the two presented colors and raised the question how the observers could *know* the second color already in the middle of the "way," that is, *before* the second color was actually presented on the screen (Goodman, 1978, see also Dennett, 1992).

The philosopher Daniel Dennett used, among other "filling-in" phenomena, the "color phi" experiment to illustrate his "multiple draft concept" of the brain and to underline that the brain does practically no "filling-in" at all (Dennett, 1991, S. 114). The general question of perceptual "filling-in" is of broad interest to philosophers as well as to scientists and is observed in a range of situations: for example filling-in at the blind spot or at scotoma. Following Dennett, the whole idea of perceptual "filling-in" is misleading and relies on the concept of "Cartesian materialism," that is, the assumption of an "audience" or a central area within the brain to which images are "presented" or where all the sensual information "comes together" (Dennett, 1991, 1992). However, possible neuronal mechanisms underlying especially blind spot and scotoma filling-in phenomena have been investigated in a range of animal studies and are subject to ongoing discussion and theorizing (Pessoa et al., 1998; Komatsu, 2006 for an overview).

With respect to the perceptual illusion of apparent motion a range of studies investigated the contribution of primary visual cortex (V1) and higher visual areas (hMT+/V5) using functional magnetic resonance imaging (Liu et al., 2004; Muckli et al., 2005; Sterzer et al., 2006) and repetitive transcranial magnetic stimulation ("rTMS," Matsuyoshi et al., 2007). These studies provided some evidence that the perceptual illusion of the moving spot is closely related to the perception of a real moving spot by demonstrating that comparable areas in the primary visual cortex are activated during both conditions (Muckli et al., 2005). The similarity between the perceived path of an apparently moving stimulus and a continuously ("real") moving stimulus was demonstrated as well by Siori et al. (2000). In the case of apparent motion, however, the activation of the primary visual cortex is not caused by the visual stimuli themselves, as in the real motion condition, but seems to be initiated by feedback connections from the higher visual area hMT+/V5 (Sterzer et al., 2006). The cortical area hMT+/V5 is thought to be involved in the perception of moving visual objects as demonstrated by case studies of Akinetopsia (Zihl et al., 1983). The (feedback) activation of the primary visual cortex by these higher cortical areas would thus provide a direct illustration of a cortical "filling-in" during the perception of apparent motion. However, Liu et al. (2004) detected no corresponding activation of the primary visual cortex in their study but confirmed solely the contribution of hMT+/V5. Matsuyoshi et al. (2007) demonstrated a lower probability to perceive apparent motion when hMT+/V5 is "disabled" using rTMS. Hence, there is a range of research with respect to the phenomenon of apparent motion also addressing philosophical aspects with respect to the question of "filling-in": How are differences between (subjective) perceptions and (objective) stimulus presentations explainable? Whereas there is at least some evidence with respect to the general phenomenon of apparent motion available, the question how the perceived color of an apparently moving stimulus might change in the middle of an illusory path *before* the observers are able to know the second color has not been addressed so far.

To illustrate how neurodynamical modeling can be used to understand and further investigate phenomena as "color phi" without the necessity to dig into philosophical speculations we would like to outline the design and testing of a neurodynamical model for an example (more complex set shifting) task. We describe how data regarding neuronal behavior and cortical organization are useable for the generation of basic modeling principles (see section "How to Build a Suitable Neuronal Model for a Psychological Experiment"). Afterwards we demonstrate how neuroimaging data can be used to verify the design of the model for a set shifting task (see section "Calculating the fMRI Signal for an Example Set Shifting Model"). The calculation of response times and error rates in an example task for human participants and for the neurodynamical model is outlined next and thus a solid base regarding the neurobiological plausibility of the neurodynamical model is obtained (see section "Response Times and Error Rates in an Example Set Shifting Task"). Lastly we sketch how a model, similar to the one used and tested in the previously mentioned tasks, through its spiking dynamics might account in a very natural way for "color phi" (see section "Summary and Back to 'Color Phi'").

HOW TO BUILD A SUITABLE NEURONAL MODEL FOR A PSYCHOLOGICAL EXPERIMENT

The first question to address when looking for a suitable neuronal model for a specific task is the question of modeling detail and the achievable neurobiological plausibility. Especially for set shifting tasks there are a range of neuronal models which use rather abstract neuronal elements (Dehaene & Changeux, 1991; Berdia & Metz, 1998; Amos, 2000; Rougier & O'Reilly, 2002; Rougier et al., 2005). However, the use of more abstract neuronal units has the disadvantage that several functional assumptions regarding the neuronal operations are necessary and, furthermore, it was demonstrated that these models are not able to account for a range of experimental data such as the appropriate discrimination of different error types (see Stemme, 2007, Chapter 4).

By contrast rather detailed biophysical descriptions of neuronal behavior are available and it was demonstrated that models using these description are able to simulate the recorded behavior of single neurons in "delayed match to sample" tasks (Deco & Rolls, 2005b) and might account as well for available neuroimaging data (Deco et al., 2004).

These models describe the neuronal behavior using the "Integrate-and-Fire" model, first introduced by Lapicque (1907) and further extended by Tuckwell (1988). Amit and Brunel (1997) converted this approach to a system of synaptic currents and spiking rates and Wang (1999) further extended the model to incorporate synaptic connections using different receptor types. Important model parameters were adjusted according to neurophysiological findings. Finally, Brunel and Wang (2001) suggested a framework comprising "Integrate-and-Fire" neurons to form a "cortical module." This module was further extended to a modular biased competition and cooperation approach (Deco & Lee, 2002; Deco & Rolls, 2003; Deco & Rolls, 2005a). Within this framework only minimal assumptions regarding neuronal operations are used, which were verified against neurophysiological findings, and different neuronal as well as synaptic connection types are considered. Hence this framework represents a rather optimal base for a neuronal modeling approach with a high degree of neurobiological orientation.

Following the "Integrate-and-Fire" model, the development of the neuron membrane potential can be represented by an equivalent electric circuit consisting of a capacity in parallel to a resistance and a battery (compare Figure 12.2). The battery represents the resting potential of a neuron, $V_{rest} = -70\,\mathrm{mV}$.

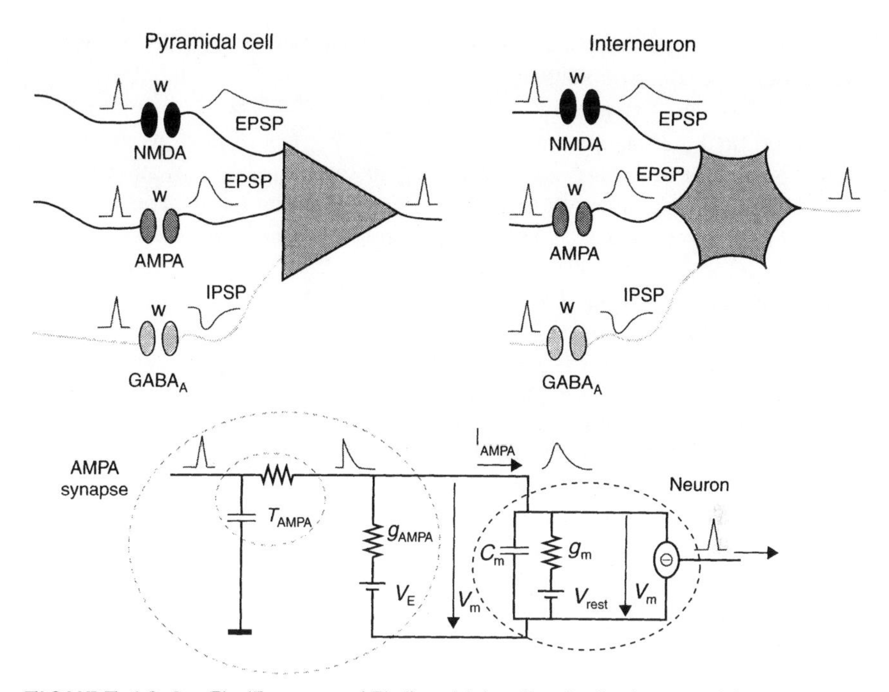

FIGURE 12.2 The "Integrate-and-Fire" model describes the development of the subthreshold membrane potential for neurons by an equivalent electrical circuit. Different neuronal and synaptic connection types are possible. EPSP—excitatory postsynaptic potential. IPSP—inhibitory postsynaptic potential. Top left: schematic diagram of a pyramidal cell and different connections types. Top right: schematic diagram of an interneuron and connections. Bottom: The equivalent circuit used to calculate the subthreshold development of the membrane potential for either of the neurons with an exemplary AMPA synapse connected. (compare also well Stemme, 2007, p. 73, Figure 6.1) (See color plate)

When the membrane potential reaches a threshold θ an action potential is generated, which means that the neuron "spikes." Afterwards the membrane potential is set to V_{reset} and not updated until a refractory period τ_{rp} passed. Below threshold, the membrane potential evolves as a function of the synaptic input current I_{syn} and the membrane parameters R_m and C_m. The subthreshold development of the membrane potential is given by:

$$C_m \frac{dV_m(t)}{dt} = -\frac{1}{R_m} V_m(t) - V_{rest} I_{syn}(t)$$

The product of capacitance and resistance is also called "membrane time constant," $\tau_m = C_m \times R_m$. Without any input the membrane potential decays as a function of τ_m to V_{rest}. The parameter values necessary to describe the behavior of two important neuron types within the cerebral cortex, excitatory pyramidal cells, and inhibitory interneurons, differ and are based again on neurophysiological data

(McCormick et al., 1985). In a similar way, the contributions of different synaptic connection types are computable.

The framework Brunel and Wang suggested (also Figure 12.3) represents a limited cortical area and consists of two different types of neurons, 80% excitatory pyramidal cells and 20% inhibitory interneurons, consistent with neurophysiological findings (Abeles, 1991). These neurons are grouped into two appropriate types of pools: excitatory pools and one inhibitory pool. The neurons are modeled as "Integrate-and-Fire" neurons considering three different synaptic connection types: two excitatory—AMPA and NMDA connections—and one inhibitory—GABA. As the cerebral cortex is highly connected and thus the simulation of a "closed" cortical area would be unrealistic, every neuron receives a certain "background" input from neurons outside the network. For the approximation of this "background" input, it is taken into account that neurons always show a certain level of activity, that is a spiking rate of approximately 3 Hz for

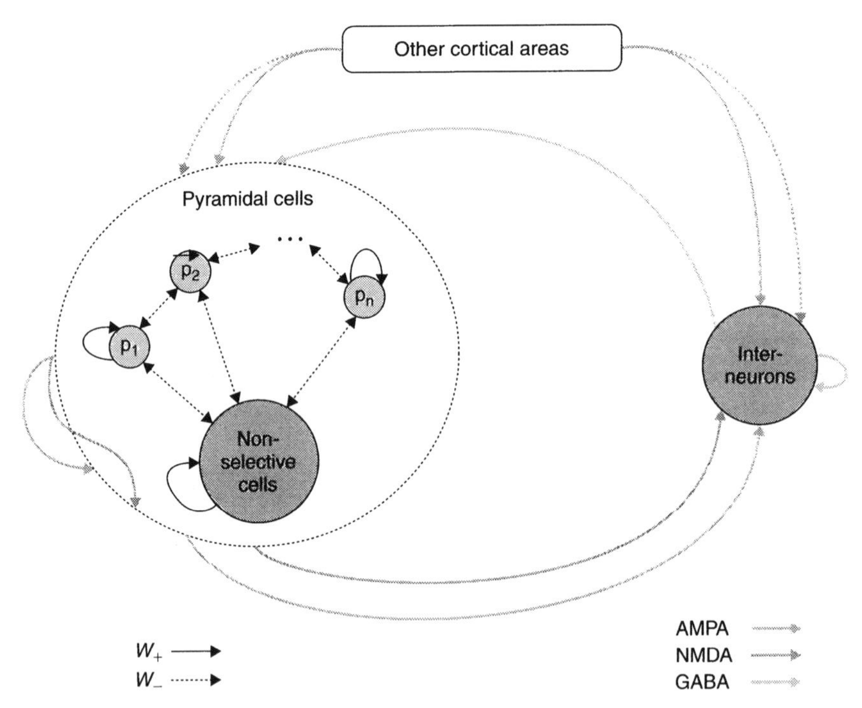

FIGURE 12.3 Overview of the cortical module. Repainted according to Brunel & Wang (2001, Figure 2, compare also Stemme, 2007, Figure 6.2). Interneurons send inhibitory connections to the pyramidal cells and to the other interneurons within the inhibitory pool. Pyramidal cells in turn send AMPA and NMDA connections to the interneurons and are also interconnected with the relatively strong strength w_+ within a selective pool or a comparatively weaker weight w_- between different selective pools, respectively. External glutamatergic input from other cortical areas can be mediated by AMPA and NMDA receptors. (See color plate)

pyramidal cells and 9 Hz for interneurons (Wilson et al., 1994), which is called the "spontaneous rate." Accordingly, the external background input is modeled as an AMPA-mediated Poisson train of spikes arriving from $N_{ext} = 800$ neurons with a rate of 3 Hz. Thus, the total background noise every neuron receives comes out to $\nu_{ext} = 800 \times 3\,\text{Hz} = 2.4\,\text{kHz}$.

The excitatory neuronal pools are considered to be selective to a certain task (i.e., an abstract response rule in a set shifting task) or to a visual stimulus, an assumption that is supported by single cell recordings with behaving monkeys (White & Wise, 1999; Wallis et al., 2001; Rainer & Miller, 2002). To simulate this "selectivity," the synaptic coupling strength ("weight") represents a major modeling parameter. It is assumed that the connections were established by Hebbian learning, that is, the coupling will be strong if a pair of neurons shows correlated activity, and weak if two neurons are activated in an uncorrelated way. Consequently, neurons within a specific pool are coupled with the relatively strong weight $w_+ = 2.1$ whereas connections between different pools are comparatively weak. Furthermore, the external input to these pools might be increased representing the presentation of a certain visual stimulus to the model, for example. For this purpose, the external AMPA-mediated input to the neurons within the specific pool is increased to $\upsilon_{ext} + \lambda_{stimulus}$. Compared to the background noise, $\nu_{ext} = 2.4\,\text{kHz}$, the stimulus specific input is rather low; a typical value might be $\lambda_{stimulus} = 0.15\,\text{kHz}$.

So far we described a neuronal framework which provides a high degree of detail with respect to the neuronal operations in combination with a rather minimal set of functional assumptions: equations and parameter values are oriented on neurophysiological observations. In the next section, we will demonstrate how this framework can be used to design an example set shifting model and how an fMRI signal for this model is computable, enabling the comparison with experimentally determined values.

CALCULATING THE fMRI SIGNAL FOR AN EXAMPLE SET SHIFTING MODEL

The "Wisconsin Card Sorting Test" (WCST, Grant & Berg, 1948) or the Stroop task (Stroop, 1935) are traditional set shifting tasks, which test the ability of the participants to switch attention from one aspect of an object to another. In these tasks subjects are required to attend to a selected property of a presented visual stimulus and select a feature-specific response. The different stimulus properties might interfere with each other as in the Stroop task: written colored words serve as stimulus displays and the participants are instructed to switch between the response rules "color naming" and "word reading." The valid response rule is usually indicated by an explicit task cue or a predefined task order. In the WCST, participants are required to sort cards according to one of three possible sorting criteria: color, form or number. The stimulus properties do not interfere with each other as in the Stroop task but the valid sorting rule is changed without explicit notice and the participants are required to detect the new valid rule by a trial and error procedure. Various phenomena have been observed with set shifting tasks: increased response times following set shifts as well as following the presentation of congruent visual stimuli. Also, for incongruent stimulus displays compared to congruent ones, response times are increased (Monsel, 2003 for an overview). Furthermore, increased error rates were detected following an uninstructed change of the valid rule, especially for patients suffering from lesions of the prefrontal cortex (Milner, 1963; Barceló & Knight, 2002).

Neuroimaging studies revealed an increase in the fMRI signal following an uninstructed shift of attention in WCST-like tasks. Nakahara et al. (2002) detected that fMRI signals from certain areas within the prefrontal cortex were transiently increased by 0.5–0.6% approximately 6 s after the set shifting event for human participants as well as for monkeys. Konishi et al. (1999) calculated fMRI signals for two variants of a WCST. They detected that the increase in the fMRI signal was significantly lower when the participants were explicitly instructed by a specific cue to shift their attention, than in the uninstructed, original WCST variant. In Stemme et al. (2005), we described in detail how a neurodynamical model for these tasks might be designed and how a theoretical fMRI signal is computable for the model. We will summarize the main steps in the following.

The first step in building a suitable neuronal model for these tasks is to identify the specific pools. As already mentioned above, single cell recordings exemplified that single neurons show rule specific (White & Wise, 1999; Wallis et al., 2001) as well as object specific (Rainer & Miller, 2002) activity in a range of behavioral tests. These results led us to the assumption that groups of neurons (i.e., the pools) code for specific stimulus features as well as for abstract rules in set shifting tasks.

Hence, the model (compare also Figure 12.4) comprises two pools serving as "rule pools," representing two different, possible active rules and four stimulus specific pools, representing the different stimulus properties (e.g., two different shapes and two different colors). A stimulus with certain features is presented to the model (e.g., shape number two, color number one) by adding an extra Poisson input to the specific pools (i.e., to shape pool No. 2 and color pool No. 1). An appropriate set of weights for the model can be identified using a mean field analysis (compare Stemme et al., 2005).

To raise and hold competition, the rule pools receive continuously a low attentional biasing input, λ_{bias}. During the simulations, the spiking dynamics of the different pools indicate which answer the model would give to a presented series of two stimuli. At the end of the trial we introduce an unspecific extra external input representing the feedback the model would receive to the previously given answer. The feedback input is provided simultaneously to both of the rule pools, thus ν_{ext} is increased by λ_{bias} and $\lambda_{feedback}$. In case of a correct answer, we refer to the feedback input as "positive feedback" and "negative feedback" in case of

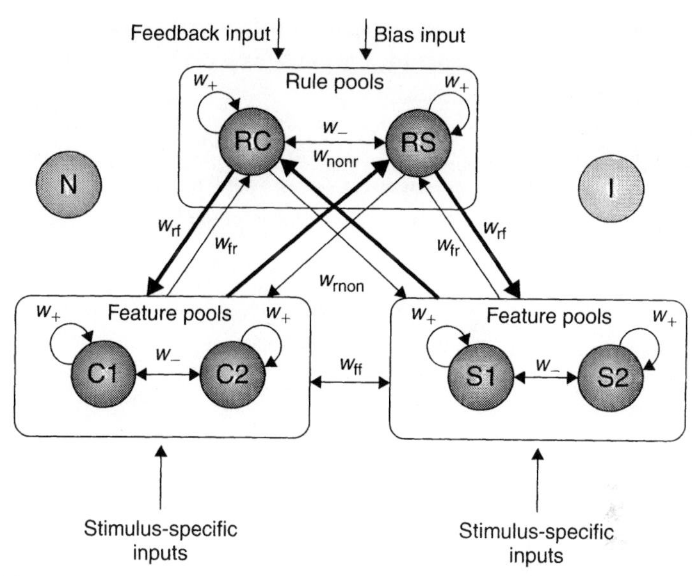

FIGURE 12.4 Neurodynamical model for the set shifting task (compare also Stemme, 2007, p. 123, Figure 8.5). The model comprises two pools responsible for different rules; RC color rule; RS shape rule. Furthermore, there are two pools representing two different colors (C1 and C2) and two pools representing different shapes (S1 and S2). Every rule pool supports "its" feature pools via the weight "w_{rf}." In the opposite direction, the weight "w_{nonr}" is comparatively stronger than the weight "w_{rf}," a configuration that enables the rule change. The neuronal pool named "N" comprises a pool of non-selective neurons. This neuronal pool accounts for the circumstance that not all neurons within a given cortical area are engaged in a specific task. The pool named "I" comprises the inhibitory neurons. Altogether, we chose to model $N_E = 1600$ excitatory neurons and $N_I = 400$ inhibitory neurons. The network of neurons is fully connected with different connection strengths as indicated. (See color plate)

an incorrect answer. However, the feedback input itself is in both cases an external, unspecific AMPA-mediated input to both of the rule pools, differing just in the amount of the value: a lower (positive) feedback value acts as a kind of "strengthener" of the currently active rule whereas a higher (negative) feedback value destabilizes the rule pool activity allowing for the change of the currently valid rule.

The design of the model reflects so far the main aspects of the WCST used by Konishi et al. (1999) and Nakahara et al. (2002). These are: (a) a certain set of visual stimuli requires a specific response of the participant based on a given rule. Hence we consider a task setup where a first stimulus consisting of two different feature dimensions (color and shape) is presented to the model followed by a second stimulus; (b) following the presentation of the second stimulus a response is required: "Match," if the presented pair of stimuli matches with respect to the currently relevant rule, "Non-Match" otherwise; (c) the relevant rule might change with or without notice, requiring a different response for the same set of stimuli. As the major objective is currently to calculate the fMRI signal theoretically emitted by the model and to compare it to experimentally determined values, we use a simple model design always providing correct answers except for the cases where a rule change occurred.

As a second step, we are now able to run the neurodynamical simulations and to compute the fMRI signal. For this purpose, it is necessary to use a hemodynamic response function. Following Glover (1999), a good estimation of the hemodynamic response to a certain brain event is given by:

$$h(t) = c_1 t^{n_1} e^{-t/t_1} - \alpha_2 c_2 t^{n_2} e^{-t/t_2}$$

with

$$c_i = \frac{1}{\max t^{n_i} e^{-t/t_i}}$$

where $n_1 = 6.0$, $t_1 = 0.9\,\text{s}$, $n_2 = 12.0$, $t_2 = 0.9\,\text{s}$ and $\alpha_2 = 0.2$ as verified in a range of experiments (Glover, 1999). The fMRI signal change for the model is computable as a convolution of this hemodynamic response function with the absolute sum of the different synaptic currents occurring during the simulation:

$$S_{\text{fMRI}}(t) = \int_0^{\infty} h(t - t') I_{\text{syn}}(t') \mathrm{d}t'$$

with

$$I_{\text{syn}} = \text{abs}(I_{\text{AMPA}}) + \text{abs}(I_{\text{NMDA}}) + \text{abs}(I_{\text{GABA}})$$

I_{syn} is considered for 100 ms intervals and normalized with the mean value of synaptic activity occurring during the simulation. A similar approach was used successfully by Deco et al. (2004). The results obtained during the simulation of instructed and uninstructed set shifts are presented in Figure 12.5. They provide a very good approximation to the results obtained for human and monkey subjects as determined by Konishi et al. (1999) and Nakahara et al. (2002).

There is an ongoing discussion whether the fMRI signal reflects synaptic activity or the spiking rate in the measured brain region (Logothetis et al., 2001; Heeger & David Ress, 2002 for a review). In our approach we used the synaptic activity, calculated as the absolute sum of the three different synaptic currents (AMPA, NMDA, and GABA) occurring in the model neurons, as the basis for the computation of the resulting fMRI signal. In Stemme et al. (2005), we also calculated the fMRI signal based on the spiking rate of the different pools. The obtained results supported the assumption that the fMRI signal is more closely related to the synaptic activity of a certain region than to the spiking rate of that same region.

In the next section, we outline how behavioral data, in terms of response times and error rates, are computable for the neurodynamical model.

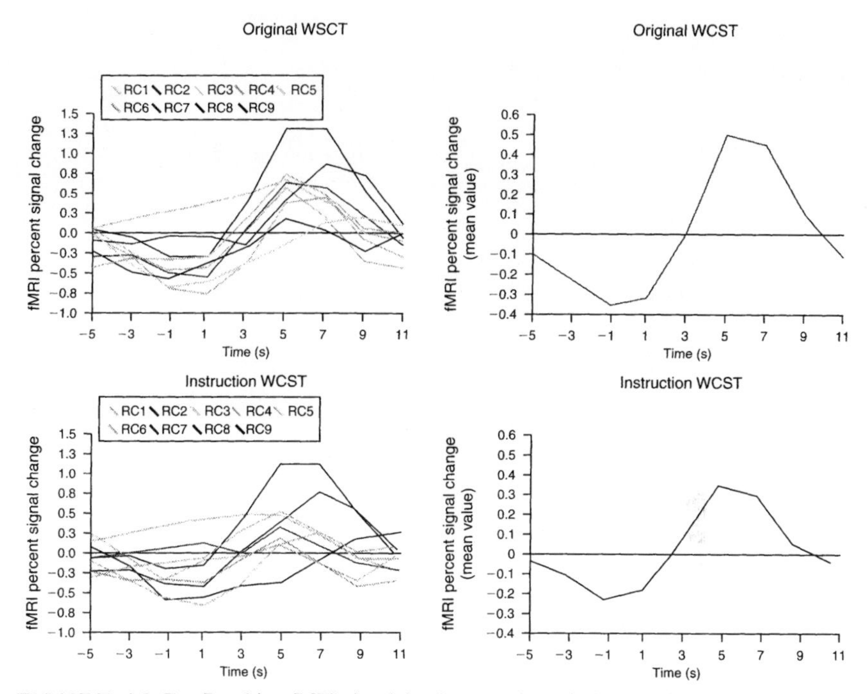

FIGURE 12.5 Resulting fMRI signal for the neurodynamical simulations. The top left diagram shows the resulting fMRI signal for nine single rule changes for the original WCST condition. The right diagram shows the corresponding mean value. The bottom diagrams show resulting fMRI signals for the instruction variant of the WCST. For all cases there is a close match with the experimentally determined values (compare also Stemme et al., 2005). (See color plate)

RESPONSE TIMES AND ERROR RATES IN AN EXAMPLE SET SHIFTING TASK

To get the necessary experimental data we performed a rather simple set shifting task using two different rules (compare also Stemme, 2007; Stemme et al., 2007a, b). The experiments were conducted with 40 healthy participants. The task setup used a combination of a "delayed match to sample" task and a Wisconsin-like paradigm and is depicted in Figure 12.6.

A sample display was shown for 500 ms, followed by a fixation delay of 1000 ms, followed by a test display which was presented until the participants responded by a key press ("y"—yes—sample and test display matched with respect to the valid rule; or "n"—no—sample and test display did not match according to the currently valid rule). Afterwards, a feedback message informed the participants whether their response was "correct" or "wrong." The feedback times varied and represented as well the inter-trial time. Participants were to discriminate between two different possible rules: Same object presented in sample and test display or same position on the screen of the visual stimuli. After an

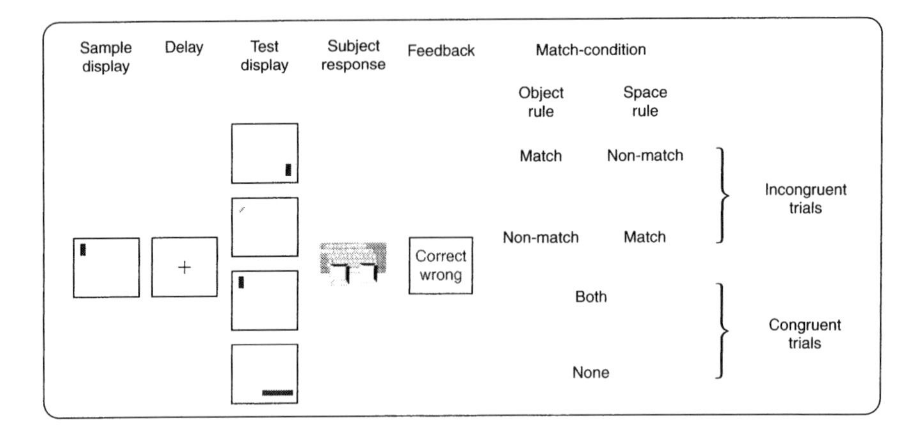

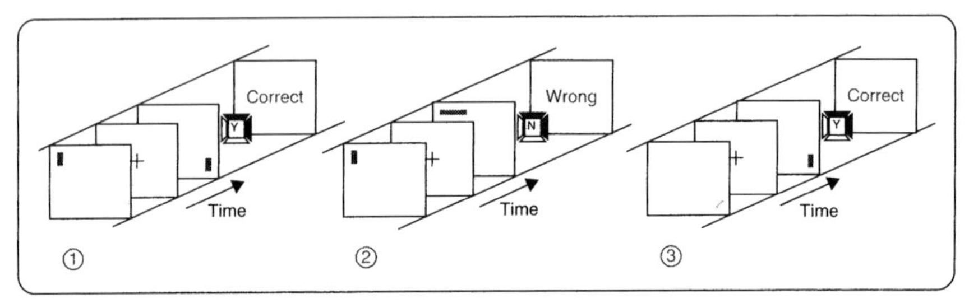

FIGURE 12.6 Task setup for the WDMS experiments. Setup of the experiments to collect experimental response data (WDMS experiment—"Wisconsin Delayed Match to Sample" experiment). Top: Timing dynamics for the single trials and possible stimulus (match) conditions. Bottom: Example trial sequence including a rule change in the second trial (feedback message: "wrong"). (See color plate)

arbitrary number of correct trials, the valid rule was changed without notice; in this case, the participants received the feedback message "wrong" although they responded correctly according to the previously valid rule.

Within this setup, four different match conditions were possible for sample and test display: "both"—the stimuli match with respect to both relevant dimensions; "match"—the stimuli match only with respect to the *currently* relevant stimulus dimension; "none"—the stimuli do not match in either dimension; "nonmatch"—the stimuli do *not* match with respect to the relevant dimension but only with respect to the *irrelevant* stimulus dimension. We calculated the response times for these different match conditions as well as relative to the rule change. Furthermore we differentiated various kinds of errors: Errors in conjunction with the rule change and attentional errors occurring during the maintenance phase of an active rule. The results are presented in Figure 12.7. Following a set shift, response times were moderately increased; response time for the different match conditions reflected the circumstance that both feature dimensions were memorized during the trials. In addition errors in conjunction with a rule change

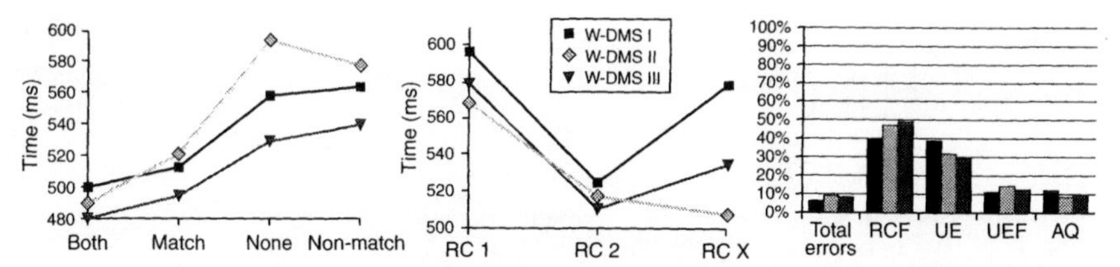

FIGURE 12.7 Experimental results. The results were obtained in three variants of the WDMS experiments (WDMS I, II, III). For these variants the feedback times and hence the intertrial time was varied (1500, 1000, 500 ms), which did not produce any significant effect on the results. Left: response times for the different match conditions. Middle: response times relative to the rule change. RC1: first trial after a rule change; RC2: second trial after a rule change. RC X: all other trials. Right: average error rates and types for the experiments. RCF: errors in the context of a rule change (rule change follow-up). UE: unmotivated, that is, attentional errors. UEF: errors following a previously unmotivated or attentional error; AQ: rule acquisition errors, which occur at the beginning of an experiment or an experimental block to determine the first valid rule. (See color plate)

("perseverative" errors, RCF in Figure 12.7), participants conducted a range of attentional errors as well.

For the neurodynamical simulations, we used a similar model to the one depicted in the previous section with an adjusted weight set accounting for the circumstance that both the relevant and the irrelevant feature dimensions influence the response times of the participants. The rule pools now represent a "space" rule for the position on the screen and a general "object" rule whereas the feature pools represent two different objects and two different locations (O1, O2, S1, S2; see also Stemme et al., 2007a). The model response is computable on the basis of spiking dynamics: If the summed spiking rate of all feature pools stays for a certain time above a threshold, the model response is considered to be "yes"; if, on the other hand, this summed spiking rate stays below a certain threshold, the model response is considered to be "no" (compare also Figure 12.8). The thresholds and the number of time intervals to pass or fail these thresholds represented further important model parameters.

The simulation results with respect to response times and error rates are depicted in Figure 12.9. We obtained an almost perfect match with respect to relative response times for the different match conditions and following a rule change. A constant factor, thought to represent a not-modeled explicit motor response, accounts for the differences of experimental to simulation response times. We were further able to demonstrate that it is important for the analysis of experimental results to consider not only experimental average values but also individual results and that different threshold sets are able to account for and thereby explain varying participant behavior (compare Stemme et al., 2007a, b). Thus, the neurodynamical simulations allowed us to consider average experimental results as well as individual response time *distributions*; in addition, the model was able to account for different types of errors as opposed to previous modeling approaches using rather simplified neuronal units (Rougier & O'Reilly, 2002; Rougier et al., 2005).

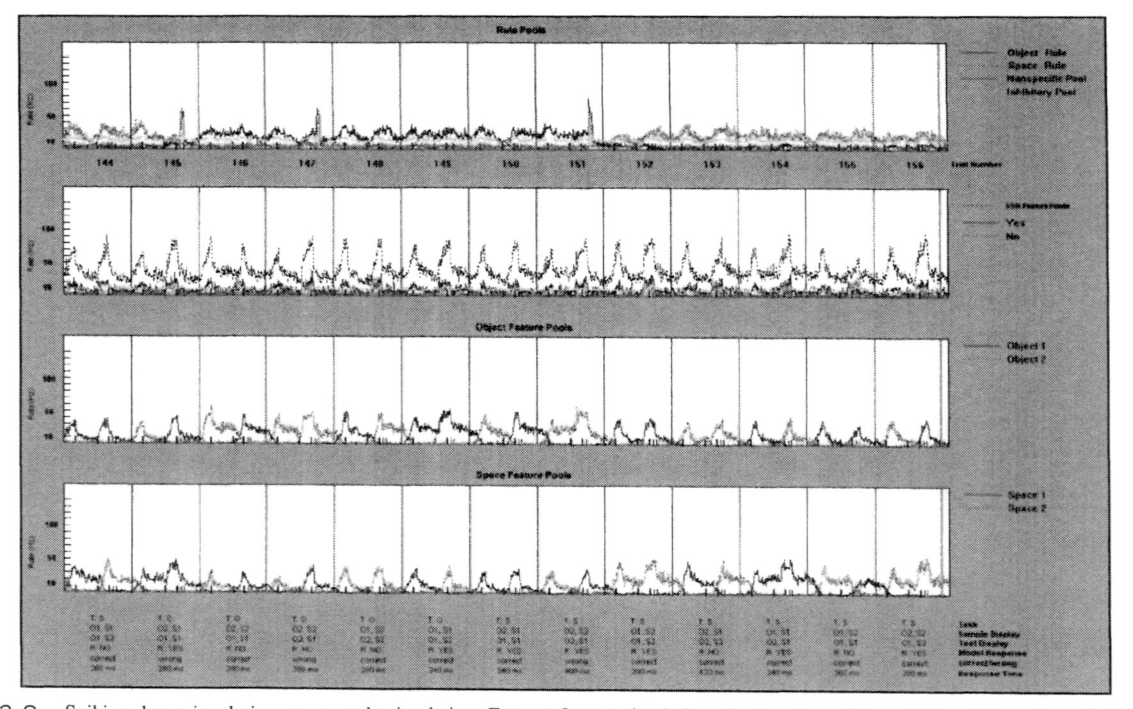

FIGURE 12.8 Spiking dynamics during an example simulation. Excerpt from a simulation showing the spiking activity of the rule pools (top diagram), the object feature pools (third diagram), the space feature pools (bottom diagram), and the summed spiking rate (ssr) of all feature pools. In other words, the time course of the ssr as depicted in the second diagram was obtained by adding the spiking activity of O1 (in Figure 12.4 labeled "C1"), O2 (in Figure 12.4 labeled "C2"), S1 and S2 for any given time point. As response times depend on both stimulus dimensions, the ssr constitutes the input provided to optional response pools of the model. Thus, a "yes" or "no" response of the model has to be based necessarily on differences in this input. The resulting model responses as presented in Figure 9 were calculated on the basis of ssr where an answer of the model was considered to be "yes" if the ssr passed a certain threshold and "no" if the ssr failed the threshold for a certain period of time. The threshold values constituted important model parameters (compare also Stemme, 2007; Stemme et al., 2007a). The bottom lines in the diagram indicate the individual trial setup, the calculated model response, the response time, and the feedback the model received (i.e., "correct" or "wrong"). (See color plate)

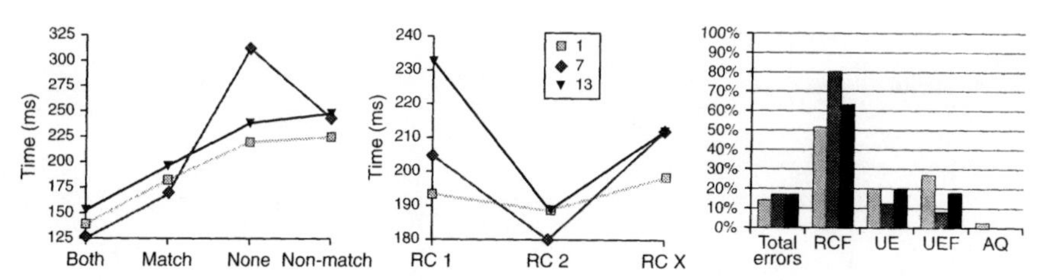

FIGURE 12.9 Simulation results. Results of the WDMS simulations with respect to response times and error rates for three example model configurations using different parameter values. A constant factor accounts for the response time differences compared to the experimental results (see Figure12.7), as an explicit motor response was not considered. Further investigations revealed the importance of considering individual participant results rather than averaged experimental data. (See color plate)

SUMMARY AND BACK TO "COLOR PHI"

With the usage of detailed biophysical descriptions of neuronal behavior it is possible to design realistic neuronal models for behavioral experiments; the calculation of fMRI signals theoretically emitted by these models allows the verification of the model design by a comparison with fMRI signals experimentally determined for human participants during the tasks. Furthermore, it is possible to calculate response times during the neurodynamical simulations and explain various facets of participant behavior. Also, the presented neurodynamical model was able to account for different types of errors. These are important aspects especially with respect to the examination of patient behavior.

The WCST is well known to be especially sensitive to dysfunctions of the prefrontal cortex (PFC) which seems to be reflected in corresponding experimental results, that is, a comparatively high amount of perseverative errors (Miller & Cohen, 2001). These findings appear to apply for patients with frontal lobe damages (first examined by Milner, 1963) as well as for patients suffering from Schizophrenia or Parkinson's disease (Owen et al., 1993; Everett et al., 2001). However, recent research revealed as well that errors more related to attentional issues might to a significant degree be responsible for the impaired performance of patients with frontal lobe damages (Barceló & Knight, 2002). Moreover, experimental results, especially for schizophrenic patients, do not always agree; schizophrenics might show perseverative behavior (Kolb & Wishaw, 1983; Everett et al., 2001) or might not (Goldstein et al., 1996; Landro et al., 2001). This divergence stresses the importance of developing detailed neuronal models considering different synaptic transmission types which are able to account for different error types.

Considering the spiking dynamics of the presented model it is possible to outline an explanation for the "color phi" phenomenon (compare also Figure 12.10). The spatial and timing conditions within the experiment might meet the requirements which indicate a moving object in the visual field, for example a moving spot or circle, respectively. As moving objects are processed via different cortical paths than

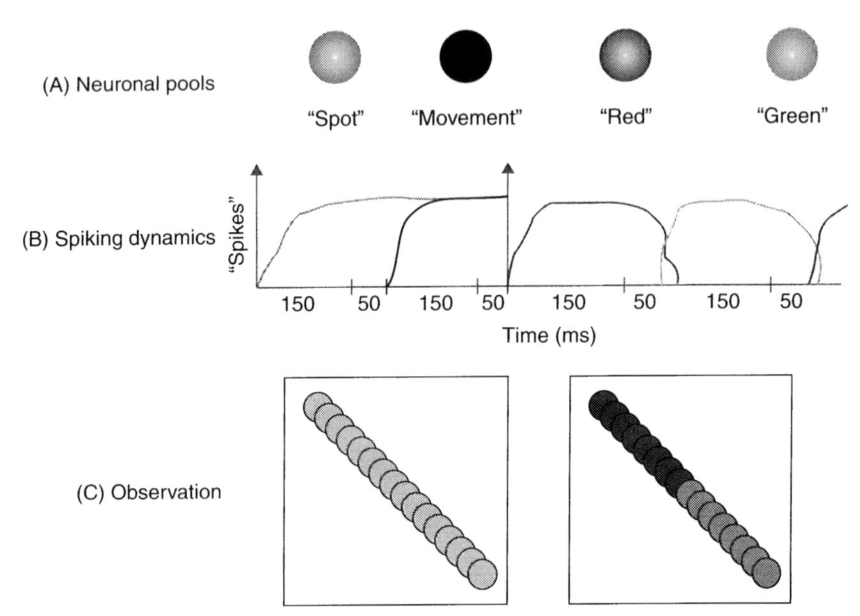

FIGURE 12.10 Illustration of possible neurodynamics underlying "color phi." (A) We might assume for example four neuronal pools selective or "responsible," respectively, for a spot presented in the visual field, for a specific movement occurring within the visual field, and for the colors red and green, in a way very similar to that described for Figure 12.4. Neuronal selectivity for colors and forms follows the neurophysiological findings, as outlined for the design of the set shifting model. Similarly, it has been demonstrated that neurons respond to specific movements within the visual field. (B) Estimated spiking dynamics of the neuronal pools. It is assumed that neurons respond according to the visual stimuli presented, which implicates that the "movement" neurons start spiking with the presentation of the second stimulus in the first trial and continue spiking for the remainder of the experiment, as do the "spot" neurons (both pools enter a state of persistant activity). The "color" neurons respond in an alternating manner according to the presented color of the stimulus. (C) Under the assumption that the activity of a neuronal pool leads to a certain (subjective) perception, that is the activity of the neuronal pool responsible for "red" leads to a "red perception", we are able to provide a rather easy explanation for "color phi" in considering the spiking dynamics; these indicate why observers see a moving spot (activity of the corresponding pools) which abruptly changes the color (for 150 ms "red" is presented and for another 150 ms "green"), at least starting with the presentation of the second stimulus in the first trial and thereafter for the remainder of the continuous experiment. (See color plate)

stationary objects, we might assume that the moving circle object is represented by a single pool of neurons that keep their activity throughout the trial. If we consider now that for a *time* frame of 150 ms a red color was presented to the observers followed by a short delay followed by the presentation of a green color, we are able to detect an easy and rather natural explanation for "color phi": The observers do not need to know the second color by some miraculous brain processes in the middle of the *way* but they surely do know the color of the second spot approximately after half of the presentation *time*. Thus, taken together, the differences between the visual presentations and the perceived moving object is easy to explain: the timing conditions in the experiments activated hMT+/V5 indicating a moving object. The color of this moving object changes because two different colors were presented in an alternating manner (150 ms for each color in the referenced experiment).

With respect to the perception of apparent motion Kolers (1984) reported a range of tested inter-stimulus intervals using different shapes and different colors. Whereas the color of the illusory moving item changes abruptly, the shape changes smoothly from triangle to square, for example, a circumstance which indicates a kind of "intermediate" interpretation of visual stimuli in the case of geometric shapes opposed to colors. Furthermore, the authors discriminated the perception of motion with replacement of the shapes from the perception of flicker and superposition. These findings generate a range of constraints regarding the neuronal organization, which are usable for the design of neurodynamical models.[1] The rather narrow timing constraints which distinguish the perception of pure compared to optimal movement, as indicated by Steinman et al. (2000), deliver further aspects to be considered in a neurodynamical model and offer also an explanation for divergent experimental results with respect to the activation of primary visual cortex: it is well possible that "pure" movements activate only hMT+/V5 whereas "optimal" movements of stationary stimuli lead as well to an activation of primary visual cortex. Thus, it seems necessary to differentiate "pure" from "optimal" movements and to determine the exact timing constraints for each perception. These constraints represent important aspects to be considered in the design of a neurodynamical model and are able to provide explanations for both phenomena. Taken together the referenced studies form a promising base for further experimental investigation accompanied by neurodynamical modeling work and thus represent a promising research path to gain a deeper understanding about the nature of the brain's "filling-in" in the case of apparent movement.

Thus, in summary, we have demonstrated the power of neurodynamical modeling work in conjunction with behavioral experiments to, on the one hand, explain human response behavior (set shifting tasks). On the other hand, we demonstrated how the use of neurodynamical models is able to provide explanations for rather philosophical questions (subjective perception of movement) and how these models might well guide experimental research in posing important questions. Hence, the combination of neurodynamical modeling work with psychological and neuroimaging experiments represents a promising path to gain a deeper understanding of the general operation of the human brain *and* the relationship to subjective perceptions. Most importantly, this research path might even be able to relinquish extensive and expensive single cell recordings with behaving monkeys by using already available knowledge with respect to the neuronal and cortical organization of the human brain. Single cell recordings

[1]Details will be outlined in a separate study currently in preparation: "The perception of visual movement—overview and neurodynamical modeling" by Stemme and Deco.

can be criticized in several respects: from an ethical point of view, from a financial point of view, and from an operational point of view; it is always necessary to "condition" the monkeys to perform a certain task contrary to the instruction of human participants in experiments. Also, it is difficult for monkeys to report their perceptions.

REFERENCES

Abeles, A. (1991). *Corticonics*. New York: Cambridge University Press.

Amit, D. J. & Brunel, N. (1997). Model of global spontaneous activity and local structured activity during delay periods in the cerebral cortex. *Cerebral Cortex, 7,* 237–252.

Amos, A. (2000). A computational model of information and processing in the frontal cortex and basal ganglia. *Journal of Cognitive Neuroscience, 12,* 505–519.

Barceló, F. & Knight, R. T. (2002). Both random and perseverative errors underlie WCST deficits in prefrontal patients. *Neuropsychologia, 40,* 349–356.

Berdia, S. & Metz, J. T. (1998). An artificial neural network stimulating performance of normal subjects and schizophrenics on the Wisconsin card sorting test. *Artificial Intelligence in Medicine, 13,* 123–138.

Brunel, N. & Wang, X.-J. (2001). Effects of neuromodulation in a cortical network model of object working memory dominated by recurrent inhibition. *Journal of Computational Neuroscience, 11,* 63–85.

Deco, G. & Lee, T. (2002). A unified model of spatial and object attention based on inter-cortical biased competition. *Neurocomputing, 46,* 775–781.

Deco, G. & Rolls, E. T. (2003). Attention and working memory: A dynamical model of neural activity in the prefrontal cortex. *European Journal of Neuroscience, 18,* 2374–2390.

Deco, G., Rolls, E. T., & Horwitz, B. (2004). What and where in visual working memory: A computational neurodynamical perspective for integrating fMRI and single-neuron data. *Journal of Cognitive Neuroscience, 16,* 683–701.

Deco, G. & Rolls, E. T. (2005a). Neurodynamics of biased competition and cooperation for attention: A model with spiking neurons. *Journal of Neurophysiology, 94,* 295–313.

Deco, G. & Rolls, E. T. (2005b). Synaptic and spiking dynamics underlying reward reversal in the orbitofrontal cortex. *Cerebral Cortex, 15,* 15–30.

Dehaene, S. & Changeux, J. P. (1991). The Wisconsin card sorting test: Theoretical analysis and modeling in a neuronal network. *Cerebral Cortex, 1,* 62–79.

Dennett, D. C. (1991). *Consciousness Explained.* Boston, Toronto, London: Little, Brown and Company.

Dennett, D. C. (1992). Filling in versus finding out: A ubiquitous confusion in cognitive science. In Van den Broek & P. Knill (Eds.), *Cognition, Conception, and Methodological Issues.* American Psychological Association. APA Print. http://cogprints.org/267/

Everett, J., Lavoie, K., Gagnon, J. F., & Gosselin, N. (2001). Performance of patients with schizophrenia on the Wisconsin card sorting test (WCST). *Journal of Psychiatry and Neuroscience, 26,* 123–130.

Glover, G. H. (1999). Deconvolution of impulse response in event-related BOLD fMRI. *Neuroimage, 9,* 416–429.

Goldstein, G., Beers, S. R., & Shemansky, W. J. (1996). Neuropsychological differences between schizophrenic patients with heterogenous Wisconsin card sorting test performance. *Schizophrenia Research, 21,* 13–18.

Goodman, N. (1978). *Ways of Worldmaking*. Indianapolis, Hackett Publishing Company.

Grant, D. A. & Berg, E. A. (1948). A behavioral analysis of degree of reinforcement and ease of shifting to new responses in a Weigl-type card sorting problem. *Journal of Experimental Psychology, 38,* 404–411.

Heeger, D. J. & David Ress, A. (2002). What does fMRI tell us about neuronal activity? *Nature Reviews Neuroscience, 3,* 142–151.

Kolb, B. & Wishaw, I. Q. (1983). Performance of schizophrenic patients on tests sensitive to left or right frontal, temporal, or parietal function in neurological patients. *The Journal of Nervous and Mental Disease, 171,* 435–443.

Kolers, P. A. (1984). Motion from continuous or discontinuous arrangements. *Computer Graphics*12–16.

Kolers, P. A. & von Grünau, M. (1976). Shape and color in apparent motion. *Vision Research, 16,* 329–335.

Komatsu, H. (2006). The neuronal mechanism of perceptual filling-in. *Nature Reviews Neuroscience, 7,* 220–231.

Konishi, S., Kawazu, M., Uchida, I., Kikyo, H., Asakura, I., & Miyashita, Y. (1999). Contribution of working memory to transient activation in human inferior prefrontal cortex during performance of the Wisconsin card sorting test. *Cerebral Cortex, 9,* 745–753.

Landro, N. I., Pape-Ellefsen, E., Hagland, K. O., & Odland, T. (2001). Memory deficits in young schizophrenics with normal intellectual function. *Scandinavian Journal of Psychology, 42,* 459–466.

Lapicque, L. (1907). Recherches quantitatives sur l'excitation electrique des nerfs traitée comme une polarisation. *Journal of Physiology (London), 9,* 620–635.

Liu, T., Slotnik, S. D., & Yantis, S. (2004). Human MT+ mediates perceptual filling-in during apparent motion. *NeuroImage, 21,* 1772–1780.

Logothetis, N. K, Pauls, J., Augath, M., Trinath, T., & Oeltermann, A. (2001). Neurophysiological investigation of the basis of the fMRI signal. *Nature, 412,* 150–157.

Matsuyoshi, D., Hirose, N., Mima, T., Fukuyama, H., & Osaka, J. (2007). Repetitive transcranial magnetic stimulation of human MT+ reduces apparent motion perception. *Neuroscience Letters, 429,* 131–135.

McCormick, D., Connors, B., Lighthall, J., & Prince, D. (1985). Comparative electrophysiology of pyramidal and sparsely spiny stellate neurons in the neocortex. *Journal of Neurophysiology, 54,* 782–806.

Miller, E. K., Jonathan, D., & Cohen, (2001). An integrative theory of prefrontal cortex function. *Annual Review of Neuroscience, 24,* 167–202.

Milner, B. (1963). Effects of different brain lesions on card sorting. *Archives of Neurology, 9,* 90–100.

Monsel, S. L. (2003). Task switching. *Trends in Cognitive Science, 7,* 134–140.

Muckli, L., Kohler, A., Kriegeskorte, N., & Singer, W. (2005). Primary visual cortex activity along the apparent-motion trace reflects illusory perception. *PLOS Biology, 3,* 1501–1510.

Nakahara, K., Hayashi, T., Konishi, S., & Miyashita, Y. (2002). Functional MRI of macaque monkeys performing a cognitive set-shifting task. *Science, 295,* 1532–1536.

Owen, A. M., Roberts, A. C., Hodges, J. R., Summers, B. A., Polkey, C. E., & Robbins, T. W. (1993). Contrasting mechanisms of impaired attentional set-shifting in patients with frontal lobe damage or Parkinson's disease. *Brain, 116,* 1159–1175.

Pessoa, L., Thompson, E., & Noe, A. (1998). Finding out about filling-in: A guide to perceptual completion for visual science and the philosophy of perception. *Behavioral and Brain Science, 21,* 723–748.

Rainer, G. & Miller, E. K. (2002). Timecourse of object-related neural activity in the primate prefrontal cortex during a short-term memory task. *European Journal of Neuroscience, 15,* 1244.

Rougier, N. P. & O'Reilly, R. C. (2002). Learning representations in a gated prefrontal cortex model of dynamic task switching. *Cognitive Science, 26,* 503–520.

Rougier, N. P., Noelle, D. C., Braver, T. S., Cohen, J. D., & O'Reilly, R. C. (2005). Prefrontal cortex and flexible cognitive control: Rules without symbols. *Proceedings of the National Academy of Sciences of the United States of America, 102,* 7338–7343.

Siori, S., Cavanagh, P., Miyamoto, T., & Yagushi, H. (2000). Tracking the apparent location of targets in interpolated motion. *Vision Research, 40,* 1365–1376.

Steinman, R. M., Zygmunt Pizlo, A., & Pizlo, F. J. (2000). Phi is not beta, and why Wertheimer's discovery launched the Gestalt revolution. *Vision Research, 40,* 2257–2264.

Stemme, A. (2007). *Neuronal principles underlying cognitive flexibility – A biophysical model for set shifting tasks.* Germany: BoD Verlag, Norderstedt.

Stemme, A., Deco, G., Busch, A., & Schneider, W. X. (2005). Neurons and the synaptic basis of the fMRI signal associated with cognitive flexibility. *NeuroImage, 26*(2), 454–470.

Stemme, A., Deco, G., & Busch, A. (2007a). The neuronal dynamics underlying cognitive flexibility in set shifting tasks. *Journal of Computational Neuroscience, 23*(3), 313–331.

Stemme, A., Deco, G., & Busch, A. (2007b). The neurodynamics underlying attentional control in set shifting tasks. *Cognitive Neurodynamics, 1*(3), 249–259.

Sterzer, P., Haynes, J. D., & Rees, G. (2006). Primary visual cortex activation on the path of apparent motion is mediated by feedback from hMT + /V5. *NeuroImage, 32,* 1308–1316.

Stroop, J.-R. (1935). Studies of interference in serial verbal reactions. *Journal of Experimental Psychology, 18,* 643–662.

Tuckwell, H. C. (1988). *Introduction to theoretical neurobiology I: linear cable theory and dendritic structure*. Cambridge University Press. Cambridge.

Wallis, J. D., Anderson, Kathleen C., & Miller, Earl. K. (2001). Single neurons in prefrontal cortex encode abstract rules. *Nature, 411,* 953–956.

Wang, X.-J. (1999). Synaptic basis of cortical persistent activity: The importance of NMDA receptors to working memory. *Journal of Neuroscience, 19,* 9503–9587.

White, I. M. & Wise, S. P. (1999). Research article: Rule-dependent neuronal activity in the prefrontal cortex. *Experimental Brain Research, 126,* 315–335.

Wilson, F. A. W., Scalaidhe, S. P. O., & Goldman-Rakic, P. S. (1994). Functional synergism between putative-aminobutyrate-containing neurons and pyramidal neurons in prefrontal cortex. *Proceedings of the National Academy of Science of the United States of America, 91,* 4009–4013.

Zihl, J., von Cramon, D., & Mai, N. (1983). Selective disturbance of movement vision after bilateral brain damage. *Brain, 106,* 313–340.

13

Dynamic Field Theory as a Framework for Understanding Embodied Cognition

Sebastian Schneegans and Gregor Schöner

Institut für Neuroinformatik, Ruhr-Universität Bochum, Bochum, Germany

Textbooks of cognitive psychology will talk a lot about such things as memory, thinking, deciding, or language, typically with some opening chapters on perception. Movement is often quite secondary in such accounts, and is considered to be a somewhat "low-level" activation of organisms. Yet, all behavior of an organism is ultimately motor behavior. Through motor actions do organisms reveal that they remember something and they have planned something. Visual perception is most commonly supported by motor action that controls where our eyes are pointing or actively supports visual exploration when we take an object into our hands. Conversely, even simple motor acts seem to require the sorts of things that are the stuff of cognition, such as when we must select one of many objects which we want to grasp, or when we must turn our body to bring into our visual array a desired object which we remember is to the right of where we currently look.

Embodied cognition is an approach to cognition that has roots in motor behavior. This approach emphasizes that cognition typically involves acting with a physical body on an environment in which that body is immersed. The approach of embodied cognition postulates that understanding cognitive processes entails understanding their close link to the motor surfaces that may generate action and to the sensory surfaces that provide sensory signals about the environment. To a certain extent, the embodiment stance implies a mistrust of the abstraction inherent in much information processing thinking, in which the interface between cognitive processes and their sensorimotor support is drawn at a level that is quite removed from both the sensory and the motor systems.

The roots in motor behavior of the embodiment stance manifest themselves also in the emphasis on the real-time autonomy of cognitive processes. These are not typically controlled or triggered by specific inputs to which an "answer" must be generated. Instead, cognition always happens on a background of ongoing behavior. The state of an organism's nervous systems comes from somewhere and goes somewhere. There is hardly any cognition that does not in some way depend on the recent behavioral and stimulation history as well as the concurrent environmental context. In relation to the environment, this context sensitivity of cognition is sometimes referred to as "situatedness," a concept we subsume here under embodiment.

Finally, for some (and for us), the embodiment stance also postulates that an understanding of cognition must be based on concepts that are consistent with the fundamental principles of neuronal organization that govern our nervous systems. This means, in particular, that cognition happens in a temporally continuous and asynchronous fashion, without a central controller that clocks computational steps. This also means that a homogeneous language is spoken within the neuronal networks which our nervous system consists of. Neurons interact through their activation levels, be they assessed by firing rates, levels of synchronicity, or intra-cellular potentials. What neurons transmit through their axons and the synapses they form is always the same type of variable. Neurons do not transmit messages beyond these physical signals. The processing of neurons is largely homogeneous across the higher nervous system, and is based essentially on weighted integration. Only through the structure of the neuronal networks, of which neurons are part, may the different functionally relevant states of neurons be brought about. Note, however, that there are no signatures of the temporal discreteness of neuronal spiking events or of the spatial discreteness of individual neurons in cognition or behavior. So the level at which the neuronal substrate provides constraints for an understanding of cognition must be identified rather than fixed a priori. In our review, that level will consist of spatio-temporally continuous neuronal activation patterns. The radical stance within the approach of embodied cognition is that the link to the sensory and motor surfaces, the constraints imposed by the physical body and the structured environment in which it is immersed, the constraints of temporal continuity and autonomy, and the constraints provided by the neuronal substrate are relevant not only for the subset of cognitive processes that control action and perception. Instead, in the radical view, all cognition is hypothesized to be of this kind. Remoteness of cognitive processes from the sensorimotor domain, independence of physical instantiation, forward computation only from given inputs, and abstraction from the neuronal substrate are all illusory. Even the highest form of cognition, thinking, is viewed as a form of motion, characterized by similar constraints as motor behavior, if not always directly acted out by the motor system (Port & Gelder, 1995). These claims cannot be considered proven at this time but provide a very stimulating research program for a fresh understanding of cognition.

It is clear, that new theoretical tools are needed to address cognition within the embodiment perspective. This chapter reviews one set of theoretical concepts which we believe to be particularly suited to address the constraints of embodiment and situatedness. We refer to this set of concepts as *Dynamical Systems Thinking* or *DST*. The concepts are based on the mathematical theory of dynamical systems, but are not identical with that theory, of course (which is why we resist the term "Dynamical Systems Theory" that is sometimes used to describe this approach). In shortest form, DST is the proposition that the states of the nervous system from which cognition emerges can be described by ensembles of continuous state variables that evolve continuously in time. That evolution is characterized by dynamical laws. Functional states of the neuronal dynamics are attractors, whose stability enables them to persist in the face of perturbations and fluctuating inputs. New solutions and qualitative functional change emerge from instabilities of the neuronal dynamics.

Stability, a core concept of DST, has obvious roots in motor behavior. As every engineer knows, stability is of the essence whenever the control of a physical effector is continuously linked to sensory information as it is during the execution, but also the planning of motor behavior (Goodale et al., 1986). This need to stabilize functional states generalizes, however, to nervous activation other than overt motor behavior, because continuous links to sensory information as well as other, ongoing neural processes is a pervasive feature of neural function. Given the high degree of functional connectivity within the central nervous system, any neuronal subpopulation engaged in a particular functional state receives signals from many other neuronal subsystems that are not contributing to this function. In effect, these signals represent perturbations of the ongoing functional state, against which the state must be stabilized. This is true even for perceptual processes, for which feedforward computation would at first sight seem a reasonable framework. Stability is required, however, to form coherent percepts from the continuous stream of inherently ambiguous sensory signals (Hock et al., 2003).

Once we recognize that functional states of neural systems have stability properties, the question arises how systems may change state to approach the flexibility that characterizes cognition. In the motor domain, such flexibility may appear limited, but cognition and perception are inherently time varying and highly responsive to changing inputs. Flexibility requires that functional states be released from stability. This happens in instabilities (or bifurcations), at which the neuronal dynamics go through qualitative change, leading to new functional states (Schöner & Kelso, 1988; Schöner, 2008).

In the motor domain, the notion of a dynamic state of the neural control systems is easily grounded in biomechanics and physiology. In fact, muscle–tendon systems contribute through their elasticity and viscosity to the stability of effector systems as do peripheral and central reflex loops (Feldman, 1986; Bizzi & Mussa-Ivaldi, 1990;

Hogan, 1990). Applying DST beyond the motor domain requires that those neuronal principles be identified that may endow representations with stability. Stability, as we will illustrate later, requires a metric which distinguishes small from large perturbations and within which resistance to and recovery from perturbations can be defined. Representations can be embedded in metric state spaces through the concept of activation fields that span the continuous, potentially high dimensional spaces of possible percepts, memory states, or action plans (Shepard, 1980). Much of this chapter will review a class of neuronal dynamics, originally inspired by the homogeneous, layered structure of cortical anatomy (Wilson & Cowan, 1973; Amari, 1977), which provides the key to endowing representations with dynamical stability properties as well as the potential for instabilities from which elementary forms of cognition emerge (Spencer & Schöner, 2003; Schöner, 2008). We refer to the conceptual framework that result from combining the concepts of DST with this class of neuronal dynamics of activation fields as *Dynamic Field Theory* or *DFT*.

The major part of this chapter will review DFT, providing first foundations, discussing the units of representation as stable localized patterns of activation, and illustrating some of the instabilities through which different forms of elementary cognition emerge. We will show how this framework connects the graded sensorimotor representations underlying estimation, detection, and motor planning to the seemingly discrete representations underlying categorical behavior. To examine the extent to which DFT is consistent with the embodiment stance, and as a pointer to the achievable complexity of cognitive function, we will review a robotic application of these neuronal ideas to object recognition. Before we start, however, we will ground the ideas in the mathematical theory of dynamical systems through a brief and quite elementary tutorial.

DYNAMICAL SYSTEMS

The theory of dynamical systems has its origins in classical physics, where it was used to understand how physical systems evolve in time. Much of physics involves so-called conservative systems that do not have (asymptotically) stable states, so that perturbations affect the long-term behavior of the systems for ever. A textbook example of a conservative system would be a frictionless pendulum. If hit somewhere along its orbit, its future time course is forever changed. Real pendulums, in contrast, are affected by friction and ultimately come to rest. The resting state is stable, which makes it a point attractor.

Dynamical systems that have stable states are called dissipative by physicists and form a special subclass, relevant to understanding neural and behavioral systems. The investigation of the stability properties of such dissipative systems has received considerable attention in mathematics, and we shall introduce some of the most basic terms and interrelationships here. (There are very many textbooks on this field of mathematics. Two examples are Braun (1993) at an elementary level and Perko (1991) for a more advanced level.) To visualize a dissipative dynamical system, imagine a ball rolling in a smoothly sloped landscape under the influence of gravity. The ball may be coated with something sticky and so it experiences a lot of friction. This will ensure that its movements will be dampened quickly, generating stable states. The state of the system is then completely described by the ball's position in the horizontal plane. The change of this state—the movement of the ball—is determined only by the slope of the landscape locally at the position of the ball. If the slope is zero at the ball's position, the system is at a fixed point, and its state does not change anymore. There are a number of different kinds of fixed points: A stable fixed point or *attractor* is reached if the ball is at the lowest point of a valley. If the ball's position is disturbed within certain bounds by an external force, it will return to this point. Stability is this property of converging to a fixed point from any point in the immediate vicinity of the fixed point. (In mathematical language this is actually called *asymptotic stability* and differs from a weaker condition that mathematicians define as stability. Like most physicists and engineers, we continue to use the term "stability" for the stronger condition of asymptotic stability.)

A stronger disturbance may cause the system to leave the fixed point's *basin of attraction* (a term that can be taken literally in our example), and the ball will come to rest in some other valley, thus putting the system into a new attractor state. Another kind of fixed point, a *repellor*, would exist at the very top of a hill. Exactly on the top the slope is zero, but any disturbance, even a very small one, will cause the system to move away from this fixed point as the ball rolls downhill toward some other valley that might be available. A repellor is not a stable state. There are other ways in which a fixed point could not be stable; for instance by lying exactly at a saddle of the landscape: the system would be attracted only along one route down the saddle but would run away from the fixed point along all other directions. Still another way in which stability could fail would be observed if there was a direction in which the landscape was exactly flat. Along a valley with a perfectly horizontal floor, any point would be a fixed point that would not be perfectly stable because a perturbation would shift the system along the valley. It would still be more stable than a repellor or saddle point, however, because it would stay close to the original position.

Formally, a dynamical system can be described by one or more differential equations of the form $\mathrm{d}x/\mathrm{d}t = f(x)$, in which the rate of change, $\mathrm{d}x/\mathrm{d}t$, depends on the current state, x. For a given initial condition, this equation makes it possible to determine the system's state for all later points in time by integrating the rate of change along time. How attractors and repellors emerge from such equations can be visualized for a one-dimensional state space by plotting the function $f(x)$ as a function of x (Figure 13.1). By definition, fixed points are zero crossings of this function. Fixed point attractors are zero crossings at which the dynamic function, f, has a negative slope: The rate of change is positive for states smaller than the fixed point, leading to increase and thus movement toward the fixed point. The rate of change is negative for states larger than the fixed point, leading to decrease and thus likewise movement toward the fixed point. A zero crossing with a positive slope of the dynamic function is a repellor, at which the analogous logic explains why the system is pushed away from the fixed point. It may be intuitive from these

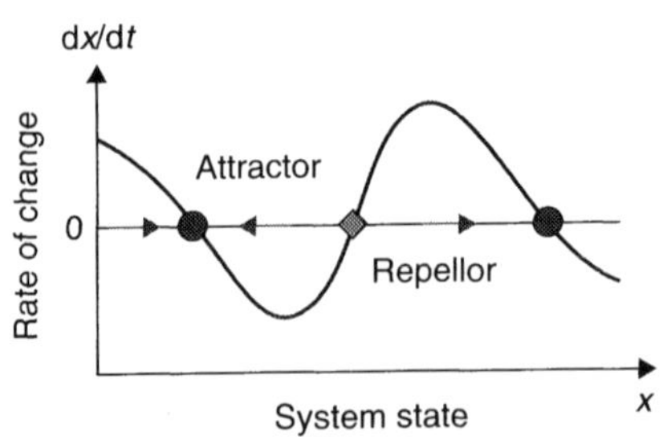

FIGURE 13.1 A one-dimensional dynamical system is described by how the rate of change dx/dt depends on the state, x, of the system. Zero crossings are fixed points that can be either attractors or repellors.

considerations that if f is a continuous function, two attractor states are always separated by a repellor (and vice versa). Thus, to switch from one attractor state to another, some external force must be exerted on the system (represented by a temporary deformation of the dynamic function, f) that is strong enough to move the system to the other side of the repellor. The repellor, therefore, demarks the boundary of the basins of attraction of the two attractors.

A change of the system's state may also occur if some external parameter–one that is not included in the state x–alters the system's dynamics. In the rolling ball metaphor, imagine that the landscape is tilted from the horizontal by a certain angle. This will cause the attractors (minima) and repellors (maxima) to shift (actually, the maxima will move in the opposite direction to the minima). When the incline reaches a critical point, some minima may stop being local minima; typically because they collide with a local maximum (try this out for a one-dimensional landscape!). The ball will track the lowest point of the valley it is in, until that minimum disappears. At this point, the attractor undergoes instability. With just a little more increase in the incline, the ball will move way from the former valley, until it reaches some other valley, which still contains a stable state (this example allows for the unfortunate outcome in which the ball runs off to infinity when no other valleys are left in the downhill direction).

This is a rather abstract view of dynamical systems. How could these terms be used to talk about the evolution in time of patterns of neural activation? What may stable states look like in such neural dynamical systems? How may they arise or disappear through instabilities? We will discuss next how Dynamical Systems Thinking can be combined with neural principles in DFT.

DYNAMIC NEURAL FIELDS AND PEAKS AS UNITS OF REPRESENTATION

The architecture of the *Dynamic Neural Field* or *DNF* is based on the finding that in the central nervous systems of vertebrates metric information is

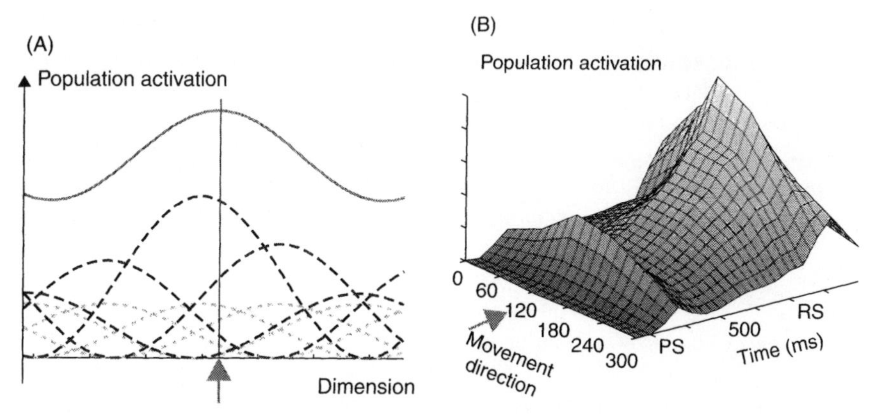

FIGURE 13.2 (A) Neurons tuned to a metric dimension have tuning curves (short dashed lines) with a single hump. These are weighted with the current firing rate of each neuron (long dashed lines) and superposed, generating the distribution of population activation over the metric dimension (solid line). When a specific value of the dimension is specified as in the illustrated case (arrow), the neurons with preferred values close to the specified value contribute more strongly than neurons with preferred values far from that value, because their firing is higher. (B) Time course of a distribution of population activation over the dimension of movement direction constructed in this way from the tuning curves of about 100 neurons in motor cortex (Bastian et al., 2003). The movement direction "120" is first signaled at the time marked as "PS," followed by the "go" signal at time "RS." A single peak located at that movement direction emerges.

commonly represented in the form of population codes (Erickson, 1974; Georgopoulos, 1991; deCharms & Zador, 2000). This means that dedicated populations of neurons exist whose activities, taken together, yield a representation of a certain feature. This may be the color of a visual stimulus, the pitch of a sound or a desired hand position in the planning of a motor action. Each neuron within such a population is maximally active when a certain, "preferred" feature value is presented, and its activation decreases as the feature value contained in the stimulus differs increasingly from this preferred value. The response property of a neuron can be visualized by its tuning curve, which plots the neuron's average activation against the feature dimension. The tuning curves of all neurons in a population cover the represented metric dimension or the relevant part thereof. There is usually a strong overlap between the tuning curves of neurons, so that each stimulus will cause activation in a number of neurons.

For the following theoretical considerations we will assume the neurons are ordered according to their preferred feature value even though this ordering does not necessarily correspond to the spatial layout of the neurons in the nervous system. In this perspective, the information represented by the population can be read out from the spatial distribution of activation (Figure 13.2): A single value along the feature dimension can be represented by a localized peak of activation, that is, by a group of neighboring neurons with high activation levels in an otherwise inactive population. The width and height of a peak may give additional information about the precision or certainty of this value. Ambiguous information about the feature can be represented by multi-modal distributions of activation, and the absence of information by uniformly low activation levels over the whole population.

These activation patterns are shaped not only by the input that the neurons receive from other structures but are greatly influenced by interactions within the population. A ubiquitous form of connectivity in the central nervous system can be characterized as local excitation and global inhibition in the spatial arrangement of feature sensitive neurons. Neurons that code for similar feature values excite each other, whereas neurons that code for distant feature values inhibit each other (via inhibitory interneurons). This kind of interaction promotes the emergence of localized activation peaks, as will be discussed later.

One key assumption of DNF models is that it is the distributions of activation over neural populations that convey the relevant information and not the behavior of the single neurons. Accordingly, DNFs abstract from the neurons as discrete computational units and model activation over continuous feature dimensions. The evolution of activation patterns is modeled as a continuous process in time, described by a set of differential equations. Special emphasis is put on the internal interactions in the field which are critical for establishing stable states. To model these interactions, an output is calculated over the whole field and fed back into the field as endogenous input. This output can be regarded as a correlate to the mean firing rate of a group of neurons, whereas the activation reflects their mean membrane potential. The field output is usually calculated from the activation via a sigmoid function, which is close to zero for low activation levels, rises around a threshold value and saturates at a constant value for higher levels of activation. The distribution of the endogenous input that originates from one position in the field can be described by an interaction kernel: It consists of an excitatory part, typically modeled as a Gaussian centered at the origin of the output, and an inhibitory part. The inhibitory component may be homogeneous over the whole field, but it may also be a broader Gaussian, resulting in a Mexican hat shape. This type of kernel implements the pattern of local excitation and global inhibition found in neural populations. External input can boost the activation in the field, either locally or globally. Finally, in many models random noise is added to the activation field to account for fluctuations in neural activation that cannot be captured by a deterministic differential equation.

In the absence of any input, the activation over the whole field is driven toward a preset resting level, which is usually chosen to be well below the threshold of the output function. This pattern of activation constitutes a first attractor state of the DNF: If the activation is perturbed by noise, it may fluctuate around the resting level, but it does not drift over extended periods of time, and it relaxes toward the resting level when the noise is turned off. If a weak localized input is added to the field, then the activation in the field rises toward a state reflecting the sum of resting level and input (Figure 13.3A). Here, the input strength acts as an external parameter that causes a shift of the attractor states (similar to the tilt of the landscape in the rolling ball example), and the activation distribution follows the attractor. We call this the input-driven state of the DNF.

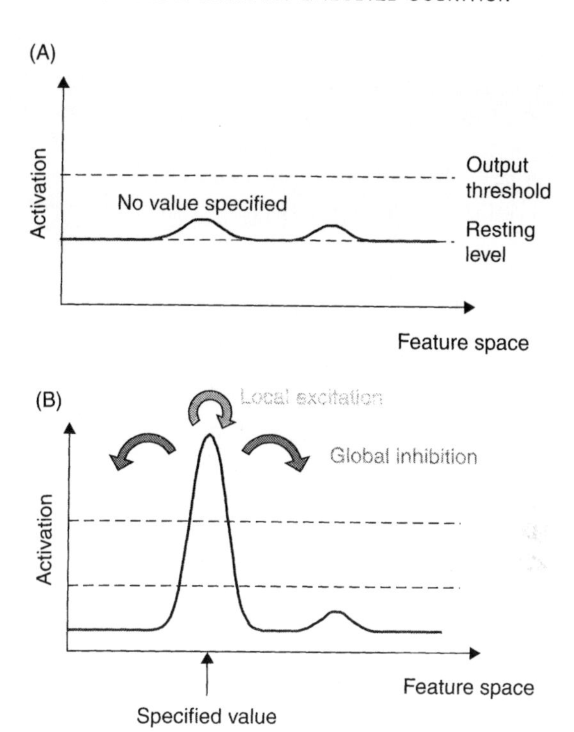

FIGURE 13.3 In DFT, metric information is represented by continuous distributions of activation over metric dimensions that span perceptual or motor feature spaces. (A) Low activation levels across the entire field index the absence of conclusive information for the feature space. (B) A single value along the feature dimension is specified by a peak of activation localized at a particular position in the field, which stands for that feature value. Such activation peaks are the units of representation in DFT and emerge as attractors from the neuronal dynamics of the activation fields.

Consider next the case of a localized input that is strong enough to lift the activation in a small section of the field above the output threshold. In this case, the interactions in the dynamic field must be taken into consideration to determine the stable state of the system. The local excitation will drive activation even higher at the location of the input and global inhibition will depress it elsewhere (Figure 13.3B). If the parameters of the interaction kernel are within a certain range, the result will be a strong localized peak of activation surrounded by a zone of inhibited activation. For this state, the endogenous excitation and inhibition as well as the exogenous input and the forces pushing the system toward resting level reach a balance at every position of the field. This constitutes another attractor state, which we refer to as a self-stabilized state and which is qualitatively different from the input-driven state.

One way to see that this is a qualitatively different state is to decrease the strength of the external input again enough, so that the combined effect of input and resting level are insufficient to reach the output threshold. The field activation in the area of the peak will, however, remain high enough to sustain output leading to self-excitation by the local excitatory interactions. At such an input level the system is bistable, that is, two attractor states co-exist: The input-driven state that is reached from low levels of activation, does not engage interaction, and mirrors the input signal. The self-excited state is reached from sufficiently high levels of activation, stabilized by interaction, but continues to be influenced by input. Which of these attractor states the system reaches at this input level is determined by the field's activation history. Weak previous levels of activation put the field into the basin of attraction of the input-driven state, strong previous levels of activation put the system into the basin of attraction of the self-excited state. For initial states near the boundary of the basins of attraction, the system may reach either state depending on stochastic perturbations (reflecting, for instance, noisy neural inputs).

The qualitative change of the attractor states when a single localized input increases in strength is illustrated in Figure 13.4. The dynamics generating the attractor solutions sketched in the left column of the figure can be illustrated by plotting the rate of change, $du(x)/dt$ of the activation level at some location, x, within the peak, as function of the activation level, $u(x)$, at that same location. Strictly speaking, this plot is not a mathematically conclusive representation of the dynamics, because the rate of change also depends on activation levels at other field sites. The intuition derived from this plot is corroborated by the correct mathematical analysis, however (Amari, 1977). The rate of change has a negative slope overall, reflecting the fundamental stability of neuronal activation. At large levels of activation, the rate of change is lifted up by the net effect of the excitatory interactions within a peak of activation. The effect of localized input is to shift the rate of change upward across all activation levels. As a result, the single attractor at low levels of activation is joined by a second attractor at high levels of activation, into which the system switches when the attractor at low activation levels becomes unstable for sufficiently strong input. When input levels are then lowered again, the system will remain in this activated state until that state becomes unstable for sufficiently weak inputs. Either switch occurs as an attractor disappears after becoming unstable.

A behaviorally relevant effect of these instabilities is that the bistable regime helps stabilize detection decisions. Consider a simple perceptual detection task and assume that a stimulus is perceived when the relevant neural population creates sufficient activation that exceeds the threshold for output to be generated. In the input-driven regime, the percept would be very unstable for a stimulus that is just strong enough to push the field to the output threshold. Due to sensory noise, the activation would fluctuate around the threshold and the output nonlinearity would produce a signal that alternates on a fast timescale. In the self-stabilized regime, the percept is stabilized once activation reaches the output threshold. The percept persists even when the input strength is reduced (within limits). Empirical support for the stabilization of detection decisions comes from psychophysical experiments demonstrating perceptual hysteresis (Hock et al., 1997).

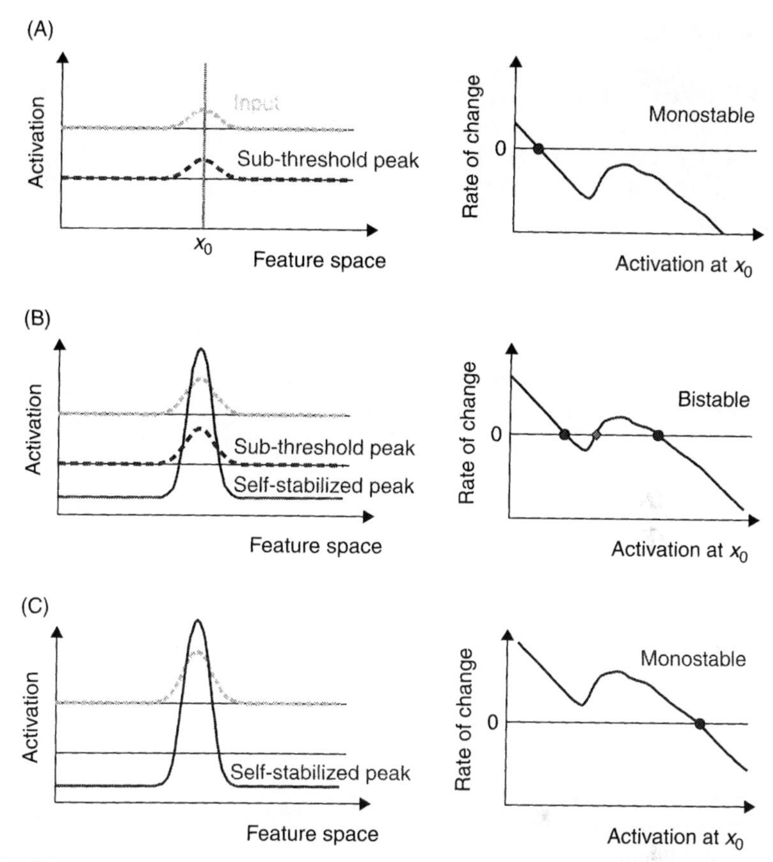

FIGURE 13.4 Stable patterns of activation induced by a single Gaussian input of varying strength are shown (left column) together with the corresponding plots of the rate of change against the current activation level, both taken at the peak position (right column). (A) For weak input (short dashes), the only stable pattern is a matching subthreshold peak (long dashes). The associated dynamics is monostable (dot marks the attractor). (B) At intermediate input strength, the system is bistable. One attractor emerges from the subthreshold peak (long dashes), which is merely shifted toward higher levels of activation (leftmost attractor in the plot on the right). The other attractor is a self-stabilized peak (solid line). It shows up as an additional attractor state of the system (rightmost attractor in the plot on the right), separated from the old attractor by a repellor (diamond). (C) At the highest input levels, the system is monostable again with only the self-stabilized peak surviving. The subthreshold peak has become unstable.

The range of input strengths for which the DNF is in the bistable regime depends on the parameters of the interaction kernel. So far, we have assumed that a DNF would be monostable in the absence of any input so that the activation would always relax to the resting level when the input is removed. In the DNF model, this is not necessarily the case, however. If the excitatory part of the interaction kernel is strong enough and is balanced by sufficient inhibition to stabilize a local peak, then a perfectly stable peak of activation may persist even without exogenous input. Such self-sustained peaks of activation have been used to model metric working memory (Zipser, 1991; Durstewitz et al., 2000; Spencer & Schöner, 2003): A peak is created by a single presentation of a localized stimulus. It is then sustained over extended periods of time, representing the former input as a memory item together with its metric value reflected by the location of the sustained peak.

We will make a few additional remarks about the relationship of DNFs to real neural populations and brain structures. Generally, DNFs can be used to describe neural systems at different levels of abstraction. Historically, DNFs were first developed to approximate the cortical neuronal architecture that is characterized by layered sheets of neurons which are relatively homogeneous along the layers with strongly overlapping dendritic trees for nearby neurons (Wilson & Cowan, 1973; Amari, 1977). DNFs can be used to model clearly identified populations of neurons using the concept of a distribution of population activation (Erlhagen et al., 1999), which frees the description of strict anatomical constraints. This has been done, for instance, for the representation of retinal location in primary visual cortex (Jancke et al., 1999), the representation of movement direction in motor cortex (Bastian et al., 1998; Cisek, 2006), and for the representation of saccadic end-points in superior colliculus (SC) (Trappenberg et al., 2001). In these cases, the parameters of DNF models can be tuned to reproduce the experimentally observed patterns of neural activation.

However, DNFs may also be used to explain the results of psychophysical experiments, such as the metrics of performance, reaction times, error rates, frequencies of responses, and other signatures of the underlying processes (Kopecz & Schöner, 1995; Erlhagen & Schöner, 2002; Schutte et al., 2003). In these cases, the feature dimensions over which the fields are defined are usually parameters of the experimental setup. It is often not known exactly where the processes modeled in the DNF take place in the brain, and it may even be doubtful whether any single neural population exists that behaves exactly as the dynamic field does. Instead, the neuronal instantiation of such DNFs could be distributed across multiple areas and populations of neurons. Such functional DNF models may properly capture the net effect of the evolution of stable activation patterns that underlie the observed behavior (Spencer et al., 2007).

INTERACTIONS BETWEEN MULTIPLE ACTIVATION PEAKS

Up to here we looked at the input-driven attractor and at a single, localized peak that is stabilized by interaction and forms a second attractor (see Amari (1977) for a complete mathematical analysis). More complex attractor configurations arise if two or more localized inputs are applied to the field. In such a case, multiple peaks of activation may emerge that influence each other due to the excitatory and inhibitory interactions. In this section we review the different effects that these interactions generate and show how the resultant solutions can be used to explain experimental results linked to sensorimotor decision-making. The metrics and timing of saccadic eye movements have been extensively investigated and much is known about the underlying neuronal substrate. The neuronal specification of such movements thus provides an excellent model system, which we can use to illustrate key ideas of DFT. We build on detailed DNF models of saccade specification (Kopecz & Schöner, 1995; Trappenberg et al., 2001; Wilimzig et al., 2006).

Saccades are the abrupt eye movements that we use to change the fixation point of our eyes from one location in visual space to another. Saccades are ballistic movements, that is, each saccade's trajectory is determined before the movement starts and normally it remains unaltered during the execution of the saccade. Furthermore, saccadic eye movements are highly stereotyped so that it is sufficient to specify the horizontal and vertical distance of the saccadic target in retinal coordinates, that is, relative to the current fixation point. These two spatial dimensions of the saccadic end-point can thus be considered relevant feature dimensions (for simplicity we will think of only a single dimension in what follows). A peak in an activation field defined over these dimensions thus indicates the metrics of a planned saccade. Such a DNF may be interpreted as a functional description of relevant neural populations, in particular, those in the Superior Colliculus (SC), a mid-brain structure that is involved in saccade planning and initiation. The SC features a topographic map of saccade target positions, in which activation peaks arise before a saccade is initiated. The SC integrates both sensory and cortical inputs and it is assumed that it is in the SC that the final decision about the initiation of a saccade is made. (A more detailed model points to multiple zones within SC and to different layers playing different roles in the specification and initiation of a saccade as well as the opposing function of fixation.)

The presence of a stimulus somewhere in the visual field is modeled by localized input to the corresponding position in the DNF. If the input is strong enough, the system goes through the detection instability and a peak emerges, indicating the metrics of a saccade to the visual target. Under natural conditions, of course, there is never a single unique visual target in the visual array. Instead, typical visual environments provide a rich selection of potential targets of saccadic eye movements, which are most commonly characterized by some high-energy local contrast, edge or corner point. Specifying a saccade under such conditions necessarily involves selection (Ottes et al., 1984). DNFs and their interactions afford such selection.

Consider first a case in which two identical inputs are presented to two field sites that are at a large distance from each other (Figure 13.5A). When these inputs are sufficiently strong, they induce levels of activation in the field that reach the output threshold and thus engage the neuronal interaction in the field dynamics. For perfect symmetry and in the absence of noise, two identical peaks may arise. Mutual inhibition may reduce the total activation in these peaks as compared to a peak induced by a single localized input. This is because

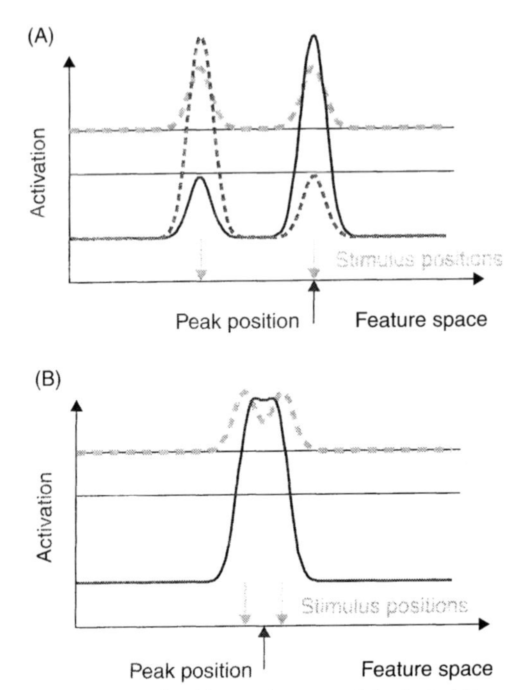

FIGURE 13.5 (A) If two localized inputs (arrows and fat dashed line on top) are applied at distant positions, the emerging peaks compete with each other due to self-excitation and mutual inhibition. This results in the selection of one peak and the suppression of the other. Any of the inputs may be selected depending on their respective strengths and the field's history, allowing two possible stable states for the dynamic field (dashed and solid line below the input). (B) If input positions are close to each other, local excitatory interactions bring about a fusion of the activation peaks, leading to a monostable response at an averaged location.

both peaks contribute to inhibitory interaction that impacts on the entire field, whereas excitatory interaction is local to each peak. The two-peak state is a fixed point of the system but not generally a stable state. If the activation level of one peak is slightly increased by noise or stimulus asymmetry, that peak generates more supra-threshold activation leading to stronger self-excitation and stronger global inhibition. This will diminish the activation in the other peak and, in turn, it reduces the amount of self-excitation within that peak as well as its inhibitory influence on the other peak. An imbalance between the two peaks will arise and grow, which may lead, for sufficiently strong interactions, to the complete suppression of one peak by the other one. Given that all output from the inhibited peak is suppressed, the remaining peak has the same shape and strength as a peak with only one localized input.

Thus, for strong bimodal input, the system is again in a bistable state: If a single peak has been established and activation at the other input location has been suppressed, this pattern is stabilized against noise and also against moderate increase of the suppressed input. Which one of the two possible peaks is realized depends on the system's activation (and thus stimulation) history as well as on random stochastic fluctuations (e.g., from many uncorrelated neural inputs). If one location receives stronger input than the other then that location has a greater chance at generating a peak, which is suppressed only if a strong, and thus rare, stochastic perturbation favors the other location. The same mechanism of competition takes place for more than two inputs, leading to the selection of one location in many, which is typically the location with strongest input.

So far, we have analyzed the case in which the locations receiving input are distant from each other, so that the associated activation peaks only inhibit rather than excite each other. What about closely spaced inputs? We first remind the reader of experimental observations for metrically close saccadic targets (Ottes et al., 1984). If multiple visual stimuli are presented in proximity to each other, but distant from the current fixation point, the result is typically an averaging saccade made to the center of the group, not to a single item. A smaller saccade to fixate a specific target may follow in a second step.

This averaging behavior can be understood in terms of DNFs as well (Figure 13.5B). Two Gaussian inputs to a dynamic field will overlap if they are close to each other, so that both input sources contribute activation to the area between the two locations. Trivially, this may result in a single localized input to the field, centered already over the averaged input locations. An averaging peak may even emerge, however, when the two inputs do not overlap so strongly that a single-humped input distribution results. Input induced activation at two locations that are close enough to experience mutual excitatory interaction will tend to fuse into a single peak. Supra-threshold activation at either location propagates toward the center. This converges to a merged peak at an averaged position, similar to one that would be created by a single broad input.

For two localized inputs that are applied very close to each other, the merged peak is the only stable activation pattern. If the distance between the inputs is increased continuously, this pattern will remain stable over a certain distance, whereas the same input may create a selection behavior when applied to a previously inactive field. If the distance is increased further, the merged peak attractor is destabilized and the peak quickly shifts to one of the stimulus positions (this is called the fusion/selection instability). Because excitatory interaction can take a direct route between excitatorily coupled neurons, whereas inhibition requires inhibitory interneurons, this account predicts that early saccades tend to fuse inputs, whereas later saccades that occur after more time has been available for the neuronal dynamics to settle, tend to select one target (Wilimzig et al., 2006). This is empirically true.

The time course of sensorimotor decisions has been studied using DFT ideas in a variety of other settings. The timed–movement–initiation–paradigm (Erlhagen & Schöner, 2002) provides access to the preparation of goal-directed hand and arm movements. Infants show reliable patterns of selection when confronted with multiple possible reaching targets in the famous A-not-B task of Jean Piaget

(Thelen et al., 2001). Here the delay across which competition between a motor habit and a cued new movement target occurs can be varied experimentally. More generally, tasks involving metric working memory provide access to the temporal evolution of selection decisions, exposing time-dependent metric biases due to the influence of competing influences. In all of these cases, prior experience within the task plays a critical role in how the selection process unfolds. To understand sensorimotor decision-making, we need to look more carefully, therefore, at how prior experience may have an impact on selection decisions.

PRESHAPE IN DYNAMIC NEURAL FIELDS

We have just seen how small inhomogeneities in a field, a little more input at one location than at another, can have a critical influence on selection decisions. The activation fields underlying the perceptual or motor decisions cannot be generally expected to be perfectly neutral, clean slates. Whenever a particular input arrives that drives the field toward a decision, the activation pattern in the field may be preshaped by other inputs that have been around longer. One source of such preshaping input is the sensed environment, in which there may be rich visual structure including potential movement targets such as graspable objects. Decisions typically take place on such a background of prior activation.

One particular source of such preshaping of activation fields is the recent activation history. Habit formation is perhaps the simplest form of learning in which an organism builds a tendency to repeat behaviors that have been successful before. Habits may be accounted for in DFT by assuming that patterns of activation leave a memory trace, which then in turn contributes to preshaping the field. A simple mathematical formalization is based on an additional layer of activation, in which such a memory trace results from a slow dynamics. This memory layer in turn provides input to the proper activation field. The resultant preshaped activation is generally subthreshold, so that it does not by itself induce decisions. Preshape may, however, exert a great influence on the activation patterns that emerge when stimulus input is added.

The concept of preshape is not meant to model one specific neural mechanism. Instead, preshape is a general functional account for a variety of neural mechanisms that contribute to activation prior to an imperative or specific signal, which triggers an instability leading to a peak being formed and a decision being made. Long-term memory and associations from other cortical areas may represent expectations, predictions, or attention directed at certain parts of the feature space. In other cases, the preshape may be thought of as residual activation from previous behavior. Learning mechanisms may involve changes in synaptic efficacy, either in the afferent or in the lateral connections. In either case, the functional effect is to facilitate the induction of a peak, which we may conceptualize as inhomogeneity or preshape of the field.

How preshape influences selection decisions can be illustrated again in the preparation of saccadic eye movements. In the laboratory, only a limited number of potential target locations are typically used. Participants may acquire prior knowledge about the possible eye movements they will need to perform to acquire these targets. The effects of this knowledge were investigated by Dorris et al. (2007) in rhesus monkeys who were trained to fixate on a light point and then make a saccade to a visual target appearing at a predictable position in the visual array. In electrophysiological measurements in the SC, they found a localized hill of activation in the area representing the target location before the stimulus was actually presented. Furthermore, they investigated the behavioral effect of this preparatory activation by presenting distracters (that differed from the actual targets by their color) at different locations in the monkey's visual field. Distracters presented close to the usual target location tended to attract the eye, leading to erroneous saccades, whereas the metrically distant distracters did not.

These observations can be understood within the DNF model of saccade preparation. In the preshape layer, a constant broad hill of activation is created at the trained target position, moderately increasing the activation level in the associated dynamic field. The distracters may be modeled as transient inputs that are weaker than the target input (because they are not reinforced by the neural systems that performs the target recognition). Such an input will be sufficient to create a peak if applied to a preactivated region of the field, but not in the other regions, explaining the different rates of erroneous saccades. A further effect of the preshape is that peaks are created faster in response to target presentation, predicting shorter reaction times for saccade initiation in situations where the target position is known in advance. This is in accordance with a large range of experimental results (as reviewed in Erlhagen & Schöner, 2002). The same effect of pre-information on neural activation levels and reaction times has been shown for motor cortex when pointing movements were prepared (Bastian et al., 2003).

How preshape may be acquired by a simple learning process is illustrated in Figure 13.6. To this end, the output of the dynamic field is fed into a memory trace layer. Thus, whenever there is supra-threshold activation in the field, a memory trace is laid down at the matching locations. Because the memory trace evolves over a much slower timescale, the pattern within the memory trace layer reflects the statistics of activation in the field, with more activation built in those locations that have repeatedly and consistently been activated.

The memory trace is thus a mechanism through which probability distributions reflecting the activation history can be autonomously acquired and neuronally instantiated, much in the manner of the prior distributions of the Bayesian framework. Metric biases may arise from the preshape pattern induced by such memory traces, which are again consistent with Bayesian estimation. This effect is illustrated in Figure 13.7, where a tiny amount of bias toward one of the preshaped locations can be seen (the effect can be larger under appropriate circumstances (see Erlhagen & Schöner, 2002); here we aim to contrast this effect with

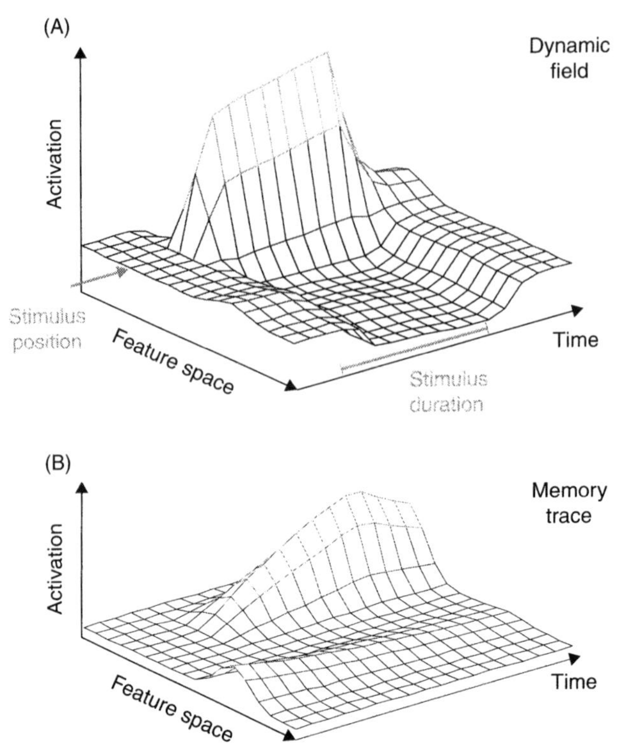

FIGURE 13.6 The evolution in time of the activation pattern in a dynamic field (A) and an associated memory trace field (B) is shown. As long as a peak is present in the activation field (here due to the presence of a stimulus marked by the bar labeled "stimulus duration"), the memory trace slowly builds up at the corresponding field location. Memory traces passively decay in the absence of such activation. In this illustration, a memory trace at a different feature value was assumed to exist initially from earlier peak events. That trace preshapes the activation field at the matching location but then slowly decays because it is not further stimulated by input from the activation field.

a different mode of integration discussed in the next section). One manifestation of the preshaping of the choice seen here is that the time needed to build a peak is shorter when the peak is consistent with the memory trace than when it is not. Note, however, how the DNF goes beyond the fusion of prior and sensory input. The field dynamics suppresses any influence from the other, metrically remote location that has also accumulated preshape. This amounts to something like robust estimation and is one aspect of the stabilization of decisions.

In DFT, memory traces reflect not only the probabilities of different peak events, but also their metrics. Probabilities are essentially encoded by the activation level within a preshaped location, whereas the metrics of prior experience is encoded by the location of the preshaped activation. An experiment and associated simulations highlight, how probability and metrics interact (McDowell et al., 2002). Human participants were asked to make center-out pointing movements to visual targets. In each block of trials, only two movement directions occurred,

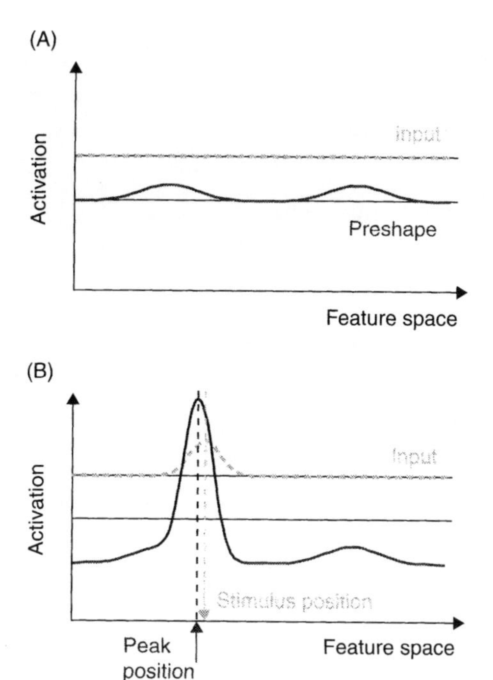

FIGURE 13.7 (A) An activation field (solid line) is preshaped by a memory trace that reflects that two locations have frequently seen activation peaks. This decreases the amount of excitation that is required to generate a peak for the preactivated field locations. (B)When a localized input is applied to the preshaped field, the resulting peak position is slightly biased away from the input specified location toward the closest location specified by the preshape. Metrically distant preactivation does not matter, as it is only within the range of 19 inhibitory interactions.

one of which was elicited frequently and the other rarely. When the two movement directions were metrically far from each other (120°), reaction times to the rare target were longer than reaction times to the frequent target, consistent with the Hyman law (which says that choice reaction time increases with decreasing probability of a choice). When the two targets were metrically close (5°), reaction times for both movement directions were equally fast. The rare movement direction was actually shared across two different blocks. Reaction time to this target was long when it was paired with a metrically far frequent target and short when it was paired with a metrically close frequent target.

Figure 13.8 illustrates the DFT account for this effect. Movement direction is the feature dimension and movement is initiated in the direction encoded by the location in the field at which a self-stabilized peak is generated. Reaction time is predicted by the rate at which activation within the peak rises, shown in the figure through the activation level at the location of the peak. Whenever a peak is created and a response made, the memory trace at the associated location is updated. The frequent movement direction is thus represented by a more strongly preactivated field location than the rare movement direction. This leads to faster buildup of a peak from the

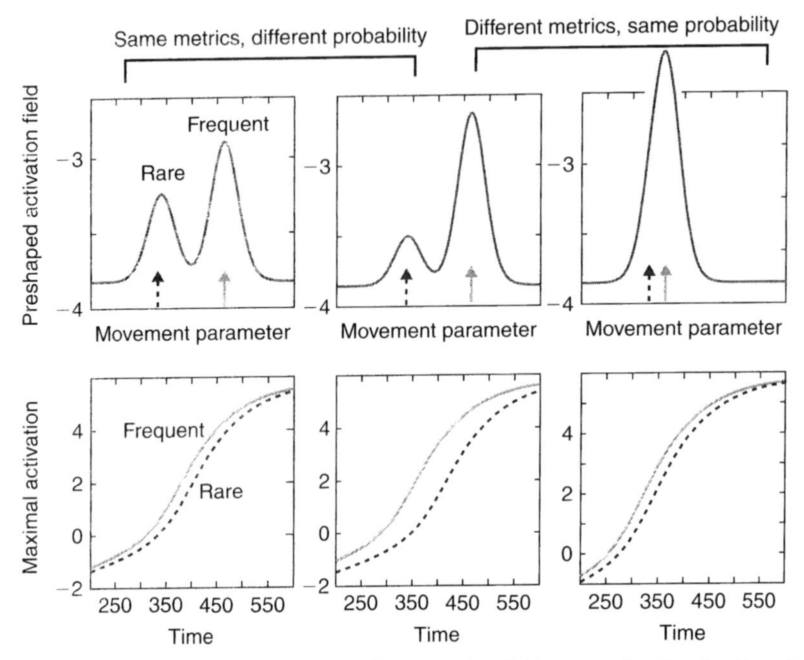

FIGURE 13.8 On top, the preshape of an activation field representing the direction of an upcoming movement is shown. The level of preshape activation at two possible movement directions (arrows) reflects the probability of each movement, higher levels arising for more probable choices (left and middle). When the two movements are metrically close, preshapes overlap, lifting the level of preactivation for the less probably choice. At the bottom, the rise of the maximal level of activation is shown for the case that the frequent (solid) and the rare movement is specified. That rise is faster for the frequent than the rare movement when probabilities and metrics are disparate (middle) but not when probabilities are similar (left) or metrics are close (right).

more strongly preshaped location, explaining the Hyman law (Erlhagen & Schöner, 2002). When the two locations are very close, however, the preshape at the frequent location spills over to the rare location, boosting buildup there and leading to similarly fast buildup time for both choices. The wider implication is that the amount of information (e.g., probability, number of choices, sensory precision) is not the only predictor of choice behavior. The contents of the selection decisions, their metrics, also matter. It is not possible to abstract from the "what," the specific, embodied and substantive contents of mental representations by focusing only on the "how much," on the abstract processing of information and its capacity limits.

CATEGORICAL BEHAVIOR FROM CONTINUOUS REPRESENTATIONS

Up to this point we have talked primarily about sensorimotor tasks, in which decisions about continuous feature dimensions needed to be made and the values of these dimensions estimated. This may entail forms of cognition such as when such estimates need to be stabilized in working memory or when a selection among different possible values is required. Much of cognition, however, may seem to involve primarily categorical behavior and the associated categories. Categorical behavior is required if the environment offers discrete objects such as when we select and reach for one object rather than another, when we name one object rather than another. Words appear categorical in nature and many language tasks seem to require the selection of one of a discrete set of choices, such as when we name an object using one word rather than another.

In the laboratory, categorical behavior is often imposed by asking participants to act on discrete objects (such as pressing one key when stimuli of a certain kind are presented and another when stimuli of another kind are presented) or use discrete responses. Although the sensorimotor tasks reviewed up to now may, in the laboratory, also involve only a small set of discrete possible choices, these are naturally embedded in a continuum (e.g., of possible movement directions). A classification task, in contrast, seems to involve inherently discrete response categories. If we are asked to recognize faces by labeling, we may be compelled to make a discrete selection rather than interpolating between two possibilities (although such interpolation may appear possible at some level of representation, we will come to that). Another way to characterize categorical tasks is to examine the kind of errors that participants can make: are errors graded and metric in nature or are they inherently categorical, for example, "right" or "wrong."

How does DFT deal with inherently categorical behaviors? The key idea is to think of such categories as embedded in underlying continua. In many cases, these may be thought of as arising from the lower level perceptual feature spaces, within which an object may be described. At a neuronal level, cortical feature maps provide a substrate for such an embedding. Population coding has been found in cortical areas as high as IT exactly for the presentations of objects (Young & Yamane, 1992) suggesting that the notion of overlapping neuronal connectivity that gives rise to the notion of peaks along a continuum applies. Psychophysically, most perceptual representations are not strictly categorical, giving access to graded information about the particulars of any given instance of a category (as is true even for the most famous case of categorical perception, the perception of the phonemes of speech, see Massaro, 1987).

Once we recognize that categories may be embedded in this way, the question is how categories may arise from such underlying continua and how categorical behavior may be generated on the basis of continuous DNFs. We have to begin with the latter question to then know what the first question entails. It turns out, there is a simple answer that requires no new mechanisms over those used up to here. Discrete categorical responses may arise from multi-peaked preshape within a continuous activation field. From such graded, subthreshold patterns of preactivation, self-excited peaks can be generated through the same detection instability discussed earlier. Figure 13.9 illustrates that a simple boost, a homogeneous excitatory input that lifts the activation across the entire field, may push the field

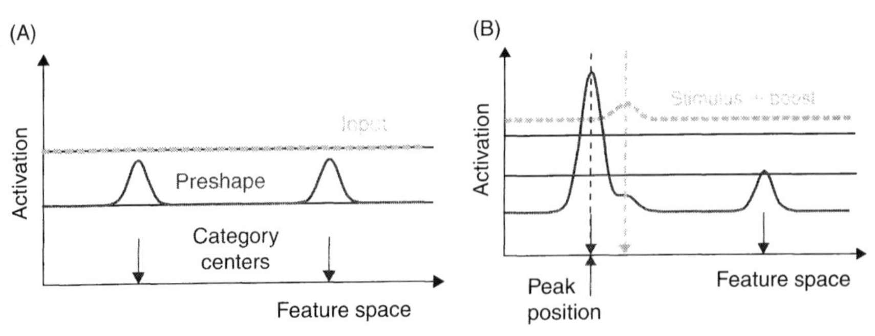

FIGURE 13.9 (A) An activation field (solid line) is preshaped around feature values that correspond to associated categories. In the absence of input (dashed line is at the zero level), the preshaped field remains below the output threshold. (B) Categorical responses are generated by combining a weak localized stimulus input with a homogeneous boost of activation that can be viewed as a "go" signal (dashed line). This lifts all preshape hills above the output threshold and engages both local self-excitation and competition among the potential peaks. The small localized input component biases the competition toward the preshaped category with which it has the greatest overlap. The position of the resulting self-stabilized peak (solid line) is largely determined by the categorical preshape pattern, not by the localized stimulus component.

through the detection instability at one of the preshaped locations. If multiple locations are preshaped, then selection may happen through the same mechanism of lateral inhibition evoked earlier to understand selection in sensorimotor tasks. The input may not be perfectly homogeneous, containing instead some localized structure and thus favoring the selection of the preshaped location that overlaps most with this input. The peak in the field, and thus, the associated behavior, is localized largely over the preshaped locations, however, when this localized component of input is small compared to the amount of preactivation. Comparing the categorical response mode to the response mode used in the sensorimotor scenarios (Figure 13.7), the roles of stimulus input and preshape are reversed. In the latter case, the localized stimulus is dominant, largely determining the peak location and being causal for the initiation of a response (so that there is no need for a separate "go" signal). The preshape makes a minor contribution to the metrics of the representation, biasing the peak toward the preshaped regions. In the categorical response model, in contrast, preshape determines the metrics of the response while the time of response initiation is determined by the homogeneous boost, the "go" signal. The specific, localized stimulus merely biases the competition between the different preshapes, thus selecting the category that will be activated. These two modes are, however, merely limit cases of a continuum, in which the relative strength of preshape and localized stimulus input takes on any intermediate value. The mechanism proposed here to explain categorical responses explains how categorical errors may arise even for unambiguous stimuli. Such errors may arise if the wrong preshape hill, which is not metrically close to the current stimulus, wins the competition. Because the field goes through an instability when the peak is brought up from preshape, it is sensitive to noise and such an outcome may result due to a fluctuation. This is more likely for smaller differences in input at different possible locations (e.g., because the localized part of the stimulus is weak) or for stronger overlap between preshape patterns (e.g., because the categories are metrically close).

This account also explains why trials in which an error occurs tend to have longer reaction time than trials with a correct response (Luce, 1986). In the dynamic field, the reaction time is determined by the time course of the selection process. If one peak is clearly stronger than all others, it can quickly suppress those others and win the competition. If on the other hand several candidates are almost equally strong, the process of competition starts more slowly, as there is little difference between the forces that act on the single peaks. Even if one of them gains a little advantage, a small amount of noise is sufficient to nullify it. In our account, the strength of different peaks is on average more similar in error trials than in correct ones (as trials with a close competition are more likely to produce errors). Thus, the DNF model will produce longer reaction times for error trials. Furthermore, reaction times of the DNF model will tend to be longer as the number of response categories increases: As more preshape hills compete with each other, the total inhibition gets stronger, slowing down the rise of a single peaks (Erlhagen & Schöner, 2002). Such an influence of the number of categories on the reaction time is experimentally well studied and is captured by Hick's law (Luce, 1986).

The generalization of this result is the Hyman law, of course, according to which reaction time increases with decreasing probability of a choice. We showed earlier how the Hyman law interacts with the metrics of choices when peaks are generated from a localized input representing an imperative signal. In that case, metrically close choices have faster reaction times irrespective of their probability (Erlhagen & Schöner, 2002). This is actually a somewhat counter-intuitive result. More common is the distance effect, in which the decision between two choices takes longer if the choices are more similar and metrically close (Anderson, 1995). As illustrated in Figure 13.10, the distance effect falls out of the DNF account of selection in the categorical mode dominated by preshape (Wilimzig, 2006). Only when the preshape is bimodal are categorical responses possible. Everything else being the same, the stimulus specifying either choice overlaps more with the other choice, so that both choices are activated to a larger extent. Mutual inhibition is more strongly engaged and slows responding down until one of the choices falls below the output level.

The detection instability driven by a homogeneous boost to the field is capable of amplifying small graded inhomogeneities into macroscopic stable states that can begin to impact behavior. This has far reaching implications for what is required to learn categories: Essentially, acquiring a graded preshaping along a feature dimension with local maxima near the centers of categories is sufficient to respond categorically to graded inputs. Categories thus emerge naturally from underlying continuous feature representations through graded, incremental learning rules such as the memory trace mechanism described earlier (Figure 13.6) or a generic Hebbian rule.

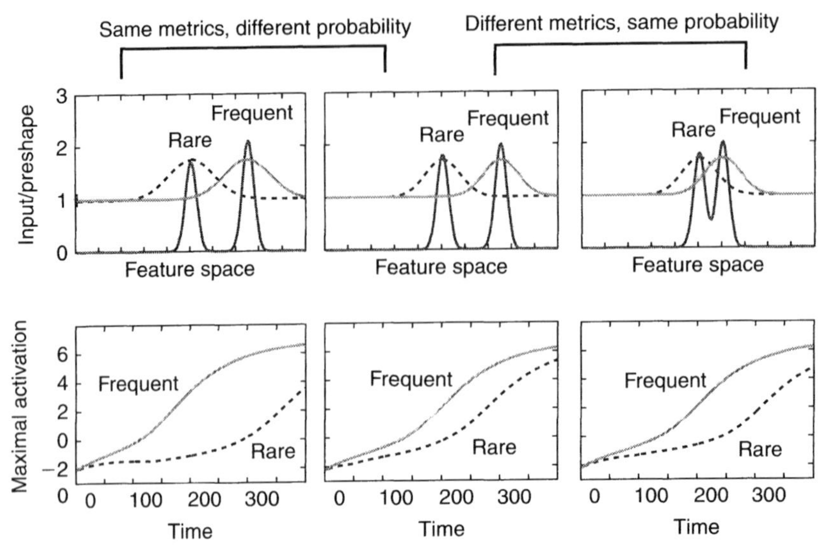

FIGURE 13.10 This figure is analogous to Figure 13.8 but now based on the DNF model in the categorical response mode. Top: Preshape (fat solid line) reflecting a frequent and rare choice along a feature dimension is shown when the difference between the probabilities is small (left), or large (middle and right). The two choices are either metrically far (left and middle) or metrically close. The stimulus specifies either the rare (dashed) or frequent choice (thin solid) and contains a homogeneous boost component. Bottom: The maximal activation at the site specified by the stimulus rises faster for the frequent than for the rare choice. This difference is larger when the probability difference is larger (left compared to middle), but increases again when the choices are metrically close, the opposite effect compared to Figure 13.8 (Figure adapted from Wilimzig (2006)).

EMBODYING DYNAMIC NEURAL FIELDS ON AUTONOMOUS ROBOTS

We have emphasized the concept of stability as a prerequisite for understanding how cognition may emerge in embodied and situated systems that are continuously linked to structured environments through sensory inputs. But how do the concepts fare when a real body is controlled based on real sensory information? One way to evaluate is to implement DNF models on autonomous robots and investigate how DNFs cope with continuously changing and noisy input and how DFT architectures generate consistent and flexible behaviors. If simple, neuronally plausible sensory and motor processes are sufficient to enact a DNF model, then this proves that there are no hidden problems in the interface between the DNFs and the sensory and motor surfaces. This is not trivial. Many a model of cognition makes strong demands on both ends of sensation and motor control. Some connectionist models, for instance, postulate that a specific neuron represents a particular kind of object (e.g., see Munakata, 1998). Recognizing objects on the basis of visual information is, however, a well-known and nontrivial problem. So there is something hidden in the interface here (which may seem particularly relevant for a model that addresses object permanence such as Munakata, 1998). Another aspect of such implementations is that they probe the real-time autonomy of behavior. Is the robot capable of behaving continuously, going from one state to another, propelled by its inner dynamics and the sensory information it actively acquires from its environment? This entails not only the issue of closed loop control in the real world but also the continuous, asynchronous operation of cognitive processes. This may be contrasted with information processing models, in which behavior is generated only as a response to stimulation, so that time is (implicitly or explicitly) parsed into input–output cycles.

There is more to be gained from robotic implementations beyond such feasibility proof. As a heuristic device, robots may reveal to us all that needs to be specified, and all that can go wrong when a particular behavior is generated. This may motivate new research questions. Examples in kind are calibration and homeostasis, both of which are often left unaddressed in more abstract models of cognitive function. Heuristics also works the other way round: Robot demonstrations of a particular function may be possible without invoking a particular concept. For instance, perseverative reaching can be modeled without using an explicit object representation (Schöner & Dineva, 2006). This does not prove that babies do not have object representations; however, it means that perseverative reaching is not necessarily an index of such representations.

Robotic implementations of DFT may also be pursued simply as a competitive approach to autonomous robotics, evaluated based on the performance of the solutions, on their robustness, ease of design, and so on. It is in this most applied sense that the first robotic demonstrations of DNF models were made (Engels & Schöner, 1995; Schöner et al., 1995; Bicho et al., 2000). Here, we illustrate how DFT can work in a real-world setting using an example close to the issues discussed in the last section, that is, the visual recognition of objects (Faubel & Schöner, in press). This is anchored in a scenario, in which a service robot interacts with a human user within a shared workspace. The robot system learns to recognize a number of objects from a single or a small number of views, associating the object with a label. The ultimate goal is to interact with the user, recognize and name objects, reach for them, manipulate them, and so on. To simplify the task, it is assumed that the number of objects to be memorized is limited, and that the environment is uncluttered and known to the robot.

As first step of the object recognition, a simple segmentation algorithm is applied to the visual input to detect objects on the table surface. Then, several low-level features are extracted for each object: Its size, the aspect ratio as a measure of its shape, and the color, described by a histogram of hue values (Figure 13.11). Each of these features serves as input to a two-dimensional label feature field. In these fields, an association between a feature value and a matching label is realized through hills of preshape: The features are represented along the first dimension of the field, whereas the labels are represented along the second one. For each label, a hill of preshape is created during the teaching procedure around the appropriate feature values. If an object is presented for recognition, its feature

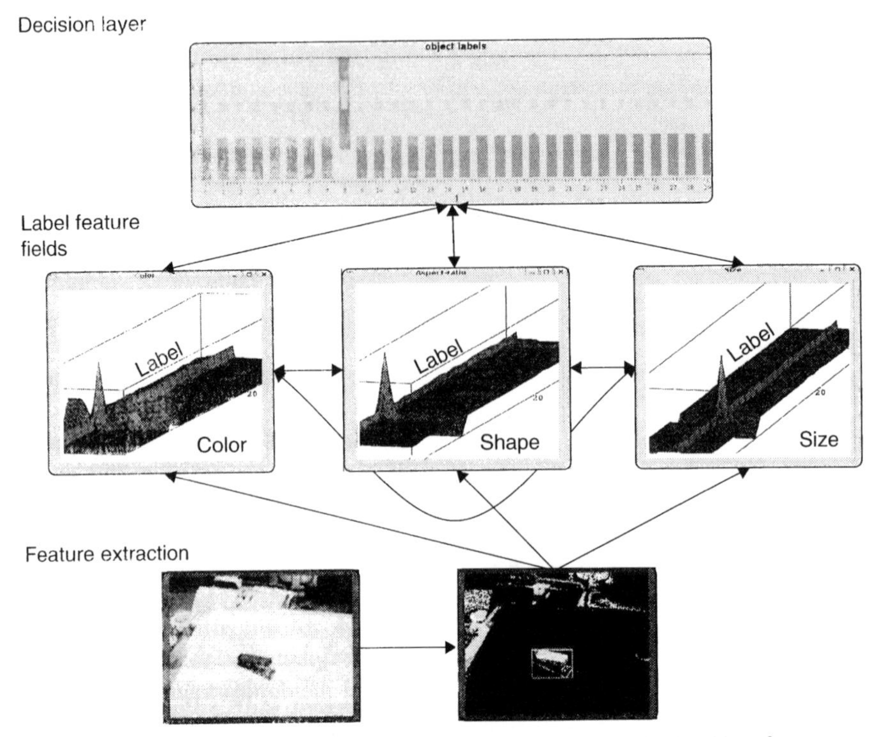

FIGURE 13.11 For the object recognition procedure, three simple object features are extracted from an image and fed into a network of coupled dynamic fields. The interactions within and between these fields result in the selection of one label, which can be read out from the decision layer. (See color plate)

values are fed into the two-dimensional field along the first dimension, creating a ridge of activation that will overlap with some of the preshapes. As for the categorization behavior, the preshape peaks compete with each other through local excitation and global inhibition after the field receives a "go" signal (a homogeneous boost of activation throughout the field), and the preshape that has the greatest overlap with the input is most likely to win the competition. A two-dimensional interaction kernel is used to implement these field interactions. As no metrics is defined for the different labels, the kernel profile is simplified along this dimension such that each label excites only itself and homogeneously inhibits all others.

The output of all label feature fields is fed into a one-dimensional decision layer that has one node for each label. The same simplified interactions are used in this layer to enforce a decision for one label, which is returned as the result of the recognition process once the activation of one node passes a preset threshold. This feedforward processing is augmented by lateral and feedback connections between the fields: The same labels in different label feature fields excite each other, such that the selection of a certain label in one field gives it a competitive advantage in the other fields. In addition to this, there is inhibitory input from the decision layer: The output of one label in this layer suppresses all other labels in the label feature fields. This ensures that after a decision, all peaks that do not match the selected label are extinguished, which is important for learning.

To teach a new object, its label is associated with one node in the decision field, the corresponding columns in the label feature fields are preactivated, and the object is presented to the robot's camera. Peaks will emerge in the label feature fields for the feature values extracted from the visual input and preshape is laid down at these positions. If similar features are found in later recognition trials, the learned label will be activated through the association mechanism of the two-dimensional fields. It is possible to continue the learning process, that is the buildup of preshape, in later trials, either after a correct recognition or after corrective input from the user. This way, the distribution of the feature values for different views of the same object can be reproduced in the preshapes.

The DFT-based object learning system was tested on 30 objects, presented in several different positions on the table and different orientations during teaching and recognition trials. A mean recognition rate of 88% was achieved after teaching each object in eight different views. In every teaching trial, the object recognition procedure was performed as well, with the correct label being given to the robot if the recognition failed. It is noteworthy that with this setting, on average only 2.8 user interventions (corrections of wrong responses) were necessary per learned object.

One important aspect of this model is the role of the lateral and feedback projections between the fields. Without them, each label feature field would independently select the label that best matches the current input and the final decision would be made between these candidates. With the lateral projections, a label that yields a good match in several features receives extra input, and thus it can win the competition in all fields even if it does not yield the best match in any of them. A second effect aids the selection of the correct label: For those feature dimensions where several labels are closely competing with each other, the selection process is slowed down (as discussed earlier for categorical responses in general). The selection is also slow in those fields where the candidate labels have a broad and flat preshape, which results from a high variance in that feature under different views. Due to these effects, a decision is first made in those fields where the stimulus input is unambiguous, and the other fields are then pushed to select the same feature by inter-field excitation. Once more, this desirable behavior emerges directly from the field interactions, and no superordinate structure is needed to select those features that are most significant in the current situation.

CONCLUSIONS

At the same time as it illustrates the embodied and situated nature of DFT accounts, the preceding example provided an outline for how the ideas reviewed in this chapter may scale up from the most elementary forms of cognition toward more complete, "higher" acts of cognition. Other work, not reviewed here, has similarly established the scalability of the concepts by accounting for the emergence of working memory for planned actions (Thelen et al., 2001), spatial working memory (Schutte et al., 2003) and spatial cognition more generally (Simmering et al., 2007), visual working memory (Johnson et al., 2008) and infant habituation (Schöner & Thelen, 2006).

In many of these cases, DNFs have to be combined across different feature dimensions. The dynamic ideas of coupling and stabilization work seamlessly, supporting complex architectures of DNFs. This includes the transformation of sensory into motor representations. Associating different feature dimensions is a natural task of neuronal networks and the dynamics of neural fields accommodate that basic neuronal functionality.

We are expressing some confidence that DFT concepts scale up to forms of cognition more traditionally at the core of concern of cognitive scientists. That said, a subtle, but fundamental issue must be recognized. The complete dynamical system characterizing an organism and its nervous system in a given environment and task context has rich internal structure and includes coupling through the outer world as well. The functionally significant states of such complete systems emerge as attractor solutions from these dynamics under the appropriate circumstances, depending on the behavioral (or activation) history and on an appropriately structured environment (Schöner & Dineva, 2006). These functions are not fixed and they do not "sit somewhere" until activated. They are simply emergent properties of the dynamical system. Individuals may differ in the circumstances that are required to bring about such functions. Individual differences may initially arise from chance events but may then become amplified over time due to the adaptive and learning capacity of dynamical systems. Because dynamical systems can amplify small graded differences into qualitatively different states, this implies a limitation of predictive power. Conversely, the same system may behave differently, exhibiting or not exhibiting a particular function, depending on the task context. Learning does not necessarily install function in a definite and fixed way. Learning may more appropriately be viewed, in Dynamical Systems Thinking, as a process that eases the constraints on the environmental and task conditions under which a function may emerge. Thus, the very nature of Dynamical Systems Thinking makes that the accounts delivered may differ from expectations built on the tradition of information processing or even connectionist thinking. It appears unlikely that there would be something like the ultimate dynamical systems model of the mind, a fixed, if complicated architecture from which behavior can be predicted. A metaphor closer to what Dynamical Systems Thinking may provide is the notion of the brain as a very high dimensional, complex dynamical system, built neuronally, but potentially coupled so closely to the environment through its own effector and sensor systems that these become part of the dynamical system. Understanding such a system amounts to understanding the constraints on its inner structure, flexible though it may be. What dynamicists try to do is find the tasks and elementary behaviors, in which such structure comes to light and in which projection from the high dimensional state space into much lower dimensional subsystems is possible. In such exemplary situations they identify the principles that describe how such subsystems form, stabilize, and adapt. Those principles provide the basis for extrapolating to the myriad and open-ended ways in which the mind may shape and reshape.

REFERENCES

Amari, S. (1977). Dynamics of pattern formation in lateral-inhibition type neural fields. *Biological Cybernetics, 27,* 77–87.

Anderson, J. R. (1995). *Cognitive Psychology and its Implications.* New York: W. H. Freeman and Company.

Bastian, A., Riehle, A., Erlhagen, W., & Schöner, G. (1998). Prior information preshapes the population representation of movement direction in motor cortex. *Neuroreports, 9,* 315–319.

Bastian, A., Schöner, G., & Riehle, A. (2003). Preshaping and continuous evolution of motor cortical representations during movement preparation. *European Journal of Neuroscience, 18,* 2047–2058.

Bicho, E., Mallet, P., & Schöner, G. (2000). Target representation on an autonomous vehicle with low-level sensors. *The International Journal of Robotics Research, 19,* 424–447.

Bizzi, E. & Mussa-Ivaldi, F. A. (1990). Muscle properties and the control of arm movement. In D. N. Osherson, S. M. Kosslyn, & J. M. Hollerbach (Eds.), *Visual Cognition and Action* (pp. 213–242). Cambridge, MA: The MIT Press.

Braun, M. (1993). *Differential Equations and Their Applications*, 4 ed. New York: Springer Verlag.

Cisek, P. (2006). Integrated neural processes for defining potential actions and deciding between them: A computational model. *Journal of Neuroscience, 26*(38), 9761–9770.

deCharms, R. C. & Zador, A. (2000). Neural representation and the cortical code. *Annual Reviews of Neuroscience, 23,* 613–647.

Dorris, M. C., Olivier, E., & Munoz, D. P. (2007). Competitive integration of visual and preparatory signals in the superior colliculus during saccadic programming. *Journal of Neuroscience, 27*(19), 5053–5062.

Durstewitz, D., Seamans, J. K., & Sejnowski, T. J. (2000). Neurocomputational models of working memory. *Nature Neuroscience Supplement, 3,* 1184–1191.

Engels, C. & Schöner, G. (1995). Dynamic fields endow behavior-based robots with representations. *Robotics and Autonomous Systems, 14,* 55–77.

Erickson, R. P. (1974). Parallel "population" neural coding in feature extraction. In F. O. Schmitt & F. G. Worden (Eds.), *The Neurosciences—Third Study Program* (pp. 155–169). Cambridge, MA: MIT Press.

Erlhagen, W. & Schöner, G. (2002). Dynamic field theory of movement preparation. *Psychological Review, 109,* 545–572.

Erlhagen, W., Bastian, A., Jancke, D., Riehle, A., & Schöner, G. (1999). The distribution of neuronal population activation (DPA) as a tool to study interaction and integration in cortical representations. *Journal of Neuroscience Methods, 94,* 53–66.

Faubel, C. & Schöner, G. (2008). Learning to recognize objects on the fly: A neurally based dynamic field approach. *Neural Networks, 21,* 562–576.

Feldman, A. G. (1986). Once more on the equilibrium point hypothesis (λ-model) for motor control. *Journal of Motor Behavior, 18,* 15–54.

Georgopoulos, A. P. (1991). Higher order motor control. *Annual Reviews of Neuroscience, 14,* 361–377.

Goodale, M. A., Pélisson, D., & Prablanc, C. (1986). Large adjustments in visually guided reaching do not depend on vision of the hand or perception of target displacement. *Nature, 320,* 748–750.

Hock, H. S., Kogan, K., & Espinoza, J. K. (1997). Dynamic, state-dependent thresholds for the perception of single-element apparent motion: Bistability from local cooperativity. *Perception and Pschophysics, 59,* 1077–1088.

Hock, H. S., Schöner, G., & Giese, M. A. (2003). The dynamical foundations of motion pattern formation: Stability, selective adaptation, and perceptual continuity. *Perception and Psychophysics, 65,* 429–457.

Hogan, N. (1990). Mechanical impedance of single- and multi-articular systems. In J. M. Winters & S. L.-Y. Woo (Eds.), *Multiple Muscle Systems* (pp. 149–164). New York: Springer Verlag.

Jancke, D., Erlhagen, W., Dinse, H. R., Akhavan, A. C., Giese, M., & Steinhage, A. et al. (1999). Parametric population representation of retinal location: Neuronal interaction dynamics in cat primary visual cortex. *Journal of Neuroscience, 19,* 9016–9028.

Johnson, J. S., Spencer, J. P., & Schöner, G. (2008). Moving to higher ground: The dynamic field theory and the dynamics of visual cognition. *New Ideas in Psychology, 26,* 227–251.

Kopecz, K. & Schöner, G. (1995). Saccadic motor planning by integrating visual information and pre-information on neural, dynamic fields. *Biological Cybernetics, 73,* 49–60.

Luce, R. D. (1986). *Response Times.* New York: Oxford University Press.

Massaro, D. W. (1987). *Speech Perception by Eye and Ear: A Paradigm for Psychological Inquiry.* Hillsdale, NJ: Lawrence Erlbaum Associates.

McDowell, K., Jeka, J. J., Schöner, G., & Hatfield, B. D. (2002). Behavioural and electro-cortical evidence of an interaction between probability and task metrics in movement preparation. *Experimental Brain Research, 144,* 303–313.

Munakata, Y. (1998). Infant perseveration and implications for object permanence theories: A pdp model of the ab task. *Developmental Science, 1*(2), 161–184.

Ottes, F. P., Gisbergen, J. A. M. van, & Eggermont, J. J. (1984). Metrics of saccade responses to visual double stimuli: Two different modes. *Vision Research, 24,* 1169–1179.

Perko, L. (1991). *Differential Equations and Dynamical Systems.* Berlin: Springer Verlag.

Port, R. & Gelder, R. van (Eds.) (1995). *Mind as Motion: Explorations in the Dynamics of Cognition.* Cambridge, MA: MIT Press.

Schöner, G. (2008). Dynamical systems approaches to cognition. In R. Sun (Ed.), *Cambridge Handbook of Computational Cognitive Modeling.* Cambridge, UK: Cambridge University Press.

Schöner, G. & Dineva, E. (2006). Dynamic instabilities as mechanisms for emergence. *Developmental Science, 10,* 69–74.

Schöner, G. & Kelso, J. A. S. (1988). Dynamic pattern generation in behavioral and neural systems. *Science, 239,* 1513–1520.

Schöner, G. & Thelen, E. (2006). Using dynamic field theory to rethink infant habituation. *Psychological Review, 113*(2), 273–299.

Schöner, G., Dose, M., & Engels, C. (1995). Dynamics of behavior: Theory and applications for autonomous robot architectures. *Robotics and Autonomous Systems, 16,* 213–245.

Schutte, A. R., Spencer, J. P., & Schöner, G. (2003). Testing the dynamic field theory: Working memory for locations becomes more spatially precise over development. *Child Development, 74,* 1393–1417.

Shepard, R. N. (1980). Multidimensional scaling, tree-fitting, and clustering. *Science, 210*(4468), 390–398.

Simmering, V. R., Schutte, A. R., & Spencer, J. P. (2007). Generalizing the dynamic field theory of spatial cognition across real and developmental time scales. *Brain Research,* (doi:10.1016/j.brainres.2007.06.081).

Spencer, J. P. & Schöner, G. (2003). Bridging the representational gap in the dynamical systems approach to development. *Developmental Science, 6,* 392–412.

Spencer, J. P., Simmering, V. R., Schutte, A. R., & Schöner, G. (2007). What does theoretical neuroscience have to offer the study of behavioral development? Insights from a dynamic field theory of spatial cognition. In J. Plumert & J. P. Spencer (Eds.), *The Emerging Spatial Mind.* New York, NY: Oxford University Press.

Thelen, E., Schöner, G., Scheier, C., & Smith, L. (2001). The dynamics of embodiment: A field theory of infant perseverative reaching. *Brain and Behavioral Sciences, 24,* 1–33.

Trappenberg, T. P., Dorris, M. C., Munoz, D. P., & Klein, R. M. (2001). A model of saccade initiation based on the competitive integration of exogenous and endogenous signals in the superior colliculus. *Journal of Cognitive Neuroscience, 13*(2), 256–271.

Wilimzig, C. (2006). *Dynamische Feldtheorie der kognitiven Informationsverarbeitung.* Unpublished doctoral dissertation, Ruhr–Universität Bochum.

Wilimzig, C., Schneider, S., & Schöner, G. (2006). The time course of saccadic decision making: Dynamic field theory. *Neural Networks, 19,* 1059–1074.

Wilson, H. R. & Cowan, J. D. (1973). A mathematical theory of the functional dynamics of cortical and thalamic nervous tissue. *Kybernetik, 13,* 55–80.

Young, M. P. & Yamane, S. (1992). Sparse population coding of faces in the Inferotemporal cortex. *Science, 256,* 1327–1331.

Zipser, D. (1991). Recurrent network model of the neural mechanism of short-term active memory. *Neural Computation, 3,* 179–193.

14

A Lazy Brain? Embodied Embedded Cognition and Cognitive Neuroscience

Pim Haselager[1], Jelle van Dijk[2] and Iris van Rooij[1]

[1]*Nijmegen Institute for Cognition and Information, Radboud University Nijmegen, The Netherlands*
[2]*Mediatechnology, Utrecht University of Applied Sciences, Utrecht, The Netherlands*

Introduction

The *E. coli* shines in its simplicity. This single-cell organism can locate food in its environment without having any plan on how to look for it, nor does it have any beliefs about the world it finds itself in. Instead it finds food, and avoids toxics, by moving its flagella in one of two ways: either it tumbles about randomly or it swims straight ahead. Without specific stimulation it changes between these two modes every few seconds, thereby engaging in a random exploration of its environment. Once a chemical gradient in its environment is sensed (e.g., an increase in sugar level or a decrease in toxic substances), it increases the amount of swimming and decreases the random tumbling resulting in a process called *chemotaxis*. In effect, the bacterium swims upward along a stream of increasing nourishment toward a food source and downward along a stream of decreasing toxics (Cairns-Smith, 1996, pp. 90–94). The behavior of the *E. coli* could in principle be described in terms of the folk psychological concepts of beliefs, desires, and intentions (Jonker et al., 2001), but these would be superfluous metaphorical ascriptions at best. It seems implausible and unnecessary to attribute such mental states to the *E. coli* as its behavioral success is readily explained in terms of the direct perception–action couplings in which chemical gradients are sensed that trigger different behavioral patterns (tumbling or swimming).

Humans are much more complicated organisms than *E coli*. Humans have much richer behavioral repertoires than the *E. coli* does, and humans can apply this repertoire with an exceptionally high degree of flexibility and sensitivity to environmental conditions, both past, present, and future. It is this flexibility that is seen as a mark of human intelligence and what has proven so difficult to replicate in robots. On the one hand, the increased flexibility makes humans' lives easier, as it allows them to survive under wider environmental conditions than *E. coli*. But, on the other hand, it also seems to make things more difficult for humans, because it confronts them with a challenging control task. The challenge seems to be that humans need to *decide what* to do and *when*. The *E. coli* does not have this problem (these bacteria do exactly what is triggered by the gradient of chemicals in their environment).

The received view in cognitive science and artificial intelligence is that cognitive systems can come to display the kind of intelligent behavior that is characteristic of human beings only by maintaining more or less accurate mental representations of the world (i.e., *beliefs*), which they derive from perceptual information. Based on their beliefs about the states of the world, humans are assumed to make plans (i.e., *intentions*) with the aim of guiding motor behaviors in a way that meets certain goals (i.e., *desires*). This internalist, cognitivist view of the relationship between cognizing and behavior is inherited by much of contemporary cognitive neuroscience, resulting in the explanation of intelligent cognitive behavior as the product of powerful brains that can maintain world models and devise plans. In other words, contemporary cognitive neuroscience tends to see cognizing as something that the *brain* does.

We think that by construing the control problem posed to the brain in this way, cognitive neuroscience, like artificial intelligence, may be making a mistake. Maintaining a stable and approximately a correct set of beliefs about the world that is sufficient for programming more or less successful behavior in situations of real-world complexity seems to pose a computationally too demanding a task for a human (or any kind of) brain to perform. This computational intractability problem, long known to plague cognitivist models of cognition (Pylyshyn, 1987; Haselager, 1997), clashes with the observation that people make split second decisions in everyday contexts, typically with good results. Hence, cognitive neuroscience may do well to consider alternative views of the control architecture of humans.

In this chapter, we make the case for one such alternative control structure. Our control structure is inspired by the theoretical framework of *Embodied Embedded Cognition*, or EEC for short (Chiel & Beer, 1997; Clark, 1997; Brooks, 1999). EEC proposes that cognition and behavior emerge from the bodily interaction of an organism with its environment. According to EEC, the physical structure of the body, the physical and social structure of the world, and the internal milieu of the organism's body, all provide important constraints that govern behavioral interactions. From this perspective, behavior is best explained by a system of interacting components, where the brain is only one such component. In other words, the brain is best viewed not as a commander or director of behavior but rather as only one of the players among equally important others (i.e., the body and the world). As a result, according to EEC, in a great number of cases, the processes subserving cognitive behavior cannot be directly mapped onto brain structures.

We are well aware of the apparent tension between an EEC perspective of the brain and contemporary cognitive neuroscience research (see also van Dijk et al., 2008). Much of the current cognitive neuroscience's methodology (e.g., brain imaging and single-cell recordings) is built on the idea that the brain implements an encapsulated mechanism for cognizing that can be understood by studying the brain in almost complete isolation, independent from any realistic bodily interaction with the world. Accordingly, much experimental effort in cognitive neuroscience is devoted to figuring out *which* of the cognitive subprocesses (perception, abduction, planning, deciding) are performed *where* in the brain and *how* these processes are neurally implemented. This research aim makes sense if one presupposes that the body and the world are merely external factors (related to the input and output) to cognition. But it is exactly this presupposition that is questioned by EEC.

In this chapter, we review arguments *against* the exclusive adoption of the cognitivist conception in cognitive neuroscience, and *for* extending it with an EEC view. Of course, empirical researchers are not easily swayed by theoretical or philosophical argumentation alone, nor should they be. If EEC is to inspire cognitive neuroscience to extend its research methodology, so that it aligns with an EEC view of the role of the brain in cognitive behavior, then EEC may do well to formulate concrete research questions that are amenable to empirical testing by cognitive neuroscientists in the near future. In this chapter, we therefore try to take some steps toward the generation of such concrete questions.

OVERVIEW

The chapter is organized as follows. First, we start by explaining the computational intractability problem, why it poses a formidable problem for cognitivism, and why we think that existing attempts fail or are unlikely to overcome the problem within a cognitivist framework. Second, we put forth some arguments for, and speculations about, how organisms can come to inhabit, and be adaptive in, relatively complex environments without the need for continuous high-level world modeling, planning, and decision-making. Basically we argue that because of a natural fit between organism and environment, organisms can be "ignorantly successful" in their "user-friendly" environments most of the time.

Third, we outline the contours of a "minimalistic" control structure that could suffice for such ignorant successfulness by introducing the metaphor of traffic facilitation as a way of conceiving the main task for higher level control mechanisms in the brain. According to this view, the brain does not primarily produce (through modeling, planning, and deciding) behavior but rather, at least most of the time, inhibits or disinhibits perception-action loops that are constitutive of ongoing behavior. We discuss the types of questions this traffic facilitator metaphor could imply for empirical research in cognitive neuroscience experimentation as well as robotics.

THE COMPUTATIONAL UNFEASIBILITY OF A BRAIN IN COMPLETE CONTROL

We examine the computational demands of the task attributed to the brain by the cognitivist. Cognitivist accounts typically assume that central control systems work in two general stages: first, based on the information provided by the sensory input systems, "higher" cognitive processes (sometimes referred to as "central systems") form beliefs about how the world is; and second, the central system selects from the entire repertoire of possible actions, a sequence of actions that when performed in the world, as it is believed to be, will lead to the realization of certain goals. Both stages can be shown to run into the problem of computational intractability (Joseph & Plantinga, 1985; Bylander, 1994), but for ease of presentation we will focus on the computational task posed by the first stage only.[1] Clearly, the beliefs generated in the first stage cannot be guaranteed to be true, always and everywhere, but assuming that behavioral success is to be explained by plans based on these beliefs they cannot be arbitrarily false either. It seems then that for a cognitivist account of adaptive behavior to work, one needs to assume that brains have a capacity for forming more or less accurate beliefs, at least sufficiently accurate to support the success of planned behavior most of the time. We present the following quote as just one example of the cognitivist idea that higher processes are involved in trying to make sense of the world on the basis of imperfect information to decide on action:

> Action selection is a fundamental decision process for us, and depends on the state of both our body and the environment. Because signals in our sensory and motor systems are corrupted by variability or noise, the nervous system needs to estimate these states. …

[1] Alternatively, one may assume that the two steps are collapsed into one, in the sense that the probability of plan success is being evaluated by the central system for all possible plans against the background of all possible worlds consistent with current perceptions, and the plan that has the largest (or a large enough) probability of success is selected. For our purposes, the simplified two-step scenario suffices to make our points about the computational intractability of centralized (disembodied) inference, planning, and decision-making. The same points would also apply to the collapsed-steps scenario explained here, since its computational complexity is at least that of the two steps considered separately.

> The approach of Bayesian statistics is characterized by assigning probabilities to any degree of belief about the state of the world ... Bayesian statistics defines how new information should be combined with prior beliefs and how information from several modalities should be integrated.
>
> Körding and Wolpert (2007, p. 319)

On this view, then, higher cognitive processes involved in planning and decision-making are engaged in generating abductive hypotheses that make (the most) sense of the perceived information, given everything else the cognitive system knows (Rock, 1983; Shanahan, 2005; see also Fodor, 1983, 2000). The word "abduction" is not often used in neuroscientific literature. However, to ensure that one's beliefs about the world correspond more or less to what is actually the case in the world, one seems to minimally require a capacity for domain-general abduction. Here, by "abduction" is meant an inferential process that takes as input partial information about the world as input, or data (as produced by sensation and perception) and generates as output hypotheses about which states of the world are believed to currently hold and which ones not. For example, if an object looks like a duck (vision) and quacks like a duck (audition), then we might (or might not—depending on what else we perceive and believe) abduce that the object in front of us is a duck. We furthermore, might or might not abduce that the object is eatable, a bird, 2 feet long, etc. By "domain-generality" is meant *both* that the abduction process can be informed by information coming from all of the input systems (vision, audition, olfaction, proprioception, etc.) *and* that the entertained hypotheses, and the information relevant to maintaining them, can span all kinds of content domains that are potentially relevant for human activity (the hypotheses can be about ducks, people, atoms, weather, etc.). This domain-generality is also expressed sometimes by saying that human abduction processes are not informationally encapsulated (Pylyshyn, 1980, 1984).

It is the requirement of domain-generality that in a sense causes trouble when one wishes to devise computational procedures for abduction. The reason is that it implies that we cannot, in general, have good abductions by considering only a handful of observational facts and only a handful of relevant beliefs. In contrast, whether or not one should entertain belief p, given observational facts $d_1, d_2, \ldots, d_m$, depends also on one's whole system of background beliefs about the world, $p_1, p_2, \ldots, p_n$. Such belief systems may contain hundreds or thousands of beliefs, and hence $n >> m$. Moreover, these beliefs are not set in stone (neither are the observational facts by the way, which may be abduced to be misperceptions or illusions; see e.g., Thagard, 2000) and each and everyone of them is a potential candidate for updating when new observations are made. Given that the number of possible updates of beliefs (i.e., combinations of held beliefs) grows exponentially as a number of potentially held beliefs, efficient updating of the whole web of beliefs seems computationally prohibitive for minds/brains with finite computational resources.

To give a numerical illustration of the problematic nature of such an exponential growth, assume that there are in total, say, $n = 100$ beliefs in one's entire system of beliefs (a gross underestimation, we would think). Then there are already $2^n = 2^{100} > 10^{30}$ many possible truth assignments ("true" or "false") possible; allowing values of believability between "true" and "false," as preferred by probabilists, makes the number of possibilities even larger. Clearly, exhaustively searching this space to find which truth assignment is supported by the observations at hand is impossible. Even if a brain (or a supercomputer) had at its disposal as many parallel computational channels as there are neurons in the human brain (about 10^{14}), and if each such channel were capable of considering millions (10^6) of possible truth assignments per second, still the computation would require more than 10 centuries to complete ($>10^{10}$ seconds). More importantly, there seems to be no other possible way to ensure that updating results in a stable and more or less accurate set of beliefs. This follows from the observation that all attempts to formally define the computational problem underlying abduction have resulted in a problem that is *NP-hard* (Bylander et al., 1991; Abdelbar & Hedetniemi, 1998; Thagard, 2000). We next explain what this means.

NP-hard problems are problems for which no practicable (i.e., polynomial-time) algorithm is known and it is strongly conjectured that no such algorithm can ever exist. In other words, it is conjectured that NP-hard problems can only be solved by some variant of exhaustive search (i.e., exponential-time) algorithms, which is why these problems are considered *computationally intractable* (Garey & Johnson, 1979, p. 8). Although the conjecture is so far unproven, it has strong empirical support.[2] There are currently hundreds of NP-hard problems known (see e.g., the available online compendia). Moreover, it is known that if any one of these problems were computable in polynomial-time then all of them would be. Despite sustained efforts by mathematicians and computer scientists over the last four decades, nobody to this day has succeeded in devising a polynomial-time algorithm for an NP-hard problem—hence, the conviction that no such algorithm exists. Unless one would want to ascribe oracle-computing powers to central brain systems (something that would be akin to

[2] As an aside, we note that the conjecture is also strongly supported by mathematical intuition. The mathematical intuition derives from the believed inequality of two problem classes, called NP and P. Here, informally, NP can be thought of as a class of problems whose solutions can be easily checked, and P can be thought of as a class of problems whose solutions can be easily found. Now, the mathematical intuition (and perhaps the layperson intuition as well) says that NP may contain problems that are not in P (not all easily checkable problems need be easily solvable). To assist the non-mathematician's intuition, think of crossword puzzle or a game like Sudoku. For each such puzzle it is easy to check if a proposed solution is correct, but it is not clear that a solution is also always easy to find, that is, there may be *hard* puzzles. Now, for technical reasons we cannot go into here, it is known that if an NP-hard problem would be computable without some form of exhaustive search, then this would imply that NP = P, which would violate mathematical intuition (see Garey & Johnson, 1979, for more details).

the avowed "homunculus" in psychological explanation), it seems implausible that central brain systems have the capacity for efficiently computing NP-hard problems.

The theoretical obstacle posed by the computational intractability of abduction is greater than many cognitivists seem to realize. First of all, choosing a different formalism for modeling abduction cannot detract the problem. Oaksford and Chater (1998), for example, argued for a switch from non-monotonic logics to Bayesianism for modeling human abductive inference based on the computational intractability of the former. But such a move seems in vain given that Bayesian models of abduction are as computationally intractable (if not more) than all other existing models of abductive inference—such as, non-monotonic logics, covering models, constraint satisfaction models, and neural network models (Bruck & Goodman, 1990; Cooper, 1990; Bylander et al., 1991; Abdelbar & Hedetniemi, 1998; Thagard & Verbeurgt, 1998; Thagard, 2000).

Second, loosening the quality of the abductions also cannot detract the computational intractability problem. It is often suggested in the cognitive science literature that computationally intractable problems can be approximately computed efficiently (e.g., Love, 2000; Chater et al., 2003, 2006), but this seems at best a misrepresentation of the state of the art. It is well known that many NP-hard problems cannot be efficiently approximated (Yoa, 1992; Arora, 1998), and almost all are inapproximable if only a constant sized error is allowed (Garey & Johnson, 1979). Moreover, models of abduction are NP-hard to approximate even for quite liberal criteria of approximation (Roth, 1996; Abdelbar & Hedetniemi, 1998), and where claims are made of polynomial-time "approximation" algorithms for abduction problems (e.g., Thagard & Verbeurgt, 1998), those algorithms do not approximate the required solution itself (i.e., the truth assignment), but instead its associated value, for example, coherence or probability (see Hamilton et al., 2007, for a discussion).

Third, computationally intractable problems cannot be rendered tractable by a divide and conquer strategy. For example, in the cognitive science literature, it is sometimes suggested that the computationally intractability problem plaguing a single, central abduction/planning system can be overcome by postulating the existence of a large set of "modules," each being able to efficiently update beliefs, or make plans, for a specific domain of situations or "contexts" (cf. the "massive modularity" of Cosmides & Tooby, 1994, the "toolbox of heuristics" of Todd & Gigerenzer, 2000, and the "multiple models" of Wolpert & Ghahramani, 2000; see also Carruthers 2003a, b, and Sperber, 2002, for discussions). If each such module implements a tractable computation, then it may seem that the whole system could tractably update our beliefs, and make plans, in all psychologically relevant contexts. However, even granting the number of required modules could be efficiently stored in the human brain (think of the potentially quite large number of possible contexts), a modular system cannot tractably compute any computationally intractable problem at risk of contradiction. If a problem were to be tractably computable by a modular system, then this would imply that that problem does not belong to the class of computationally intractable problems.[3] If a problem is intractable, as seems to be the case for domain-general abduction, then *no* algorithm for tractably computing it can exist.

Much more can be said about the topic of computational intractability, its proposed solutions, and their failings (see, e.g., van Rooij, 2008) but for our purposes the point is merely this: The computational intractability problem is not going to go away for cognitivist models of planning and action control. The only way to achieve tractability of control, so it seems, is to assign an easier computational task to control processes than domain-general abductive inference, and in effect make the explanation of success of behavior to a large extent independent of the success of our abductions of beliefs about the world. The question, of course, is how the successfulness of behavior in the world can be explained if not by an appeal to a control system that plans on the basis of beliefs about the world. The next section will put forth an argument for why it may be plausible to assume that organisms with control structures that maintain no internal model of the world can nevertheless behave successfully and adaptively in the world. Moreover, we argue, that such organisms may very well come to inhabit the most complex or challenging worlds that their control structures can successfully handle or approximations thereof.

IGNORANTLY SUCCESSFUL IN A USER-FRIENDLY ENVIRONMENT

No animal is behaviorally adapted to react in appropriate ways to all possible changes of all possible variables existing in the "world out there." Consider an ant, stamped upon by a casual pedestrian: this poor creature "has no idea" what hit him, and, more importantly, it has no means whatsoever to counteract such occasions. The ant is either extremely lucky or it dies. From the perspective of the ant, a passing pedestrian is a true Deus ex Machina. Still ants are successful creatures. On the whole, every organism seems to get by pretty successfully, using the behavioral capacities it possesses. So how is it that organisms can be successful in a complex and unpredictable world? The speculative idea we pursue in this section is based on the assumption that the *local* or *personal* environment in which an intelligent creature is situated is not formed independent from the organism's own behavioral and evolutionary history. "Environments" are not simply pre-given, arising out of nothing. Organisms do not wake up to find themselves in completely new, unfamiliar, and hostile worlds. In a confined

[3]It would also mean that the problem is tractably computable by a single non-modular system, because a non-modular system could tractably compute it by (i) simulating the process by which the modular system selects the right module for the current context, and (ii) simulating the workings of the selected module. If both steps are tractable for the modular system, then the simulation is also tractable for the non-modular system.

region of the global chaos we call reality, each creature "makes a living," based on its sensory capacities and its behavioral repertoire, thereby creating its own *Umwelt* (Von Uexkull, 1934; Ziemke & Sharkey, 2001). In the words of Varela et al. (1991) one might say that the organism, by its own actions, *brings forth*, or *enacts*, a world. In yet other words, organisms and their environments can be said to *co-evolve* (Chiel & Beer, 1997; Deacon, 1997) or as Mead (1934) put it:

> The sort of environment that can exist for the organism, then, is one that the organism in some sense determines. If in the development of the form there is an increase in the diversity of sensitivity there will be an increase in the responses of the organism to its environment, that is the organism will have a correspondingly larger environment. ... In this sense it selects and picks out what constitutes its environment. It selects that to which it responds and makes use of it for its own purposes, purposes involved in its life-processes. It utilizes the earth on which it treads and through which it burrows, and the trees that it climbs; but only when it is sensitive to them.
>
> Mead (1934, p. 245), quoted in Jarvilehto (1999)

Our suggestion is that it is this interdependency between organisms and their environments that makes these environments generally facilitative to interaction. It is this intimate "fit," we speculate, which ensures that actions, once taken, will generally prove to be successful/adaptive. Moreover, under most, ordinary, circumstances, inappropriate actions will generally prove to be repairable: we are allowed to make mistakes in our Umwelt, so that we may even learn something along the way. For example, think of the way in which parents provide safe environments for their offspring to explore and learn in. In other words, the naturally emerging embodied embedded behaviors of an organism generally tend to be quite effective for the survival of *that* organism in *that* Umwelt. Consider that most ants live their lives successfully, without knowing about, nor having had to deal with, stamping feet. Most ants are ignorantly successful. And so, we claim, are we humans.

An ignorantly successful interaction with a by and large user-friendly environment might very well be an apt description of what takes place during ongoing behaviors of individual human beings in daily life. As an illustration of this, consider a situation in which a human being is in need of locating an often used object in the kitchen during cooking, for example, a milk-beater. In cognitivist theory this is a problem of search, involving not only inspecting the visual scene, but also memory, as when we try to remember (or form hypotheses about) where we may have put the object. In practice, however, memory search is often not needed. In many situations the structure of the environment naturally constrains the kind of actions that can be performed, and one may question whether the brain needs to search through mental models and memory stores at all. For instance, in a kitchen, some drawers and shelves are more easily reached by the human agent than others. Such drawers and shelves will be among the first to be inspected, that is, if the natural flow of body–world interaction is followed. Note that this is a physical constraint that exists because of the bodily characteristics of the person and the physical organization of the kitchen, and independent from any potential deliberation in the person's mind. Chances are that a daily used object like a milk-beater is also put on one of those easily reachable shelves or drawers, perhaps even by the same human agent that is searching for it later. So when we experience ourselves doing a seemingly random inspection of drawers and shelves instead of a rational search, we are actually being constrained both by body and world, leading to higher chances of behavioral success even if these actions in isolation would seem to be senseless or random. The example illustrates how success of behavior can follow from the "fit" between the person's behaviors and the local environment. Where you can put objects most easily is also where you can look for them most easily, which in turn is where you have a high chance of finding the object that you where looking for.

Environments can thus be "user-friendly," not unlike a well-designed interface. An agent's natural tendencies for action can tend to match the environmental structure in ways that turn out to be functional with respect to the agent's needs. This would be the case, for example, when the agent's behavioral repertoire and the structure of the situated environment co-developed with one another.[4] In such a process of co-evolution organism and situated environment (Umwelt) are mutually affected by one another. Evolution is sometimes seen as a one-way effect in which an animal adapts to changes in its environment. What is less often recognized is the reversed process, in which changes in an organism's structure might also lead to changes in the (situated) environment. Have polar bears turned white because their environment became snowy? Or did the white colored polar bears use their skin-color to their advantage, leading them to travel even further up north into snowy territory? Or consider the human eye. From a traditional perspective, the eye would be viewed as the animal's *solution* to an environmental *problem*. An evolutionary explanation might begin by stating that, at some point, due to a change in the environment, the ability to detect the visible spectrum of light became relevant for survival (where previously it had not been). How to acquire the capacity to use light can be seen as the *problem*. Selection forces then procure a sensor that is able to detect light, in humans the eye. This is the *solution*. We think that such a view need not be correct. For one thing, it has been argued that sometimes structural properties of organisms emerge and persist (over numerous generations) long before the property in question becomes adaptive (Goodwin, 1994). In other words, evolution creates exaptations (Gould & Vrba, 1982; Gould, 1991), which in a way can be seen as "solutions" for problems that do not even exist (yet). A perhaps even more fundamental question is why the visible spectrum of light became relevant for survival in the first place. In many situations, it is not unreasonable to suggest that such aspects of the environment co-evolve with changes in the behavioral repertoire of the organism itself. Consider, as a hypothetical example, a blind creature that has developed

[4]Incidentally, such a fit between organism and environment might, at least for human beings, emerge not only for a species on an evolutionary timescale but also, for an individual, from the ongoing interaction with the environment during his or her lifetime.

the means to move significantly faster than before. Now speed may be a useful adaptation, but it also presents dangers, such as a fatal collision. For this animal, sensitivity to distal (e.g., visual) rather than proximal sources now becomes adaptive, whereas its slow ancestor would have had no use for it. Hence, once the eye has evolved, the system relaxes into a stable relation between animal and environment, in which its new eyes team up nicely with its fast legs. But that is not the end of it. Once there is vision, the environment "broadens up" once more. A "visual environment" might help the animal in dealing with the dangers of going fast (the original "problem"), but it also creates new challenges. As Lock (2003, p. 105) states:

> Simpler organisms can handle their simpler worlds by less complex means, but once evolution has come up with the where-withal for simpler organisms to handle their somewhat simpler selection problems, then it effectively creates for itself a new problem. That is, as organisms find ways of sustaining themselves, they create new potential sources of energy that can be preyed upon. And as new sources of energy, they present more complex worlds for their possible prey to operate in.

That is, when compared to its blind ancestors, the eyed creature faces some challenges of its own: How to cross that distant river, how to climb that far-away tree, how to fight that approaching competitor, and so on. The idea is thus that behavioral capacities co-evolve with changes in the organism's environment in a corresponding manner. New capacities enable the animal to be adaptive in that new environment. But the new situation has both "advantages" and "disadvantages." The advantage is that the new extension to the behavioral repertoire helps the organism in dealing "better" with some aspects of the environment than before. However, the disadvantage is that the animal has now projected itself into a new environment and this environment poses new cognitive challenges as compared to the previous situation. The development of new capacities, seen as a means to resolve some tension between organism and environment, can therefore also be seen as *generating* new challenges as well: new kinds of behaviors lead to an extension of the environment, which poses new demands. Therefore, instead of saying that animals become *more* adaptive with each step in evolution, we would rather formulate it as animals becoming *equally adaptive again and again*, at each new critical equilibrium (Goodson, 2003), albeit in a broader range of (more complex) environments. For a related view of the co-evolution of psycho-linguistic capacities and sociolinguistic environments, see Deacon (1997).

Overall, we propose that the local, situated environments in which organisms are embedded are relatively comfortable and safe environments. Organisms and their environments co-develop, making environments generally "user-friendly" life-worlds. We argued that success of behavior follows from the "fit" between the embodied embedded repertoire of the organism and the structure of the situated environment. Next, we showed how new capacities in effect broaden up the situated environment, which has both upsides as well as downsides: new possibilities for action and perception may be useful in dealing with certain existing challenges, but they also generate new challenges as well.

GENERATING RESEARCH QUESTIONS FOR COGNITIVE NEUROSCIENCE AND ROBOTICS

As indicated in the introduction, it would be highly desirable for EEC to formulate concrete research questions that can provide the basis for research in cognitive neuroscience. A first apparent obstacle is that in the current neuroimaging methodology the movements of subjects have to be restricted almost completely to reduce noise. This prevents the occurrence of the natural organism–environment fit, discussed earlier, that forms the basis for the view on brain control to be outlined in this section. Another problem is that the perspective of EEC tends to get formulated at a rather abstract, philosophical, or even generally descriptive, level. Hence, most statements (including our own so far) about the value of EEC tend to be far removed from concrete empirical research questions in cognitive neuroscience. A third problem is that existing theories and models of EEC commonly deal with relatively low-level organism–environment interactions, usually as far removed from the complexity of daily life behaviors as the research of the often scorned cognitivist perspective (hence, e.g., Clark & Toribio's (1994) challenge to deal with "representation-hungry" cases of behavior; see also van Rooij et al., 2002, and Haselager et al., 2003). These problems are indeed formidable and cannot be solved within one chapter. However, we do feel that there are enough ingredients available, from the area of robotics as well as from neuroscience, to at least tentatively outline a view that might lend itself to empirical testing. In this section, then, we will try to work our way from a metaphorical depiction of high-level brain functioning during common sense behavior to its consequences for empirical research in robotics and cognitive neuroscience.

Brooks (1999, p. 81) suggested that it is fundamental for an organism to have "the ability to move around in a dynamic environment sensing the surroundings to a degree sufficient to achieve the necessary maintenance of life and reproduction." He modeled this capacity by means of his well-known layered architecture: reactive creatures consisting of behavioral layers that each instantiate a direct input–output coupling. According to Brooks, it is a major advantage of his approach that no intermediate (in between input and output) world modeling, planning, and decision-making takes place. Instead, layers compete for dominance on the basis of the input received by the system. From this perspective, a creature can be seen as a repertoire of behavioral dispositions and the environment selects from it. A creature is inclined, in virtue of its bodily possibilities and its history of interactions with its environment, to respond to stimuli in specific ways without high-level thought or planning. Perception, action, and world are structurally coupled to form a temporarily stable behavioral pattern that is functional with respect to the task. We call this structural coupling a "basic interaction cycle." A creature carries its set of potential behaviors with it across contexts, and if these contexts fit with the creatures' behavioral repertoire (as well may the case, as indicated in the section "Ignorantly Successful in a User-Friendly Environment") its overall conduct may be satisfactory for a long time.

The fit between environment and behavioral repertoire might to a large extent underlie the relative success of most of our common sense behavior in daily life, such as having a drink in a bar, going home, or making dinner, and so on. Common sense behavior actually consists in quite complicated sequences of behavior, even though it does not require the type of planning and decision-making characteristic of say playing a chess game or buying a house. Instead we seem to operate more or less on "autopilot," our behavior flows naturally out of the stimulations from the environment.

In the reactive robots of the early 1990s, the number of distinct behavioral layers was typically small and the precedence relations between them were set beforehand and were hardwired into the system. This resulted in creatures not unlike the *E. coli* discussed earlier. However, once the set of basic behavioral capacities of a creature becomes larger, and its sensorimotor capacities quite rich, a more flexible and integrated way of setting up behavioral layers and their interrelations becomes necessary. To illustrate, consider the following: If an organism has n basic behavioral layers available, then it could come to display, in principle, as many as 2^n distinct behavioral patterns by simply turning "on" some layers and turning others "off." With even as few as 32 layers this could result in as many as $2^{32} = 10^{10}$ distinct potential behaviors, which would, to quote Wolpert and Kawato, be "sufficient for a new behavior for every second of one's life" (1998, p. 1318). If additionally quantitative adjustments are possible—that is, states in between "on" and "off," possibly implementing dominance relations—then the same organism would have the capacity for displaying an even larger number of possible behaviors. To help in regulating the selection (or dominance relations) of behavioral layers, we suggest, is the main task of the high-level control function of the brain. In other words, instead of interpreting the brain's control system as the driver or pilot of the body, we see it as a *traffic regulator* (van Dijk et al., 2008)—it is (merely, but importantly) assisting the environment-driven selection from the behavioral repertoire. Notably, we do not propose that this traffic facilitation is achieved by computing the best (or even, a good enough) behavior from the set of possible behaviors given the current context (see, e.g., Wolpert & Kawato, 1998; Wolpert. & Ghahramani, 2000; Körding & Wolpert, 2007), because doing so would lead us right back to the computational intractability problem discussed in the section "The Computational Unfeasibility of a Brain in Complete Control." Therefore, contrary to the traditional view of the control system as involved in world modeling, planning, and decision-making, we would like to hypothesize that the control function of the brain works in a, dare we say, more "lazy" way.

There may be several ways in which one could conceive of a "lazy" control system. We will describe just one such possibility here, drawing on an analogy with the control system of the *E coli*. Recall that the *E coli* can perform two modes of behaviors (tumbling or swimming), and the probability with which it switches between these two modes depends on chemicals (food or poison) it picks up from the environment. In a similar vein, our lazy control system may work by stochastically sampling from the set of behavioral options with a non-uniform *bias*, that is, not every behavioral option is equally likely to be selected. The bias can be represented by a probability distribution P over the set of possible behaviors (e.g., combinations of "on" and "off" layers and/or combinations of dominance relations), where $P(t, i)$ would denote the probability that behavioral disposition i is sampled at time t. Here, the bias P may be fixed, but more likely it is variable over time, for example, as a function of experience and the organism's internal (homeostatic) milieu. This proposal raises several (more or less) concrete questions for cognitive neuroscience: How is P implemented in the human brain? What is the shape of the distribution P for humans? Is P fixed or variable? If P is variable, what is it a function of? If P is a function of experience and/or homeostatic states, how do these factors contribute to changes in the distribution P over time, both descriptively and mechanically?

It seems to us that these questions can in principle be researched using (existing and developing) cognitive neuroscientific methods. Consider for instance the question of how such a lazy control system could be implemented in the brain. A concept that could help to elucidate how the brain might be involved in the temporary creation of a relevant behavioral repertoire is Edelman's (1992; Edelman & Tononi, 2000) notion of functional clusters. A functional cluster consists of "elements within a neural system that strongly interact among themselves but interact much less strongly with the rest of the system" for a certain amount of time (Edelman & Tononi, 2000, p. 120, see also pp. 184–185). Several neuronal groups form a strongly integrated assembly for brief periods (most likely to be measured in the range of 50–100 milliseconds). In other words, functional clusters exist only temporarily, consist of various contributing areas that are recruited for the specific occasion and are changeable over contexts. A similar concept, that of neuronal assemblies, is discussed by Chakraborty et al., (2007, p. 491):

> Large-scale, coherent, but highly transient networks of neurons, 'neuronal assemblies', operate over a sub-second time frame. Such assemblies of brain cells need not necessarily respect well-defined anatomical compartmentalisation, but represent an intermediate level of brain organisation

Functional clusters or neuronal assemblies can be assumed to implement short-lived changes in the organisms behavioral dispositions. In that case, the nature of the postulated bias P with which behavior dispositions are sampled could be experimentally investigated by studying the stochastic dependencies between different possible functional clusterings over time. We may observe that of the many different ways in which neural systems may cluster in principle, only relatively few cluster types happen with high frequency in practice over long periods of time under constant conditions. If so, this would suggests that P is relatively high peaked, implementing a stronger bias than when P would be flat throughout. Also, the hypothesis of non-constancy of P could be investigated by trying to fit a constant model to the observed stochastic dependencies and see if it fails to account for the observations. Following this, different *P*s, each a different hypothesized function of internal conditions and environmental factors, can be formulated and tested for their ability to explain observed stochastic dependencies of clustering over time and under variable conditions. Of particular interest and relevance for the latter type of experimental investigation would be to consider internal homeostatic states as variables for the function *P*, since by analogy with the *E coli* we hypothesize that much (if not all) of the bias in our sampling of behavioral dispositions is a function of such states.

Our proposal of a "lazy" traffic facilitator control system also raises a question that we think may be answered using robotic simulation: How can humans, or any other complex organism, come to have a bias *P* that works well enough for the organism to get around the world on "autopilot," without giving the selection of behaviors much thought, most of the time? We think that the answer lies in the type of co-evolution of control systems (in this case the bias *P*) with the life world of the organism, as described in the section "Ignorantly Successful in a User-Friendly Environment." This explanation may be tested, or at least a proof of concept may be given, using robotic simulation. For example, a robotic simulation could start by endowing robotic systems with a "lazy" control system P_0 and letting it evolve for *n* generations through $P_1, P_2, \ldots, P_n$ in interaction with its life world. By systematically manipulating (i) the set of layers available to the robot, (ii) the nature of the initial P_0, (iii) the way each P_i depends on internal and external conditions of life world *i*, (iv) aspects of the environment, and (v) the nature of the evolution process, one could get a better understanding of how these factors (i)–(v) interrelate. Observations of the interrelations generated in the simulation process may serve as hypotheses for how these factors relate in (higher) organisms and could be subjected to cognitive neuroscience testing.

Although we realize that our suggestions for experimentation in cognitive neuroscience and robotics need to be worked out more concretely to result in actual simulations and experiments, we do feel that they indicate that the traffic facilitation metaphor and the general view of EEC underlying it are not too far removed from empirical investigations.

CONCLUSION

Compared to for instance the *E. coli*, humans have an exceptionally rich behavioral repertoire that gets applied with great flexibility and sensitivity to environmental conditions. We argued against the received view in cognitive neuroscience, that is, that cognitive systems can display this behavior only by maintaining mental representations of the world on the basis of which plans are made to achieve specific goals. We explained how such a position leads to the problem of computational intractability. We proposed that effective control may be possible for a more tractable, even "lazy," control system that does not maintain any

internal models of the world, assuming that such "lazy" control systems co-evolve with the bodies and environments of organisms. This co-evolution ensures a certain degree of fit between the control system of an organism and its life world. The "lazy" control mechanism that we postulated raises several interesting questions, each of which we think is amenable to experimental investigation using brain measuring methods. Also, our claim that "lazy" control systems can plausibly evolve, even for quite complex organisms in quite complex environments, can be directly investigated using the methods of robotic simulation. In all, we hope to have shown that an EEC view on the higher level control functions of the brain is not only possible, but that it can be made precise enough to suggest experimental investigation in cognitive neuroscience, as well as robotics.

REFERENCES

Abdelbar, A. M. & Hedetniemi, S. M. (1998). Approximating MAPs on belief networks is NP-hard and other theorems. *Artificial Intelligence, 102,* 21–38.

Arora, S. (1998). The approximability of NP-hard problems. Survey based upon a plenary lecture at the *ACM Symposium on Theory of Computation*, 1998.

Brooks, A. (1999). *Cambrian Intelligence: The Early History of the New AI.* Cambridge, MA: MIT Press.

Bruck, J. & Goodman, J. W. (1990). On the power of neural networks for solving hard problems. *Journal of Complexity, 6*(2), 129–135.

Bylander, T. (1994). The computational complexity of propositional STRIPS planning. *Artificial Intelligence, 69,* 165–204.

Bylander, T., Allemang, D., Tanner, M. C., & Josephson, J. R. (1991). The computational complexity of abduction. *Artificial intelligence, 49,* 25–60.

Cairns-Smith, A. G. (1996). *Evolving the Mind: On the Nature of Matter and the Origin of Consciousness.* Cambridge: Cambridge University Press.

Carruthers, P. (2003a). On Fodor's problem. *Mind & Language, 18*(5), 502–523.

Carruthers, P. (2003b). Is the mind a system of modules shaped by natural selection? In C. Hitchcock (Ed.), *Contemporary Debates in the Philosophy of Science.* Oxford: Blackwell.

Chakraborty, S., Sandberg, S., & Greenfield, S. A. (2007). Differential dynamics of transient neuronal assemblies in visual compared to auditory cortex. *Experimental Brain Research, 192,* 491–498.

Chater, N., Oaksford, M., Nakisa, R., & Redington, M. (2003). Fast, frugal and rational: How rational norms explain behavior. *Organizational Behavior and Human Decision Processes, 90,* 63–86.

Chater, N., Tenenbaum, J. B., & Yuille, A. (2006). Probabilistic models of cognition: Conceptual foundations. *Trends in Cognitive Sciences, 10*(7), 287–291.

Chiel, H. J. & Beer, R. D. (1997). The brain has a body: Adaptive behavior emerges from interactions of nervous system, body and environment. *Trends in Neurosciences, 20,* 553–557.

Clark, A. (1997). *Being There: Putting Brain, Body and World Together Again.* Cambridge, MA: MIT Press.

Clark, A. & Toribio, J. (1994). Doing without representing? *Synthese, 101,* 401–431.

Cooper, G. F. (1990). The computational complexity of probabilistic inference using Bayesian belief networks. *Artificial Intelligence, 42*(2–3), 393–405.

Cosmides, L. & Tooby, J. (1994). Origins of domain specificity: The evolution of functional organization. In L. Hirschfeld & S. Gelman (Eds.), *Mapping the Mind: Domain Specificity in Cognition and Culture.* New York: Cambridge University Press.

Deacon, T. (1997). *The Symbolic Species: The Co-evolution of Language and the Human Brain.* London: Penguin Press.

Edelman, M. & Tononi, G. (2001). *Consciousness: How Matter Becomes Imagination.* London: Penguin Books.

Edelman, G. M. (1992). *Brilliant Air, Brilliant Fire: On the Matter of Mind.* New York: Basic Books.

Fodor, J. (1983). *The Modularity of Mind.* Cambridge, MA: The MIT Press.

Fodor, J. (2000). *The Mind Doesn't Work that Way: The Scope and Limits of Computational Psychology.* Cambridge, MA: The MIT Press.

Garey, M. R. & Johnson, D. S. (1979). *Computers and Intractability: A Guide to the Theory of NP-Completeness.* New York: Freeman.

Goodson, F. (2003). *The Evolution and Function of Cognition.* Mahwah, NJ: Lawrence Erlbaum.

Goodwin, B. (1994). *How the leopard changed its spots: The evolution of complexity.* Princeton: Princeton University Press.

Gould, S. J. (1991). Exaptation: A crucial tool for evolutionary psychology. *Journal of Social Issues, 47,* 43–65.

Gould, S. J. & Vrba, E. S. (1982). Exaptation—a missing term in the science of form. *Paleobiology, 8*(1), 4–15.

Hamilton, M., Müller, M., van Rooij, I., & Wareham, T. (2007). Approximating solution structure. In E. Demaine, G. Z. Gutin, D. Marx, & U. Stege (Eds.), *Structure Theory and FPT Algorithmics for Graphs, Digraphs and Hypergraphs.* Dagstuhl Seminar Proceedings (Nr. 07281). Internationales Begegnungs- und Forschungszentrum für Informatik (IBFI), Schloss Dagstuhl, Germany.

Haselager, W. F. G. (1997). *Cognitive Science and Folk Psychology: The Right Frame of Mind.* London: Sage.

Haselager, W. F. G., Bongers, R. M., & van Rooij, I. (2003). Cognitive science, representations and dynamical systems theory. In W. Tschacher & J.-P. Dauwalder (Eds.), *The Dynamical Systems Approach to Cognition* (pp. 229–242). Singapore: World Scientific.

Jarvilehto, T. (1999). The theory of the organism-environment system: III. Role of efferent influences on receptors in the formation of knowledge. *Integrative Physiological and Behavioral Science, 34,* 90–100.

Jonker, C. M., Snoep, J. L., Treur, J., Westerhoff, H. V., & Wijngaards, W. C. (2001). Putting intentions into cell biochemistry: An artificial intelligence perspective. *Journal of Theoretical Biology, 214,* 105–134.

Joseph, D. A. & Plantinga, W. H. (1985). On the complexity of reachability and motion planning problems. *Proceedings of the First ACM Symposium on Computational Geometry* (pp. 62–66). New York: ACM Press.

Kording & Wölpert, D. (2006). Bayesian decision theory in sensorimotor control. *Trends in Cognitive Sciences, 10*(7), 320–326.

Lock, A. (2003). *Book Review of the Evolution and Function of Cognition by Felix Goodson.* Mahwah, NJ: Lawrence Erlbaum Associates, Inc. 2003. *Human Nature Review, 3,* 104–107.

Love, B. C. (2000). A computational level theory of similarity. *Proceedings of the Cognitive Science Society, 22,* 316–321.

Oaksford, M. & Chater, N. (1998). *Rationality in an Uncertain World: Essays on the Cognitive Science of Human Reasoning.* Hove, UK: Psychology Press.

Mead, G. H. (1934). *Mind, self, and society.* Chicago: Chicago University Press.

Pylyshyn, Z. W. (1980). Computation and cognition. *Behavioral and Brain Sciences, 3,* 111–169.

Pylyshyn, Z. W. (1984). *Computation and cognition: Towards a foundation for cognitive science.* Cambridge, MA: MIT Press.

Pylyshyn, Z. W. (Ed.) (1987). *The Robot's Dilemma: The Frame Problem in Artificial Intelligence.* Norwood, NJ: Ablex Publishing.

Rock, I. (1983). *The Logic of Perception.* Cambridge, MA: MIT Press.

Roth, D. (1996). On the hardness of approximate reasoning. *Artificial Intelligence, 82,* 273–302.

Shanahan, M. P. (2005). Perception as abduction: Turning sensor data into meaningful representation. *Cognitive Science, 29,* 103–134.

Sperber, D. (2002). In defense of massive modularity. In E. Dupoux (Ed.), *Language, Brain and Cognitive Development: Essays in Honor of Jacques Mehler* (pp. 47–57). Cambridge, MA: MIT Press.

Thagard, P. (2000). *Coherence in Thought and Action*. Cambridge, MA: MIT Press.

Thagard, P. & Verbeurgt, K. (1998). Coherence as constraint satisfaction. *Cognitive Science, 22*, 1–24.

Todd, P. M. & Gigerenzer, G. (2000). Precis of simple heuristics that make us smart. *Behavioral and Brain Sciences, 23*, 727–780.

Varela, F. J., Thompson, E., & Rosch, E. (1991). *The embodied mind: Cognitive science and human experience*. Cambridge, MA: MIT Press.

van Dijk, J., Kerkhofs, R., van Rooij, I., & Haselager, P. (2008). Can there be such a thing as embodied embedded cognitive neuroscience? *Theory & Psychology*, *13*(8), 297–316.

van Rooij (2008). The tractable cognition thesis. *Cognitive Science*, *32*(6).

van Rooij, I., Bongers, R. M., & Haselager, W. F. G. (2002). A non-representational approach to imagined action. *Cognitive Science, 26*(3), 345–375.

Von Uexkull, J. (1934). A stroll through the worlds of animals and men. In C. Schiller (Ed.), *Instinctive Behavior*. New York: International Universities Press, 1957.

Wolpert, D. M. & Ghahramani, Z. (2000). Computational principles of movement neuroscience. *Nature Neuroscience, 3*(supp), 1212–1217.

Wolpert, D. M. & Kawato, M. (1998). Multiple paired forward and inverse models for motor control. *Neural Networks, 11*, 1317–1329.

Yoa (1992). Finding approximate solutions to NP-hard problems by neural networks is hard. *Information Processing Letters, 41*, 93–98.

Ziemke, T. & Sharkey, N. E. (2001). A stroll through the worlds of robots and animals: Applying Jakob von Uexküll's theory of meaning to adaptive robots and artificial life. *Semiotica, 134*(1–4), 653–694.

SECTION V

EMBODIED MEANING

15

THE ROLE OF SENSORY AND MOTOR INFORMATION IN SEMANTIC REPRESENTATION: A REVIEW

LOTTE METEYARD[1,2] AND GABRIELLA VIGLIOCCO[2]

[1]*MRC Cognition and Brain Sciences Unit, Cambridge, UK*
[2]*Language Processing Laboratory, Department of Psychology, University College London, London, UK*

INTRODUCTION

Embodied theories of cognition propose that simulation is the basis for cognitive representation (Barsalou, 1999; Jeannerod, 2001; Hesslow, 2002; Gallese & Lakoff, 2005). Simulation is assumed to use the same sensory–motor systems that are engaged during real experience; when this principle is applied to the representation of linguistic meaning (semantics) theories propose that semantic content is achieved by recreating, usually in weaker form, the sensory and motor information produced when the referent of a word or sentence is actually experienced. Simulations are content-specific; for example, words referring to motion, such as rise and fall, are thought to recruit sensory systems involved in perceiving motion, and words referring to motor actions, such as kick and walk, are thought to recruit the motor systems used for those actions.

Therefore, embodied theories of semantic representation focus on semantic content, rather than the structure of the semantic system as a whole (e.g., distributed or localist, Dell, 1986 vs. Levelt et al., 1999), how words are related to one another (e.g., via associative connections or featural similarity, Collins & Loftus, 1975 vs. Smith et al., 1974), or how categories are represented (e.g., via modal similarity or a priori categories, Farah & McClelland, 1991 vs. Caramazza & Shelton, 1998). At a basic level, embodiment extends the non-controversial idea that we learn from experience, so semantics must be grounded in our sensations and actions; however, it appears to be a departure from the commonly accepted view that semantic content is amodal and thus not dependent on sensory–motor information (Fodor, 1987; Levelt, 1989; Jackendoff, 2002). This chapter reviews different theories of semantic representation placing them on a continuum as regards to their proposals about the role of sensory and motor information (see Figure 15.3). A brief review of behavioral and neuroscientific evidence is then presented and we end the chapter with a discussion of what the implications are for semantic representation.

DIRECT VERSUS INDIRECT ENGAGEMENT

Strong versions of embodiment assume what we refer to as the *direct engagement* hypothesis: to achieve representation, semantic content *necessarily* and *directly* recruits the sensory and motor systems used during experience. The *necessity condition* states that primary sensory and motor systems are essential for the semantic representation for concrete objects and events. The *directness condition* states that sensory and motor systems are engaged during semantic access without being mediated by other cognitive processes. One important idea here is modulation; semantic representation modulates activity in sensory or motor areas because those areas simulate the experience of the referent. Since the two share a common substrate, effects should be observed bilaterally, from language to perception/action and vice versa (see Figure 15.1).

There are several theories that subscribe to strong embodiment (see Figure 15.3). The most extreme of these is Gallese and Lakoff (2005) in which everything needed for representation (e.g., decomposition or abstraction) is considered to be present in sensory–motor systems and simulations within these modal systems underpin semantic representation. Thus, most (if not all) cognitive functions

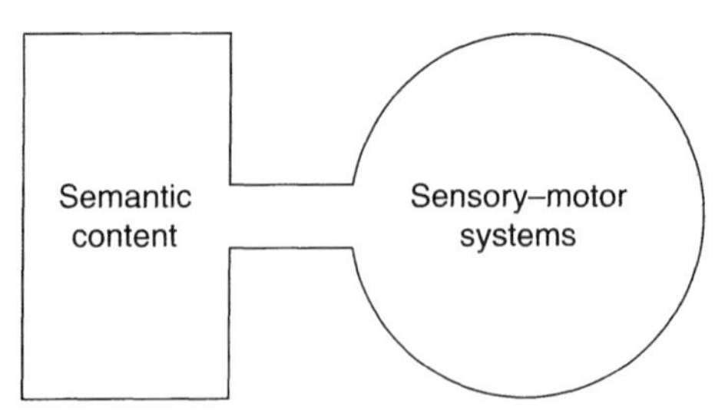

FIGURE 15.1 A schematic of direct engagement.Content-specific elements of semantic representation are isomorphic with sensory–motor systems.

are carried out *within* modal systems, creating multi-modal conceptual representation (as opposed to supra-modal systems where some information is "collapsed" across modality). Pulvermüller (1999, 2001) proposes that Hebbian learning produces embodied content: activity related to a word form occurs alongside sensory–motor activity corresponding to the word's referent, therefore, the two become associated, and sensory–motor activations become the semantic representation for a particular word. Barsalou (1999) presents a comprehensive theory of representation-as-simulation. Here, a more traditional cognitive model is presented where representations are schematic re-enactments of sensory and motor experience. However, the central tenet is the same with simulations taking place within the sensory and motor systems themselves (i.e., multi-modality). Finally, Glenberg and colleagues (Glenberg & Robertson, 2000; Glenberg & Kaschak, 2002, 2003) and Zwaan (2004) refer to the theories of Barsalou (1999) and Pulvermüller (1999), respectively when fleshing out their own theories of sentence/narrative comprehension, therefore adopting the same strong assumptions. These theories deal with both word and sentence level representations so simulation at all levels is proposed (single word, sentence, and narrative) and details of the integration of individual words, syntactic structures, and the existing context are provided. All of these theories make the following two assumptions:

1. Semantic processing automatically recruits low-level sensory and motor systems.
2. Semantic processing necessarily recruits these low-level processes (modulation), so effects should be consistent across tasks.

A weaker version of embodiment is what we will call the *indirect engagement* hypothesis. There are several possible formulations of this hypothesis, but in terms of necessity and directness it can be summarized as follows: to achieve representation, semantic content requires close contact to sensory and motor systems but activation of those systems is not necessary. The *non-essential condition* states that sensory and motor systems are implicated in semantic processing because of stable associative relationships between the semantic representation for concrete objects and events and the experience of those events. However, sensory and motor content is not necessary for semantic representation (at least once semantic representations are stable). The *indirect condition* states that sensory and motor systems are engaged during semantic access in a task-dependent manner, being mediated by cognitive processes, such as attention or perceptual learning. An important idea here is mediation; the impact of semantic representation is equivalent to an external system influencing activity in sensory or motor areas. Mediation means that bilateral effects will not always be present as the connection between semantic and sensory–motor systems is variable. Weak versions of embodiment differ from amodal theories because they assume non-arbitrary connections between semantic processing and sensory–motor systems. In contrast, amodal theories assume no direct connection: semantic processing is completely independent from sensory and motor systems (see Figure 15.2).

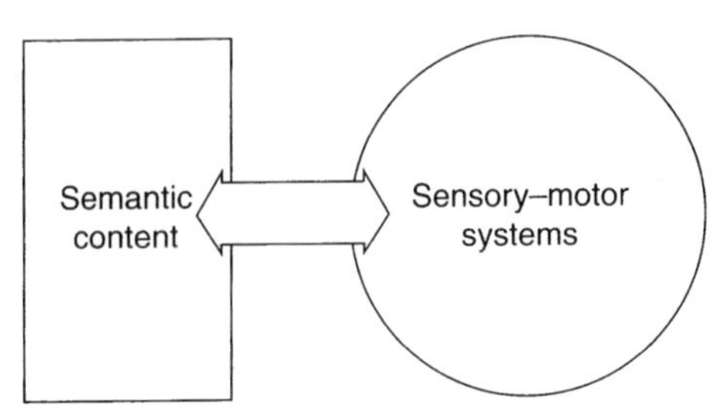

FIGURE 15.2 A schematic and indirect engagement.Content-specific elements of semantic representation are linked to sensory–motor systems in a non-arbitrary way.

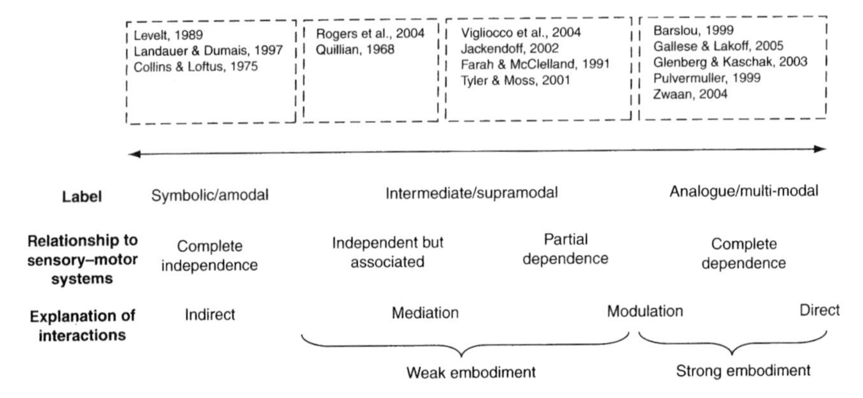

FIGURE 15.3 Schematic of theories' position along the continuum from amodal to modal.Theories are divided into four broad groups, the leftmost being symbolic/amodal theories (complete independence from and indirect interactions with modal content); intermediate supra-modal theories (associated to modal content with interactions via mediation, or partial dependence on modal content with interactions via mediation or modulation) and analog/multi-modal theories (complete dependence on and direct, modulatory interactions with modal content). Weak embodiment is claimed by intermediate supra-modal theories and strong embodiment by analog/multi-modal theories. McRae et al. (1997) and Smith et al. (1973) not included as no clear assertions are made.

There are a number of theories that adopt (or could adopt) some weak version of embodiment. Vigliocco et al. (2004) state that semantic representations are supra-modal representations that bind together modality-related conceptual features, hence they would only be partially dependent on modal systems. Jackendoff (2002) also proposes that modality-specific features are grounded in their respective modal systems whilst maintaining that much of language processing is based in an abstract, amodal conceptual structure. Other featural theories implicitly subscribe to embodiment (e.g., if their "visual" and "functional" features were grounded in the visual and motor system respectively, e.g., Farah & McClelland, 1991; Tyler & Moss, 2001). Here, the assumption would be partial dependence (i.e., supra-modality), although the precise mechanism is ambivalent between modulation and a strong form of mediation (Figure 15.3). One further step away from embodiment are theories that propose an amodal, abstract semantic system with associations to sensory–motor content (Quillian, 1968; Collins & Quillian, 1969; Rogers et al., 2004). Rogers et al. (2004) are explicit that semantic representations do not carry any content at all, but act as links to the relevant conceptual information. Quillian (1968) makes a brief reference to a common representational level between semantic and perceptual content whilst presenting a network of nodes (derived from modal content) as the basis for the semantic system. These two theories propose an independent but associative relationship where mediation is the only mechanism by which semantic and sensory–motor content can interact. For example, modality specific information could be recruited by association areas that integrate the modal information (supra-modality) and therefore have access to it. Despite the many differences across theories, they make the following weak assumptions:

1. Semantic processing is linked to sensory and motor systems but low-level sensory and motor processing is not necessarily recruited (e.g., semantic representations may be derived from higher order object concepts).
2. Modality-related effects of semantic representation are mediated so effects will vary depending on task demands (e.g., recruitment of attention and task parameters).

Finally, there are theories that propose a completely independent, amodal semantic store (Collins & Loftus, 1975; Levelt, 1989; Landauer & Dumais, 1997). Here, the link between sensory–motor and semantic content is formed outside the semantic system, by designatory processes (i.e., basic cognitive processes that link perceptions to internal representations, Pylyshyn, 1985). Here, interactions would be explained via indirect mechanisms (coming via other cognitive processes such as working memory or attention) or produced by the connection between semantic representations and the level at which designation occurs (a theoretically opaque process). Figure 15.3 summarizes where all these theories lie on the continuum from modal to amodal and which fall under weak or strong embodiment.

Below we present a review of the available evidence in light of the assumptions of direct versus indirect engagement. The strong prediction is direct engagement: to achieve representation, semantic content *necessarily* and *directly* recruits the sensory and motor systems in a simulation of the on-line experience of the referents. A weaker prediction is that semantic content recruits sensory and motor systems through association, rather than simulation. Here, the recruitment of semantic content may not be necessary but it may still be direct. Interestingly, this still predicts consistent interactions between semantic and sensory–motor information. Sensory and motor information may be recruited routinely during semantic access because of intimate ties that develop as a result of experience, but these ties do not equate to simulation.

The best way to assess the necessity constraint of direct engagement is through neuropsychological evidence or directly suppressing sensory or motor information (e.g., through transcranial magnetic stimulation, TMS).

Neuropsychological evidence is limited and currently presents a mixed picture (see e.g., Neininger & Pulvermüller, 2001; Spatt et al., 2002; Bak et al., 2006; Boulenger et al., 2008; Mahon & Caramazza, 2005, 2008) so the current review will focus on the more abundant behavioral and neuroscientific evidence that explore the directness of the connection. Associative connections (weak embodiment) should be open to more mediation than simulation (strong embodiment), so the apparent directness of the connection still allows us to distinguish someway between stronger and weaker versions of embodiment.

A BRIEF REVIEW OF THE EVIDENCE

BEHAVIORAL EVIDENCE

In a classic study, Tucker and Ellis (1998) presented pictures of objects, with affordances (potential interactions between the body and an object, for example, a handle) on the left or right. The judgments of the pictures (is it upright or inverted) were faster, and fewer errors were made, when the hand making the "upright" response and the affordance were congruent, compatible with the idea that seeing a picture of an object activates the motor actions associated with using it. Richardson et al. (2001) extended this study by presenting participants with a rapid serial visual presentation (RSVP) of eight pictured objects, with left or right affordances, followed by a decision about whether a named object was or was not in the sequence, and they found that responses were faster when the hand making the "yes" response was opposite to affordance, suggesting that the object name re-activated the motor affordances triggered when the referent object was perceived. However, it could be that the semantic content of the object name only accesses task-based affordances produced by object–picture perception. Myung et al. (2005, Experiment 1) addressed this issue by using an auditory lexical decision task in which primes did or did not share affordances with the target word, for example, a typewriter and a piano are both manipulated through fine pressing movements of the fingers. Lexical decisions were faster when the prime shared affordances with the target, supporting the automatic activation of motor plans upon semantic access. Siakaluk et al. (2007) provided further support for this by using target words that were previously rated on how easy or hard they were to physically interact with (a Body–Object Interaction score, BOI). Participants performed either a lexical or a phonological decision task and results showed that in both the tasks, decisions were faster for high BOI words as compared to low BOI words. The authors concluded that semantic representations include information about sensory–motor experience, on the assumption that high BOI words have "more" of this information. Two studies reinforce the inference that motor information accessed during semantic processing is based in the motor system itself by demonstrating that single-word comprehension of action verbs interacts with the motor system. Tseng and Bergen (2005) used

American Sign Language and showed that signs with semantic or metaphorical motion mirrored in a physical movement toward or away from the body were judged faster when the decision response was in a congruent direction (toward or away from the body). Signs with only phonological motion did not show the congruency effect. Boulenger et al. (2006) showed that reaching responses required to make a lexical decision, with the word appearing once the movement had been initiated, had smaller acceleration peaks when the word referred to an action as compared to a concrete nouns (suggesting interference). In contrast, when the item was presented as the go signal, peak accelerations were earlier for action verbs than for nouns (suggesting facilitation). Thus, motor information was activated early in comprehension, interfering with motor actions when concurrent and facilitating actions when precedent.

Klatzky and colleagues (1989) found that presentation of a congruent action cue speeded sensibility judgments for action sentences, for example, the cue for a flat palm followed by the sentence "rub your stomach." The same priming effects were found when subjects made a button press or verbalized their response for the sensibility judgment. Crucially, the effects were removed when motor tapping, but not syllable repetition, was used as a secondary task: that is, the preparation for the tapping task abolished the facilitation from preparing the hand shape (McCloskey et al., 1992). Similarly, reaction times to judge sentences that describe motor actions toward or away from the comprehender were found to be faster when the response was congruent with the described action; this is known as the Action Sentence Compatability Effect (Glenberg & Kaschak, 2003), and it has been shown to be dependent on timing, such that it is only present when preparation of the response is concurrent with sentence comprehension (Borreggine & Kaschak, 2006). Expanding on these results, two studies in Italian have shown interactions between the effector used for responding (hand, foot, or mouth) and the judgment of action sentences specifying those effectors. Buccino et al. (2005) used a hand (button) or foot (pedal) response, demonstrating that reaction times were slower when the response and effector described in the sentence were congruent. Scorolli and Borghi (2007) used a mouth (verbal) or foot (pedal) response and found sensibility judgments for pairs of nouns and verbs were faster when the effectors were congruent. The difference between interference and facilitation may be due to the response being prepared during the sentence (Buccino et al., 2005) or after it has been presented (Scorolli & Borghi, 2007) as in Boulenger et al. (2006). Finally, in a set of elegant experiments, Zwaan and Taylor (2006) found that sensibility judgments for sentences containing implied manual rotation, for example "Jane started the car," were faster when responses were made by turning a knob in the same direction as the implied rotation. Congruent facilitation was also found (but only at the verb region where the direction of rotation is specified) when participants smoothly turned a knob to progress through implied rotation sentences in self-paced reading. The use of "motor resonance" between the visual and motor domains extended these results when it was shown that the perception of congruent visual rotation speeded the judgment and reading (at the verb) of manual rotation sentences (see Fischer & Zwaan, in press, for an extensive discussion on motor resonance in comprehension). Thus there is cogent support for a semantic system that has access to effector specific motor information that is relevant to particular referents (i.e., manipulable objects and body actions); this does support a direct connection between motor semantics and motor information.

Outside the motor domain a similar picture is beginning to emerge but the evidence is more variable. Zwaan and Yaxley (2003) presented pairs of words for speeded similarity judgments; the critical items were pairs that referred to objects with a canonical spatial relation where one is above the other (e.g., root–branch, floor–ceiling). Reaction times were faster when the visual presentation of the words was congruent with the canonical relation (e.g., root at the bottom and branch at the top) rather than incongruent (e.g., root at the top and branch at the bottom). A series of experiments with a similar motivation explored whether visual attention was similarly affected by comprehension (Estes et al., in press): participants were presented with a category word (e.g., cowboy) and a part word (e.g., boot or hat) that was located at the top or bottom of the object. Following the presentation of the part word, participants identified a target letter presented in a congruent or incongruent location to the part word (i.e., top or bottom of the visual field). They found that letter identification was slower when the target location was congruent; this finding was replicated when the part-words were presented alone. Thus, the semantic content of words referring to concrete nouns with a canonical location interacts with visuo-spatial processing. In a different manipulation of visual properties, Pecher et al. (1998) manipulated the relationship between the prime and the target in a semantic priming task according to the similarity of their referents' visual form (e.g., pizza–coin, honey–glue). A small perceptual form priming effect was found for word naming when the prime referent shared visual form with the target, but only when the naming task was preceded by a perceptual decision task, which made the item's form salient (i.e., whether the word referred to an oblong object). Hence, these results do not support a direct link. Studies using sentence stimuli have supported interactions with visuo-spatial and motion processing. Richardson et al. (2003) displayed pairs of object pictures centrally as they were concurrently described in an aurally presented sentence that described a vertical ("The ship sunk in the ocean") or horizontal ("The mechanic pulled the chain") event; subsequent recognition of the pairs (i.e., were these two pictures seen in the same sentence?) was faster when they were presented in a spatial orientation congruent with the sentence. Post hoc analyses showed that the effect was significant for concrete, but not abstract, sentences. When the same sentences were presented before categorization of a shape in the vertical (top or bottom) or horizontal (left or right) meridian, categorization was slower when the visuo-spatial location and sentence orientation were congruent. These results have been replicated for concrete sentences describing upward and downward motion or objects with a canonical location "up" or "down" (e.g., ceilings or cellars) (Bergen et al., 2007). This suggests that visuo-spatial attention may only be influenced when a concrete location (provided by a concrete object) is defined, providing a potential target. This explanation contrasts with the one where a necessary simulation of all semantic representations (concrete and abstract) is similar to conscious visual imagery (Richardson et al., 2003; Bergen et al., 2007).

Stronger support for the connection between visual and semantic motion comes from studies showing that the comprehension of aurally presented sentences that describe vertical or egocentric motion is slowed when congruent visual motion is perceived (Kaschak et al., 2005). In an extension of the original experiment, Kaschak et al. (2006) found that when the motion and sentence stimuli (now describing auditory motion events) were presented in the same modality (e.g., both aurally), reaction times showed congruent facilitation. However, when presented in different modalities (i.e., visual sentences via RSVP and auditory motion stimuli), reaction times showed congruent interference. The authors proposed that when the two are presented in the same modality they are processed serially, producing congruent facilitation (priming), whereas presentation in a different modality results in concurrent processing and congruent interference (taxing the same resources). In contrast to this explanation, but in line with embodiment, Meteyard et al. (2007) showed that comprehension of blocked single words referring to motion (e.g., rise, climb, ascend) impaired the detection of concurrent motion signals (set at the threshold of conscious perception) when the two were incongruent (i.e., upward and downward motion), as shown by lower d' values. Crucially, no effects were found in reaction times and a measure of decision threshold (c) showed reduced values under congruent conditions (suggesting decision priming). These results show incongruent interference at low levels of perception (d'), and congruent facilitation at higher levels (c). In support of incongruent interference between semantic motion and low-level motion processing, Meteyard et al. (in press) found that motion patterns at the threshold of perception (assumed to be obligatorily processed) produced longer reaction times for lexical decision on motion words when the two were incongruent (i.e., "rise" with a downward motion pattern). Crucially, salient motion signals produced no congruency effects (supporting the inference that they are suppressed by top-down mechanisms; Tsushima et al., 2006). These results suggest that the relationship between semantic and perceptual information is complex, supporting a mediated connection and weaker versions of embodiment.

Several studies have used property verification, for example, is "feathers" a property of "pigeon"? and it is assumed that conceptual representations are accessed for this task to be performed and when the stimuli are words, we assume that the semantic representation of the word is accessed too. The data from property verification is quite consistent, showing effects of the perceptual modality and spatial location of the property. Reaction times are faster when the current trial modality (e.g., blender-loud) was the same as for the previous trial (e.g., leaves-rustle) rather than different (e.g., soap-perfumed) (Pecher et al., 2003). This modality switching cost is also present when the same *concept* is presented in successive trials with properties in the same or different modality (Pecher et al., 2004). Marques (2006) showed that the modality switching cost was present even when category (living vs. non-living) was kept constant. These results support simulation as there is a cost analogous to the cost of attending to events in different modalities during an on-line task (Spence et al., 2001). Property verifications are also faster when the concept is presented in a sentence, which implies a particular perspective (e.g., standing near the front or back of a car), and the property is salient given that perspective (e.g., the hood or trunk of the car respectively). This was also found when the property was at the top or bottom of a concept (e.g., the hair or shoes of a doll) and the response was made with a congruent response action, that is, pressing the top or bottom button on a response box (Borghi et al., 2004). Finally, a regression analysis showed that more variance in reaction times for property verification was accounted for by perceptual (read embodied) properties of a concept, when filler items precluded the use of simple word association strategies (Solomon & Barsalou, 2004). One criticism of these studies is that all property verification tasks may invoke imagery or more conscious processing than is typically required for semantic access: participants have to make an explicit judgment. Therefore, these tasks may not be representative of normal semantic processing. However, it does provide support for a somewhat direct connection between semantic and sensory–motor information.

A series of experiments, mostly conducted by Zwaan and colleagues, have used picture judgments to explore perceptual simulation in language comprehension. All experiments present target pictures following the comprehension of sentences. The consistent finding is that when pictures are congruent with the preceding sentence, their recognition or naming is facilitated. This has been demonstrated for congruence in object orientation (Stanfield & Zwaan, 2001), object form (Zwaan et al., 2002), and apparent motion (Zwaan et al., 2004). When color congruence was manipulated, slower responses were found for congruent conditions, this was tentatively explained by the instability of color representations in visual processing (Connell, 2006). In addition to the evidence from sentence comprehension, there is substantial evidence that narrative comprehension engages analog visuo-spatial and temporal representations (Zwaan, 1999 or Zwaan & Radvansky, 1998 for a review; Rinck et al., 1997; Rinck & Bower, 2000; Horton & Rapp, 2003; Kaup & Zwaan, 2003; Kaup et al., 2006). Nevertheless, there is only preliminary evidence that visuo-spatial processing interferes with narrative comprehension (Fincher-Kiefer, 2001), so whilst it is clear that situation models can be seen as embodied simulations, it is still possible to explain the results with an associative amodal network (Rinck & Bower, 2000).

Stronger evidence for the activation of perceptual information during comprehension comes from eye movement studies that we only briefly summarize here: when participants are asked to actively imagine or to simply listen to scene descriptions, their eye movements reflect the implied location of events (Spivey & Geng, 2001; Matlock & Richardson, 2004). This is in line with an embodied interpretation where the eyes move as if those events were being observed. Eye movements also provide evidence that perceptual and motor features of individual words are active during comprehension (Dahan & Tanenhaus, 2005; Myung et al., 2005). This provides support for perceptual and motor features being a part of semantic representation (see also Spivey et al., 2000; Chambers et al., 2002, 2005; Laeng & Teodorescu, 2002). Eye movements are an increasingly useful tool to explore comprehension (for a review see Henderson & Ferreira, 2004) and through the use of inventive methodologies, they are also supporting the role of perceptual and motor information in semantic representations. One small caveat is that eye movements may not be a veridical mirror of the mind, directly reflecting the immediate contents of cognitive processing. The mechanisms that influence oculo-motor movements, such as attention, imagination and task-demands, need to be better understood before eye movement studies can provide strong evidence for embodiment.

NEUROSCIENTIFIC EVIDENCE

Neuroscience has held "embodied" views for a long time: concepts are defined by sensory and motor attributes that arise from experience, when we see, hear, touch, and manipulate things in the environment. Distributed feature networks of sensory and motor attributes will be reflected in sensory and motor cortices of the brain; for example, the ventral occipital cortex (fusiform gyrus) supports knowledge about object form and the lateral temporal cortex (MT) supports knowledge about object motion (Martin & Chao, 2001). Neuropsychological and neuroimaging studies typically use verbal labels as one of several access routes to conceptual information (similar to pictures, e.g., Martin & Chao, 2001; Plaut, 2002; Rogers et al., 2004) rather than exploring intermediate semantic representations (Damasio et al., 1996). As such, neuroscientific theories do not typically explore the division between conceptual and semantic information (Damasio, 1989; Farah & McClelland, 1991; Martin & Chao, 2001; Tyler & Moss, 2001; Plaut, 2002; for an exception see Damasio et al., 1996, 2004), allowing modality specific content into "semantics" without much consternation about embodiment. This contrasts with cognitive/psycholinguistic theories that build on a classical cognitive heritage (Newell, 1980; Pylyshyn, 1985; Fodor, 1987) and typically propose some division between conceptual and semantic information that allows amodal semantics to be extracted from modal concepts.[1] Although there is necessary overlap (Farah & McClelland, 1991; Tyler & Moss, 2001), it is outside the scope of this chapter to explore the debate in neuroscience and neuropsychology about the organization of the conceptual system[2]; but the debate does

[1]For a marginal separation see McRae et al. (1997); partial separation see Jackendoff (2002) and Vigliocco et al. (2004); and complete separation see Levelt (1989) and Levelt et al. (1999).

[2]Briefly, the debate rages over where to draw the major fault lines in conceptual content; is it by modality (e.g., visual and functional features), domain (e.g., the categories of animals and fruit/vegetables), or some other systematic structure (e.g., Warrington & McCarthy, 1983, 1987; Warrington & Shallice, 1984; Caramazza et al., 1990; Shallice, 1993; Humphreys & Forde, 2001; Martin & Chao, 2001; Caramazza & Mahon, 2003).

show us that sensory and motor information has been implicated in (at least) conceptual representation for a long time. For example, in Convergence Zone Theory (Damasio, 1990; Damasio & Damasio, 1994) primary sensory cortices contain featural components and basic combinatorial arrangements of those features (parts, shape, color, movement, etc.). More complex combinatorial codes, which define the perception of events (spatial and temporal relationships) are "inscribed" (p. 127) in higher order association areas (frontal and temporal cortices), called convergence zones (Cz). Thus, the physical properties of experience are represented in the primary cortices, but their synchronized activation and co-ordination depends on feedback connections from Cz. Thus, embodied theories represent a strengthened version of existing ideas in neuroscience; for example, proposing multi-modal direct engagement without the need for higher order, progressively supra-modal associations (Barsalou, 1999; Pulvermüller, 1999; Gallese & Lakoff, 2005) or building on Cz theory with embodied neural principles (Simmons & Barsalou, 2003[3]).

The strength of the neuroscientific evidence for embodiment depends on modality (but see Kemmerer et al., in press). The premotor and motor cortices are consistently activated across studies and methods. These cortical areas are not only seen for language referring to body actions (Pulvermüller et al., 2000, 2001, 2004; Tettamanti et al., 2005; Aziz-Zadeh et al., 2006; Vigliocco et al., 2006), but also for tool actions and tools/manipulable objects (Grabowski et al., 1998; Chao & Martin, 2000; Gerlach et al., 2002). TMS studies provide converging evidence that lexical and sentential items with motor associations activate motor areas of the cortex (Oliveri et al., 2004; Buccino et al., 2005) and localized motor cortical areas corresponding to the specific effector of an action (Buccino et al., 2005; Pulvermüller et al., 2005). The timing of the TMS, early in the time-course of comprehension and production, supports the argument that modality specific activations are part of the early lexico-semantic processes. For most of the studies, the motor activation is left lateralized, although there is some evidence that the right hemisphere is implicated for tool action generation (e.g., Damasio et al., 2001). This strongly suggests that the motor cortex plays a role in the semantic representation of objects and actions with salient motor associations. Simulation during comprehension is supported by effector specific manipulations (Pulvermüller et al., 2005; Aziz-Zadeh et al., 2006), which suggest that the motor cortex is selectively recruited depending on the content of the language. Alongside the motor cortex, MT activity is repeatedly seen for body and tool actions as well as tool objects[4] (Martin et al., 1995, 1996; Damasio et al., 2001; Phillips et al., 2002; Tettamanti et al., 2005). When activity in this area is observed for tools and tool actions, it is usually explained as a reflection

[3]Simmons and Barsalou (2003) extend Cz theory with the Similarity in Topography (SIT) principle (see Plaut, 2002, for a very similar, but computational, approach). Here, the actual cortical proximity of convergence zones is dictated by their similarity, which is in turn dictated by the modalities (visual, motor, etc.) and/or properties (shape, color, movement) of the features they conjoin.

[4]This may be related to stimuli in these experiments being pictures rather than word stimuli.

of knowledge about the movement of objects during their use (Phillips et al., 2002). This is also in line with accounts that propose modality specific areas (in this case, those processing visual motion) are implicated in the representation of knowledge from that modality. Despite the fact that MT is typically understood as a motion processing area, there is only one study that has used motion sentences (both literal and fictive) and the active area in this case was proximal, but not isomorphic, with MT (Wallentin et al., 2005). Finally, the Amygdala supports "modality" specific representation, being active for threat words (Isenberg et al., 1999). Evidence from the motor/action domain is, therefore, in line with strong and weak embodied theories where sensory–motor features are represented multi-modally or via associations with the sensory–motor cortices; the TMS evidence lends weight to the directness of this connection, and possibly its necessity.

Beyond body actions, tool actions, and tools, the evidence is considerably less coherent. EEG data have shown attenuation of the N400 (typically interpreted as responding to semantic incongruence) when targets where preceded by visual-form related primes (Kellenbach et al., 2000). The fusiform gyrus is documented as playing a role in the representation of object form (Chao et al., 1999; Vuilleumier et al., 2002) and different areas of the fusiform have been implicated for different categories, that is lateral fusiform for animals and medial fusiform for tools (Martin & Chao, 2001). Fusiform activity was observed for tool and animal names relative to a nonsense object baseline (Martin et al., 1996), for conceptual access during property verification for objects (Kan et al., 2003), and for words related to form and color (Martin et al., 1995; Pulvermüller & Hauk, 2005). For sensory words in general (e.g., darken, darkness), an area proximal to the fusiform was observed (Vigliocco et al., 2006). These results support the role of the fusiform in representing the visual attributes of known objects, and more generally this area of the cortex as involved in higher order visual association; combining features from different modalities (Vigliocco et al., 2006). As regards to embodiment, fusiform activation is not that informative. It can be taken as a predominantly visual area, therefore supporting modality specific representation (multi-modality), but its role as an area that represents objects regardless of idiosyncratic variations in appearance (e.g., Vuilleumier et al., 2002) suggests that it responds to combinations of features or attributes to provide a more abstract representation of objects: thus, being a supra-modal rather a multi-modal area.

It is of course crucial whether the cortical areas implicated in semantic representation are isomorphic with the cortical areas involved in experience, as this is the strong version of embodied simulation when it is applied to neural structures. But higher order association areas are problematic for strong embodiment, which predicts the concurrent activation of different modality specific areas rather than concentrated activity in one area that is connected to these modal systems. It is an open question whether supra/hetero-modal areas that combine information across modalities still constitute embodied representations, or whether they indicate a progression from modality specific to modality invariant (and ultimately modality independent) representations (see Kemmerer et al., in press, for some evidence for multi-modal activations).

It is clear that language referring to objects and actions with a salient modality (e.g., tools or body actions) activate cortical areas involved in the experience of that modality. However, this can be taken as support for weaker versions of embodied theories that do not necessitate simulation, or full embodiment. It is always possible that sensory and motor cortices become active in a secondary manner, incidental to necessary processing in semantics; however, the evidence of early modal activity shown in TMS and EEG speaks against this conclusion (Pulvermüller et al., 2001, 2005). The isomorphism between the cortical areas used during real-world experience and semantic representation is supported for the motor cortex, but it is less clear what the literature shows for non-motor information. It is worth noting that the motor cortex has a special status as an efferent area that responds to top-down commands (such as verbal instruction), therefore what applies there may not apply for afferent sensory areas that respond to sensory stimuli.

CONCLUSIONS

As stated in the "Introduction" paragraph, the strong prediction from embodied theories of semantic representation is the *direct engagement* hypothesis: to achieve representation, semantic content *necessarily* and *directly* recruits the sensory and motor systems used during experience. The necessity condition states that without the support of sensory and motor systems, semantic representation for concrete objects and events is impaired. The directness condition states that sensory and motor systems are engaged during semantic access without being mediated by other cognitive processes. So, what can be concluded about the necessity and direct engagement of sensory and motor systems in semantic representation?

Neuroscientific evidence reliably shows motor cortex activation for tools, tool actions and body actions (Gerlach et al., 2002; Tettamanti et al., 2005), but the evidence for other domains is less consistent (e.g., Pulvermüller & Hauk, 2005; Vigliocco et al., 2006; Kemmerer et al., in press). However, brain activity (particularly in fMRI/PET) is always correlational rather than causal. Sensory and motor activity could be the result of the high association between particular semantic domains and particular modalities, rather than the result of direct engagement in representation.

One important correlate of automaticity is speed: the faster the access to sensory and motor information, the more likely it is to be a typical and elemental part of semantic processing (Pulvermüller, 2001). There is evidence of fast access to motor information during comprehension (Pulvermüller et al., 2000, 2001; Boulenger et al., 2006) and behavioral studies as the motor domain do support timing as a crucial element (Borreggine & Kaschak, 2006; Zwaan & Taylor, 2006). Numerous reaction time studies show the influence of sensory–motor semantic content on sensory or motor processing, and neuroscientific evidence shows sensory and motor activation following both active and passive comprehension. However, unless low-level processes are directly tapped (Meteyard et al., 2007, in press), results could still be contaminated by decision or some other mediating processes (such as imagery, attention, or a task-set, which sets up implicit relationships between linguistic and sensory/motor manipulations). In contrast, if sensory or motor activity is shown to affect comprehension, it is harder to explain away these effects by mediating processes. Such evidence is available for the motor domain (Glenberg & Kaschak, 2003; Zwaan & Taylor, 2006) and TMS studies show that direct activation of the motor system affects the comprehension of motor words (Pulvermüller et al., 2005); but evidence is limited for the senses (Kaschak et al., 2005, 2006; Meteyard et al., in press).

So far, a few studies have directly manipulated low-level sensory processes. These results show that the perception of motion affects the comprehension of motion sentences (Kaschak et al., 2005) and words (Meteyard et al., in press) and another which shows influences of motion words on low-level motion perception (Meteyard et al., 2007). But there is preliminary evidence that the influence of motion perception on comprehension may be task dependent (Meteyard et al., submitted), suggesting an automatic but mediated connection.

The evidence for sensory–motor information in semantic representation is growing, with increasing evidence that there is a direct connection between systems involved in sensory–motor experience and the representation of sensory–motor content in language, but the question of necessity is unanswered. This argues strongly against theories of semantics, which propose complete independence between semantic and sensory–motor information. More complex questions about the precise nature of the connection remain to be mapped out, with strong and weak embodiment holding equal explanatory potential at the present time.

REFERENCES

Aziz-Zadeh, L., Wilson, S. M., Rizzolatti, G., & Iacoboni, M. (2006). Congruent embodied representations for visually presented actions and linguistic phrases describing actions. *Current Biology, 16*, 1818–1823.

Bak, T. H., Yancopoulou, D., Nestor, P. J., Xuereb, J. H., Spillantini, M. G., Pulvermüller, F., & Hodges, J. R. (2006). Clinical, imaging and pathological correlates of a hereditary deficit in verb and action processing. *Brain, 129*(2), 321–332.

Barsalou, L. W. (1999). Perceptual symbol systems. *Behavioral and Brain Sciences, 22,* 577–660.

Bergen, B. K., Lindsay, S., Matlock, T., & Narayanan, S. (2007). Spatial and linguistic aspects of visual imagery in sentence comprehension. *Cognitive Science, 31*(5), 733–764.

Borghi, A. M., Glenberg, A. M., & Kaschak, M. P. (2004). Putting words in perspective. *Memory & Cognition, 32*(6), 863–873.

Borreggine, K. L. & Kaschak, M. P. (2006). The action-sentence compatibility effect: It's all in the timing. *Cognitive Science, 30*, 1097–1112.

Boulenger, V., Roy, A. C., Paulignan, Y., Deprez, V., Jeannerod, M., & Nazir, T. A. (2006). Cross-talk between language processes and overt motor behaviour in the first 200 msec of processing. *Journal of Cognitive Neuroscience, 18*(10), 1607–1615.

Boulenger, V., Mechtouff, L., Thobois, S., Broussolle, E., Jeannerod, M., & Nazir, T. (2008). Word processing in Parkinson's disease is impaired for action verbs but not for concrete nouns. *Neuropsychologica 46*(2), 743–756.

Buccino, G., Riggio, L., Melli, G., Binkofski, F., Gallese, V., & Rizzolatti, G. (2005). Listening to action-related sentences modulates the activity of the motor system: A combined TMS and behavioural study. *Cognitive Brain Research, 24*, 355–363.

Caramazza, A. & Mahon, B. Z. (2003). The organisation of conceptual knowledge: The evidence from category-specific semantic deficits. *Trends in Cognitive Sciences, 7*(8), 354–361.

Caramazza, A., Hillis, A. E., Rapp, B. C., & Romani, C. (1990). The multiple semantics hypothesis: Multiple confusions? *Cognitive Neuropsychology, 7*(3), 161–189.

Carmazza, A. & Shelton, J. R. (1998). Domain specific knowledge systems in the brain the animate-inanimate distinction. *Journal of Cognitive Neuroscience, 10*, 1–34.

Chambers, C. G., Tanenhaus, M. K., Eberhard, K. M., Filip, H., & Carlson, G. N. (2002). Circumscribing referential domains during real-time language comprehension. *Journal of Memory and Language, 47*, 30–49.

Chambers, C. G., Tanenhaus, M. K., & Madden, C. J. (2004). Actions and affordances in syntactic ambiguity resolution. *Journal of Experimental Psychology: Learning, Memory and Cognition, 30*(3), 687–696.

Chao, L. L. & Martin, A. (2000). Representation of manipulable man-made objects in the dorsal stream. *NeuroImage, 12*, 478–484.

Chao, L. L., Haxby, J. V., & Martin, A. (1999). Attribute-based neural substrates in temporal cortex for perceiving and knowing about objects. *Nature Neuroscience, 2*, 913–919.

Collins, A. C. & Quillian, M. R. (1969). Retrieval time from semantic memory. *Journal of Verbal Learning and Verbal Behavior, 12*, 240–247.

Collins, A. M. & Loftus, E. F. (1975). A spreading-activation theory of semantic processing. *Psychological Review, 82*(6), 407–428.

Connell, L. (2006). Representing object colour in language comprehension. *Cognition, 102*(3), 476–485.

Dahan, D. & Tanenhaus, M. K. (2005). Looking at the rope when looking for the snake: Conceptually mediated eye movements during spoken-word recognition. *Psychonomic Bulletin and Review, 12*(3), 453–459.

Damasio, A. R. (1989). The brain binds entities and events by multiregional activation from convergence zones. *Neural Computation, 1*, 123–132.

Damasio, A. R. (1990). Category-related recognition defects as a clue to the neural substrates of knowledge. *Trends in Neuroscience, 13*, 95–98.

Damasio, A. R. & Damasio, H. (1994). Cortical systems for retrieval of concrete knowledge: The convergance zone framework, Ch. 4. In C. Koch, & J. L. Davis (Eds.), *Large-Scale Neuronal Theories of the Brain.* London, UK: MIT Press.

Damasio, H., Grabowski, T. J., Tranel, D., Hichwa, R. D., & Damasio, A. R. (1996). A neural basis for lexical retrieval. *Nature, 380*, 499–505.

Damasio, H., Grabowski, T. J., Tranel, D., Ponto, L. L. B., Hichwa, R. D., & Damasio, A. R. (2001). Neural correlates of naming actions and of naming spatial relations. *NeuroImage, 13*, 1053–1064.

Damasio, H., Tranel, D., Grabowski, T. J., Adolphs, R., & Damasio, A. R. (2004). Neural systems behind word and concept retrieval. *Cognition, 92*, 179–229.

Dell, G. S. (1986). A spreading activation theory of retrieval in sentence production. *Psychological Review, 93*, 283–321.

Estes, Z., Verges, M., & Barsalou, L. W. (2008). Head up, foot down: Object words orient attention to the objects' typical location. *Psychological Science, 19*(2), 93–97.

Farah, M. J. & McClelland, J. L. (1991). A computational model of semantic impairment: Modality specificity and emergent category specificity. *Journal of Experimental Psychology: General, 120*(4), 339–357.

Fincher-Kiefer, R. (2001). Perceptual components of situation models. *Memory & Cognition, 29*(2), 336–343.

Fischer, M. H. & Zwaan, R. A. (2008). Embodied language: A review of the role of the motor system in language comprehension. *Quarterly Journal of Experimental Psychology, 61(6)*, 825–850.

Fodor, J. A. (1987). *Psychosemantics: The problem of meaning in the philosophy of mind.* London, UK: MIT Press.

Gallese, V. & Lakoff, G. (2005). The brain's concepts: The role of the sensory–motor system in conceptual knowledge. *Cognitive Neuropsychology, 22*(3/4), 455.

Gerlach, C., Law, I., & Paulson, O. B. (2002). When action turns to words. Activation of motor-based knowledge during categorisation of manipulable objects. *Journal of Cognitive Neuroscience, 14*(8), 1230–1239.

Glenberg, A. M. & Kaschak, M. P. (2002). Grounding language in action. *Psychonomic Bulletin and Review, 3*(9), 558–565.

Glenberg, A. M. & Kaschak, M. P. (2003). The body's contribution to language. In B. H. Ross (Ed.), *The Psychology of Learning and Motivation* (Vol. 43, pp. 93–126). San Diego, CA: Academic Press.

Glenberg, A. M. & Robertson, D. A. (2000). Symbol grounding and meaning: A comparison of high-dimensional and embodied theories of meaning. *Journal of Memory and Language, 43*, 379–401.

Grabowski, T. J., Damasio, H., & Damasio, A. R. (1998). Premotor and prefrontal correlates of category-related lexical retrieval. *NeuroImage, 7*, 232–243.

Henderson, J. M. & Ferreira, F. (Eds.) (2004). *The Interface of Language, Vision, and Action: Eye Movements and the Visual World.* London, UK: Routledge.

Hesslow, G. (2002). Conscious thought as simulation of behaviour and perception. *Trends in Cognitive Sciences, 6*, 242.

Horton, W. S. & Rapp, D. N. (2003). Out of sight, out of mind: Occlusion and the accessability of information in narrative comprehension. *Psychonomic Bulletin and Review, 10*(1), 104–110.

Humphreys, G. W. & Forde, E. M. E. (2001). Hierachies, similarity and interactivity in object recognition: "Category-specific" neuropsychological deficits. *Behavioral and Brain Sciences, 24*, 453–509.

Isenberg, N., Silbersweig, D., Engelien, A., Emmerich, S., Malavade, B., Beattie, B. et al. (1999). Linguistic threat activates the human amygdala. *Proceedings of the National Academy of Sciences, 96*, 10456–10459.

Jackendoff, R. (2002). *Foundations of Language: Brain, Meaning, Grammar, Evolution.* New York: Oxford University Press.

Jeannerod, M. (2001). Neural simulation of action: A unifying mechanism for motor cognition. *NeuroImage, 14*, S103–S109.

Kan, I. P., Barsalou, L. W., Solomon, K. O., Minor, J. K., & Thompson-Schill, S. L. (2003). Role of mental imagery in a property verification task: fMRI evidence for perceptual representations of conceptual knowledge. *Cognitive Neuropsychology, 20*, 525–540.

Kaschak, M. P., Madden, C. J., Therriault, D. J., Yaxley, R. H., Aveyard, M., Blanchard, A. A. et al. (2005). Perception of motion affects language processing. *Cognition, 94*, B79.

Kaschak, M. P., Zwaan, R., Aveyard, M., & Yaxley, R. H. (2006). Perception of auditory motion affects language processing. *Cognitive Science, 30*(4), 733–744.

Kaup, B. & Zwaan, R. (2003). Effects of negation and situational presence on the accessibility of text information. *Journal of Experimental Psychology: Learning, Memory and Cognition, 29*(3), 439–446.

Kaup, B., Ludtke, J., & Zwaan, R. (2006). Processing negated sentences with contradictory predicates: Is a door that is not open mentally closed? *Journal of Pragmatics, 38*, 1033–1050.

Kellenbach, M. L., Wijers, A. A., & Mulder, G. (2000). Visual semantic features are activated during the processing of concrete words: Event-related potential evidence for perceptual semantic priming. *Cognitive Brain Research, 10*, 67–75.

Kemmerer, D., Castillo, J. G., Talavage, T., Patterson, S., Wiley, C. (in press). Neuroanatomical distribution of five semantic components of verbs: Evidence from fMRI, *Brain and Language*.

Klatzky, R. L., Pellegrino, J. W., McCloskey, B. P., & Doherty, S. (1989). Can you squeeze a tomato? The role of motor representations in semantic sensibility judgements. *Journal of Memory and Language, 28*, 56–77.

Laeng, B. & Teodorescu, D.-S. (2002). Eye scanpaths during visual imagery re-enact those of perception of the same visual scene. *Cognitive Science, 26*, 207–231.

Landauer, T. K. & Dumais, S. T. (1997). A solution to Plato's problem: The latent semantic analysis theory of acquisition, induction and representation of knowledge. *Psychological Review, 104*(2), 211–240.

Levelt, W. J. M. (1989). *Speaking: From Intention to Articulation*. London, UK: MIT Press.

Levelt, W. J. M., Roelofs, A., & Meyer, A. S. (1999). A theory of lexical access in speech production. *Behavioral and Brain Sciences, 22*, 1–75.

Mahon, B. Z. & Caramazza, A. (2005). The orchestration of the sensory–motor systems: Clues from neuropsychology. *Cognitive Neuropsychology, 22*(3/4), 480–494.

Mahon, B. Z. & Caramazza, A. (2008). A critical look at the embodied cognition hypothesis and a new proposal for grounding conceptual content. *Journal of Physiology*, Paris, *102*(1–3), 59–70.

Marques, J. F. (2006). Specialization and semantic organisation: Evidence for multiple semantics linked to sensory modalities. *Memory & Cognition, 34*(1), 60–67.

Martin, A. & Chao, L. L. (2001). Semantic memory and the brain: Structure and processes. *Current Opinion in Neurobiology, 11*, 194–201.

Martin, A., Haxby, J. V., Lalonde, F. M., Wiggs, C. L., & Ungerleider, L. G. (1995). Discrete cortical regions associated with knowledge of color and knowledge of action. *Science, 270*(5233), 102–105.

Martin, A., Wiggs, C. L., Ungerleider, L. G., & Haxby, J. V. (1996). Neural correlates of category-specific knowledge. *Nature, 379*, 649–652.

Matlock, T. & Richardson, D. C. (2004). Do eye movements go with fictive motion? *Proceedings of the 26th Annual Conference of the Cognitive Science Society*, NJ, USA.

McCloskey, B. P., Klatzky, R. L., & Pellegrino, J. W. (1992). Rubbing your stomach while tapping your fingers: Interference between motor planning and semantic judgements. *Journal of Experimental Psychology: Human Perception and Performance, 18*(4), 948–961.

McRae, K., de Sa, V. R., & Seidenberg, M. S. (1997). On the nature and scope of featural representations of word meaning. *Journal of Experimental Psychology: General, 126*(2), 99–130.

Meteyard, L., Bahrami, B., & Vigliocco, G. (2007). Motion detection and motion verbs: Language affects low-level visual perception. *Psychological Science, 18*(11).

Meteyard, L., Zokaei, N., Bahrami, B. & Vigliocco, G. (in press) Now you see it: visual motion interferes with lexical decision on motion words. Current Biology.

Myung, J., Blumstein, S. E., & Sedivy, J. C. (2005). Playing on the typewriter, typing on the piano: Manipulating knowledge of objects. *Cognition, 98*, 223–243.

Neininger, B. & Pulvermuller, F. (2001). The right hemisphere's role in action word processing: A double case study. *Neurocase, 7*, 303–317.

Newell, A. (1980). Physical symbol systems. *Cognitive Science, 4*, 135–183.

Oliveri, M., Finocchiaro, C., Shapiro, K., Gangitano, M., Caramazza, A., & Pascual-Leone, A. (2004). All talk and no action: A transcranial magnetic stimulation study of motor cortex activation during action word production. *Journal of Cognitive Neuroscience, 16*(3), 374–381.

Pecher, D., Zeelenberg, R., & Raaijmakers, J. G. W. (1998). Does pizza prime coin? Perceptual priming in lexical decision and pronunciation. *Journal of Memory and Language, 38*, 401–418.

Pecher, D., Zeelenberg, R., & Barsalou, L. W. (2003). Verifying different-modality properties for concepts produces switching costs. *Psychological Science, 14*(2), 119–124.

Pecher, D., Zeelenberg, R., & Barsalou, L. W. (2004). Sensorimotor simulations underlie conceptual representations: Modality-specific effects of prior activation. *Psychonomic Bulletin and Review, 11*, 164–167.

Phillips, J., Noppeney, U., Humphreys, G. W., & Price, C. J. (2002). Can segregation within the semantic system account for category-specific deficits? *Brain, 125*, 2067–2080.

Plaut, D. C. (2002). Graded modality-specific specialisation in semantics: A computational account of optic aphasia. *Cognitive Neuropsychology, 19*(7), 603–639.

Pulvermüller, F. (1999). Words in the brain's language. *Behavioral and Brain Sciences, 22*, 253–336.

Pulvermüller, F. (2001). Brain reflections of words and their meaning. *Trends in Cognitive Sciences, 5*(12), 517–524.

Pulvermüller, F. & Hauk, O. (2005). Category-specific conceptual processing of color and form in left fronto-temporal cortex. *Cerebral Cortex, 16*, 1193–1201.

Pulvermüller, F., Harle, M., & Hummel, F. (2001). Walking or talking? Behavioral and neurophysiological correlates of action verb processing. *Brain and Language, 78*, 143–168.

Pulvermüller, F., Hauk, O., Nikulin, V. V., & Ilmoniemi, R. J. (2005). Functional links between motor and language systems. *European Journal of Neuroscience, 21*, 793–797.

Pylyshyn, Z. W. (1985). *Computation and Cognition: Toward a Foundation for Cognitive Science*, 2nd ed.. London, UK: MIT Press.

Quillian, M. R. (1968). Semantic Memory. In M. Minsky (Ed.), *Semantic Information Processing*. (pp. 216–271). London, UK: MIT Press.

Richardson, D. C., Spivey, M. J., & Cheung, J. (2001). Motor representations, *Memory and Mental Models: Embodiment in Cognition*, NJ, USA.

Richardson, D. C., Spivey, M. J., Barsalou, L. W., & McRae, K. (2003). Spatial representations activated during real-time comprehension of verbs. *Cognitive Science, 27*, 767.

Rinck, M. & Bower, G. H. (2000). Temporal and spatial distance in situation models. *Memory & Cognition, 28*(8), 1310–1320.

Rinck, M., Hahnel, A., Bower, G. H., & Glowalla, U. (1997). The metrics of spatial situation models. *Journal of Experimental Psychology: Learning, Memory and Cognition, 23*, 622–637.

Rogers, T. T., Lambon Ralph, M. A., Garrard, P., Bozeat, S., McClelland, J. L., Hodges, J. R. et al. (2004). Structure and deterioration of semantic memory: A neuropsychological and computational investigation. *Psychological Review, 111*(1), 205–235.

Scorolli, C. & Borghi, A. M. (2007). Sentence comprehension and action: Effector specific modulation of the motor system. *Brain Research, 1130*, 119–124.

Shallice, T. (1993). Multiple semantics: Whose confusion? *Cognitive Neuropsychology, 10*, 251–261.

Siakaluk, P. D., Pexman, P. M., Aguilera, L., Owen, W. J., & Sears, C. R. (2008). Evidence for the activation of sensorimotor information during visual word recognition: The body-object interaction effect. *Cognition, 106*(1), 433–443.

Simmons, K. W. & Barsalou, L. W. (2003). The similarity-in-topography principle: Reconciling theories of conceptual deficits. *Cognitive Neuropsychology, 20*(3–6, 4 p. 3/4), 451–486.

Smith, E. E., Shoben, E. J., & Rips, L. J. (1974). Structure and process in semantic memory: A featural model for semantic decisions. *Psychological Review, 81*(3), 214–241.

Solomon, K. O. & Barsalou, L. W. (2004). Perceptual simulation in property verification. *Memory & Cognition, 32*, 244–259.

Spatt, J., Bak, T., Bozeat, S., Patterson, K., & Hodges, J. R. (2002). Apraxia, mechanical problem solving and semantic knowledge: Contributions to object usage in corticobasal degeneration. *Journal of Neurology, 249*(5), 1432–1459.

Spence, C., Nicholls, M. E. R., & Driver, J. (2001). The cost of expecting events in the wrong sensory modality. *Perception & Psychophysics, 63*(2), 330–336.

Spivey, M. J. & Geng, J. J. (2001). Oculomotor mechanisms activated by imagery and memory: Eye movements to absent objects. *Psychological Research, 65*, 235–241.

Spivey, M. J., Tyler, M. J., Richardson, D. C., & Young, E. E. (2000). Eye movements during comprehension of spoken scene descriptions, *Proceedings of the 23rd annual meeting of the cognitive science society*. Mawhah, NJ: Erlbaum.

Stanfield, R. A. & Zwaan, R. A. (2001). The effect of implied orientation derived from verbal context on picture recognition. *Psychological Science, 12*(2), 153–156.

Tettamanti, M., Buccino, G., Saccuman, M. C., Gallese, V., Danna, M., Scifo, P. et al. (2005). Listening to action-related sentences activates fronto-parietal motor circuits. *Journal of Cognitive Neuroscience, 17*(2), 273–281.

Tseng, M. J. & Bergen, B. K. (2005). Lexical Processing Drives Motor Simulation, *Proceedings of the 27the annual meeting of the cognitive science society*. Mawhah, NJ: Erlbaum

Tucker, M. & Ellis, R. (1998). On the relation between seen objects and components of potential actions. *Journal of Experimental Psychology: Human Perception and Performance, 24*, 830.

Tsushima, Y., Sasaki, Y., & Watanabe, T. (2006). Greater disruption due to failure of inhibitory control on an ambiguous distractor. *Science, 314*, 1786–1788.

Tyler, L. K. & Moss, H. E. (2001). Towards a distributed account of conceptual knowledge. *Trends in Cognitive Sciences, 5*(6), 244–252.

Vigliocco, G., Vinson, D. P., Lewis, W., & Garrett, M. F. (2004). Representing the meaning of object and action words: The featural and unitary semantic space hypothesis. *Cognitive Psychology, 48*, 422–488.

Vigliocco, G., Warren, J., Arcuili, J., Siri, S., Scott, S., & Wise, R. (2006). The role of semantics and grammatical class in the neural representation of words. *Cerebral Cortex, 16*, 1790–1796.

Vuilleumier, P., Henson, R. N., Driver, J., & Dolan, R. J. (2002). Multiple levels of visual object constancy revealed by event-related fMRI of repetition priming. *Nature Neuroscience, 5*(5), 491–499.

Wallentin, M., Ellegaard, T., Ostergaard, S., Ostergaard, L., & Roepstorff, A. (2005). Motion verb sentences activate left posterior middle temporal cortex despite static context. *NeuroReport, 16*(6), 649–652.

Warrington, E. K. & McCarthy, R. (1983). Category specific access dysphasia. *Brain, 106*, 859–878.

Warrington, E. K. & McCarthy, R. (1987). Categories of knowledge: Further fractionations and an attempted integration. *Brain, 110*, 1273–1296.

Warrington, E. K. & Shallice, T. (1984). Category-specific semantic impairment. *Brain, 107*, 829–854.

Zwaan, R. A. & Radvansky, G. A. (1998). Situation Models in Language Comprehension and Memory. *Psychological Bulletin, 123*(2), 162–185.

Zwaan, R. (1999). Situation models: The mental leap into imagined worlds. *Current Directions in Psychological Science, 8*(1), 15–18.

Zwaan, R. (2004). The immersed experiencer: Toward an embodied theory of language comprehension. In B. H. Ross (Ed.), *The Psychology of Learning and Motivation 44* (p. 35). San Diego, CA: Academic Press.

Zwaan, R. & Taylor, L. J. (2006). Seeing, acting, understanding: Motor resonance in language comprehension. *Journal of Experimental Psychology: General, 135*(1), 1–11.

Zwaan, R. & Yaxley, R. H. (2003). Spatial iconicity affects semantic relatedness judgements. *Psychonomic Bulletin and Review, 10*(4), 954–958.

Zwaan, R., Stanfield, R. A., & Yaxley, R. H. (2002). Language comprehenders mentally represent the shapes of objects. *Psychological Science, 13*(2), 168–171.

Zwaan, R., Madden, C. J., Yaxley, R. H., & Aveyard, M. (2004). Moving words: Dynamic mental representations in language comprehension. *Cognitive Science, 28*, 611–619.

16

EMBODIED CONCEPT LEARNING

BENJAMIN BERGEN[1] AND JEROME FELDMAN[2]

[1]*Department of Linguistics, University of Hawaii at Manoa, Honolulu, HI, USA*
[2]*Department of Electrical Engineering and Computer Science, University of California, Berkeley, CA, USA*

HOW CONCEPTS ARE LEARNED

We address the question "How do people learn new concepts?" from the perspective of Unified Cognitive Science. By Unified Cognitive Science, we simply mean the practice of taking seriously all relevant findings from the diverse sciences of the mind, and here we are focusing on the question of concept learning. The particular perspective on concept learning advocated here grows out of the Neural Theory of Language project (www.icsi.Berkeley.edu/NTL), but is compatible with most cross-disciplinary work in the field.

Leaving aside for now Fodor's (1998) argument that concepts cannot be learned (which turns on disputable definitions of *learn* and *concept*), concept learning poses an ancient and profound scientific question. If we exclude divine intervention, then there are only two possible sources for our mental abilities: genetics and experience. There is obviously something about our genetic endowment that enables people, but not other animals, to become fluent language users and possessors of human conceptual systems. As nothing can enter our minds without intervention of our senses, which are themselves in large part the product of genetics, nature must provide the semantic basis for all the concepts that we acquire. So, in some sense, people really cannot learn any concepts that go beyond the combinatorial possibilities afforded by genetics.

At the same time, the conceptual systems of individual humans are profoundly marked by their experience—from maternal vocalization while still in the womb (Moon et al., 1993) to experience with culture-specific artifacts like baseball, chairs, or bartering practices. Evidence for relativistic effects of language on conceptual categories (Majid et al., 2004; Boroditsky, 2003) shows how conceptual systems are shaped by linguistic and other cultural experience. The scientific question confronting the field is how conceptual systems, which are so profoundly constrained by genetics, can at the same time be shaped by experience such that they display the great breadth of cultural diversity that they do.

A coherent and plausible picture of human concept learning is arising from combining biological, behavioral, computational, and linguistic insights. This account is similar in form to the solution to another biological question—a question for which the answer is now understood in great detail. That is the question of immunology. Animal immune systems are remarkably good at generating antibodies to combat novel antigens that invade the body. The raging question used to be whether this is a process where the killer antibody is selected from a fixed innate repertoire or whether the system somehow manufactures a custom antibody, instructed by the intruder. The full answer is beyond the scope of this chapter (and our knowledge) but the basic idea is clear. The immune system works because of a large number of primitive molecules that, in combination, can cover an astronomical number of possible antigens. These immunological primitives also evolve but not fast enough to attack a new intruder. Gerald Edelman, who won the 1972 Nobel Prize for his research on the selection/instruction problem in immunology, has worked for decades to show how the same combinatorial principles can help explain the mind (Edelman, 1987).

A "primitives plus composition" account of conceptual structure is appealing but requires further specification. We need an account of how primitive concepts arise, and an account of how the processes of conceptual composition work to generate new concepts. Details are emerging from a unified approach to cognitive science, and the story goes as follows. There is indeed an internal foundation for our concepts and it is we. As part of our animal heritage, we have a wide range of perceptual, motor, emotional, and social capabilities all expressed in our neural circuitry. This neural circuitry forms the basis for primitive concepts, which are in turn grounded in these neural structures. Furthermore, like our primate cousins, we have considerable competence at combining existing concepts to achieve desired goals, through binding, conjunction, and analogy, among other mechanisms.

In what follows, we will be using facts about language in our discussion of concepts and thought. Language is a particularly clear conduit to mental organization. Words express a speaker's concepts and evoke them in the listener. Much of conscious internal thought appears to be self-talk and, as we will discuss, there are many well-established empirical findings demonstrating how words are linked to concepts. This embodied view of language is hardly novel. Pinker and Jackendoff (2005) present a wide range of current evidence for the evolutionary continuity of language and thought. And within traditional philosophy, the American Pragmatists[1] stressed the continuity of all human activity and our evolutionary continuity. Thus, for our present purposes, a *concept* is the meaning of a word or phrase. This includes both basic, embodied words like *red* and *grasp* as well as abstract and technical words like *goal* and *continuity*. We will not address the possibility that there are concepts that cannot be described in words.

We will first provide an outline of the modern view of concepts as embodied, then outline how concrete concepts are learned, and discuss some known mechanisms for constructing new concepts from previously known ones.

EVIDENCE FOR EMBODIED CONCEPTS

Using concepts—accessing their features, imagining them, recalling them, and processing language about them—makes extensive use of their perceptual, motor, social, and affective substrates. The picture that has emerged from the broad range of convergent evidence surveyed below shows that when people use concepts, they perform mental simulations—internal enactments—of their embodied content.

Let us start with an example. Can you say how many windows there are in your current living quarters? Almost everyone simulates a walk-through to count them. Or consider a novel question—could you make a jack-o-lantern out of a grapefruit? To access the concept of a grapefruit—to reflect on its actual or hypothetical properties or to compare or combine it with other entities—you make use of detailed, encyclopedic and modality-specific knowledge. Subjectively, accessing this knowledge takes the form of sensory and motor experiences associated with the concept; reflecting on the carvability of a grapefruit involves creating internal motor and sensory experiences of carving a jack-o-lantern out of a grapefruit. Any time we use concepts, whether in performing categorization tasks, processing language about concepts, or reflecting on their features, we use mental simulation—the internal creation or recreation of perceptual, motor, and affective experiences. And we can simulate these experiences from different perspectives—it is quite different to imagine pushing, being pushed, or observing a third party pushing.

The notion that mental access to concepts is based on the internal creation of embodied experiences is supported by recent brain research, which shows that motor and pre-motor cortex areas associated with specific body parts (i.e., the hand,

[1] From The Internet Encyclopedia of Philosophy: "The basis of Dewey's discussion in the *Logic* is the continuity of intelligent inquiry with the adaptive responses of pre-human organisms to their environments in circumstances that check efficient activity in the fulfillment of organic needs. What is distinctive about intelligent inquiry is that it is facilitated by the use of language, which allows, by its symbolic meanings and implicatory relationships, the hypothetical rehearsal of adaptive behaviors before their employment under actual, prevailing conditions for the purpose of resolving problematic situations."

leg, and mouth) become active in response to motor language referring to those body parts. Using behavioral and neurophysiological methods, Pulvermüller et al. (2001) and Hauk et al. (2004) found that verbs associated with different effectors activate appropriate regions of motor cortex. In particular, Pulvermüller and colleagues had subjects perform a lexical decision task—they decided as quickly as possible whether a letter string was a word of their language—with verbs referring to actions involving the mouth (e.g., *chew*), leg (e.g., *kick*), or hand (e.g., *grab*). They found that the motor cortex areas responsible for mouth, leg, and hand motion exhibited more activation, respectively, when people were processing mouth, leg, and hand words. This result has been corroborated through transcranial magnetic stimulation work (Buccino et al., 2005). Tettamanti et al. (2005) have also shown through imaging that passive listening to sentences describing mouth versus leg versus hand motions activates corresponding parts of pre-motor cortex (as well as other areas).

Behavioral studies also offer convergent evidence for the automatic and unconscious use of perceptual and motor systems during language processing. Work on spatial language (Richardson et al., 2003; Bergen et al., 2007) has found that listening to sentences with visual semantic components can result in selective interference with visual processing. While processing sentences that encode upward motion, like *The ant climbed*, subjects take longer to perform a visual categorization task in the upper part of their visual field (deciding whether a shape is a circle or a square). The converse is also true—downward-motion sentences like *The ant fell* interferes with shape categorization in the lower half of the visual field. These results suggest that understanding spatial language evokes visual simulation that interferes with visual perception.

A second behavioral method (Glenberg & Kaschak, 2002) tests the extent to which motor representations are activated during language understanding. When subjects hear or read a sentence that describes someone performing a physical action, and are then asked to perform a physical action themselves, such as moving their hand away from or toward their body in response to a sentence, it takes them longer to perform the action if it is incompatible with the motor action described in the sentence. For example, if the sentence is *Andy gave you the pizza*, subjects take longer to push a button requiring them to move their hand away from their body than one requiring them to move their hand toward their body, and the reverse is true for sentences indicating motion away from the subject, like *You gave the pizza to Andy*. This interference between understanding language about action and performing a real action with our bodies suggests that, while processing language, we use neural structures dedicated to motor control.

A third method, used by Stanfield and Zwaan (2001) and Zwaan et al. (2002), investigates the nature of visual object representations during language understanding. Zwaan and colleagues have shown that the implied orientations of objects in sentences (like *The man hammered the nail into the floor* vs. *The man hammered the nail into the wall*) affect how long it takes subjects to decide whether an image of an object (such as a nail) was mentioned in the sentence.

When the image of an object is seen in the same orientation as it was implied to have in the sentence (e.g., when the nail was described as having been hammered into the floor and was depicted as pointing downward), it takes subjects less time to perform the task than when it was in a different orientation (e.g., horizontal). The same result is found when subjects are just asked to name the object depicted. Zwaan and colleagues also found that when sentences imply that an object would have different shapes (e.g., an eagle in flight vs. an eagle at rest), subjects once again responded more quickly to images of that object that were coherent with the sentence—images of that objects that have the same shape as they would have as described in the sentence.

A final method investigates whether sentences take longer to process when the scenes they describe take longer to mentally scan. Matlock (2004) demonstrates that the time subjects take to understand fictive motion sentences (sentences like *The road runs through the* desert or *The fence climbs up to the house*) is influenced by how quickly one could move along the described paths. For example, a sentence like *The path followed the creek* is processed faster when it follows a paragraph describing an athletic young man who jogs along the path than when it follows one describing an old man who has difficulty walking all the way down the path. Similarly, characteristics of the path itself like its distance or difficulty to navigate influence processing time in the same direction—the longer it would take the mover to travel the path, the longer it takes subjects to process the fictive motion sentence. This work once again implies that processing language makes use of a dynamic process of mental simulation.

These convergent results suggest a major role for embodied perceptual and motor experiences in language understanding. Language understanders automatically mentally imagine or simulate the scenarios described by language. The mental simulations they perform can include motor detail at least to the level of the particular effector that would be used to perform the described actions, and perceptual information about the trajectory of motion (toward or away from the understander; up or down) as well as the shape and orientation of described objects and paths. The neural imaging studies cited above suggest that these simulations involve some of the very brain mechanisms responsible for perceiving the same percepts or performing the same actions.

Mental simulation has an equally important role in other higher cognitive functions like memory and imagery. Behavioral evidence shows that recalling motor experiences recruits cognitive mechanisms responsible for performing the same motor actions (Barsalou, 1999). Several recent neural imaging studies show that this cognitive overlap mirrors a neural overlap; recalling motor experiences makes use of motor-control-specific neurocognitive structures (Wheeler et al., 2000; Nyberg et al., 2001). Similarly to recall, the performance of mental imagery involving motor control or visual or auditory perception yields activation of appropriate motor or perceptual brain areas (Porro et al., 1996; Lotze et al., 1999; Kosslyn et al., 2001; Ehrsson et al., 2003). It thus seems that recalling, imagining, or understanding language about actions and percepts recruit brain structures responsible for performing the actions or perceiving the percepts that appear in the mind's eye.

Even purely conceptual tasks involve the activation of modality-specific knowledge. For instance, in performing a property verification task (e.g., Is *mane* a property of *horse*?), subjects make use of mental simulation. This is demonstrated through longer times to correctly identify more perceptually difficult (e.g., smaller or physically peripheral) properties (Solomon & Barsalou, 2001, 2004). Using the same property verification task, Pecher and colleagues (2003, 2004) showed that verifying properties for the same concept from different sensory modalities (e.g., Apple-Green and Apple-Shiny) entailed a cost in processing time, relative to verifying properties from the same modality (e.g., Apple-Tart and Apple-Shiny). Both of these sets of findings imply that subjects performing mundane property verification are accessing modal mental simulations.

Other conceptual tasks also require mental simulation. One of the most important of these for conceptual processes is the use of covert or inner speech. At more or less frequent intervals, most people report the subjective experience of hearing a voice in their mind's ear, and also of feeling themselves articulating speech, especially when they are performing or preparing for cognitively difficult tasks. Talking to oneself internally, even without producing any speech or speech gestures, is itself demonstrably a sort of mental simulation. Empirical measures confirm that the motor and auditory systems are activated during inner speech. For example, covert speech results in brain activation whose lateral localization correlates with that of overt, actual speech (Baciu et al., 1999). In addition, covert speech, which results in no visible facial movement, nevertheless yields significantly greater electrical activity in the oral articulators than non-linguistic tasks like visualization (Livesay et al., 1996). And finally, activation of brain areas responsible for actual language production can be shown to be critical for covert speech through evidence that suppressing activity in these areas through transcranial magnetic stimulation results in decreased performance in both overt and covert speech tasks (Aziz-Zadeh et al., 2005). Inner speech is a sort of mental simulation of a particularly interesting variety, as it can itself drive mental simulation of a second sort. Suppose that one is taking care to correctly attach jumper cables to start a car with a dead battery. If one says to oneself *First attach one red clip to the positive post of the dead battery, then the other red clip to the positive post of the good one*, then this internally generated language, like language that a hearer might perceive, drives a enactment of the described events. This simulated experience thus facilitates simultaneous or future performance of the same task.

All these lines of research point to a common conclusion. Conceptual processes make use of the internal execution of imagery, qualitatively similar to the past experiences it is created or recreated from. As such, using concepts is qualitatively similar in some ways to experiencing the real-world scenarios they are built from. It is important to note that motor and perceptual experiences hold a privileged position in the study of mental simulation only because their basic mechanisms and neural substrates are relatively well understood. Other dimensions of experience are also relevant to simulation: anything that is experienced, including affect, social interactions, subjective judgments, and other imagined scenarios can be recruited to form part of a simulation. For example, recent work suggests that processing language about scenarios in which a protagonist would be likely to experience a particular emotion yields the internal recreation of similar affective experience on the part of the understander (Glenberg et al., 2005).

There are obviously limits to the extent to which previous experience can define simulation. If conceptual knowledge, as argued here, involves the activation of motor and perceptual (and other) representations of past experiences, then how can counterfactual or previously unexperienced meanings be understood? After all, one of the "design features" of human language is the possibility of describing things that do not exist (Hockett, 1960), for example, "the Easter Bunny" or "the current King of France." Moreover, because language is so important in helping children (and adults) learn about the world, it cannot be the case that linguistic meaning simply associatively reflects past experiences—if this were the case, then we could never learn anything new through language. However, a mental simulation-based account of meaning does not imply a purely behaviorist or empiricist perspective. In fact, as Kosslyn and colleagues (2001) argue, there is good reason to believe that "mental images need not result simply from the recall of previously perceived objects or events; they can also be created by combining and modifying stored perceptual information in novel ways" (p. 635). Mental simulation involves the active construction by the conceiver of novel perceptual, motor, and affective experiences, on the basis of previous percepts, actions, and feelings. Although it is constrained and informed by these experiences, compositional and other creative capacities allow departures from them.

One class of these is counterfactual or hypothetical situations, like those described through negation or conditionals (Fauconnier, 1985; Dancygier & Sweetser, 2005). For instance, an utterance like *If you hadn't painted your wall red, you wouldn't have gotten grounded* describes two scenes, neither of which actually happened (the non-painting of the wall and the non-grounding). There is evidence that suggests that language like this, and the corresponding reasoning, evokes simulations of the counterfactual or hypothetical scenes, though more transiently than factually presented content (Kaup & Zwaan, 2003).

There is also a significant literature on the computational modeling of actions and how such models can be learned and used. The most relevant work employs models of action that are themselves executable; that is, the models specify in detail how the action (say grasping) is carried out. Our work on the Neural Theory of Language uses a Petri-net based formalism called X-schemas (Bailey, 1997; Narayanan, 1999). The same X-schema can be used for carrying out an action, planning it, recognizing the action, or understanding language about it. The X-schema computational mechanism antedates the discovery of mirror neurons (Rizzolatti & Craighero, 2004) but obviously fits those data. The same formalism has proved its utility in simulation-based programs for understanding stories such as those found in newspapers (Narayanan, 1999).

As other authors have presented more detailed accounts of how neurally embodied concepts exhibit the behaviors traditionally ascribed to concepts, such as compositionality, internal structure, and so on (Barsalou, 1999; Gallese & Lakoff, 2005), we will forgo further discussion of those issues here. Instead, we will focus in the next section on how embodied concepts are learned.

LEARNING BASIC WORDS/CONCEPTS

From birth, children exhibit imitation and other social skills (Meltzoff & Prinz, 2002). They develop sophisticated methods of communication and joint attention well before they produce any language (Hoff, 2001). So we know that children have a rich set of conceptual and communication skills before they produce any language (Mandler, 1992, 2004).

Children learning about the world (and how to communicate about it) start first with concepts and words that are grounded in their direct perceptual and motor experiences. First words vary significantly across individuals, but most English-speaking children's first words (Figure 16.1) consist predominantly of concrete nouns, like *truck* and *ball* and social-interactional words, like *up* and *more* (Bloom, 2000; Tomasello, 2000). It is relatively obvious that concrete nouns are grounded in direct experience, but importantly, social-interactional words are equally bound to embodied experience. A child who utters *up!* is not soliloquizing on the existence of "upness" in the universe—he is using the word to label (often to bring about) a particular type of experience, where he

	cow								
apple	ball								yes
juice	bead		girl				down		no more
bottle	truck		baby	woof	yum	go	up	this	more
spoon	hammer	shoe	daddy	moo	whee	get	out	there	bye
banana	box	eye	mommy	choo-choo	uhoh	sit	in	here	hi
cookie	horse	door	boy	boom	oh	open	on	that	no
food	**toys**	**misc.**	**people**	**sound**	**emotion**	**action**	**prep.**	**demon.**	**social**

FIGURE 16.1 The words learned by most 2 year olds in a play school (Bloom, 1993).

is lifted. Often children also acquire concrete verbs like *get* and *sit*. It is only once they are far along in their development of these words that they begin to develop language for abstract, distant, or general concepts (Johnson, 1999). Conceptual development progresses in the same way, with concrete and directly experienced concepts leading the way for greater complexity. In addition to concepts that directly label their experience, children have pre-linguistic organizing schemas, such as support, containment, and source-path-goal (Mandler, 1992, 1994).

If all children acquired words and concepts identically, with concrete words and concepts being learned first—object before actions—and then abstract ones coming an afterward, then concept development could plausibly be accounted for as the progressive maturation of innate concepts. However, across languages and cultures, systematic differences in the character of children's experience, including linguistic differences and others, yield systematic variation in the course of word and concept acquisition. For instance, Korean and Chinese are languages in which verbal arguments can be omitted if they are obvious from context. Thus, if it is clear to both interlocutors that they are talking about what the doll is doing to the cake, the speaker would not have to say the equivalent of *The doll is throwing the cake* or even *She is throwing it*—it would suffice to say the equivalent of *Is throwing*. As a result, children growing up learning Korean and Chinese, and other languages like them, hear fewer nouns than their English-learning counterparts, and their order of word acquisition differs accordingly; significantly more of their early words are concretely grounded verbs (Choi, 2000). There is no universal order of word or concept acquisition—the only one is that children start by labeling concepts that are directly accessible to them through experience, whatever their experience happens to be.

The account we present here, then, is quite straightforward. Children learn their early words and concepts on the basis of perception, action, and other aspects of their embodied experience. Early words, and their conceptual meanings, are schematic representations of experiences, which abstract away from certain details, but still remain tightly bound to the modality-specific experiences they are based on. Using a concept thus involves reactivating a subset of those neural structures that underlay the experience in the first place. Language learning is closely integrated with conceptual learning, as a learner comes to associatively pair two aspects of experience—the perceptuo-motor schemas responsible for the perception and articulation of a particular piece of language, together with the schemas corresponding to its meaning. Moreover, language directs a learner to attend to certain aspects of his perceptual and motor experiences to make categorical linguistic distinctions (McDonough et al., 2003).

A strong test of this account is to build a computational model that realizes its claims, and see if it exhibits the right behavior. Bailey (1997) did just this when he built a program that was meant to learn the meanings of a subset of hand action words. To do this, the model needed to capture the full range of the conceptual space of potential hand actions, as described in any of the world's languages. Building in too many assumptions would preclude learning some languages, whereas leaving everything unspecified would gives the program no chance of learning at all. Bailey's (1998) solution was to base his solution on the body and on neural control networks. The idea is that all people share globally similar neural circuitry and bodies and thus exhibit the same semantic potential.

But there seems to be a complexity barrier. How could the meaning of an action word be the activity of a vast distributed network of neurons? The key to solving this in Bailey's (1998) model, and also in the brain, is *parameterization*. A motor action such as grasping involves many coordinated neural firings, muscle contractions, and so on, but we have no awareness of these details. What we can be aware of (and talk about) are certain parameters of the action—force, direction, effector, posture, repetition, and so on. The crucial hypothesis is that languages only label those action properties of which we can be aware. That is, there is a fixed set of embodied features that determine the semantic space for any set of concepts, such as motor actions.

Figure 16.2 presents an overview of Bailey's model for learning words that describe one-hand actions. The first thing to notice is that there is an intermediate set of features, shown as a large rectangle in the middle of the figure. These are the parameters just discussed—those aspects of actions that we can consciously know about can be described by a relatively small number of features. People do not have direct access to the elaborate neural networks that coordinate

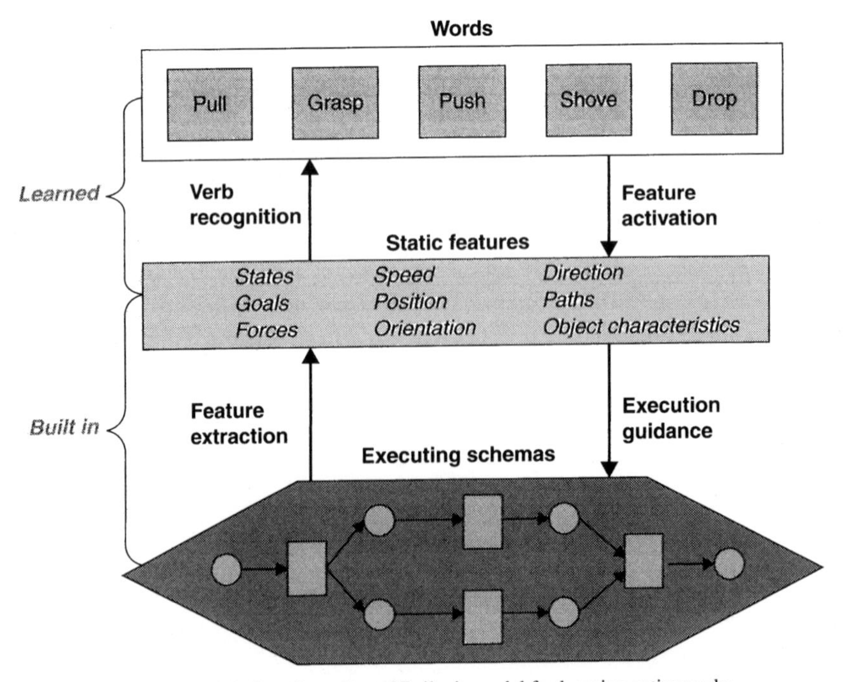

FIGURE 16.2 Overview of Bailey's model for learning action verbs.

our actions and neither does the model. This parameterization of action is one key to the success of the program.

A second critical feature of the model is the schematic representation of actions, called executing schemas (X-schemas) as shown at the bottom of Figure 16.2. In addition to parameters like force, actions are characterized by control features. For example, some actions are repetitive, some conditional, and so on. Depicted in Figure 16.2 is a generic control diagram showing an action followed by a test that causes branching to one of two alternatives, either of which leads to the final state. This kind of abstract action schema is common in the motor control literature and has also been used effectively in various computational models. The X-schema computational formalism for actions has considerable independent interest (Narayanan, 1997). The crucial point here is that control of action can also be parameterized and thus he made available to language learning. Even with these representational insights, the computational problems involved in embodied language learning are significant. The key to Bailey's success was approximating best-fit neural computation with Bayesian MDL (minimum description length) learning algorithms (Bailey, 1997).

In Figure 16.2 we note that the arrows are bi-directional. The system not only learns to label actions with words but will also carry out requests expressed using the words that it has learned. The upward arrows on the left describe the labeling pathway—features are extracted from executing schemas (bottom right arrow) and then these features are used to decide which verb is the most appropriate label for the action. The corresponding two-step path from word to parameters to action is depicted on the right of the figure.

Bailey's program learned the appropriate words for hand actions for a range of different languages, including Farsi and Spanish. A somewhat similar program by Regier (1996) learned spatial relation terms across languages that conceptualize these quite differently, including English, Russian, and Mixtec, a language that bases a large part of its spatial language on body parts. In principle, and as demonstrated by models like these, in practice as well, there seems to be no barrier against explaining in detail how children could learn those words of their language whose semantics is directly embodied. Projecting beyond existing models, these should also include words based on emotional and social cognition as well as perception, action, and goal seeking. Basic words and their concepts label instances and combinations of core neural capabilities. In the next section, we suggest how these mechanisms are extended in the learning and use of words for abstract and technical concepts.

LEARNING AND USING ABSTRACT AND TECHNICAL WORDS AND CONCEPTS

We have argued that language about directly experienced aspects of the world and the related concepts derive from generalization over concrete, embodied experiences. Abstract language and concepts—those with a less direct basis in experience—are built up from these conceptual primitives, by combining them using a modest set of productive mechanisms.

Existing concepts are used to produce novel ones through composition mechanisms like the following: conjunction (a narwhal is easily learned to be like a beluga with a long unicorn-like tusk); modification (a llama is like a camel without a hump); abstraction (a vehicle is anything that can be used for transportation); and mapping (ideas are like objects) among others. These productive mechanisms can function through direct perceptual or motor experience (e.g., seeing an image of a narwhal). But language can also indirectly ground conceptual learning. As discussed earlier, language drives perceptual, motor and affective simulation. This simulation itself constitutes experience that can form the basis for new concepts. Thus, one's only experience with flamingos being used as croquet mallets might be through reading about it (Carroll, 1865), but that still might be part of one's conceptual knowledge about flamingos. The mental experience driven by language, and reproduced using the relevant neural circuits, is a sufficient basis for conceptual reorganization.

In fact, because of the brain's massive connectivity and spreading activation, concepts are never learned or activated in isolation as each of us boasts richly interrelated concepts. We are also continuously composing or "blending" concepts. For example, quite different hues are suggested by "red hair", "red pencil", "red light", and so on. We easily understand and image novel combinations like "mauve marzipan narwhale." Fauconnier and Turner (2002) are particularly interested in blends that combine different domains through mapping to a common space like "trashcan basketball." They suggest that the human ability for complex conceptual integration was the key evolutionary advance that gave rise to language and thought.

The best studied of mechanisms for grounding abstract concepts is through mappings to them from concrete source domains. Abstract conceptual domains have long been known to be talked about in terms of concrete source domains, through linguistic metaphor. For instance, English speakers (and speakers of many other related and unrelated languages) talk about ideas in terms of objects and knowledge in terms of object manipulation. For instance, *I'm running out of ideas*, *I'm in the market for some new ideas*, *Now that we've deconstructed the proposal, let's see if we can reassemble it*, and *I'm having trouble grasping the gist of the sermon.* Close analysis of texts reveals that for most abstract domains, non-expert language users exploit very little, if any, non-metaphorical language. The domain of ideas is a case in point. Ideas can be possessed, acquired, shared, chewed on, swallowed, recast, and worn out, among many other metaphorical construals.

A large body of research spanning the past 30 years provides convergent evidence that abstract conceptual domains are not only talked about in terms of these concrete ones but are also actually thought about in terms of them as well. Early work in the Cognitive Linguistics framework (Lakoff & Johnson, 1980; Lakoff, 1993) provides three main types of evidence that metaphor is not just describing-as but conceptualizing-as. First, metaphorical language is systematic—when ideas are described as objects, considering the idea is always manipulating the object; the considerer is always the manipulator, and the idea is always the object (and never the reverse). Second, this metaphorical language is productive. It is not just due to a set of conventionalized metaphorical meanings associated with particular words. Instead, concrete language is regularly used in novel, metaphorical ways, like the word *disintegrated* in *The new human stem cell research disintegrated under the light of scrutiny.* Third, not just language but also reasoning transfers from a concrete conceptual domain to an abstract one through a metaphor. So if *this theory is hard to get a grip on*, then we infer that this is due to a property of the theory itself—it is slippery or bulky—or to a property of the understander—they do not have sufficient mental skills to get their head around it. More recently, an important fourth type of evidence has appeared, behavioral evidence using tools from cognitive psychology, showing that language users activate concrete source domains when thinking about abstract target domains (Gibbs et al., 1997; Boroditsky, 2000, 2001; Tseng et al., 2005).

How do learners come to understand an abstract domain in terms of a concrete source domain? In the simplest cases, the two domains are aligned in experience and can thus become associated (Lakoff & Johnson, 1980; Grady, 1997). For instance, quantity is a relatively abstract domain, especially when applied to concepts like power, love, and social capital. But in early childhood experiences, as throughout life, quantity of physical entities varies systematically with concrete, perceptible correlates. Perhaps most pervasive of these is relative height. In general, the more liquid in a container, the higher the level of the liquid; the more objects in a pile, the higher the pile. The systematic correlation between a concrete, perceptible cue (physical height) and a more abstract and subjective one (quantity) leads the learner to scaffold the conceptual and linguistic structure on top of the former. As the learner subsequently develops, the two domains are pulled apart—adults know that abstract quantity does not always correlate with physical height. But the conceptual and linguistic links between the two domains persist, as shown in the four types of evidence described earlier.

The case of conceptual metaphor shows not only how abstract concepts can be built up on the basis of concrete ones, but also how existing conceptual structures can be productively combined. It is clear that the metaphorical grounding account sketched out above is insufficient to completely deal with some cases, like *Theories are buildings* (*Modularity is a foundation of the theory of generative grammar; These observations buttress the theory of natural selection, Under the weight of conflicting evidence, the Newtonian physics came crashing down*, etc.). There is no experiential correlation between the creation and structure of buildings on the one hand and the invention and organization of theories on the other. But Grady (1997) has shown that the actual mappings by which theories are described and understood as buildings are partial—only certain aspects of buildings are mapped onto theories. These include the physical structure of buildings (foundation, support, and buttresses), and their persistent erectness but not plumbing. The metaphor *Theories are buildings* is thus best seen as instantiating a combination of two primary metaphors—*Persistent functioning is remaining erect*, and *Abstract organization is physical structure*. Each of these has a clear basis in experience. Many physical objects, like buildings, trees, chairs, and so on, function persistently only while erect. Many objects with complex physical structure also have associated organization—the legs are not only at the bottom of a table but also serve to the function of support. Put together through composition, these two primary metaphors produce a mapping whereby *Persistently functioning entities with abstract organization are erect objects with physical structure*. Buildings happen to be a good example of concrete objects with physical structure that saliently remain erect, and theories happen to be a good example of abstract entities with organization that persists.

Concrete concepts are learned through schematization over direct experiences and abstract concepts are indirectly grounded through co-experience with concrete ones, or through compositional mechanisms that produce them on the basis of previously grounded ones.

CONCLUSIONS

We have provided an outline of how people learn and use new concepts. The account provides a plausible theory that is supported by a broad range of linguistic, computational, behavioral, and brain imaging data. It goes something like as follows:

1. Our core concepts are based on the neural embodiment of all our sensory, motor, planning, emotional, and social abilities, most of which we share with other primates. This yields a huge, but not unbounded, collection of primitives.
2. We can only be aware of or talk about a limited range of parameters over these abilities and human languages are based on these parameterizations, plus composition. Composition can give rise to additional abilities and parameters.
3. The meanings of all new words and concepts are formed by compositions of previously known concepts. We use a wide range of compositional operations including conjunction, causal links, abstraction, analogy, and metaphor.
4. Domain relations, particularly conceptual metaphors, are the central compositional operations that allow us to learn technical and other abstract concepts.
5. We understand language by mapping it to our accumulated experience and imagining (simulating) the consequences.

We could end this chapter here, but there is a related a priori contention that we can address with the same basic line of reasoning—the postulated innateness of grammar. The logical argument from the "poverty of the stimulus" (Chomsky, 1980) proposes that children do not get a rich enough training to enable them to learn the grammar of their native language(s). The reasoning summarized above provides part of the answer to the grammar learning problem, a solution one might call the "opulence of the substrate." This alternative states that children come to language learning with a very rich collection of conceptual primitives, rules for composing them, and breadth of embodied experiences. None of these is specifically tailored to language.

The only additional insight required is that grammar is itself constituted of mappings from linguistic form to meaning. A rule of grammar is what linguists call a *construction*, a form–meaning pair. We can combine the idea of linguistic constructions with the notion of embodied meaning outlined above and define Embodied Construction Grammar or ECG (Bergen & Chang, 2005). In ECG, a word like "into" maps to its conceptual meaning—a source–path–goal schema with its goal role bound to the interior role of a container schema. Larger constructions at the phrasal level would map a phrase like "into the house" into a conceptualization where the house was assigned as the conceptual container.

Given that language is embodied and that grammar maps from sound to experience, the child's problem in learning grammar is not overwhelming. They learn basic words as labels for their experience, as pointed out in the section on Learning basic words/concepts. The key insight for learning compositional rules of grammar is that the job of a grammar rule is to specify conceptual composition. A child who already understands a scene conceptually and hears a sentence about it only needs to hypothesize what about the linguistic form licenses the known conceptual composition. Of course, these early hypotheses about grammar rules are sometimes wrong, and the usual learning processes of testing, refinement, and abstraction are also involved. This is a short version of a fairly long and complex story, but a full and computationally tested account is available in theses by Chang and Mok (2006). Some additional descriptions of ECG and its applications can be found in Chang et al. (2002) and Bergen and Chang (2005).

An account of concept learning based on cognitive and evolutionary continuity triggers an obvious question: what is unique about the human mind that enables us to become fluent language users and conceptual thinkers? This is a subject of considerable current research, most notably in Michael Tomasello's group in Leipzig. There is unlikely to be a single feature that explains all unique human mental attributes, but Tomasello has identified one feature that is clearly important—the ability to understand other minds. From our perspective, mind reading appears to be a special case of a more general capability for mental simulation. As we have seen, there is converging evidence that people understand language and other behaviors at least in part by simulation (or imagination). This ability to think about situations not bound to the here and now (*displacement*) is also obviously necessary for evaluating alternatives, for planning, and for understanding other minds.

More speculatively, there is a plausible story about how a discrete evolutionary change could have given early hominids a simulation capability that helped

start the process leading to our current mental and linguistic abilities. Mammals in general exhibit at least two kinds of involuntary simulation behavior—dreams and play. While a cat is dreaming, a center in the brainstem (the locus coeruleus) blocks the motor nerves so that the cat's dream thoughts are not translated into action. If this brainstem center is disabled, the sleeping cat may walk around the room, lick itself, catch imaginary mice, and otherwise appear to be acting out its dreams. There is a general belief that dreaming is important for memory consolidation in people and this would also be valuable for other mammals. Similarly, it is obvious that play behaviors in cats and other animals have significant adaptive value.

Given that mammals do exhibit involuntary displacement in dreams, it seems that only one evolutionary adaptation would have been needed to achieve our ability to imagine situations of our choice. Suppose that the mammalian involuntary simulation mechanisms were augmented by brain circuits that could explicitly control what was being imagined. This kind of overlaying a less flexible brain system with one that is more amenable to control is a hallmark of brain evolution. Now, hominids who could do detached simulations could relive the past, plan for the future, and would be well on their way to simulating other minds. Understanding other minds would then provide a substrate for richer modeling and communication, just as Tomasello and others have suggested.[2]

And what about Fodor's contention that people cannot learn new concepts? We have suggested a slight variant: people can only learn new concepts that map to things they already know. This is not as exciting as Fodor's version, but it has two significant advantages. First of all, it is true. In addition, it provides a framework for studying individual and cultural development as the interplay of genetics and experience. For people who take the science of the mind seriously, a unified approach to cognitive science is the only game in town.

REFERENCES

Aziz-Zadeh, L., Cattaneo, L., Rochat, M., & Rizzolatti, G. (2005). Covert speech arrest induced by rTMS over both motor and non-motor left hemisphere frontal sites. *Journal of Cognitive Neuroscience, 17*(6).

Baciu, M. V., Rubin, C., Décorps, M. A., & Segebarth, C. M. (1999). fMRI assessment of hemispheric language dominance using a simple inner speech paradigm. *NMR in Biomedicine, 12*(5), 293–298.

Bailey, D. (1997). *When Push Comes to Shove: A Computational Model of the Role of Motor Control in the Acquisition of Action Verbs*. Ph.D. Thesis, UC Berkeley.

Barsalou, L. W. (1999). Perceptual symbol systems. *Behavioral and Brain Sciences, 22*, 577–609.

Bergen, B. & Chang, N. (2005). Embodied construction grammar in simulation-based language understanding. In J.-O. Östman & M. Fried (Eds.), *Construction Grammars: Cognitive Grounding and Theoretical Extensions* (pp. 147–190). Amsterdam: Benjamin.

[2]Notice how close this is to the pragmatist view of Note 1.

Bergen, B., Lindsay, S., Matlock, T., & Narayan, S. (2007). Spatial and linguistic aspects of visual imagery in sentence comprehension. *Cognitive Science, 31*(5): 733–764.

Bloom, L. (1993). *The Transition from Infancy to Language: Acquiring the Power of Expression.* New York: Cambridge University Press.

Bloom, P. (2000). *How Children Learn the Meanings of Words.* Cambridge, MA: MIT Press.

Boroditsky, L. (2000). Metaphoric structuring: Understanding time through spatial metaphors. *Cognition, 75*(1), 1–28.

Boroditsky, L. (2001). Does language shape thought? English and Mandarin speakers' conceptions of time. *Cognitive Psychology, 43*(1), 1–22.

Boroditsky, L. (2003). Linguistic Relativity. In Nadel, L. (Ed.) *Encyclopedia of Cognitive Science* (pp. 917–921). London: MacMillan Press.

Buccino, G., Riggio, L., Melli, G., Binkofski, F., Gallese, V., & Rizzolatti, G. (2005). Listening to action-related sentences modulates the activity of the motor system: A combined TMS and behavioral study. *Cognitive Brain Research, 24,* 355–363.

Carroll, L. (1865). *Alice's Adventures in Wonderland.*

Chang, N., Feldman, J., Porzel, R., & Sanders, K. (2002). Scaling Cognitive Linguistics: Formalisms for Language Understanding. *Proceedings of 1st International Workshop on Scalable Natural Language Understanding.* Heidelberg, Germany.

Chang, N., & Mok, E. (2006). A structured context model for grammar learning. The 2006 International Joint Conference on Neural Networks. Vancouver, B.C.

Choi, S. (2000). Caregiver input in English and Korean: Use of nouns and verbs in book-reading and toy-play contexts. *Journal of Child Language, 27,* 69–96.

Chomsky, N. (1980). *Rules and Representations.* Oxford: Basil Blackwell.

Dancygier, B. & Sweetser, E. (2005). *Mental Dpaces in Grammar: Conditional Constructions.* New York: Cambridge University Press.

Edelman, G. (1987). *Neural Darwinism. The Theory of Neuronal Group Selection.* New York: Basic Books.

Ehrsson, H. H., Geyer, S., & Naito, E. (2003). Imagery of voluntary movement of fingers, toes, and tongue activates corresponding body-part specific motor representations. *Journal of Neurophysiology, 90,* 3304–3316.

Fauconnier, G. (1985). *Mental Spaces: Aspects of Meaning Construction in Natural Language.* Cambridge, MA: MIT Press.

Fauconnier, G. & Turner, M. (2002). *The Way We Think: Conceptual Blending and the Mind's Hidden Complexities.* New York: Basic Books.

Fodor, J. (1998). *Concepts: Where Cognitive Science Went Wrong.* Oxford: Oxford University Press.

Gallese, V. & Lakoff, G. (2005). The brain's concepts: The role of the sensory–motor system in reason and language. *Cognitive Neuropsychology.*

Gibbs, R. W., Bogdanovich, J. M., Sykes, J. R., & Barr, D. J. (1997). Metaphor in idiom comprehension. *Journal of Memory and Language, 37,* 141–154.

Glenberg, A. & Kaschak, M. (2002). Grounding language in action. *Psychonomic Bulletin and Review, 9,* 558–565.

Glenberg, A. M., Havas, D., Becker, R., & Rinck, M. (2005). Grounding language in bodily states: The case for emotion. In R. Zwaan & D. Pecher (Eds.), *The Grounding of Cognition: The Role of Perception and Action in Memory, Language, and Thinking.* Cambridge, MA: Cambridge University Press.

Grady, J. E. (1997). Theories are buildings revisited. *Cognitive Linguistics, 8*(4), 267–290.

Hauk, O., Johnsrude, I., & Pulvermüller, F. (2004). Somatotopic representation of action words in human motor and premotor cortex. *Neuron, 41*(2), 301–307.

Hockett, C. F. (1960). The origin of speech. *Scientific American, 203,* 88–96.

Hoff, E. (2001). *Language Development,* 2nd ed. Pacific Grove, CA: Brooks/Cole.

Johnson, C. R. (1999). *Constructional Grounding: The Role of Interpretational Overlap in Lexical and Constructional Acquisition.* Ph.D. dissertation, University of California, Berkeley.

Kaup, B. & Zwaan, R. A. (2003). Effects of negation and situational presence on the accessibility of text information. *Journal of Experimental Psychology: Learning, Memory, and Cognition, 29.*

Kosslyn, S., Ganis, G., & Thompson, W. (2001). Neural foundations of imagery. *Nature Reviews Neuroscience, 2,* 635–642.

Lakoff, G. (1993). The contemporary theory of metaphor. In A. Ortony (Ed.), *Metaphor and Thought* (2nd ed.). Cambridge, MA: Cambridge University Press.

Lakoff, G. & Johnson, M. (1980). *Metaphors We Live by.* Chicago: University of Chicago Press.

Livesay, J., Liebke, A., Samaras, M., & Stanley, A. (1996). Covert speech behavior during a silent language recitation task. *Perceptual Motor Skills, 83*(3 Pt 2), 1355–1362.

Lotze, M., Montoya, P., Erb, M., Hülsmann, E., Flor, H., Klose, U., Birbaumer, N., & Grodd, W. (1999). Activation of cortical and cerebellar motor areas during executed and imagined hand movements: An fMRI study. *Journal of Cognitive Neuroscience, 11*(5), 491–501.

Majid, A., Bowerman, M., Kita, S., Haun, D., & Levinson, S. (2004). Can language restructure cognition? The case for space. *Trends in Cognitive Sciences, 8*(3), 108–114.

Mandler, J. M. (1992). How to build a baby: II. Conceptual primitives. *Psychological Review, 99*(4), 587–604.

Mandler, J. M. (2004). *The foundations of mind: The origins of the conceptual system.* New York: Oxford University Press.

Matlock, T. (2004). Fictive motion as cognitive simulation. *Memory & Cognition, 32*(8), 1389–1400.

McDonough, L., Choi, S., & Mandler, J. (2003). Understanding spatial relations: Flexible infants, lexical adults. *Cognitive Psychology, 46,* 229–259.

Meltzoff, A. N. & Prinz, W. (2002). *The Imitative Mind: Development, Evolution, and Brain Bases.* Cambridge, England: Cambridge University Press.

Moon, C., Cooper, R. P., & Fifer, W. P. (1993). Two-day olds prefer their native language. *Infant Behaviour and Development, 16,* 495–500.

Narayanan, S. (1997). *KARMA: Knowledge-based Action Representations for Metaphor and Aspect.* Ph.D. Thesis, UC Berkeley.

Narayanan, S. (1999). Moving right along: A computational model of metaphoric reasoning about events. *Proceedings of the National Conference on Artificial Intelligence AAAI-99.* Orlando, Florida.

Nyberg, L., Petersson, K.-M., Nilsson, L.-G., Sandblom, J., Åberg, C., & Ingvar, M. (2001). Reactivation of motor brain areas during explicit memory for actions. *NeuroImage, 14,* 521–528.

Pecher, D., Zeelenberg, R., & Barsalou, L. W. (2003). Verifying properties from different modalities for concepts produces switching costs. *Psychological Science, 14,* 119–124.

Pecher, D., Zeelenberg, R., & Barsalou, L. W. (2004). Sensorimotor simulations underlie conceptual representations: Modality-specific effects of prior activation. *Psychonomic Bulletin and Review, 11,* 164–167.

Pinker, S. & Jackendoff, R. (2005). What's special about the human language faculty? *Cognition, 95*(2), 201–236.

Porro, C., Francescato, M., Cettolo, V., Diamond, M., Baraldi, P., Zuian, C., Bazzocchi, M., & diPrampero, P. (1996). Primary motor and sensory cortex activation during motor performance and motor imagery: A functional magnetic resonance imaging study. *Journal of Neuroscience, 16,* 7688–7698.

Pulvermüller, F., Haerle, M., & Hummel, F. (2001). Walking or talking?: Behavioral and neurophysiological correlates of action verb processing. *Brain and Language, 78,* 143–168.

Regier, T. (1996). *The Human Semantic Potential: Spatial Language and Constrained Connectionism.* Cambridge: MIT Press.

Richardson, D., Spivey, M., McRae, K., & Barsalou, L. (2003). Spatial representations activated during real-time comprehension of verbs. *Cognitive Science, 27,* 767–780.

Rizzolatti, G. & Craighero, L. (2004). The mirror-neuron system. *Annual Review of Neuroscience, 27,* 169–192.

Solomon, K. O. & Barsalou, L. W. (2001). Representing properties locally. *Cognitive Psychology, 43,* 129–169.

Solomon, K. O. & Barsalou, L. W. (2004). Perceptual simulation in property verification. *Memory and Cognition, 32,* 244–259.

Stanfield, R. & Zwaan, R. (2001). The effect of implied orientation derived from verbal context on picture recognition. *Psychological Science, 12,* 153–156.

Tettamanti, M., Buccino, G., Saccuman, M. C., Gallese, V., Danna, M., Scifo, P., Fazio, F., Rizzolatti, G., Cappa, S. F., & Perani, D. (2005). Listening to action-related sentences activates fronto-parietal motor circuits. *Journal of Cognitive Neuroscience, 17,* 273–281.

Tomasello, M. (2000). First steps in a usage based theory of first language acquisition. *Cognitive Linguistics,* (11), 61–82.

Tseng, M., Hu, Y., Han, W.-W., & Bergen, B. (2005). "Searching for happiness" or "Full of joy"? Source domain activation matters. *Proceedings of the 31st Annual Meeting of the Berkeley Linguistics Society.*

Wheeler, M., Petersen, S., & Buckner, R. (2000). Memory's echo: Vivid remembering reactivates sensory specific cortex. *Proceedings of the National Academy of Science of the USA, 97,* 11125–11129.

Zwaan, R., Stanfield, R., & Yaxley, R. (2002). Do language comprehenders routinely represent the shapes of objects? *Psychological Science, 13,* 168–171.

17

Mathematics, the Ultimate Challenge to Embodiment: Truth and the Grounding of Axiomatic Systems

Rafael E. Núñez

Department of Cognitive Science, University of California, San Diego, CA, USA

The human body is an animal body. A body that has evolved over millions of years coping with real-world properties such as temperature, gravity, humidity, color, space, texture and so on. With this same body humans have been able to create concepts—and think with them—in a way that transcends immediate bodily experience. Today, millions of modern humans effortlessly operate in everyday life with abstract notions like "democracy," "black humor," "inflation," and the "flow of time." In technical domains, like mathematics, humans have created abstract concepts, such as "square root of minus one" and "transfinite numbers"—rich and precise entities that lack any concrete instantiation in the real world. These entities are the product of the human imagination, and exist in the realm of mental abstractions and social practices. How do humans achieve this with the body of a primate? In what sense are the abstract ideas humans create *embodied*? And then there is a question of what *is* mathematics in the first place? What is the nature of this body of knowledge that appears to be timeless, eternal, absolute, and effective to the point that many scholars firmly believe it is part of the very fabric of the universe, transcending human existence?

I must make it clear, right up front, that to address these questions within embodied cognitive science, we must go beyond a concrete understanding of embodiment—"material embodiment" as I called it nearly a decade ago (Núñez, 1999)—a view that centers primarily on *physical corporality*. Moreover, we need to go beyond usual views in embodied cognition that tend to focus on individual perception, action, motor control, emotional states, and neural correlates of given phenomena. As important as this work is, I suggest that relative to the above questions, beyond individuals' performances, behavioral observations, and biological measurements, this time the focus should be on non-material supra-individual cognitive products and their genesis—*conceptual systems*. And by concepts I do not mean pre-defined notions as evoked or thought by someone, but the concepts *themselves*, with their semantic properties and inferential organization. The idea is to explain what *is* mathematics, what makes it possible, and what brings it into being, rather than how individual people learn about *it* (as a pre-existing entity), what people feel in their bodies when they think about it, what parts of people's bodies help them think mathematically, and so on. George Lakoff and I called such an endeavor the Cognitive Science *of* mathematics (Lakoff & Núñez, 2000). In this chapter, we will be looking at the embodiment of stable inferential patterns created and sustained by communities of individuals, which exist beyond the individuals themselves. The approach we take here is comparable to the study of, say, speech accents, in that although they are created, manifested, and sustained *by* individuals, they—the speech accents themselves—constitute distinctions we make at a supra-individual level. In the same way that we speak of Welsh or Jamaican accents, or of John as having an Australian accent but not of John-ian or Sally-ian accents (i.e., the Welsh accent is the one we observe among Welsh *people*) here we will talk about abstract concepts—not individual conceptualizations but concepts that constitute collective domains of knowledge. In sum, we will analyze the embodied cognitive mechanisms that make human abstraction and their supra-individual crystallization possible, and we will see how they bring mathematics, its concepts, and inferential organization, into being.

My goal in this chapter is to provide a brief overview of what is the nature of mathematics from the perspective of embodied cognitive science and conceptual systems. I want to show how the inferential organization of mathematics emerges from everyday cognitive mechanisms of human imagination realized via embodied conceptual mappings such as metaphor, metonymy, conceptual blends, and so on. This, of course, is a vast enterprise, so here I will only concentrate on the fundamental concept of *axiom*, which modern formal mathematics takes to be an essential building block for the study of the foundations of mathematics itself. Contrary to the widespread belief among mathematicians and logicians who see axioms as meaningless formal statements, I want to show that axioms are the product of embodied cognitive mechanisms. Through the analysis of hypersets—a specific branch of contemporary set theory—I intend to show how the quintessential abstract conceptual system we call mathematics (1) emerges from embodied cognitive mechanisms for imagination such as conceptual metaphor; (2) that truth and objectivity comes out of the collective use of these mechanisms; (3) that it can have domains that are internally consistent but mutually inconsistent; (4) and that these domains built on corresponding axiom systems that while grounded in embodied meaning provide different "truths" and inferential organization. Finally, I want to show that these properties are not unique to mathematics but that they exist in everyday abstract conceptual systems as well. I will illustrate this point with empirical observations from my investigation contrasting spatial construals of time in the western world with that in the Aymara culture of the Andes' highlands. I will defend the idea that everyday conceptual systems possess elementary embodied forms of "truth," "axioms," and "theorems" (i.e., true statements derived within a logical system) that are "objective" within the communities that operate with them. These properties of ordinary human imagination serve as grounding for developing more complex and refined forms of abstraction, which find the most sublime form in mathematics.

MATHEMATICS, A REAL CHALLENGE TO EMBODIMENT

Mathematics is a unique body of knowledge. The very entities that constitute what it is are idealized mental abstractions, which cannot be perceived directly through the senses. The empty set, for instance—the simplest entity in set theory—cannot be actually perceived. We cannot physically observe collections with no members. Or take the simplest entity in Euclidean geometry, the point. As defined by Euclid, a point is a dimensionless entity, which has only location but no extension! The empty set and the Euclidean point, with their precision and clear identity, are idealized abstract entities that do not exist in the real physical world, and therefore they are not available for empirical investigation. Yet, they are fundamental building blocks for the construction of set theory and Euclidean geometry, respectively. But nowhere can the imaginary nature of mathematics be seen more clearly than in concepts involving infinity. Because of the finite nature of our bodies and brains, no direct experience can exist with the infinite itself! Yet, infinity in mathematics is essential. It lies at the very core of many fundamental domains such as projective geometry, infinitesimal calculus, point-set topology, mathematical induction, and set theory, to mention a few. Taking infinity away from mathematics would mean the collapse of this extraordinary edifice, as we know it.

Moreover, mathematics has a unique collection of features. It is (extremely) precise, objective, rigorous, generalizable, and, of course, applicable to the real world. It is also extraordinarily stable, in that a theorem once proved, stays proved forever! Any attempt to address the nature of mathematics must explain these features. What is, then, the nature of mathematics? What makes it possible? What is the cornerstone of such a fabulous objective and precise logical edifice? Such questions have been treated extensively in the realm of the philosophy of mathematics, which are becoming, in the 20th century, specific subject matters for rather technical fields of formal logic and metamathematics. Ever since, the foundations of mathematics have taken to be intramathematical (i.e., inside mathematics proper), as if the tools of formal logic alone are to provide the ultimate answers about the nature of mathematics. But can the foundations of mathematics be, themselves, mathematical entities? Or do they lie outside of mathematics? And if they do, where do these entities come from? And what forms do they have? As we will see, within the formalist approaches, the quest for *axioms* and the study of their deductive power have become a fundamental issue in the investigation of the nature of mathematics.

In an attempt to answer these questions, another (very influential) approach has come from good old Platonism, which relying on the existence of transcendental worlds of ideas beyond human existence, sees mathematical truths and entities as existing independently of human beings. This view, however does not have any support based on scientific findings and does not provide any link to current empirical work on human ideas and conceptual systems (although, paradoxically, as a matter of faith it is supported by many mathematicians, physicists, and philosophers). For scholars who endorse a socio-cultural view (often along the lines of postmodernism), the question of the nature of mathematics is relatively straightforward: Mathematics, much like art, poetry, architecture, music, and fashion, is a "social construction" (Lerman, 1989). Although I endorse the relevance of socio-cultural dimensions in mathematics (e.g., see Lakoff & Núñez, 2000, pp. 355–362), I defend the idea that mathematics is *not just* the result of socio-cultural practices. It is not clear, in a purely socio-cultural constructivist view, what makes mathematics so special. What distinguishes mathematics from other forms of social constructions, say, fashion or poetry? Any precise enough explanatory proposal of the nature of mathematics should give an account of the peculiar collection of features that make mathematics so unique: precision, objectivity, rigor, generalizability, stability, and applicability to the real world. This is what makes the scientific study of the nature of mathematics so challenging: mathematical entities (organized ideas and stable concepts) are abstract and imaginary, yet they are realized through the biological and social peculiarities of the human animal. For those studying the human mind scientifically, the question of the nature of mathematics is indeed a real challenge, especially for those who endorse an embodied oriented approach to cognition. The crucial question is: *How can an embodied view of the mind give an account of an abstract, idealized, objective, precise, sophisticated, and powerful domain of ideas if direct bodily experience with the subject matter is not possible?* In *Where Mathematics Comes From*, Lakoff and I propose some preliminary answers to this question (Lakoff & Núñez, 2000). Several basic elements of this proposal are analyzed in the next section.

EVERYDAY EMBODIED MECHANISMS FOR HUMAN IMAGINATION

Building on findings in mathematical cognition and the neuroscience of numerical cognition, and using mainly methods from cognitive linguistics, a branch of cognitive science, Lakoff and I asked, what cognitive mechanisms are used in structuring mathematical ideas? And more specifically, what cognitive mechanisms can characterize the inferential organization observed in mathematical ideas themselves? We suggested that most of the idealized abstract technical entities in mathematics are created via everyday human cognitive mechanisms that extend the structure of bodily experience while preserving inferential organization. Such "natural" mechanisms are, among others, conceptual metaphors (Lakoff & Johnson, 1980, 1999; Johnson, 1987; Sweetser, 1990; Lakoff, 1993; Lakoff & Núñez, 1997; Núñez & Lakoff, 1998, 2005), conceptual metonymy (Lakoff & Johnson, 1980), and conceptual blends (Fauconnier & Turner, 1998, 2002; Núñez, 2005). Using a technique we called *Mathematical Idea Analysis* we studied in detail many mathematical concepts in several areas in mathematics, from set theory to infinitesimal calculus, to transfinite arithmetic, and showed how, via these everyday embodied mechanisms, the inferential patterns drawn from bodily experience in the real world get extended in very specific and precise ways to give rise to a new emergent inferential organization in purely imaginary domains.

Consider the following two everyday linguistic expressions: "The election is *ahead* of us" and "the long Winter is now *behind* us." Literally, these expressions do not make any sense. "An election" is not something that can physically be "ahead" of us in any measurable or observable way, and the "Winter" is not something that can be physically "behind" us. Hundreds of thousands of these expressions, whose meaning is not literal but *metaphorical*, can be observed in human everyday language: "he is a *cold* person," "she has *strong* opinions," "the market is quite *depressed*." Metaphor, in this sense, is not just a figure of speech, or an exceptional communicational tool in the hands of poets and artists. It is an ordinary mechanism of thought, which, usually operating unconsciously and effortlessly, permeates nearly every aspect of human everyday (and technical) language, making imagination possible.

Cognitive linguistics (and more specifically, cognitive semantics) has studied this phenomenon in detail and has shown that the meaning of these hundreds of thousands metaphorical linguistic expressions can be modeled by a relatively small number of conceptual metaphors. These conceptual metaphors, which are inference-preserving cross-domain mappings, are cognitive mechanisms that allow us to project the inferential structure from a grounded *source domain*, for instance, thermic experience, into another one, the *target domain*, usually more abstract, say, affection. As a result, specific temperature-related notions like "cold" and "warm" get mapped onto "lack of affection" and "presence of affection," respectively, and open up an entire world of inferences where the relatively abstract domain of "affection" is conceived and understood in terms of a more concrete one, namely, thermic experience. A crucial component of what is modeled is inferential organization, the network of inferences that is generated via the mappings.

We can illustrate how the mappings work with the above temporal examples where events are conceived as being in front of us or behind us. Note that although the expressions use completely different *words* (i.e., one refers to a location *ahead* of us, whereas the other to a location *behind* us), they are both linguistic manifestations of a single general conceptual metaphor, namely, TIME EVENTS ARE THINGS IN SAGGITAL UNIDIMENSIONAL SPACE.[1] As in any conceptual metaphor, the inferential structure of target domain concepts (time, in this case) is created via a precise mapping drawn from the source domain (in this case saggital unidimensional space: the linear space in front and behind an observer). The general mapping of this metaphor is shown in the following table[2]:

Source Domain Saggital unidimensional space relative to ego		Target Domain Time
Objects in front of ego	→	Future times
Objects behind ego	→	Past times
Object co-located with ego	→	Present time
The further away in front of ego an object is	→	The "further away" an event is in the future
The further away behind ego an object is	→	The "further away" an event is in the past

The inferential structure of this mapping accounts for a number of linguistic expressions, such as "The summer is still *far away*," "The end of the world is *near*" and "Election day is *here*." Many important entailments—or truths—follow from the mapping. For instance, transitive properties applying to spatial relations between the observer and the objects in the source domain are preserved in the target domain of time: if, relative to the front of the observer, object *A* is further away than object *B*, and object *B* is further away than object *C*, then object *C* is closer than object *A*. Via the mapping, this implies that time *C* is in a "nearer" future than time *A*. The same relations hold for objects behind the observer and times in the past. Also, via the mapping, time is seen as having extension, which can be measured; and time can be extended, like a segment of a path, and conceived as a linear bounded region, and so on.

Of course spatial construals of time and conceptual mapping, in general, present many more subtleties and complexities. We will come back to some of them later (in section "Everyday Abstraction: the Embodiment of Spatial Construals of Time and Their 'Axioms'"). For the moment, let us stop here and

[1] Following a convention in cognitive linguistics, small capitals here serve to denote the name of the conceptual mapping as such. Particular instances of these mappings, called metaphorical expressions (e.g., "she has a great future in front of her") are not written with capitals.

[2] There are two main forms of this general conceptual metaphor defined according to the nature of the moving agent—the relative motion of ego with respect to the objects, or the objects with respect to ego (as in *Easter is approaching* vs. *We're approaching Easter*). Their analysis goes beyond the scope of this chapter. For details see Lakoff (1993), Núñez (1999), and Núñez & Sweetser (2006).

see how this theoretical framework applies to more sophisticated imaginary ideas such as the ones constituting mathematical concepts and their axiomatic systems.

MATHEMATICAL ABSTRACTION: THE EMBODIMENT OF AXIOMS, SETS, AND HYPERSETS

Axioms are the modern, and more technical manifestation of the old Euclidean idea of *postulate*. The great mathematician Euclid (ca. 325–265BC) is best known for having systematized the knowledge of geometry and developed what is known as Euclidean geometry today. Although Greek thinkers had been developing geometry at least since the time of pre-Socratic philosophers, three centuries before Euclid, they generated a body of knowledge that was far from constituting a unified discipline. It was Euclid who put together the results and advancements in geometry in a systematic manner. He organized this body of knowledge in such a way that he could derive, through logical deduction, all the known facts of geometry from a few simple and fundamental facts. He took these essential facts to be trivial, intuitive, and too self-evident to be deduced from other facts (e.g., "A straight line may be extended to any finite length"). He called these facts *postulates*. Euclid claimed that only five postulates were required to characterize, using a ruler and compass only, the essence of the entire domain of plane geometry as a subject matter, and believed that from these essential facts all other geometric truths could be derived by deduction alone (i.e., theorems). From this came the idea that every subject matter in mathematics could be characterized in terms of a few essential facts—a short list of postulates, taken as truths, from which all other truths about the subject matter could be deduced. The rest is history. From over two millennia, from Euclid until Kurt Gödel in the 20th century, it has been assumed that an entire mathematical subject matter should follow from a small number of logically independent postulates or axioms. Following this influential view, axioms became the deductive source of all the properties of a given mathematical system (the theorems). When written symbolically in formal logic, a collection of axioms symbolically represents, in compact form, the essence of an entire mathematical system. Euclid's deductive method is still, today, the backbone of mathematics.

For more than a century now, axiomatizing mathematical subject matters has become a crucial enterprise in mathematics, serving as an engine for developing new mathematics: the axiomatization of different forms of geometry, number systems, different types of set theory, of statistics, and so on. Generations of mathematicians have developed entire careers seeking to find the smallest number of logically independent axioms for specific subject matters. The quest for the most appropriate and logically fruitful set of axioms for given subject matters became the ultimate goal for many who were investigating the foundations of mathematics. But, whereas Euclid understood his postulates for geometry to be meaningful to human beings, modern axiomatic mathematics has

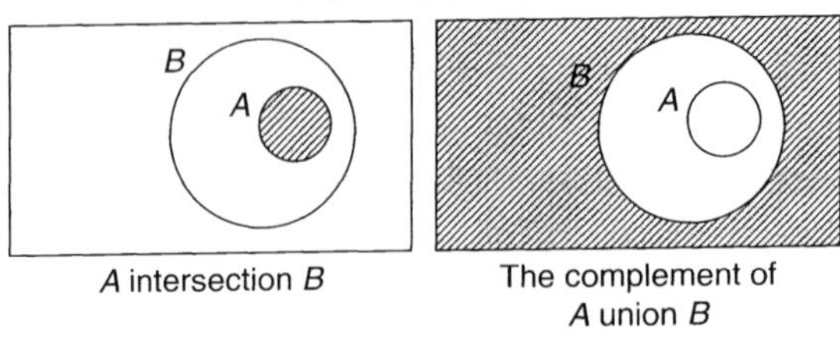

FIGURE 17.1 Venn diagrams representing the case "*A* is a subset of *B*." The diagram on the left depicts "*A* intersection *B*," and the one on the right "the complement of *A* union *B*." Venn diagrams implicitly assume a metaphorical conception of sets as container schemas, deriving their logics from the logic of conceptual container schemas. Members of set *A* are thus conceptualized as being *inside* the container or bounded region *A*, whereas non-members are *outside* of it (Lakoff & Núñez, 2000).

taken axioms as mind-free sequences of symbols. Most of modern mathematics today sees axioms as defined to be free of human conceptual systems and human understanding, characterizing the essence of each mathematical subject matter. But are axioms genuinely mind-independent? Are they simply meaningless strings of symbols? Recent developments in set theory provide rich insight into these questions.

Let us start with a simple question. What is a set? Intuitively, many people (including mathematicians) would say that a set is some kind of collection or aggregate.[3] Many authors speak of sets as "containing" their members and most students think of sets this way. Even the choice of the word "member" suggests such a reading, as do the Venn diagrams used to introduce the subject (Figure 17.1).

Implicit in this form of understanding of sets is the conceptual metaphor SETS ARE CONTAINER SCHEMAS, whose mapping and inferential organization is shown in the following table:

Source Domain Container schemas		Target Domain Sets
Display of container schemas and entities	→	The membership structure of a set
Interior of the container	→	The membership relation
The containers themselves	→	Sets
Entities inside a container	→	Members of a set

[3]The father of modern set theory, the German mathematician Georg Cantor (1845–1918) referred to it as *Menge*: "any collection into a whole (*Zusammenfassung zu einem Ganzen*) *M* of definite and separate objects *m* of our intuition or our thought" (Cantor, 1915/1955, p. 85).

If we operate with this conceptual metaphor, the understanding of Venn diagrams follows immediately. On the modern formalist view of the axiomatic method, however, a "set" is not a container but rather any mathematical structure that "satisfies" the axioms of set theory as written in symbols. The traditional axioms for set theory (the Zermelo–Fraenkel axioms) are often taught as being about sets conceptualized as containers. But if we look carefully through those axioms, we will find nothing in them that characterizes a container. The terms "set" and "member of" are both taken as undefined primitives. In formal mathematics, it means that they can be anything that fits the axioms. Here are the classic Zermelo–Fraenkel axioms, including the axiom of choice, what are commonly called the ZFC axioms.

- *The axiom of extension*: Two sets are equal if and only if they have the same members. In other words, a set is uniquely determined by its members.
- *The axiom of specification*: Given a set A and a one-place predicate, $P(x)$ that is either true or false of each member of A, there exists a subset of A whose members are exactly those members of A for which $P(x)$ is true.
- *The axiom of pairing*: For any two sets, there exists a set that they are both members of.
- *The axiom of union*: For every collection of sets, there is a set whose members are exactly the members of the sets of that collection.
- *The axiom of powers*: For each set A, there is a set $P(A)$ whose members are exactly the subsets of set A.
- *The axiom of infinity*: There exists a set A such that (1) the empty set is a member of A and (ii) if x is a member of A, then the successor of x is a member of A.
- *The axiom of choice*: Given a disjointed set S whose members are non-empty sets, there exists a set C that has as its members one and only one element from each member of S.

There is nothing in these axioms that explicitly requires sets to be containers. What these axioms do, collectively, is to *create* entities called "sets," first from elements and then from previously created sets. The axioms do not say explicitly how sets are to be conceptualized.

The point is that, within formal mathematics, where all mathematical concepts are mapped onto set-theoretical structures, the "sets" used in these structures are not technically conceptualized as container schemas. They do not have container schema structure with an interior, boundary, and exterior. Indeed, within formal mathematics, human ideas are not supposed to exist at all, and hence sets are not supposed to be conceptualized as anything in particular. They are undefined entities whose only constraints are that they must "fit" the axioms. For formal logicians and model theorists, sets are those entities that fit the axioms and are used in the modeling of other branches of mathematics. Of course, most of us do conceptualize sets in terms of container schemas (as in the case of Venn diagrams) and that is perfectly consistent with the axioms just described.

But when we conceptualize sets as container schemas, a constraint follows automatically: Sets cannot be members of themselves, as containers cannot be inside themselves. Strictly speaking, this constraint does not follow from the axioms but from our metaphorical understanding of sets in terms of containers. The axioms do not rule out sets that contain themselves. However, an extra axiom was proposed by the mathematician John von Neumann (1903–1957) that does rule out this possibility.

- *The axiom of foundation*: There are no infinite descending sequences of sets under the membership relation. That is, $\ldots S_{i+1} \in S_i \in \ldots \in S$ is ruled out.

Since allowing sets to be members of themselves would result in such a sequence, this axiom has the indirect effect of ruling out self-membership.

But despite the fact that this axiom is somewhat fixing the "self-containing problem" (by ruling out self-membership), certain model-theorists have found that for special cases they would like to preserve the possibility of allowing "self-membership." For example, consider an expression like

$$x = 1 + \cfrac{1}{1 + \cfrac{1}{1 + \cdots}}$$

If we observe carefully, we can see that the denominator of the main fraction has, in fact, the value defined for x itself. In other words, the above expression is equivalent to

$$x = 1 + \frac{1}{x}$$

Such recursive expressions are common in mathematics and computer science. The possibilities for modeling such expressions using "sets" are ruled out if the only kind of "sets" used in the modeling must be ones that cannot have themselves as members. And these mathematicians have pointed out that despite that "containment" in itself is not part of the Zermelo–Fraenkel axioms, still our implicit ordinary grounding metaphor that SETS ARE CONTAINER SCHEMAS gets in the way of modeling kinds of phenomena (especially recursive phenomena) like the one above. They realized that a new non-container metaphor (not based on what they called the "box metaphor") was needed for thinking about sets, and explicitly constructed one (see Barwise & Moss, 1991).

The idea is to use graphs, not containers, for characterizing sets. The kinds of graphs used are accessible pointed graphs or APGs. "Pointed" indicates an asymmetric relation between nodes in the graph, indicated visually by an arrow pointing from one node to another —or from one node back to that node itself (Figure 17.2). "Accessible" indicates that there is a single node, that is linked

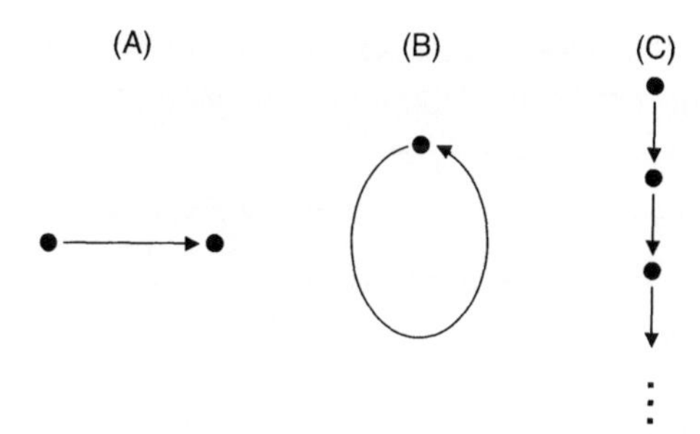

FIGURE 17.2 Hypersets: Sets conceptualized as graphs, with the empty set as the graph with no arrows leading from it. The set containing the empty set is a graph whose root has one arrow leading to the empty set (A). Illustration (B) depicts a graph of a set that is a "member" of itself, under the SETS ARE GRAPHS metaphor. Illustration (C) depicts an infinitely long chain of nodes in an infinite graph, which is equivalent to (B).

to all other nodes in the graph and can therefore be "accessed" from any other node.

From the axiomatic perspective, what has been done is to replace the Axiom of Foundation with another axiom that implies its negation, the "Anti-Foundation Axiom." But from the perspective of Mathematical Idea Analysis the creators of this new conception of "sets" have implicitly used a radically different conceptual metaphor, with graphs—not container schemas—as a source domain. The following table shows the mapping of such a powerful metaphor:

Source Domain Accessible pointed graphs		Target Domain Sets
An APG	→	The membership structure of a set
An arrow	→	The membership relation
Nodes that are tails of arrows	→	Sets
Decorations on nodes that are heads of arrows	→	Members
APG's with no loops	→	Classical sets with the Axiom of Foundation
APG's with or without loops	→	Hypersets with the Anti-Foundation Axiom

The effect of this conceptual metaphor is to eliminate the notion of containment from the concept of a "set." The graphs have no notion of containment built into them at all, and containment is not modeled by the graphs. Graphs that have no loops satisfy the ZFC axioms and the Axiom of Foundation. They, thus, work just like sets conceptualized as container schemas. But graphs that *do* have loops model sets that can "have themselves as members." They do not work like sets that are conceptualized as containers, and they do not satisfy the Axiom of Foundation. A "hyperset" is an APG that may or may not contain loops. Hypersets, thus, do not fit the Axiom of Foundation but rather another axiom with the opposite intent:

- *The anti-foundation axiom*: Every APG pictures a unique set.

With this example we can see the power of the embodied mechanism of conceptual metaphor in mathematics, playing a crucial role even at the foundational level of axiomatic systems. Sets, conceptualized in everyday terms as containers, do not have the right properties to model everything needed. So some mathematicians have metaphorically reconceptualized "sets" to exclude containment by using other more appropriate conceptual metaphors—certain kinds of graphs. The only confusing thing is that this special case of graph theory is still called "set theory" for historical reasons. Because of this misleading terminology, it is sometimes said that the theory of hypersets is "a set theory in which sets can contain themselves." From a cognitive point of view, this is completely misleading because it is not a theory of "sets" as we ordinarily understand them in terms of containment. The reason that these graph theoretical objects are called "sets" is a functional one: they play the role in modeling axioms that classical sets with the Axiom of Foundation used to play.

The moral is that mathematics has (at least) two *internally consistent* but *mutually inconsistent* metaphorical conceptions of sets: one in terms of container schemas and another in terms of graphs. And in both cases, corresponding axioms have been especially concocted to organize the inferential structure (theorems) of both kinds of "set" theory, namely, the Axiom of Foundation and the Anti-Foundation Axiom, respectively. Is one of these conceptions right and the other wrong? Are truths in one system "higher" than the truths in the other one? What axiom system is providing the ultimate truth about sets? A Platonist might want to think that there must be only one literal correct notion of a "set" transcending the human mind. But from the perspective of Mathematical Idea Analysis, these two distinct notions of "set" define different and mutually inconsistent subject matters, conceptualized via radically different human conceptual metaphors. Interestingly, in mathematics, cases like this one are more of a rule than an exception!

Let us now go back to the discussion about our ordinary forms for conceiving Time, and see how some of the embodied mechanisms that bring mathematics into being are the same that make everyday imagination possible.

EVERYDAY ABSTRACTION: THE EMBODIMENT OF SPATIAL CONSTRUALS OF TIME AND THEIR "AXIOMS"

Time, which for centuries has intrigued philosophers, physicists, and theologians, is a fundamental component of human experience. It is intimately related with everything we do, yet it is abstract, in the sense that we do not experience it directly as an isolated thing we can point to. Besides, our brains do not seem to have specific areas dedicated to process pure temporal experience in the way it does with, say, visual or auditory stimulation. Still, humans from all cultures must cope, implicitly or explicitly—with time-related entities, whether it is for cooking, dancing, hunting, traveling, or raising children. So, how do humans make up time concepts? As we saw earlier, the short answer is by treating "time" metaphorically as being *spatial* in nature, and one widespread form allows us to conceive the future as being in front of us, and the past behind us. This (mostly unconscious) way of thinking seems extremely obvious and natural, to the point that we barely notice that this is a major form of comprehension of temporal experience shared by many cultures around the globe. Even though nobody explicitly taught us this way of thinking about time, we master it effortlessly. It is simply part of who we are. This form of conceiving future and past, however, despite being spread across countless unrelated cultures around the world, is not universal! In collaboration with linguist Eve Sweetser from the University of California at Berkeley, we were able to reach this conclusion after studying in detail the conceptions of time in the Aymara people of the South American Andes (Núñez & Sweetser, 2006). This constituted the first well-documented case violating the postulated universality of the metaphorical orientation future-in-front-of ego and past-behind-ego.

Aymara, an Amerindian language spoken by nearly 2 million people in the Andean highlands of western Bolivia, southeastern Peru, and northern Chile, present a fascinating contrast to the well-known spatial-temporal mappings described earlier, and a clear challenge to the cross-cultural universals of metaphoric cognition studied so far. In Aymara, the basic word for "front" (*nayra*, "eye/front/sight") is also a basic expression meaning "past," whereas the basic word for "back" (*qhipa*, "back/behind") is a basic expression meaning "future". For example, *nayra mara*, whose literal translation is "eye/front year" means "last year," and *qhipa pacha*—"back time"—means future time. Many more temporal expressions in Aymara follow this pattern. But here is where, as cognitive scientists, we had to remain very cautious in reaching fast conclusions regarding possible exotic conceptions of time. To proceed, we needed to address two important research questions:

1. What exactly are the mappings involved in these metaphorical expressions?
2. Is there evidence of their psychological reality? That is, do Aymara people really *think* metaphorically in this manner, or are they simply using *dead* fossilized expressions with no inherent metaphorical meaning?

The first question pushed us to make further theoretical distinctions. In cases like "The election is *ahead* of us" and "the long Winter is now *behind* us" the terms "ahead," "behind", and so on, are defined relative to *ego*. In other words, ego is the *reference point* and therefore the conceptual metaphor described earlier—Time Events Are Things in Saggital Unidimensional Space—is said to be an instance of an Ego-reference-point (Ego-RP) metaphorical mapping. It is crucial not to confuse this mapping with another type called Time-reference-point (Time-RP), that underlies metaphorical expressions such as "the day *before* yesterday" or "revive your *post* summer skin," where morphemes like *fore* (front) and *post* (posterior) denote earlier than and later than relations, respectively.[4] The Time-RP mapping is the following:

Source Domain (unmarked) Unidimensional space		Target Domain Time
Objects	→	Times
Sequence of objects	→	Chronological order of times
Object *A* is in front of object *B*	→	Time *A* occurs earlier than time B
Object *A* is in behind object *B*	→	Time *A* occurs later than time B
Object *A* is co-located with object *B*	→	Time *A* occurs simultaneously with time *B*

This mapping is in many respects, simpler than the Ego-RP one. As it does not have an ego, it does not have a "now" in the target domain of time, and, therefore, it does not have built in the intrinsically deictic categories, past, present, and future. The Time-RP mapping has only earlier than and later than relationships. But when a particular moment is picked as "now," then "earlier than now" (past) and "later than now" (future) can be obtained. According to this mapping, however, "earlier than now" (past) gets its meaning from a "front" relationship, and "later than now" events (future) from a "behind" relationship. This may create confusion as in the case of the Ego-RP mapping the opposite seems to be happening: "front" (of us) means "future" and "behind" (us) means "past." The confusion, however, is immediately clarified by asking the following simple question: in front of *what*? or, behind *what*? Technically, this means identifying the underlying reference point. In "the day before yesterday," the reference point is "yesterday," "in front" of which is located the day the expression refers to. In "revive your post summer skin" the reference point is "summer," with the phrase targeting the times that follow the sunny season. To understand the Aymara case, we must keep this fundamental distinction between Ego-RP and Time-RP mappings clearly in our minds.

[4]For empirical evidence of the psychological reality of this metaphor, see Núñez & Sweetser (2006) (gestural), and Núñez et al. (2006) (priming experiments).

The crucial question we needed to address was: What are the reference points involved in the uses of *nayra* (front) and *qhipa* (back) in Aymara? That is, what is "in front of" or "behind," when these terms are used for temporal meaning? If the reference points are temporal entities such as "winter," "sunrise," "lunch time," or "rainy season," as opposed to "us" or "me," then there is absolutely nothing intriguing or exotic in the above Aymara temporal expressions. In such cases Aymara uses of "front" and "back" would be equivalent to English Time-RP cases like "the day before yesterday" or "post summer." In fact, this is what occurred with some Polynesian and African languages that had been claimed to be "special" with respect to space–time metaphors, but whose data, after proper analysis, turned out to be standard Time-RP cases (Moore, 2000). If in Aymara, however, the reference point is indeed ego, that is, "front of *us*" means past and "behind *us*" means future, then this finding would be critical since it would provide a counterexample to the largely universal Ego-RP mapping.

The second question we needed to investigate was how people—Aymara or otherwise—actually think about time. For this, we had to go beyond the mere analysis of words and their etymological roots. We needed to investigate empirically the psychological reality of these space–time metaphors, and ask: Do people actually *think* this way? Or perhaps the expressions simply used "dead" lexical items from a distant past that lost its original metaphorical meaning? And how can we tell?

Along with my colleagues, I addressed both these questions. Regarding the first one—the question of reference points—we quickly hit some dead ends. It turns out that in Aymara, due to grammatical reasons too involved to explain here, it is not possible to simply find markers like "us" in temporal uses of "front" (*nayra*) or "back" (*qhipa*). In short, using purely linguistic methods, we could not tell whether the Aymara expressions given earlier were Ego-RP or Time-RP. This is precisely the nightmare scenario for a scientist: not being able to provide an answer to the research question with the methods at hand. We therefore looked for other methods. The best candidate—which turned out to be essential in answering the second question as well—ended up being a long forgotten dimension of human language: Embodied spontaneous gestures.

Why gestures? When speaking, humans from all cultures spontaneously produce gestures. These are effortless but complex sequences of motor activity—especially hand movements—that are co-produced with speech. The study of human gestures, after being ignored in academic circles for a long time, has made a substantial progress over the last couple of decades. Research in a variety of areas, from child development (Bates & Dick, 2002) to neuropsychology (Kelly et al., 2002; McNeill, 2005), to linguistics (McNeill, 1992; Cienki, 1998), and to anthropology (Haviland, 1993), has shown the intimate link between oral and gestural production. Moreover, it is known that linguistic metaphorical mappings are paralleled systematically in gesture (Cienki & Müller, 2008; Núñez, 2006). We, therefore, could reliably ask what kinds of gestures Aymara speakers produce when uttering temporal expressions using "front" (*nayra*) or "back (*qhipa*).

(A)

(B)

FIGURE 17.3 The speaker, at right, is referring to the Aymara expression *aka marat(a) mararu*, literally "from this year to next year." (A) When saying *aka marat(a)*, "from this year," he points with his right index finger downward and then (B) while saying mararu, "to next year," he points backwards over his left shoulder. (©2008 Rafael Núñez. Published by Elsevier Ltd. All rights reserved) (See color plate)

Where are they pointing when doing so? What is the built-in reference point of such pointings?

In order to find out, in collaboration with Chilean colleagues Manuel Mamani and Vicente Neumann from the University of Tarapacá, and Carlos Cornejo from the Pontifical Catholic University of Chile, I conducted videotaped ethnographic interviews with Aymara people from the north-easternmost tip of Chile, up in the Andes, along the border with Bolivia. As we were interested in *spontaneous* gestures (with high ecological validity), the interviews were informal and were designed to cover discussions involving reference to time. Participants were asked to talk about, make comments, compare, and explain a series of events that had happened or that were expected to happen in the context of their communities. They were also asked to talk about traditional "sayings," anecdotes, and expressions in Aymara involving time and to give examples of them. To our amazement, what we found was that Aymara speakers gestured in Ego-RP patterns! Alongside the Ego-RP spatial language used to represent time as in front (*nayra*) and in back (*qhipa*) of ego, they gesturally represented time as deictically centered space: the speaker's front surface was essentially "now," as in English speakers' gestures (Figure 17.3a). The space behind the speaker was the Future (Figure 17.3b), whereas the space in front of the speaker was the Past (Figure 17.4).

Moreover, locations in front and closer to the speaker were more recent past times, while locations in front and farther from the speaker corresponded to less recent times. For instance, speakers contrasted "last year" with "this year" by pointing first at a more distant point and then at a nearer one. When talking about wider ranges of time, rather than particular points in time, we saw speakers sweeping the dominant hand forward to the full extent of the arm as they talked about distant past generations and times. In sum, our data showed, on the one hand, that the reference point in the above temporal expressions in Aymara

FIGURE 17.4 The speaker, at left, is talking about the Aymara phrase *nayra timpu*, literally "front time," meaning "old times." When he translates that expression into Spanish, as he says *tiempo antiguo* he points straight in front of him with his right index finger. (©2008 Rafael Núñez. Published by Elsevier Ltd. All rights reserved) (See color plate)

is indeed ego centered (our first question) and on the other hand, thanks to the analysis of gestures, that for Aymara speakers the Ego-RP metaphorical spatial conception of time has genuine psychological reality (our second question). With these empirical data at hand we were thus able to characterize the actual mapping of the Aymara form of the conceptual metaphor as shown in the following table:

Source Domain Saggital unidimensional space relative to ego		Target Domain Time
Objects in front of ego	→	Past times
Objects behind ego	→	Future times
Object co-located with ego	→	Present
The further away in front of ego an object is	→	The "further away" an event is in the Past
The further away behind ego an object is	→	The "further away" an event is in the Future

This analysis of Aymara language and gesture provides the first empirically demonstrated case of a counterexample to the largely spread space–time metaphors where "future" is conceived as being "in front" of ego and "past" behind ego. Aymara has the opposite pattern (and it may not be the only such culture). Beyond its anecdotal flavor, this finding is crucial as it shows that human abstraction is not pre-wired in the brain. It tells us that there is no single way for achieving abstraction, not even for a fundamental domain such as time. Human biology is certainly fundamental in providing the basis for human imagination. But, building on universal species-specific body morphology and neural organization, different aspects of bodily experience may be recruited for the systematic construction of more abstract concepts, which allow for plasticity and cultural variation. Regarding temporal metaphorical uses of front–back relationships, we tend to profile frontal motion. Based on this, our basic postulate (or "axiom") builds on prototypical frontal motion. If we walk (forward) at any given time we will reach a location that is in front of us, leaving behind us the original location. *That* location is reached in the future relative to the moment we started the action, with the initial position where we were initially (past) located behind us. Aymara people, however, although do walk in the same way as the rest of the world does, operate with a radically different postulate (or "axiom"). They profile a fundamentally different aspect of front–back features: what is seen (and therefore known), lies in front of the observer and behind them lies what is outside the visual inspection. These features parallel essential temporal properties, namely, past events are known, whereas future events are not. In Aymara, visual perception appears to play the leading role in bringing temporal concepts to being, and several data sources support this explanation, from evidential grammatical markers to special social practices and values.

The moral is that humans have at least two forms for conceiving time along a bodily front–back axis, which are—like in set theory—internally consistent but mutually inconsistent. These forms are defined by mutually exclusive ways of orienting the body in saggital unidimensional space, providing a radically different collection of truths. By profiling different aspects of bodily grounded experience we get one case with a built-in postulate ("axiom") that puts the observer "facing" the future and the other case with the very opposite postulate with the observer "facing" the past. Once the orientation of the observer is defined, a series of theorem-like entailments follow. Which one is the correct one? Where really is the past? In front of us? Behind us? Like in mathematics, no ultimate transcendental answer can be provided. Both forms have their own postulates (or axioms), and truth rests on the underlying embodied mappings that made these very abstractions possible.

CONCLUSION

We have analyzed two types of human abstract conceptual systems, one in the realm of contemporary technical mathematics and the other in a fundamental domain of human everyday experience—time. We have seen how, ordinary embodied cognitive mechanisms that sustain human imagination, such as conceptual metaphor, are essential in structuring the meaning and the inferential organization of these abstract conceptual systems. And we have seen how even the most fundamental and abstract building blocks of modern mathematics—axioms—find their grounding on human everyday understanding and embodied sense-making. There are, of course, many differences between technical abstraction and everyday abstraction: usually the former requires writing systems, whereas the latter can evolve with oral tradition alone; the former requires explicit (usually effortful) goal-directed instruction, whereas the latter does not; the former defines conceptual systems that usually are shared by specialized communities, whereas the latter tends to be spread over entire cultures and ethnic groups; and so on. Despite these differences, both forms of human abstraction share essential properties—embodiment, supra-individuality, and truth—that show the same origin: human imagination as realized in the body/mind of the human primate.

Embodiment: In this chapter, we showed how the abstract conceptual systems we develop are possible *because* we are biological beings with specific morphological and anatomical features. In this sense, human abstraction is *embodied* in nature. It is because we are living creatures with a salient and unambiguous front and back, for instance, that we can build on these properties and the related bodily experiences we have to bring forth stable and solid concepts such as "the future in front of us." This would not be possible if we had the body of, say, a jellyfish. Similarly, we can have the experiences and the understanding of containers because we have brain mechanisms as topographic maps of the visual field, center-surround receptive fields, and gating circuitry in which container schemas appear to be realized neurally (Regier, 1996). But whereas other non-human primates share these mechanisms with us, and have fronts and backs as well, it is the modern human primate that has an embodied cognitive apparatus, such as conceptual metaphor, that can systematically extend immediate bodily experiences to create imaginary notions like future-as-front-locations and sets-as-containers. Moreover, biological properties and specificities of human bodily grounded experience impose very strong constraints on what concepts can be created. Because of this, abstract conceptual systems are not "simply" socially constructed, as a matter of convention. Although social conventions usually have a huge number of degrees of freedom, many human abstract concepts do not. For example, the color pattern of the Euro bills was socially constructed via convention (and so were the design patterns they have). But virtually any color ordering would have done the job. In the case of metaphorical construals of time, not any source domain serves the purpose: human construals of time are *spatial*. And this is an *empirical* observation, not an arbitrary or speculative statement, since, as far as we know, there is no language or culture on earth where time is conceived in terms of thermic or chromatic source domains. Human abstraction is thus not merely "socially constructed." It is constructed through strong non-arbitrary biological and cognitive constraints that play an essential role in constituting what human abstraction is, from everyday ideas to highly sophisticated mathematics. Human cognition is *embodied,* shaped by species-specific non-arbitrary constraints.

Supra-individuality: We saw that to study the embodiment of conceptual systems, the level of analysis is situated above the individuals. The primary focus is not on how *single individuals* learn to use, say, conceptual metaphors, or what difficulties they encounter when they learn them, or how they may lose the ability to use them after a brain injury, and so on. The focus is on the characterization, across hundreds of linguistic expressions and other manifestations of meaning (e.g., gestures) of the structure of the inferences that can be drawn from these metaphors, which is available for a community of people operating with such mapping. For example, when English speakers hear "the winter is behind us," they can implicitly and effortlessly infer that the previous fall is not just behind them but *further away behind* them. Similarly, if they read "the election is ahead of us," they implicitly infer that the various effects of the political climate building up to the election are not only ahead of them but also *much closer in front of* them than the election itself. And if we operate with the metaphor SETS ARE CONTAINER SCHEMAS, then we implicitly know that an object cannot be, both, inside and outside a container, and therefore, via the metaphor, that an element cannot be a member and a non-member of a set. The focus of embodied idea analysis (mathematical or other) is thus situated at a supra-individual level, at the level of the mappings and networks of metaphorical inferences. Large communities sharing networks of metaphors and mappings constitute cultures. In what concerns everyday ideas such as time these cultures may naturally coincide with ethnic groups located geographically in specific places (e.g., Aymara people in the Andes) but in mathematics, irrespective of ethnicity or geography, one could speak of the culture of mathematicians practicing set-theory with ZFC axioms and the other one practicing with hypersets and the Anti-Foundation axiom.

Truth: One of the most important morals of this chapter is that when imaginary entities are concerned, truth is always relative to the inferential organization of the mappings involved in the underlying conceptual metaphors. "Last summer" can thus be conceptualized as being *behind us* as long as we operate with the general conceptual metaphor TIME EVENTS ARE THINGS IN SAGGITAL UNIDIMENSIONAL SPACE, which determines a specific bodily orientation with respect to metaphorically conceived events in time—future as being "in front" of us and the past as being "behind" us. As we saw, this way of conceptualizing time, although spread worldwide, is not universal. For an Aymara speaker from the Andes' highlands, it is *not true* that the sentence "The Winter is *behind* us" refers to an event that has already occurred. In fact that sentence means the very opposite, namely, that the Winter has not taken place yet! Aymara people operate with a different conceptual time–space metaphor, which provides a different set of truths. The same occurs with sets and hypersets in contemporary set-theory. They have different collections of truths, characterized by different collection of axioms. The moral is that there is no *ultimate truth* regarding human imaginative structures. In the cases we saw, there is no ultimate truth about where, really, lies the ultimate metaphorical location of the future (or the past) or whether sets can allow self-membership. Truth depends on the details of the mappings of the underlying conceptual metaphor. This turns out to be of paramount importance when mathematical concepts are concerned: their ultimate truth is not hidden in the structure of the universe, but it rests on the underlying embodied conceptual mappings used to create them.

REFERENCES

Barwise, J. & Moss, L. (1991). Hypersets. *The Mathematical Intelligencer, 13*(4), 31–41.

Bates, E. & Dick, F. (2002). Language, gesture, and the developing brain. *Developmental Psychobiology, 40,* 293–310.

Cantor, G. (1915/1955). *Contributions to the Founding of the Theory of Transfinite Numbers*. New York: Dover.

Cienki, A. (1998). Metaphoric gestures and some of their relations to verbal metaphoric counterparts. In J.-P. Koenig (Ed.), *Discourse and Cognition: Bridging the Gap* (pp. 189–205). Stanford, CA: CSLI.

Cienki, A. & Müller, C. (Eds.). (2008). *Metaphor and Gesture*. Amsterdam: John Benjamins.

Fauconnier, G. & Turner, M. (1998). Conceptual integration networks. *Cognitive Science, 22*(2), 133–187.

Fauconnier, G. & Turner, M. (2002). *The Way We Think: Conceptual Blending and the Mind's Hidden Complexities*. New York: Basic Books.

Haviland, J. B. (1993). Anchoring, iconicity and orientation in Guugu Ymithirr pointing gestures. *Journal of Linguistic Anthropology, 3*, 3–45.

Johnson, M. (1987). *The Body in the Mind: The Bodily Basis of Meaning, Imagination, and Reason*. Chicago, IL: University of Chicago Press.

Kelly, S., Iverson, J., Terranova, J., Niego, J., Hopkins, M., & Goldsmith, L. (2002). Putting language back in the body: Speech and gesture on three time frames. *Developmental Neuropsychology, 22*, 323–349.

Lakoff, G. (1993). The contemporary theory of metaphor. In A. Ortony (Ed.), *Metaphor and Thought* (2nd ed.). New York: Cambridge University Press.

Lakoff, G. & Johnson, M. (1980). *Metaphors We Live By*. Chicago, IL: University of Chicago Press.

Lakoff, G. & Johnson, M. (1999). *Philosophy in the Flesh*. New York: Basic Books.

Lakoff, G. & Núñez, R. (1997). The metaphorical structure of mathematics: Sketching out cognitive foundations for a mind-based mathematics. In L. English (Ed.), *Mathematical Reasoning: Analogies, Metaphors, and Images*. Mahwah, NJ: Erlbaum.

Lakoff, G. & Núñez, R. (2000). *Where Mathematics Comes From: How the Embodied Mind Brings Mathematics into Being*. New York: Basic Books.

Lerman, S. (1989). Constructivism, Mathematics and Mathematics Education. *Educational Studies in Mathematics, 20*, 211–223.

Moore, K. E. (2000). Spatial experience and temporal metaphors in Wolof: Point of view, conceptual mapping, and linguistic practice., Unpublished doctoral dissertation, University of California, Berkeley.

Núñez, R. (1999). Could the future taste purple?. In R. Núñez & W. Freeman (Eds.), *Reclaiming Cognition: The Primacy of Action, Intention, and Emotion* (pp. 41–60). Thorverton, UK: Imprint Academic.

Núñez, R. (2005). Creating Mathematical Infinities: The Beauty of Transfinite Cardinals. *Journal of Pragmatics, 37*, 1717–1741.

Núñez, R. (2006). Do real numbers really move? Language, thought, and gesture: The embodied cognitive foundations of mathematics. In R. Hersh (Ed.), *18 Unconventional Essays on the Nature of Mathematics* (pp. 160–181). New York: Springer.

Núñez, R. & Lakoff, G. (1998). What did Weierstrass really define? The cognitive structure of natural and e-d continuity. *Mathematical Cognition, 4*(2), 85–101.

Núñez, R. & Lakoff, G. (2005). The cognitive foundations of mathematics: The role of conceptual metaphor. In J. Campbell (Ed.), *Handbook of Mathematical Cognition* (pp. 109–124). New York: Psychology Press.

Núñez, R. & Sweetser, E. (2006). With the future behind them: Convergent evidence from Aymara language and gesture in the crosslinguistic comparison of spatial construals of time. *Cognitive Science, 30*, 401–450.

Núñez, R., Motz, B., & Teuscher, U. (2006). Time after time: The psychological reality of the ego- and time-reference-point distinction in metaphorical construals of time. *Metaphor and Symbol, 21*, 133–146.

McNeill, D. (1992). *Hand and Mind*. Chicago, IL: University of Chicago Press.

McNeill, D. (2005). *Gesture and Thought*. Chicago, IL: University of Chicago Press.

Regier, T. (1996). *The Human Semantic Potential*. Cambridge: MIT Press.

Sweetser, E. (1990). *From Etymology to Pragmatics: Metaphorical and Cultural Aspects of Semantic Structure*. New York: Cambridge University Press.

18

EMBODIMENT FOR EDUCATION

ARTHUR M. GLENBERG

Department of Psychology, Arizona State University, Tempe, AZ, USA
Department of Psychology, University of Wisconsin, Madison, WI, USA

> Watching a child makes it obvious that the development of his mind comes about through his movements ... Mind and movement are parts of the same entity.
>
> ...Maria Montessori (1967)

It is clear that one of the 20th century's greatest educational thinkers believed that there is a close connection between the body and education. But why should we think in the same line? The answer that I will develop in this chapter will be in two parts. After a brief discussion of embodiment theory, I will first briefly review data showing an intimate connection between the body and simple mathematics. Second, I will spend considerably more time reviewing data from a research project investigating a reading intervention based on an embodied theory of language. This intervention has been successfully applied across various populations of young readers, and we are beginning to explore its application in learning abstract concepts in science.

WHY EDUCATION?

The essence of embodied theories of cognition is that the body, particularly bodily systems that have evolved for perception, action, and emotion, contribute to "higher" cognitive processes. Many of these cognitive processes are important to education, such as language comprehension, reading, mathematics, and scientific thinking. Thus, the classroom offers a fertile ground for observing effects of embodiment and testing theories.

But there is another reason for putting embodiment and education together. Consider why many modern societies have great belief in and respect for science; because it works. For example, why do lay people think that physicists are pursuing something worthwhile? It is not because lay people have a clear understanding of esoteric theories; instead it is because modern physics has produced tremendous achievements that lay people can use, such as computers and television, as well as tremendous achievements that we can admire, such as traveling to the moon. Similarly, why do societies credit biological sciences? Because those biologists created amazing advances in healthcare. By analogy, what will lead societies to value cognitive science? It will be a demonstration of its practical applications, and those practical applications are likely to be in education. If embodiment theory can lead to the educational equivalent of a moon landing or a polio vaccine, it will demonstrate both its worth to society and the likelihood that it is the correct approach to understanding cognition.

EMBODIED MATHEMATICS

There are good reasons to believe that there is a strong relation between embodied mechanisms and mathematics. Some of that research will be reviewed here (and see Chapter 7). Nonetheless, this section will be relatively brief because educational interventions for mathematics based on embodiment theory have not yet emerged.

MATHEMATICS AND ACTION SYSTEMS

It is not a news that the hand is used by children in learning to count. But is the association between hand and number also found in adults? And, does the hand play a role in mathematical cognition, or is the association purely epiphenomenal?

The first question can be answered in the affirmative. Several reports using transcranial magnetic stimulation (TMS) have demonstrated a close relation between mathematical and motor processes in adults. TMS uses a hand-held electromagnet that is positioned on the scalp. When pulsed, the magnetic field penetrates the scalp, the skull, and outer parts of the cortex, and it thereby induces an electrical current in neurons. Repetitive application of TMS can be used to temporarily alter the functioning of the stimulated area. Single pulses, particularly in motor areas of the brain, can be used to measure how a cognitive task modulates cortico-spinal activity. For example, when the magnetic field stimulates areas of cortex that control the hand, measurable EMG activity can be recorded from muscles in the hand (and a strong enough pulse to the magnet generates overt movement). This EMG activity is referred to as a motor evoked potential (MEP). Thus, if a cognitive task modulates the MEP evoked by TMS, it can be inferred that the task influences motor areas of cortex (or more appropriately, the cortico-spinal system).

Andres et al. (2007) used TMS to uncover a relation between adult counting and the hand. In their experiments, participants either counted the number of dots in a semi-circular array, or determined if two adjacent dots had the same color (control task). During the task, TMS pulses were delivered to hand, arm, or leg areas of cortex and MEPs were recorded from hand, arm, or leg muscles, respectively. They found that counting the dots (relative to the control task) increased MEPs measured in the hand, but not the arm or leg. Furthermore, a subsequent experiment ruled out the possibility that the effect was produced solely by subvocal articulation. Namely, mentally reciting numbers (without counting) did not affect MEPs.

The Andres et al. data clearly demonstrate an association between counting and motor system activity, particularly for the hand. Given that many children may learn to count by enumerating with their fingers, this result may not be very surprising. For this reason, Sato et al. (2007) chose a numerical task not easily associated with hand-based enumeration, namely parity (odd/even) judgments. Participants were shown single digits and responded orally with the parity. Shortly after presentation of the digit, a TMS pulse was delivered over the cortex controlling the left or right hand. The major finding was that right hand MEPs were affected by the parity task for small numbers. Thus, the data demonstrate that a mathematical task affects motor system even when there is no need for explicit counting.

Lindemann et al. (2007) also used the parity judgment task, but without TMS. Participants were required to make the parity judgment by grasping either a large (6 cm in diameter) wooden object using a power grip or a small (0.7 cm in diameter) object using a precision grip. The major finding was that parity judgments on large numbers were faster using the power grip, whereas the judgments on smaller numbers were faster using the precision grip. Thus again, it appears that there is a connection between the hand and simple mathematics in adults.

From the point of view of many theories of mathematical cognition (McCloskey, 1992; Anderson, 2005) the results are close to bizarre. That is, mathematics has been conceptualized as rule-like manipulation of abstract symbols that have no direct connections to perception or action. The data reviewed earlier indicate that this abstractionist account of mathematics must be wrong in at least some details. Nonetheless, one can still question if the embodiment effects are functional or not, and whether the action system is literally used in mathematical cognition, or do the effects simply reveal a residual activation of the hand based on early experience? Some evidence along these lines is presented in the next section.

MATHEMATICS AND GESTURE

Several studies have demonstrated causal links between action, in the form of gesture, and classroom performance (see Nathan, in press for a thorough review). Perhaps the strongest of these is the study by Wagner Cook et al. (2008). In this experiment, children learned to solve problems such as $4 + 9 + 3 = 4 + ?$ In one condition, children were taught a problem-relevant gesture (sweeping the hand under the left side and then the right side) to perform while solving the problem.

In another condition, the children were taught a verbal statement, "I want to make one side equal to the other." Four weeks later, children were tested again, and those who had been taught the gesture were significantly more likely to maintain learning gains than children who were taught the verbal statement.

MATHEMATICS AND PERCEPTION

The abstractionist account of mathematics disavows connections between mathematics and perception as well as action. According to this account, perceptual systems are used to encode the mathematical information, but then the cognitive processes are independent of any perceptual information such as modality of presentation. Several research programs demonstrate that this independence is not found.

Campbell (1994) (Campbell & Fugelsang, 2001) presented the participants with relatively simple problems for solution (e.g., $3 + 4 = ?$) and verification (e.g., $3 + 4 = 8$), using either Arabic numerals or words (e.g., three + four = ?). As might be expected, given the differential size and familiarity of the stimuli, the word format resulted in people taking longer to solve the problems. More importantly, participants reported that more calculation was needed with the words, and that this became proportionally greater as the size of the numbers increased (Campbell & Fugelsang, 2001). In addition, Campbell (1994) found that people made different calculation errors with the two formats and that the two formats resulted in different patterns of priming from one trial to the next. Thus, perceptual format appears to affect not just peripheral encoding, but also calculation processes.

A similar conclusion was reached by Goldstone et al. (in press). In their experiments, participants judged if Equations [e.g., (18.1) and (18.2)] were correct. An important component of the judgment was the order of operations (e.g., multiplication before addition). This was explained to the participants and they were given feedback on their performance that depended on the participant's proper use of the order of operations. That is, there was no ambiguity regarding the task.

$$R * E \quad + \quad L * W = L * W \quad + \quad R * E \tag{18.1}$$

$$R * \quad E + L \quad * W = L * \quad W + R \quad * E \tag{18.2}$$

Both Equations (18.1) and (18.2) are correct, but participants found it easier to confirm this when the perceptual contiguity matched that order of operations, as in Equation (18.1), than when they mismatched as in Equation (18.2). That is, in Equation (18.1) the terms that are multiplied (e.g., R and E) are closely grouped and separated from the terms that should be added. In Equation (18.2), however, the spatial arrangement suggests that E and L should be processed together.

The grouping effect can be found even with quite subtle manipulations. Consider, for example, when the order of operations is consistent with alphabetical proximity, as in Equation (18.3), or alphabetical proximity is inconsistent with the order of operations, as in Equation (18.4).

$$A * B + X * Y = X * Y + A * B \tag{18.3}$$

$$A * X + Y * B = Y * B + A * X \tag{18.4}$$

Participants were more likely to err when judging Equations such as (18.4). Again, the conclusion is that mathematical processing does not discard perceptual information.

In some ways, these studies are stronger than the TMS studies reviewed earlier. The TMS studies demonstrated that mathematical operations and actions are correlated, but not that action systems are literally used in doing mathematics. The Campbell and Goldstone studies do show a causal connection between perceptual format and mathematical problem solving.

All of these studies point to the same conclusion: Mathematics is not the cognitive manipulation of abstract symbols by rules. Instead, mathematical problem solving makes use of representations based on bodily systems of action and perception. Consequently, it seems reasonable to expect that teaching strategies that capitalize on the embodied nature of mathematics would be successful. To date, however, those strategies have not been developed and tested. The situation is different in the domain of reading comprehension, to which we turn next.

EMBODIED READING

My approach to developing a reading intervention is based on an embodied account of language comprehension, the indexical hypothesis (IH; Glenberg & Robertson, 1999, 2000; Glenberg & Kaschak, 2002). According to the IH, three processes are used to understand a sentence. The first is indexing (mapping) words and phrases in a sentence to objects in the environment or perceptual symbols. Barsalou (1999) defines a perceptual symbol as a representation based on neural activity in perceptual areas of the brain. Thus, activating a perceptual symbol provides much the same information as that apprehended during experience with the relevant objects and events. For example, in the sentence "Art stood on the chair to change the bulb in the ceiling fixture," the phrase "the chair" is mapped either to an actual chair in the comprehender's environment or to a perceptual symbol of a chair.

The second process is deriving affordances from the indexed objects. Affordances are possibilities for interaction between a particular biological system and a physical situation (Gibson, 1979). Thus, kitchen chairs afford both sitting-on and standing-on for adults, whereas beanbag chairs afford only sitting-on.

The third process is meshing (integrating) the affordances as directed by syntax. The syntax of the example sentence indicates that "Art" is standing on the chair rather than vice versa. The mesh process takes into account how affordances or actions can be combined while respecting biological constraints on action. Because a human can stand on a chair while holding a light bulb (i.e., the affordances can be meshed), the example sentence can be understood. Thus the processes of indexing, deriving affordances, and meshing the affordances ground the abstract language symbols (words and syntax) in a sensorimotor representation of what the language is about. This approach to language comprehension has received strong support (e.g., Glenberg & Robertson, 1999, 2000; Borghi et al., 2004; Chambers et al., 2004; Zwaan & Taylor, 2006).

The indexical hypothesis provides a rationale for why children have difficulty with symbolic information when reading—namely, the children have not learned to index *written* symbols to grounded representations. When children are learning a natural language, symbols are indexed and grounded immediately (Masur, 1997). For example, when a caregiver says to an infant, "Here is your bottle," invariably (in the United States) the caregiver will point to or display a bottle. [Tomasello (2003) notes that explicit indexing for infants is not part of all cultures. However, mechanisms of joint attention ensure that the infant can induce the referents of many words.]

In contrast, when learning to read, children must concentrate on the (initially) laborious process of decoding print into sound. The objects read about are not in the environment and are rarely illustrated (and if illustrations are provided, reference to them is haphazard). Even when the child succeeds in pronouncing a word, the prosody may be so different from that in conversation that the laboriously pronounced word does not strongly activate appropriate perceptual symbols. For a child in this situation, reading becomes an exercise in naming ungrounded, and hence meaningless, symbols.

A similar analysis applies to older children (and adults) reading in unfamiliar domains (e.g., science). If the words are not adequately indexed, the material will be, at best, difficult to understand. Finally, the analysis holds for mathematical operations. To the extent that the numbers and symbols of mathematics are ungrounded, children will have a hard time understanding how those symbols can be applied in situations other than rote symbol manipulation.

To summarize, I propose that experience when learning a natural language leads to indexing of words, phrases, and grammatical constructions to objects and events, thereby grounding the symbols and imbuing them with meaning. In contrast, typical experiences in learning to read (e.g., concentrating on letter-to-sound correspondences) does not encourage indexing to objects and events, and may even work against it.

PHYSICAL AND IMAGINED MANIPULATION AS A READING INTERVENTION

Our intervention is designed to directly illustrate to young readers the indexing process, how comprehension flows from indexing, and to assist them in

Breakfast on the farm

Ben needs to feed the animals.
He pushes the hay down the hole.
The goat eats the hay.
Ben gets eggs from the chicken.
He puts the eggs in the cart.
He gives the pumpkins to the pig.
All the animals are happy now.

FIGURE 18.1 The farm toys and one text. The green traffic light signaled that a sentence was to be reread or used to direct manipulation. (See color plate)

developing skill at indexing. The procedure has two main components, physical manipulation (PM) and imagined manipulation (IM). With PM, children read a text about activities in a particular situation (e.g., on a farm), and toys representing the important characters and objects (e.g., a toy barn, animals, tractor, farmer) are simultaneously available (Figure 18.1). After reading a critical sentence, the child is cued (by the image of a green traffic light) to manipulate the toys to correspond to the sentence. This manipulation ensures that the words are indexed to objects, affordances derived (the child must manipulate the toys), and the concepts meshed to simulate the sentence. Thus, PM ensures grounding of the symbols. For young readers in the first and second grades, PM produces gains in recall and comprehension of 1.5–2.0 standard deviations compared to children who read and reread the texts without PM (Glenberg et al., 2004).

Substantial reading comprehension also materialized in additional experiments with second- and third-grade Native American learners (Marley et al., in preparation). PM can also be used in small reading groups in which one child reads a sentence and manipulates followed by a different child who reads the next sentence and manipulates (Glenberg et al., 2007a). In these reading groups, the gains hold for information from sentences that the child has manipulated as well as for information from sentences the child has watched others manipulate. Thus, the procedure is applicable in classrooms where it would be impractical to have toy objects for every child.

One might suspect that the theory would necessarily predict that children in the PM condition with literal interaction with the toys would outperform children who simply observed manipulation. However, recent work on the human mirror neuron system (Rizzolatti & Craighero, 2004) has demonstrated that action systems can be activated whether one takes literal action or simply observes action. Consequently, embodiment theory need not predict a difference between manipulation and observation conditions (although I will describe a subtle difference later).

Following PM, children are trained in IM by being asked to imagine manipulating the toys. With the very young children, the training requires them to describe what they imagine so that the researcher can correct misconceptions. Thus, children are corrected if they (a) simply repeat information in the sentence, (b) describe a mental image without describing actions, or (c) do not provide details about how an action could have taken place. (The overt description of the content of IM is required only during training, not during the application of IM.) We believe that this sort of training is more effective than having the child simply imagine a static situation (i.e., form a visual image of the situation).

We have found that IM leads to effects of comparable magnitude to PM, and that for the youngest children, IM can be used effectively at least 1 week after the child has used PM (Glenberg et al., 2004). Furthermore, for third-grade children, IM is effective for at least 2 weeks after initial training, and it can be applied to texts about situations with which the child has had no PM experience (Glenberg et al., 2007b).

HOW IM WORKS

Both empirically and theoretically, IM following PM is different from simply providing visual information and instructions to make mental images. First, consider how PM works. Undoubtedly, part of the effect stems from the formation of visual and motoric representations in addition to a verbal one. These codes result in part from the process of indexing, as specified by the indexical hypothesis. Another reason PM works is that it forces the reader to consider how all of the parts of the sentence fit together. In brief, by requiring real action with real objects in real time, PM forces the reader to consider who did what to whom and when. This is the process of meshing (integrating) described by the indexical hypothesis. The success of IM, particularly following PM, trades on these processes, and is why IM is so effective, even compared to visualization strategies. For example, Marley et al. (in preparation) compared three conditions with third-grade, Native American English Language Learning (ELL) students. Two of the conditions were Reread and PM. The third condition was an Observe condition in which the children observed the experimenter manipulate the toys. When children literally manipulated or observed manipulation, children in the PM and Observe condition outperformed those in the Reread condition. When children were asked to engage in IM, however, only children who had previously engaged in PM outperformed those who observed the experimenter manipulating the toys.

Thus, IM is different from simply forming visual images in several respects. First, the instruction can be made very clear, because children are asked to imagine how they would act to create the situation, and these are actions that they have performed during PM. Second, to the extent that there is a close relation between language and action (as reviewed earlier), engaging the motor system will lead to a particularly appropriate type of encoding. Third, the images (visual and motoric) the child creates are dynamic, and thus capture many grammatical relations (e.g., who is acting on whom) and relations between narrative episodes. Finally, teaching IM appears to teach a skill, namely the skill of creating mental models from text.

COMPARISON TO OTHER WORK ON CONCRETE MANIPULATIVES

A finding in much of the developmental and educational literature is that positive effects of concrete manipulatives are inconsistently obtained (Uttal et al., 1997; Uttal, 2003). In contrast, we have found that concrete manipulatives can lead to enormous benefits in reading comprehension as well as in some mathematical problem-solving contexts. There are three reasons why this contrast is more apparent than real. First, Uttal (2003) noted that when children are taught mathematics using manipulatives, they have a difficult time transferring that knowledge to written forms of representation. In our procedures, the manipulatives and the written information are combined, so that children can integrate the firsthand and secondhand knowledge (Schwartz et al., 2005).

Second, Uttal notes that concrete objects make for ineffective learning aids when children must treat the concrete object as a symbol—for example, when using a large block to symbolize 10 units. In cases such as this, children have a hard time dissociating the concrete uses of the object (e.g., that it can be stacked with other blocks to form a tower) from its symbolic use. When using PM and IM, the concrete manipulatives are not treated as symbols. Instead, the manipulatives are the physical situation to which the symbols (i.e., words) refer. Thus, when reading about animals on a farm, children using PM are reading about the particular farm toys that are in front of them. And when children are reading using IM, they map the words onto perceptual symbols learned from interacting with the farm toys. Similarly, when children are solving story problems about animals at a zoo, the problem is about the animals in front of them. In this way, the manipulatives are not symbols but they are ground.

Third, difficulties with the use of manipulatives have been investigated in the context of mathematics. In the main, PM and IM are procedures for enhancing reading comprehension. As far as I know, use of manipulatives in reading contexts has not lead to any difficulties in symbol use or understanding.

One might suspect that the use of PM and IM would lead to rather brittle knowledge. After all, how often do we have just the right toys in front of us while reading? However, our work demonstrates otherwise (Glenberg et al., 2007b). That is, once children have learned IM in a particular context, it is relatively easily transferred to other domains. The reason appears to be that the combination of PM and IM teaches children the general skill of how to ground abstract symbols (words and mathematical symbols) in their experiences (i.e., to create embodied mental models).

RELATION TO CURRENT EDUCATIONAL PRACTICE

Many classroom teachers already use manipulatives in teaching. Thus, how does the use of PM and IM differ from business as usual in the classroom? The first answer rests on the nature of the manipulatives. For example, Glenberg et al. (2007b) compared PM using story-relevant manipulatives (e.g., toy balloons in a fair scenario) to PM using abstract manipulatives (Lego pieces) which were meant to simulate the sort of counting aids used in many classrooms. Children in both conditions were given the same instructions—namely, to act out the mathematical story problem by counting, using the manipulatives. Nonetheless, children using story-relevant manipulatives outperformed the children using the abstract manipulatives. We believe that this effect arose because in the story-relevant condition the children were likely to use the manipulatives to create a representation of the problem world that constrained the mathematical operations. In the abstract-manipulatives condition, the children simply counted the Legos without any attempt to model the problem world. These findings are consistent with previous research that has demonstrated that the experience with different types of manipulatives can have differential effects on learning (Chao et al., 2000; Martin & Schwartz, 2005).

Perhaps most importantly, although manipulatives are commonplace during the elementary math lesson, they are rarely found during the elementary reading lesson. Thus the importance of PM and IM is that they promote skill in symbol manipulation in an area essential for learning, reading comprehension.

PM AND IM WITH ENGLISH LANGUAGE LEARNING CHILDREN

Marley et al. (2007) demonstrated benefits of PM for listening comprehension for learning-disabled Native American children. Marley et al. (in preparation) extended this research to non-disabled, third-grade Native American children all of whom had been identified by their teachers as having limited English proficiency. The school district serving the children in this sample had not made adequate yearly progress, as defined by the *No Child Left Behind Act of 2001 (NCLB),* since the enactment of the act. In addition, 88% of the children in the three elementary schools received free or reduced lunch.

Children were randomly assigned to Reread, PM, or an Observe condition in which the children observed the experimenter manipulate the toys. When children literally manipulated or observed manipulation, they outperformed children in the Reread condition. Using Cohen's *d* as a measure of effect size (i.e., the number of standard deviations between the means of the two conditions), the differences are substantial, with *d*s = 1.12 and 0.84 for the contrast of Reread to PM and Observe, respectively. When children were asked to engage in IM, however, only children who had previously engaged in PM outperformed those who Reread, with $d = 1.07$ (when toys were present during IM) and $d = 1.08$ (when toys were not present during IM). These data illustrate the effectiveness of PM for ELL students, and the students' ability to transfer what they learned during PM to new stories using IM. In addition, the data demonstrate that for these unskilled readers, PM is needed to ensure the effectiveness of IM.

PM AND IM AND VOCABULARY ACQUISITION

Reading consists of many processes in addition to comprehension. Children must learn the alphabetic principle, they must develop strong phonological skills, and they must develop fluency and a sight vocabulary. Thus, an important question is whether PM and IM interfere with the development of these other skills, or whether PM and IM might benefit the development of those skills.

To study the effect of PM and IM on vocabulary acquisition, the stories from previous reading experiments were adapted to contain two new pseudo-words each: one singular noun (e.g., "thrabe") and one present tense, third-person verb (e.g., "skigs"). Before reading a story, all children were introduced to both of the new words contained in the story: The experimenter pronounced the pseudo-words, used them in context, and used PM to define them. Then the children read and pronounced the words, read the words in context, and used manipulatives to act out sentences containing each new word.

Following the introduction to the new words, all children read a total of four stories (using PM or Reread) in each of two sessions. Thus, the experimental question is whether PM will result in better retention of the new words compared to the Reread control condition. At the end of both sessions, children were tested on item and action recognition. The children were shown the novel objects and actions and asked if they had seen the objects before and whether they remembered the names. Next, the children were asked to read a sentence out loud. Each sentence included one of the pseudo-words (in bold). After reading the sentence and pronouncing the word, the child was asked if the bold word made sense in the sentence (half of the sentences made sense based on the meanings learned) and why. At the end of the second session, they were tested on memory of word definitions, where the experimenter read a word, and the child stated whether it was heard before or not. When the child said "yes," the child was asked what the word meant.

Data from 31 second-grade children confirmed that children who practiced PM showed better comprehension for story events ($M = 0.87$) than those who Reread ($M = 0.72$). Also, children who manipulated were better able to determine whether a pseudo-word was used correctly in a novel sentence ($M = 0.91$) than children who Reread ($M = 0.85$). Finally, children assigned to PM were able to more accurately define words ($M = 0.65$) than their counterparts who

Reread (M = .38). In summary, the students in the PM condition were better able to remember story events, determine correct usage of pseudo-words, and define them in context, than the children who Reread. Thus, PM appears to be a useful tool for creating better memory for new words as well as supporting reading comprehension.

PM AND IM IN SCIENCE EXPOSITION

Our previous work has demonstrated benefits of PM and IM for young children reading narratives. Is the technique also effective for older children reading exposition about abstract ideas? The answer is important for both theoretical and practical reasons. Theoretically, evidence for an embodied approach to cognition is strongest when applied to concrete concepts. Although there is some work extending the approach to more abstract ideas (Boroditsky & Ramscar, 2002; Richardson et al., 2003; Glenberg et al., 2008), the extent to which abstract concepts can be grounded in bodily experiences is still an open question.

The practical significance comes about because around the fourth grade, children experience particular difficulties in comprehending expository text (Armbruster & Anderson, 1988; Beck et al., 1997). Also, young readers face another challenge—the relative lack of expository or informational reading opportunities at the elementary level (Pressley et al., 1996; Morrow & Pressley, 1997; Duke, 2000) and a preference among teachers to select narrative texts for reading instruction (Donovan & Smolkin, 2001). Given that learning science, mathematics, and other content areas often rely on comprehension of expository text (Sweet & Snow, 2003), it is not surprising that many students are not up to these challenges. Thus, we have begun to explore the efficacy of PM and IM for understanding abstract expository text.

The application of PM and IM in exposition is based on research conducted primarily by Klahr and associates (e.g., Chen & Klahr, 1999; Toth et al., 2000; Klahr et al., 2001; Triona & Klahr, 2003) and a theoretical idea developed by Schwartz et al. (2005). Klahr has worked extensively with the control-of-variables strategy (CVS). CVS is the idea underlying experimentation, namely, that all (confounding) variables should be controlled to determine if the independent variable has an effect. Two important points that have emerged from Klahr's work are pertinent here: (1) PM is effective for learning CVS (although manipulation of pictorial representations on a computer screen can be equally effective) and (2) direct instruction on the CVS principle is more effective than pure discovery learning.

From Schwartz et al. (2005), we take the idea of the importance of combining firsthand (experiential) knowledge with secondhand (derived from language) knowledge. While recognizing the necessity of symbol grounding, Schwartz et al. also note that most formal learning is secondhand, mainly through reading. Hence, an important question is, how should firsthand and secondhand knowledge be integrated so that reading by itself becomes an effective mode for learning?

Our proposed answer is that PM provides the firsthand knowledge, and IM provides the skill needed to extend that knowledge to reading when physical objects are not available.

One additional idea is relevant for understanding the design of the experiment. Part of our explanation for the success of PM and IM is that children who are fluent in oral language use must, nonetheless, also learn to ground written words if they are to become skilled readers. This supposition that there is a difference between grounding heard words and written words is tested in the experiment.

In the experiment (Richmond et al., in preparation), we adapted the basic CVS design to answer the following questions: (a) When one is acquiring an understanding of CVS, how important is the opportunity to ground the written word (in contrast to the heard word)? (b) Is grounding necessary at all when learning an abstract principle, or are abstract principles better conceived of as rules operating on ungrounded symbols? (c) Will grounding of the abstract CVS concept during reading in one domain (e.g., how springs work) produce transfer when the children are reading to apply CVS in other domains (e.g., how plants grow)?

In the first of three sessions, fourth-grade children were given a brief oral introduction to CVS. Then, children read (or heard) texts that described how to set up experiments that conform to the CVS principle. Table 18.1 presents an example text, and Figure 18.2 illustrates one context, the ramps context, in which children carried out experiments to determine the factors that influence how far a ball will roll.

In the read and manipulate (RM) condition, children read aloud the texts describing how to set up the experiment, and they literally manipulated the experimental apparatus to conform to the text. In the listen and manipulate (LM) condition, the experimenter read the text aloud, but the children literally manipulated

TABLE 18.1 Example of a text used to study the application of PM to the learning and application of CVS.

In this experiment, we will try to find out if the ramp surface makes a difference in how far the ball travels after leaving the ramp. Circle ramp surface on your worksheet.

- Ramp A surface should be smooth, and Ramp B surface should be rough. ☐

Then, you need to be sure that all of the other variables (steepness, length, and ball type) are exactly the same for both ramps:

- The two ramps should both be steep. ☐
- The two ramp lengths should both be long. ☐
- The two ramps should both have a squash ball. ☐

Now, record on your worksheet whether or not you think this is a good experimental design to test whether the ramp surface makes a difference in how far the ball travels after leaving the ramp.

- One squash ball should be placed on Ramp A, and the gate should be lifted. ☐
- One squash ball should be placed on Ramp B, and the gate should be lifted. ☐

Note: Children put a check mark in the box when the activity was completed.

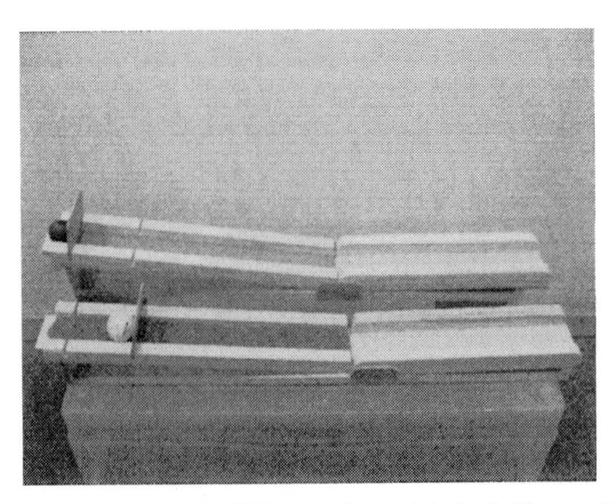

FIGURE 18.2 Ramps context for CVS. A confounded design is illustrated because the ramps differ in ball starting location, type of ball, ramp angle, and ramp surface. (See color plate)

the experimental apparatus. Note that the abstract information content is equated in these two conditions, but only in the RM condition do children have the opportunity to ground written words in their actions. Finally, in the read condition, children read the text aloud while the experimenter set up the experiment out of sight. In all conditions, after the experiment was set up, the children viewed the apparatus, observed the experiment and its outcome, and discussed again the CVS principle.

During the first session, children worked with two experiments in two of the three experimental contexts (e.g., ramps and how different types of objects sink). After an experiment was set up, but prior to conducting the experiment, the children were asked to explain and justify the experimental design (i.e., account for whether they think the design represents a "good test" of the focal variable). Then, the experiment was conducted, and children made observations of the outcome. Finally, the researcher changed the apparatus to set up confounded or noncontrastive (i.e., the same level of the independent variable was used in both conditions) experiment, and the researcher lead a discussion addressing why this was not a good experiment. The reason for this discussion was that learning the CVS principle involves not only the ability to design and execute unconfounded experiments, but also the ability to distinguish between confounded, unconfounded, and noncontrastive designs (Chen & Klahr, 1999).

Multiple types of assessments were used in the experiment. Here we report data from arguably the most important assessment, namely, how well children can set up unconfounded experiments and how well they can assess the experiments created by others. On the second day of the experiment, children were introduced to a third context for which they had had no previous experience. Table 18.2 presents the script used to introduce the "springs" context and an experiment that a child was asked to conduct. Each child (in the group of three) was asked, in

TABLE 18.2 Introduction to the "springs" context and example performance assessment.

We will use these springs to test the effect of different variables on how long springs stretch. Just like with the ramps and sinking materials, there are four things you will test on the springs to see if they make a difference in how long a spring stretches. These four things are called variables and variables are things that can change. The variables for the springs are:

1. *Spring length* – the springs can be either short or long.
2. *Spring width* – the springs can have either narrow coils or wide coils.
3. *Wire thickness* – the springs can have either thin wires or thick wires.
4. *Weight size* – the weights that you hang from the springs can be either heavy or light.

Now, each you will have the opportunity to set up an experiment. You will read all of the texts, but only one person will set up the experiment. After that person has set up the experiment, the other two students in the group will decide whether or not they think he/she has set up a good experiment and record their decision on the worksheet (just as you did yesterday). You can also make a prediction about what you think will happen in the experiment. After the experiment has been conducted, you will complete questions 2 and 3 on the worksheet.

Spring Experiment 1

In this experiment, you will try to find out if the spring width makes a difference in how long springs stretch. Circle spring width on your worksheet.
The two springs should both be long.
The two springs should both have thin wires.
The two springs should both have light weights.

Before you do this experiment, record on your worksheet your reasons for why you have set up the experiment this way (*for the child conducting the experiment*).

To other students in group: Use this time to decide whether or not you think he/she has set up a good experiment. Circle your choice on the worksheet and then write a sentence below explaining your choice.

When you have finished writing, you can make a prediction about the experiment. Then, we will do the experiment and determine if the focal variable made a difference in how the experiment turned out.

---

turn, to create an experiment to test the effect of a variable (e.g., spring width) that had not yet been investigated, and the child was asked to justify the design. The other children in the group then evaluated the adequacy of design.

The means for the various measures reported next have been statistically adjusted to take into account effects associated with the eight classrooms from which the children were sampled. In addition, means are reported only for the first performance assessment in each group of three. Because the researcher and children continued to discuss CVS after each experiment, eventually children in all of the conditions came to produce unconfounded designs. How well did students design the experiment? The proportions of unconfounded designs were 0.91, 0.75, and 0.81 for the RM, LM, and Read conditions, respectively. The effect size (d) for the statistically significant contrast between RM and LM was 0.78. When asked to justify the design, to what extent did children invoke

CVS (scored using a 0–4 rubric)? The mean scores were 2.10 1.52, and 1.83 for the RM, LM, and Read conditions, respectively (d = .81 for the RM to LM contrast). Finally, how accurately did the other children in the group evaluate the experiment? Based on the same 0–4 rubric, the means were 2.82, 2.18, and 2.15 (d = 1.01 for the RM to LM contrast). In other words, the data consistently point to the success of children in the RM condition (which gave children the opportunity to index and ground written words) compared to the children in the symbolically equivalent LM condition and the Read condition.

These data lead to several important conclusions. First, the data suggest that PM can be of benefit when learning from exposition and applying abstract knowledge. Second, the data suggest that grounding written words is not automatic. Children in the RM condition had the opportunity to ground written words in their actions, and the benefits of that grounding become apparent when the children were required to read and interpret a new text (Table 18.2). Previous experience hearing the same texts in the LM condition was not as effective. Finally, the data demonstrate strong transfer. That is, children learned the CVS in two experimental contexts, but the performance assessments were conducted using a third, newly introduced context.

CONCLUSIONS

If embodied approaches to cognition are on the right track, then they should provide key insights into educational processes. This chapter has surveyed two areas of promise, mathematics and reading comprehension. The work in mathematics suggests strong connections between the body and mathematical reasoning (Chapter 17). Nonetheless, this work has yet to produce effective interventions.

I have also provided an extensive overview of my own work applying an embodied approach to language comprehension to teaching reading comprehension. To date, the data are very encouraging. The PM intervention increases reading comprehension by 1 to 1.5 standard deviations over a Reread control. Importantly, once children have had experience with PM, they can engage in IM and thereby apply the strategy on their own. We have shown that the procedures can be applied to small reading groups, that they help with retention of vocabulary, and that they are effective when dealing with more abstract material such as CVS.

Are PM and IM the embodied educational equivalent of a successful moon shot? Clearly not. Nonetheless, the launching pad appears to be in sight.

REFERENCES

Anderson, J. R. (2005). Human symbol manipulation within an integrated cognitive architecture. *Cognitive Science, 29,* 313–341.

Andres, M., Seron, X., & Olivier, E. (2007). Contribution of hand motor circuits to counting. *Journal of Cognitive Neuroscience, 19,* 563–576.

Armbruster, B. B. & Anderson, T. H. (1988). On selecting "considerate" content area textbooks. *Remedial and Special Education, 9,* 47–52.

Barsalou, L. (1999). Perceptual symbols systems. *Behavioral and Brain Sciences, 22,* 577–609.

Beck, I. L., McKeown, M. G., Hamilton, R. L., & Kucan, L. (1997). *Questioning the Author: An Approach for Enhancing Student Engagement With Text.* Newark, DE: International Reading Association.

Borghi, A. M., Glenberg, A. M., & Kaschak, M. P. (2004). Putting words in perspective. *Memory and Cognition, 32,* 863–873.

Boroditsky, L. & Ramscar, M. (2002). The roles of body and mind in abstract thought. *Psychological Science, 13,* 185–188.

Campbell, J. I. D. (1994). Architectures for numerical cognition. *Cognition, 53,* 1–44.

Campbell, J. I. D. & Fugelsang, J. (2001). Strategy choice for arithmetic verification: Effects of numerical surface form. *Cognition, 80,* B21–B30.

Chambers, C. G., Tanenhaus, M. K., & Magnuson, J. S. (2004). Actions and affordances in syntactic ambiguity resolution. *Journal of Experimental Psychology: Learning, Memory, and Cognition, 30,* 687–696.

Chao, S., Stigler, J. W., & Woodward, J. A. (2000). The effects of physical materials on kindergarteners' learning of number concepts. *Cognition and Instruction, 18,* 285–316.

Chen, Z. & Klahr, D. (1999). All other things being equal: Acquisition and transfer of the control of variables strategy. *Child Development, 70,* 1098–1120.

Donovan, C. A. & Smolkin, L. B. (2001). Genre and other factors influencing teachers' book selections for science instruction. *Reading Research Quarterly, 36,* 412–440.

Duke, N. K. (2000). 3.6 minutes per day: The scarcity of informational text in first grade. *Reading Research Quarterly, 35,* 202–224.

Gibson, J. J. (1979). *The Ecological Approach to Visual Perception.* New York: Houghton Mifflin.

Glenberg, A. M. & Robertson, D. A. (1999). Indexical understanding of instructions. *Discourse Processes, 28,* 1–26.

Glenberg, A. M. & Robertson, D. A. (2000). Symbol grounding and meaning: A comparison of high-dimensional and embodied theories of meaning. *Journal of Memory and Language, 43,* 379–401.

Glenberg, A. M. & Kaschak, M. P. (2002). Grounding language in action. *Psychonomic Bulletin and Review, 9,* 558–565.

Glenberg, A. M., Gutierrez, T., Levin, J. R., Japuntich, S., & Kaschak, M. P. (2004). Activity and imagined activity can enhance young children's reading comprehension. *Journal of Educational Psychology, 96,* 424–436.

Glenberg, A. M., Brown, M., & Levin, J. R. (2007a). Enhancing comprehension in small reading groups using a manipulation strategy. *Contemporary Educational Psychology, 32,* 389–399.

Glenberg, A. M., Jaworski, B., Rischal, M., & Levin, J. R. (2007b). What brains are for: Action, meaning, and reading comprehension. In D. McNamara (Ed.), *Reading Comprehension Strategies: Theories, Interventions, and Technologies* (pp. 221–240). Mahwah, NJ: Lawrence Erlbaum Publishers.

Glenberg, A. M., Sato, M., Cattaneo, L., Riggio, L, Palumbo, D., & Buccino, G. (2008). Processing abstract language modulates motor system activity. *Quarterly Journal of Experimental Psychology, 61,* 905–919.

Goldstone, R., Landy, D., & Son, J. Y. (in press). A well-grounded education: The role of perception in science and mathematics. In M. DeVega, A. M. Glenberg & A. C. Graesser (Eds.), *Symbols, Embodiment and Meaning: A Debate.* Cambridge, England: Oxford University Press.

Klahr, D., Chen, Z., & Toth, E. E. (2001). Cognitive development and science education: Ships that pass in the night or beacons of mutual illumination? In S. Carver & D. Klahr (Eds.), *Cognition and Instruction: Twenty-five years of progress* (pp. 75–119). Mahwah, NJ: Lawrence Erlbaum Associates, Inc.

Lindemann, O., Abolafia, J. M., Girardi, G., & Bekkering, H., (2007). Getting a grip on number: Numerical magnitude priming in object grasping. *Journal of Experimental Psychology: Human Perception and Performance, 33,* 1400–1409.

Marley, S. C., Levin, J. R., & Glenberg, A. M. (2007). Improving Native American children's listening comprehension by means of concrete representations. *Contemporary Educational Psychology, 32,* 537–550.

Marley, S. C., Levin, J. R., & Glenberg, A. M. (in preparation). Young Native American readers benefit from simulated interaction while reading narrative texts.

Martin, T. & Schwartz, D. L. (2005). Physically distributed learning: Adapting and reinterpreting physical environments in the development of fraction concepts. *Cognitive Science, 29,* 587–625.

Masur, E. F. (1997). Maternal labeling of novel and familiar objects: Implications for children's development of lexical constraints. *Journal of Child Language, 24,* 427–439.

McCloskey, M. (1992). Cognitive mechanisms in numerical processing: Evidence from acquired dyscalculia. *Cognition, 44,* 107–157.

Morrow, L. M. & Pressley, M. (1997). The effect of a literature-based program integrated into literacy and science instruction with children from diverse backgrounds. *Reading Research Quarterly, 32,* 54–77.

Nathan, M. J. (in press). An embodied cognition perspective on symbols, gesture and grounding instruction. In M. DeVega, A. M. Glenberg & A. C. Graesser (Eds.), *Symbols, Embodiment and Meaning.* Cambridge, England: Oxford University Press.

Pressley, M., Rankin, J., & Yokol, L. (1996). A survey of instructional practices of primary teachers nominated as effective in promoting literacy. *The Elementary School Journal, 96,* 363–384.

Richardson, D. C., Spivey, M. J., Barsalou, L. W., & McRae, K. (2003). Spatial representations activated during real-time comprehension of verbs. *Cognitive Science, 27,* 767–780.

Richmond, E. K., Glenberg, A. M., & Levin (in preparation). Learning science from text: The importance of grounding written words.

Rizzolatti, G. & Craighero, L. (2004). The mirror-neuron system. *Annual Review of Neuroscience, 27,* 169–192.

Sato, M., Cattaneo, L., Rizzolatti, G., & Gallese, V. (2007). Numbers within our hands: Modulation of corticospinal excitability of hand muscles during numerical judgment. *Journal of Cognitive Neuroscience, 19,* 684–693.

Schwartz, D. L., Martin, T., & Nasir, N. (2005). Designs for knowledge evolution: Towards a prescriptive theory for integrating first- and second-hand knowledge. In P. Gardenfors & P. Johansson (Eds.), *Cognition, Education, and Communication Technology* (pp. 21–54). Mahwah, NJ: Lawrence Erlbaum.

Sweet, A. P. & Snow, C. E. (2003). Reading for comprehension. In A. P. Sweet & C. E. Snow (Eds.), *Rethinking Reading Comprehension* (pp. 1–12). New York: Guilford Press.

Tomasello, M. (2003). *Constructing a Language.* Cambridge, MA: Harvard University Press.

Toth, E. E., Klahr, D., & Chen, Z. (2000). Building research and practice: A research based classroom intervention for teaching experimentation skills to elementary school children. *Cognition and Instruction, 18,* 423–459.

Triona, L. M. & Klahr, D. (2003). Point and click or grab and heft: Comparing the influence of physical and virtual instructional materials on elementary school students' ability to design experiments. *Cognition and Instruction, 21,* 149–173.

Uttal, D. H. (2003). On the relation between play and symbolic thought: The case of mathematics manipulatives. In O. N. Saracho & B. Spodek (Eds.), *Contemporary Perspectives on Play in Early Childhood Education* (pp. 97–114). Greenwich, CT: Information Age Publishing, Inc.

Uttal, D. H., Scudder, K. V., & DeLoache, J. S. (1997). Manipulatives as symbols: A new perspective on the use of concrete object to teach mathematics. *Journal of Applied Developmental Psychology, 18,* 37–54.

Wagner Cook, S., Mitchell, Z., & Goldin-Meadow, S. (2008). Gesturing makes learning last. *Cognition, 106,* 1047–1058.

Zwaan, R. A. & Taylor, L. J. (2006). Seeing, acting, understanding: Motor resonance in language comprehension. *Journal of Experimental Psychology: General, 135,* 1–11.

SECTION

VI

SCALING UP

19

HOW DID WE GET FROM THERE TO HERE? AN EVOLUTIONARY PERSPECTIVE ON EMBODIED COGNITION

MARGARET WILSON

Department of Psychology, University of California, Santa Cruz, CA, USA

INTRODUCTION

Human cognition really is "something new under the sun." Although research in animal cognition is rapidly piling up evidence that animals have abilities that were previously thought to be impossible, nevertheless there is no serious challenge to the apparent fact that human cognition is unique. The human capacities for language, long-term plans, manipulation of abstract concepts, and accretion of knowledge and skills across generations simply have no competitors in other animals. Yet to understand how human cognition works, investigators increasingly recognize, we must tell an evolutionarily plausible story about how we got from there to here. The embodied cognition perspective is fundamentally an evolutionary one, viewing cognition as a set of abilities that built upon, and still reflects, the structure of our physical bodies and how our brains evolved to manage those bodies. But the embodied cognition literature has sometimes taken a very strong stance that cognition is fundamentally and directly bound to the body in its immediate physical environment. Instead, I argue here, the value of the embodied cognition approach is not to deny the existence of abstract and de-contextualized thought, but to explain how it grew out of previously existing sensorimotor abilities.

This chapter will consider a cluster of possibly linked capacities that may have driven human embodied cognition, including the ability to exert flexible voluntary control over particular effectors, the ability to see analogies, and the ability to imitate. The story to be told is one of escape from situation-bound cognition to a more flexible, abstract, and "general purpose" form of cognition.

ANIMAL COGNITION: WHERE WE STARTED FROM

Humans and their ancestors, the hominids, are one branch of the great apes, having diverged from the ancestors of gorillas about 8 million years ago, and the ancestors of chimpanzees and bonobos about 6 million years ago. Humans evolved in the Great Rift Valley of eastern Africa, in a drier, more open environment than that of gorillas, chimpanzees and bonobos, with different ecological demands. Humans are the sole survivors of the hominid branch, but this branch was originally much "bushier," with multiple species of genus *Australopithecus*, and later genus *Homo*.

The australopiths, although still possessing relatively small brains, developed the skeletal structure to walk upright and the ability to manufacture and use stone tools. The larger-brained genus *Homo* improved upon these abilities, creating tools according to a preconceived plan (rather than randomly chipping to create a sharp edge), and migrating from Africa to populate much of Europe and southern Asia (becoming, among others, the Neanderthals in Europe). Finally, a mere 200,000 years ago, *Homo sapiens* emerged from those hominids that had remained behind in Africa. They began to invent a much wider variety of tools, and to create decorative items. A small group left Africa to become the ancestors of all non-African modern humans, and out-competed their cousin species in Europe and Asia, leaving humans as the only surviving hominids on the planet. This new species, virtually identical to modern humans, apparently possessed the mental equipment that later made possible the "cultural explosion" of art and artifacts 50,000 years ago, and the discovery of agriculture and animal domestication 9000 years ago, leading to the first large population centers. Evolution has continued since the emergence of *Homo sapiens* 200,000 years ago, but in minor ways such as in skin tone to accommodate different amounts of sunlight, nose shape to accommodate different air temperatures, and the ability to digest different foodstuffs. (For a highly readable overview of hominid evolution for the non-specialist, see Zimmer, 2005.)

Today, we are left trying to reconstruct the cognitive past of the human lineage from a range of indirect sources, including dramatic increases in brain size, structural changes suggesting adaptation for vocal language production and manual manipulation, changes in the time course of infant and juvenile development, artifacts left behind, artifacts moved long distances from their manufacture sites, migration patterns, genetic divergence, and evidence of organized group activity such as hunting. Another important source of evidence, of course, is comparative study of other existing species, which allows us to make educated guesses about the cognitive abilities of our common ancestors, as well as the cognitive abilities demanded by various ecological niches and social structures.

The last few years have seen a flood of studies that are eroding previous dogma on what non-human animals supposedly cannot do, and the game is no longer just to find precursors to human cognition in our own closest relatives the great apes (and by implication, our mutual ancestors). Surprising cognitive abilities are also being uncovered in more distantly related primates such as old-world and new-world monkeys (e.g. Fragaszy & Cummins-Sebree, 2005; Cheney & Seyfarth, 2007), and also in animals much further removed from us on the evolutionary tree, such as birds, marine mammals, dogs, and elephants (e.g. Emery & Clayton, 2004; Kuczaj & Walker, 2006; Miklósi et al., 2007; Hart et al., 2008).

At the present moment, there is much excitement about the emergence of social cognition abilities in particular, and how these might have been the driving force behind the evolution of sophisticated cognition in general (e.g. Hare, 2007). One set of abilities receiving a great deal of attention is the ability to understand what other individuals know, see, and think. Although the details are still in dispute, non-human primates are revealing surprisingly sophisticated abilities to understand eye gaze, to deceive, and to manipulate others based on their state of knowledge (e.g. Byrne & Corp, 2004; Scerif et al., 2004; Leavens et al., 2005; Hare et al., 2006; Melis et al., 2006; Braüer et al., 2007; Hattori et al., 2007; Wood et al., 2007). Similar findings are emerging for more distantly related species as well, such as scrub jays (Dally et al., 2006), and dolphins (Pack & Herman, 2007). These abilities may be interpreted as a rudimentary theory of mind, or as precursors to such a theory.

A second set of social cognition abilities receiving much attention revolves around fairness, reciprocity, morality, and justice. Evidence is increasingly showing that the human sense of fairness and the desire to punish violators (even at a cost to oneself, and even when the victim of the unfairness is someone else) have biological bases (for review see Fehr & Camerer, 2007). Research on non-human animals is investigating whether other species will choose options that benefit another individual, and other related empathic tendencies (Stevens & Hauser, 2004; Nowak & Sigmund, 2005; Silk, 2006; Warneken & Tomasello, 2006; Jensen et al., 2007). It has been argued that these kinds of social abilities and social awareness were a driving force behind modern human brain size, intelligence, and cognition, and made possible such important abilities as explicit teaching, cultural transmission, and complex communication leading to language and symbolic thought (e.g. Tomasello & Rakoczy, 2003; Seyfarth et al., 2005; Csibra, 2007).

But these social abilities do not constitute the whole story. Various animal species reveal additional cognitive abilities that were previously thought not to exist in non-humans, and many of these appear to be precursors or prerequisites for the kinds of cognitive abilities that humans display. Following is a brief menu of topics of current interest among researchers.

Planning and "mental time travel": To what extent are non-human animals able to disconnect their behavior from the present situation, in ways that cannot be reduced to simple learned associations (e.g., delaying a behavior will lead to a future reward) or innate predispositions (e.g., caching food or migrating), and actually take into account future outcomes in choosing their behavior? A subset of this question is whether animals experience mental simulations of the future that are phenomenologically and neurologically similar to episodic memory of the past, an ability in humans that has been dubbed "mental time travel" (Suddendorf & Corballis, 1997, 2007). Although Suddendorf and Corballis (2007) argue that this latter ability has not been demonstrated in any non-human animal, nevertheless evidence of various non-simplistic planning abilities is beginning to emerge. For example, orangutans and bonobos save and transport tools for future use (Mulcahy & Call, 2008), and scrub jays preferentially cache food for the next morning in a location that they have previously learned will not have food at that time (Raby et al., 2007).

Numerosity and proto-mathematical abilities: Evidence is emerging that non-human primates are sensitive to the abstract concept of number, including both its ordinal (serial order) and its cardinal (quantity) properties (see Nieder, 2005; Cordes et al., 2007, for reviews). With respect to ordinal properties, rhesus macaques are able to correctly order displays of one to four items, but are unable to do so with an arbitrary order (Brannon & Terrace, 2000). These monkeys also generalize the ordering task to sets with greater numbers of items, without reinforcement. And with respect to quantity, macaques match number of seen monkey faces to number of heard monkey voices. That is, when they hear two monkey voices, they preferentially look at a display with two faces rather than a display with three (Jordan et al., 2005). Similarly, capuchin monkeys have shown sensitivity to quantity (Beran, 2008). It has even been shown that rhesus macaques can compute approximate addition of sets of items (Flombaum et al., 2005).

Causal reasoning. Although evidence for inferences about cause-and-effect in non-humans is by and large either absent or negative (see Penn & Povinelli, 2007, for review), two recent results suggest that rats can engage in certain forms of causal reasoning. Rats who learned through observation that a light predicted both a tone and food unsurprisingly came to expect that food should accompany the tone. However, they did not exhibit this expectation when they themselves were allowed to cause the tone by pressing a lever, suggesting that they had developed a causal model of the relations among the events, and not just mere associations (Blaisdell et al., 2006). In a different study, Beckers et al. (2006) showed that whether rats show blocking in classical conditioning (a failure to develop a new conditioned stimulus when it is paired with another previously conditioned stimulus) is sensitive to the rat's knowledge of causal factors.

Imitation. Imitation, which humans do frequently and effortlessly, is a topic of hot debate in the animal literature (for reviews see Byrne, 2005; Zentall, 2006). Some authors have argued that non-human apes do not truly imitate at all, and that behaviors that appear to be imitation can instead be explained as instances of *emulation*, that is, learning about and then reproducing the affordances of objects that another individual has been seen manipulating (e.g. Call et al., 2005). The ape does not appear to be attending to and duplicating the actual actions of the individual. More recently though, it has been reported that chimpanzees learn to pull a lever to deliver a reward when they see the apparatus used by another chimp, but not when the lever is activated by an invisible fishing line (Whiten, 2007). This appears to eliminate the emulation explanation for the learning. Furthermore, reports from observations of apes in the wild and in naturalistic captive settings support a richer view of imitation than what has been observed in laboratory experiments, although of course these observations are subject to the criticism that they do not involve strict controls (e.g. Russon & Galdikas, 1993; van Schaik et al., 2003; Byrne & Tanner, 2006; Whiten et al., 2007). In addition, vocal imitation of non-species-typical sounds has been reported in elephants (Poole et al., 2005), paralleling the well-known vocal imitation abilities of birds such as parrots; and there is a documented anecdote of a human-reared seal, Hoover, who produced imitations of the voice of the man who raised him. (A recording of Hoover saying "Get over here, come on, come on" can be found at: http://www.neaq.org/scilearn/kids/hooveronly.html.)

Tool use: Tool use is well documented in great apes, including the use of sticks to probe termite mounds, rocks to break open nuts, and broad leaves to serve as hats in the rain, as well as more sophisticated awareness of tool properties in human-enculturated chimps (Furlong et al., 2008). Tool use has also been demonstrated in capuchin monkeys for purposes of digging, cracking, and probing (Moura & Lee, 2004; Waga et al., 2006), and a recent study also documents probe-tool manufacture, transport, and use by crows (Kenward et al., 2005).

HUMAN COGNITION: WHERE WE ENDED UP

Given this impressive array of newly discovered animal abilities, with more likely to come, what makes human cognition unique? This is a case that is almost easier to make by mere looking around, rather than appeal to published data. Animals can make tools and can alter their environments through niche construction, but none has produced such a dizzying array of artifacts as even pre-agricultural humans, let alone electronic-age humans. Animals can recognize basic numerical relations, but none has discovered calculus. Animals can plan ahead, but none has conceived of a construction project like the medieval cathedrals that took longer than one person's lifetime to complete. And perhaps most obviously, animals can communicate, but none has developed language. Some of these abilities appear to slightly greater degrees in human-enculturated apes than wild apes (for review see Tomasello & Call, 2004), suggesting untapped cognitive potential, but still none comes even close to approximating participation in the human life of the mind.

Given that many of the precursors to human intelligence are present in other animals—some in virtue of related lineage, others by convergent evolution—one might ask why various unique human abilities have not arisen in other species. It is difficult to see how these kinds of abilities could *not* contribute to adaptive fitness. Why, then, are humans the oddballs? One possible reason is that the "engineering solutions" required to build such a cognitive system may not be easy for evolution to construct out of previous genetic resources (cf. Wilson, 2002, p. 627). If this is true, then humans are the beneficiaries of a lucky and unlikely evolutionary accident. An alternative explanation, and one that is gaining some empirical support, is that the caloric needs of brain tissue offset the survival advantages of smarter brains. According to this view, hominids broke through the brain-size barrier by learning to extract additional calories from their environment, through the discovery of roots and tubers ("underground storage organs") as a food source during times when other food sources were unreliable (Laden & Wrangham, 2005; Yeakel et al., 2007), or the discovery of cooking, which releases additional calories (Lucas et al., 2006; Gibbons, 2007).

But however it came about, humans undeniably possess startling cognitive abilities, with no precedent in the previous life of the planet. How did we get from there to here, and, particularly if one takes seriously the embodied view of the cognition of our ancestors, how do we explain human cognition in embodied terms?

EMBODIED COGNITION: CAN IT EXPLAIN WHERE WE ARE NOW?

Underlying most of the embodied cognition literature is an implicit evolutionary argument. Traditional abstract-symbol-processing views of cognition that were typical in the 1950s–1980s are now considered implausible precisely because there is no continuity with the cognitive skills that would have been demanded of the ancestors of the human species. Thus, the embodied cognition approach attempts to provide continuity between cognitively simpler creatures that lived in the moment, and modern humans.

This has led to interesting new perspectives on human cognition, including the study of "situated cognition," cognition that is about, entwined with, and time locked to unfolding events in the immediate physical environment. However, I have argued previously that human cognition cannot, in principle, be entirely or even largely reduced to situated cognition (Wilson, 2002). This fundamental error, resulting from an over-application of the evolutionary argument, has riddled the field of embodied cognition since its inception: the assumption that, because human cognition is *for* survival in immediate real-world situations, therefore human cognition must be entirely *about* those immediate real-world situations. Instead, an essential feature of human cognition is the ability to decouple from the present environment and represent situations and ideas that are of the past, the future, the physically distant, the imaginary, the generalized, or the abstract.

A productive science of embodied cognition, if it takes seriously the claim that much or all of human cognition has its roots in embodiment, must consider how embodied cognition can go "off-line"—decouple from situation-bound reactivity and use body-based resources for other purposes.

How do we embody abstract thought? In the early years of embodied cognition theorizing, this was a large unmet promissory note. It was assumed that progress would be made in this area, and plausible examples were proposed, but empirical data were lacking. Instead, studies tended to focus on embodying "thoughts" that were themselves very close to external activities, such as mental rotation (Kosslyn et al., 1998), playing video games (Kirsh & Maglio, 1994), and understanding mechanical diagrams (Hegarty, 1992). Recently, however, considerable progress has been made in identifying the embodied underpinnings of various cognitive domains. These include representation of language and of abstract concepts (Barsalou, 2005; Gallese & Lakoff, 2005; Zwaan & Taylor, 2006), spatializing of abstract quantities such as numbers and time (Stoianov et al., 2008), gesturing to support cognition (Broaders et al., 2007; Goldin-Meadow, 2006), and offloading information onto body-based resources in working memory (Wilson, 2001b; Wilson & Fox, 2007). The striking feature that these have in common is the way in which we expand the domain of what is "embodyable" by creative use of body resources, decoupled from immediate action on the environment. In this chapter, I advance a proposal of a related cluster of abilities that allowed this transformation to take place.

FLEXIBILITY AND RESEMBLANCE: KEYS TO OFF-LINE EMBODIMENT?

VOLUNTARY CONTROL

One important prerequisite for a flexible, general purpose, off-line embodied cognition is voluntary control over important articulators, if not the whole body. Species differ as to how much control they have, and over which muscle groups. Non-human primates appear to lack any great degree of voluntary control over their vocal apparatus, both in the sense that they possess a limited and fixed repertoire of calls, and that the calls are predictably elicited by certain situations and are not deployed flexibly or creatively. For this reason, it has been suggested that manual dexterity (Arbib & Rizzolatti, 1996; Gentilucci & Corballis, 2006; Pollick & de Waal, 2007), or non-vocal facial movements such as lip-smacking and chewing (MacNeilage, 1998), rather than vocal calls, may have been the evolutionary precursor to human speech, because these are domains in which our closest relatives (and therefore presumably our common ancestor) show greater flexibility and control.

Species more distantly related to us can also possess a remarkable degree of voluntary control over articulators that they use to manipulate their environments. Walruses, to take one example, forage for food on the bottom of the ocean by way of extremely sensitive and sophisticated movements of the mouth and snout, and also use their highly mobile mouth and throat structures to produce a large repetoire of social calls. Walruses in captivity have been observed to use these abilities for other purposes, including manipulating toys and generating novel vocalizations (Schusterman & Reichmuth, 2008). Thus, voluntary control of particular articulators appears to be an ability that evolves specifically when and where it is needed for the animal's particular survival strategy.

What is remarkable about humans is that we have managed to achieve an unprecedented degree of control over our bodies. Of notable importance to humans are the hands and the vocal apparatus, but to a great degree our entire bodies are under voluntary control. Even large and cumbersome muscle groups, such as those of the abdomen, can with practice be brought under voluntary control. Thus, we can not only choose at will from among pre-existing skills, but we can also develop a virtually unlimited range of new ones, as demonstrated by different dance styles from belly dance to ballet to hula to classical Indian dance.

How does this relate to cognition? In order to take embodied cognition off-line, it is necessary to decouple our bodily actions (or mental representation of those actions) from the demands of the immediate situation. We need to be able not only to walk, jump, turn, point, move objects, smile at people, and speak words, but also to perform or think about performing those actions at will when they are not necessarily appropriate to the situation but can assist us in our cogitations. Escaping stimulus-driven activation of our motor skills is a necessity for off-line embodied cognition.

Furthermore, it is plausibly of great importance to a broad, general cognitive system to be able to learn or invent new bodily actions at will, expanding the array of tools available for embodied thinking. A gymnast, for example, familiar with rotating the entire body not only around the vertical axis (remaining upright) but also around both horizontal axes (consider a cartwheel and a back handspring, respectively), may possess spatial imagery skills that differ from those of a ballet dancer. A person who is mathematically inclined can learn to count on their fingers in binary (a finger up is a 1, down is a 0), enabling representation of the integers up to 512 by hand configuration alone. A pianist, whose musical skill involves chords, may have greater cognitive flexibility in thinking about music than a vocalist of equal musical training, whose instrument can only produce one note at a time. According to Yale musician Joel Wizansky, "it's more or less a truism that any composer, regardless of his main instrument, must have some level of keyboard skill, so as to be able to 'think polyphonically' at the piano" (J. Wizansky, personal communication, February 19, 2008). And to take a far more mundane example, whose astounding cognitive usefulness gets overlooked in its sheer familiarity, we learn the fine motor skills of handwriting and typing in order to be able to put our thoughts on paper. For purposes of embodied cognition, this last example is perhaps most interesting not so much in terms of its archival functions (writing down information for later reference or to be read by other people) but for its functions in serving as an external memory device during ongoing cognitive processing—polishing a line of poetry, for example, or trying to remember all the names of the seven dwarfs. All of these examples can be considered instances of what has been termed "cognitive technologies" (Frank et al., in press; see also O'Connor, 1996). Rather than being universal tools of human cognition, they are inventions, made possible by creative uses of the body.

A broad degree of voluntary control over the body, then, is arguably a bonanza for the expansion of the capabilities of embodied cognition. In the next section we turn to how this voluntary elicitation of movement can be harnessed to be able to do cognitive work.

ANALOGY

The ability to exploit analogies is ubiquitous in human cognition and perception. Humans excel at recognizing and capitalizing on structural isomorphisms between different objects or different conceptual domains. When an analogy is not obvious or made explicit by others (as in the well-known tumor/fortress problem), we do not always immediately recognize the parallelism between domains; but once we are aware of the parallelism we exploit it effortlessly (Gick & Holyoak, 1980, 1983; Holyoak & Koh, 1987; Pedone et al., 2001; Didierjean & Nogry, 2004).

One example, where the analogy between two domains is very clear and nearly perfect, is the use of maps, scale models, and drawings to understand their real world counterparts. This is an ability that emerges in children around 2 years of age, when they begin to recognize the relation between, for example, "feeding" a doll with a toy spoon and feeding a person (Johnson et al., 2005). By the age of three, children master the more complex analogical skill of using a scale model of a room to find a hidden object in the real room (Troseth et al., 2007). Interestingly, the power of the analogy is so compelling that children go through a stage of making scale errors, attempting to treat smaller model objects as if they were their larger counterparts (DeLoache et al., 2004; Ware et al., 2006). The ability to understand and use scale models appears to be shared with chimpanzees, although only some individuals use the information spontaneously whereas others need to be taught (Kuhlmeier et al., 1999; Kuhlmeier & Boysen, 2002).

A second example of our ability to see analogies is our ability to recognize structural parallels between objects that differ in their superficial characteristics, such as a penguin and a canary, or a truck and a sports car. This ability to "see" structural or functional similarities is crucial to the formation of concepts and categories. Bar (2007) goes further, suggesting that it also underlies our ability to generalize from our memories of past experiences in order to successfully anticipate new situations. As such, this tendency may have precursors in other animals (e.g. Murai et al., 2005), who must also generalize beyond the particular stimulus in order for past learning to be useful in new situations. Such an ability may arise from our perceptual systems, which likewise face the problem of needing to generalize in order to do pattern recognition, which cannot be reduced to mere template matching. Around the age of 2 years, children begin to recognize and creatively invent analogies that are based on only a partial match between domains. This can be seen in pretend play, in which one object stands in for another to which it may bear only gross featural similarity (e.g., pretending that a banana is a telephone).

A special case of seeing and exploiting imperfect analogies involves recognizing isomorphisms between different parts of the human body, and using one set of body parts to represent another set of body parts, as in gesturing with the two hands to show an action with the feet, or gesturing with two fingers to show walking. More sophisticated versions of these last examples can be found in signed languages, in the form of iconic signs and classifier constructions that refer to bodies or body parts (Taub, 2001; Emmorey, 2002). These are various even within a single signed language, and also differ across signed languages. For example, in American Sign Language the action of walking can be represented by a single upright index finger indicating the whole body, by the index and middle fingers of the dominant hand representing legs, by the two index fingers of the two hands representing legs, or by the two whole hands held in a flat position to represent feet. Each of these is used systematically in different situations to convey different shades of meaning. Other examples abound: fluttering eyelashes can be represented by the fingers, closing eyes by two flat hands coming together as the upper and lower lids, widened eyes by fists opening into semi-circles, and so on.

Certain body-to-body analogies may actually have a biological basis. There is evidence that a connection between opening and closing of the mouth and opening and closing of the hands is hardwired. Gentilucci et al. (2004) have shown that grasping or observing someone else grasp a small object (a cherry) or a large object (an apple) actually alters the way that people simultaneously pronounce a syllable, resulting in a more closed or open vowel sound. This hand–mouth connection may be an exception though, with its basis in the specific brain circuits that govern these two important manipulators and which appear to overlap in Broca's area (e.g. Rizzolatti & Arbib, 1998; Gentilucci & Dalla Volta, 2007; Skipper et al., 2007). In contrast to this hardwired case, though, humans are also able to productively employ analogies between any variety of body parts, provided the analogy is apt, as seen in the wide variety of body representations by the hands in signed languages.

Further examples of the ability to recognize isomorphism become progressively more abstract. When they reach school age, children are capable of mapping an abstract domain onto a concrete, spatial one, such as a clock face to represent time, or the number line to represent sequence or quantity. This ability becomes increasingly complex as children mature into adults, and forms the basis for many of our most sophisticated cognitive abilities, including higher math, computer programming, literary symbolism, and musical composition.

This strategy of using a concrete domain to represent an abstract one also underlies much of our use of analogical mappings in everyday life. This has been extensively explored in the context of conceptual metaphor (Johnson & Lakoff, 2002), in which an extensive network of part-by-part correspondences between domains is systematically exploited. Well-known examples include conceptual metaphors such as *communication is transferring an object between containers*, which can be broken down to a fine grain of detail (minds are containers, ideas are objects, difficulties in communication are physical barriers, communicative acts are bodily actions of handing, placing, or throwing, and so on). These extended metaphors tend to be culturally shared, as shown by their systematic use in language, but can vary from culture to culture. One striking example is the observation that, unlike most cultures, which talk about the future as ahead and the past behind, the Aymaras of the Andes do the reverse, metaphorically placing the future behind the body (since it is unknown) and the past in front (Núñez & Sweetser, 2006). The grounding of abstract conceptual knowledge in concrete domains has also been explored by Barsalou and colleagues, proposing that abstract concepts are grounded in schemata based on the sensory, motor, and introspective qualities of specific experiences (e.g. Barsalou, 1999). Although less extensively explored in the empirical literature, this same principle of abstract-to-concrete is clearly at work in activities such as mathematics (cf. Lakoff & Núñez, 2000). The insight that abstract quantities and relationships could be represented graphically underlay much of ancient Greek mathematics; led to the recognition in Persia in the 11th century and Europe in the 17th century that algebra and geometry are fundamentally related, which spawned the field of analytic geometry; and contributed to the development of calculus in the 17th century. One recent study that investigates this principle in the laboratory showed that formally irrelevant aspects of how elements are physically grouped can affect accuracy of mathematical judgements (Landy & Goldstone, 2007).

In short, recognizing and exploiting analogies, particularly physical and spatial analogies, seem to be a general feature of human cognition, ranging from the perceptual to the abstractly cognitive (cf. Wharton et al., 2000).

IMITATION

How does the ability to see analogies further the cause of embodied cognition? One special case of analogizing is the ability to recognize the isomorphism between one's own body and something else. This includes recognizing the isomorphism between one's own body and another person's body, and using this for imitation and social learning (Wilson, 2001a).

Imitation is ubiquitous in human activity, and in many cases appears to be unconscious and automatic (see Wilson, 2001a; Wilson & Knoblich, 2005, for reviews). Examples that have been studied experimentally include the chameleon effect (unconscious copying of another's posture and movements); speeded reaction times when the stimulus is a human body movement that matches the required response; rudimentary imitation in neonates and the subsequent development of imitation in children; disinhibition of imitation in frontal lobe patients; activation of motor brain areas and even muscles themselves in response to perceived action; and, expanding on the remarks above about sign language, the use of "iconic" (i.e., imitative) handshapes and movements to represent a wide variety of objects and events, which occurs in all documented signed languages of the world.

In addition there is the discovery of the "mirror system," involving regions in pre-frontal and parietal cortex (for reviews see Fadiga et al., 2005; Lepage & Théoret, 2007). The mirror system is so called because it is involved both in the perception and the production of body movements. This system is sometimes incorrectly referred to in the literature on humans as *mirror neurons*. However, only with single-cell recording in monkeys have specific neurons been observed that serve both perceptual and motor functions. In humans, at most we know that certain brain *regions* are involved in both. A second difference from the animal literature on mirror neurons is that monkey mirror neurons seem to be quite limited in the range of actions to which they respond. So far, mirror neurons have been found that respond to specific movements of the hand and mouth, such as grasping, placing, and tearing of objects, communicative and ingestive mouth movements, and reaching with a tool (Gallese et al., 1996; Kohler et al., 2002; Ferrari et al., 2003; Ferrari et al., 2005). In contrast, the human mirror system has been observed to respond to a wide variety of activities, including learned skills that can in no way be considered part of the "natural" repertoire of human movement (the way that walking might), for example, playing the piano (Haueisen & Knösche, 2001). In a parallel to the previous remarks about recent research on animal cognition, research on mirror neurons has tended to focus on the social functions of such a system, with speculative links being drawn to empathy and theory of mind (Meltzoff & Decety, 2003; Iacoboni & Dapretto, 2006; Agnew et al., 2007; Braten, 2007; Kilner et al., 2007). However, there are also several ways in which the mirror system may have contributed more generally to the development of human cognition. One obvious and direct way is that the mirror system plausibly gave rise to the ability to imitate.

As noted earlier, the existence of true imitation in non-humans is a subject of intense debate. If we concede that our closest relatives do not imitate with the same ease and flexibility as humans, it becomes worthwhile to ask what role imitation has played in the emergence of uniquely human cognition. What benefits did it confer? As noted earlier, it may have played a role in developing social cognition, but it may have played other roles as well. I have argued elsewhere that, in addition to the perceptual system driving the motor system, the information may flow back the other direction (Wilson & Knoblich, 2005). Unconscious activation of the motor system in response to perceived human body action may feed back again into the perceptual system, and provide more robust processing of the perceptual signal.

I have also argued that the human capacity for imitation, that is, easy automatic translation between perceptual and motor codes of body action, may be a key factor in working memory performance (Wilson, 2001b; Wilson & Fox, 2007). Rehearsal in working memory involves repeated covert articulation of the stimuli to be remembered, closely coupled to a quasi-perceptual representation of those stimuli. As a result, stimuli that can be imitated, which is to say stimuli that can be rehearsed in this articulatory way, yield much more robust working memory performance than stimuli that cannot.

These two cases, perception and working memory, can be seen as specific examples of a more general principle, which is that imitation is crucial to embodied cognition. To represent something with the body, we need to be able to shape the body to match the thing to be represented. This takes us beyond an embodied cognition of actions we might perform to operate on the world (as in the video game and mechanical diagram examples), and allows us to represent, using bodily resources, a much wider variety of events. That is, we can represent *what someone or something else is doing*, and not just *what we ourselves might do*.

In this connection, it is important to note that the ability to imitate can go beyond imitating other people, and can give us the ability to use the body to represent other objects. This includes anthropomorphizing jointed, multi-part entities, such as machinery and animals; and also includes using body movements such as hand trajectories, rotations, and pointing to represent object motion and location. These types of uses have been documented in a variety of cognitive tasks, such as mental rotation of non-human objects (Amorim et al., 2006), memory retrieval (Dijkstra et al., 2007), simple algebra (Broaders et al., 2007), and working memory (Chieffi et al., 1999).

In sum, in order to develop an embodied cognition that is about more than just the immediate situation (or, at most, planning future actions using representations of the same actions one would use in the actual situation), requires two things: the first is the ability to exert flexible control over one's body to recruit existing motor skills at will and to take on new and diverse body shapes and actions; and the second is the ability to use that control and flexibility to represent, by resemblance, a wide variety of things in the world.

FUTURE DIRECTIONS

In this chapter, I have argued that core properties of human cognition that are radically new in the scope of evolutionary history can nevertheless still be accommodated within an embodied cognition framework. It is not necessary either to insist that we are still situation-bound, or on the other hand to abandon embodiment for an ungrounded symbol-processing view of cognition. What is needed, though, is an evolutionarily plausible account of how we broke the limits of a situation-bound sensorimotor cognition, further limited by a small repertoire of species-typical behaviors. Two well-documented human abilities—our broad flexibility in control of our bodies, and our capacity to imitate and otherwise see and exploit analogies—can explain this shift.

This account still leaves several important questions unanswered. One, of course, is how voluntary control and imitation themselves arose. Were they driven directly by the advantages conferred by taking embodied cognition off-line? Or did they emerge for other reasons (perhaps socially based reasons), to then be exploited by an increasingly sophisticated off-line cognition?

Another important question that has not been addressed here is how language fits into an off-line embodied cognition theory. In the rush to get away from traditional views of cognition as abstract, symbolic, and disembodied, there has been a tendency to discount the unavoidable fact that humans *do* use symbols. As Barsalou (2005, p. 389) puts it, "Although abstraction has gone out of fashion, it will not go away." One of the most obvious ways that we use symbols, of course, is language. Does this mean that language has allowed us to escape embodiment, that aspects of cognition related to language are in fact disembodied in the tradition of old-fashioned cognitive psychology? Not necessarily. One important piece of this puzzle is being addressed by sensorimotor accounts of meaning (e.g. Barsalou, 1999, 2005; Johnson & Lakoff, 2002; Zwaan & Taylor, 2006). Another piece may reside in the fact that language is a bodily activity, expressed with the vocal tract or, more rarely, with the hands. This allows a meaning to be offloaded into mere motor activity, from which it can be retrieved and re-converted into meaning. In this respect, it is not an exaggeration to say that the emergence of human language revolutionized working memory. Nevertheless, the scope of the problem of understanding the evolution of language, and how it drove or was driven by off-line cognition, should not be underestimated.

In spite of these unanswered questions, though, the ideas put forward in this chapter may help to explain how embodied cognition theory can account for many of the important features of human cognition that appear on the surface to be disconnected from our immediate sensorimotor experience and behavior. Further, these ideas may help to bridge the apparent gulf between human and animal cognition, and help to explain how we got from there to here. As with all of evolution, spectacular new characteristics are possible, even though, in the words of Darwin (1862, p. 348), they "use old wheels, springs, and pulleys."

REFERENCES

Agnew, Z. K., Bhakoo, K. K., & Puri, B. K. (2007). The human mirror system: A motor resonance theory of mind reading. *Brain Research Reviews, 54,* 286–293.

Amorim, M.-A., Isableu, B., & Jarraya, M. (2006). Embodied spatial transformations: "Body analogy" for the mental rotation of objects. *Journal of Experimental Psychology: General, 135,* 327–347.

Arbib, M. A. & Rizzolatti, G. (1996). Neural expectations: A possible evolutionary path from manual skills to language. *Communication and Cognition, 29,* 393–424.

Bar, M. (2007). The proactive brain: Using analogies and associations to generate predictions. *Trends in Cognitive Sciences, 11,* 280–289.

Barsalou, L. W. (1999). Perceptual symbol systems. *Behavioral and Brain Sciences, 22,* 577–660.

Barsalou, L. W. (2005). Abstraction as dynamic interpretation in perceptual symbol systems. In L. Gershkof-Stowe & D. H. Rakison (Eds.), *Building Object Categories in Developmental Time* (pp. 389–431). Mahwah, NJ: Lawrence Erlbaum Associates.

Beckers, T., Miller, R. R., De Houwer, J., & Urushihara, K. (2006). Reasoning rats: Blocking in Pavlovian animal conditioning is sensitive to constraints of causal inference. *Journal of Experimental Psychology: General, 135,* 92–102.

Beran, M. J. (2008). Capuchin monkeys (*Cebus apella*) succeed in a test of quantity conservation. *Animal Cognition, 11,* 109–116.

Blaisdell, A. P., Sawa, K., Leising, K. J., & Waldmann, M. R. (2006). Causal reasoning in rats. *Science, 311,* 1020–1022.

Brannon, E. M. & Terrace, H. S. (2000). Representation of the numerosities 1–9 by rhesus macaques (Macaca mulatta). *Journal of Experimental Psychology: Animal Behavior Processes, 26,* 31–49.

Braüer, J., Call, J., & Tomasello, M. (2007). Chiimpanzees really know what others can see in a competitive situation. *Animal Cognition, 10,* 439–488.

Braten, S. (2007). *On Being Moved: From Mirror Neurons to Empathy.* Amsterdam: John Benjamins Publishing Company.

Broaders, S. C., Cook, S. W., Mitchell, Z., & Goldin-Meadow, S. (2007). Making children gesture brings out implicit knowledge and leads to learning. *Journal of Experimental Psychology: General, 136,* 539–550.

Byrne, R. W. (2005). Detecting, understanding, and explaining animal imitation, Ch. 9. In S. Hurley & N. Chater (Eds.), *Perspectives on Imitation: From Mirror Neurons to Memes* (Vol.1, pp. 255–282). Cambridge, MA: MIT Press.

Byrne, R. W. & Corp, N. (2004). Neocortex size predicts deception rate in primates. *Proceedings of the Royal Society B, 271,* 1693–1699.

Byrne, R. W. & Tanner, J. E. (2006). Gestural imitation by a gorilla: Evidence and nature of the capacity. *International Journal of Psychology and Psychological Therapy, 6,* 215–231.

Call, J., Carpenter, M., & Tomasello, M. (2005). Copying results and copying actions in the process of social learning: Chimpanzees (*Pan troglodytes*) and human children (*Homo sapiens*). *Animal Cognition, 8,* 151–163.

Chieffi, S., Allport, D. A., & Woodin, M. (1999). Hand-centred coding of target location in visuo-spatial working memory. *Neuropsychologia, 37,* 495–502.

Cheney, D. L. & Seyfarth, R. M. (2007). *Babboon metaphysics.* Chicago: University of Chicago Press.

Cordes, S., Williams, C. L., & Meck, W. H. (2007). Common representations of abstract quantities. *Current Directions in Psychological Science, 16,* 156–161.

Csibra, G. (2007). Teachers in the wild. *Trends in Cognitive Sciences, 11,* 95–96.

Dally, J. M., Claton, N. S., & Emery, N. J. (2006). The behavior and evolution of cache protection and pilferage. *Animal Behaviour, 72,* 13–23.

Darwin, C. (1862). *On the various contrivances by which British and foreign orchids are fertilized by insects.* London: John Murray.

DeLoache, J. S., Uttal, D. H., & Rosengren, K. S. (2004). Scale errors offer evidence for a perception-action dissociation early in life. *Science, 304,* 1029–1047.

Didierjean, A. & Nogry, S. (2004). Reducing structural-element salience on a source problem produces later success in analogical transfer: What role does source difficulty play? *Memory & Cognition, 32,* 1053–1064.

Dijkstra, K., Kaschak, M. P., & Zwaan, R. A. (2007). Body posture facilitates retrieval of autobiographical memories. *Cognition, 102,* 139–149.

Emery, N. J. & Clayton, N. S. (2004). The mentality of crows: Convergent evolution of intelligence in corvids and apes. *Science, 306,* 1903–1907.

Emmorey, K. (2002). *Language, Cognition, and the Brain: Insights from Sign Language Research.* Hillsdale, NJ: Lawrence Erlbaum Associates.

Fadiga, L., Craighero, L., & Olivier, E. (2005). Human motor cortex excitability during the perception of others' action. *Current Opinion in Neurobiology, 15,* 213–218.

Fehr, E. & Camerer, C. F. (2007). Social neuroeconomics: The neural circuitry of social preferences. *Trends in Cognitive Sciences, 11,* 419–427.

Ferrari, P. F., Gallese, V., Rizzolatti, G., & Fogassi, L. (2003). Mirror neurons responding to the observation of ingestive and communicative mouth actions in the monkey ventral premotor cortex. *European Journal of Neuroscience, 17,* 1703–1714.

Ferrari, P. F., Rozzi, S., & Fogassi, L. (2005). Mirror neurons responding to observation of actions made with tools in monkey ventral premotor cortex. *Journal of Cognitive Neuroscience, 17,* 212–226.

Flombaum, J. I., Junge, J. A., & Hauser, M. C. (2005). Rhesus monkeys (*Macaca mulatta*) spontaneously compute addition operations over large numbers. *Cognition, 97,* 315–325.

Fragaszy, D. M. & Cummins-Sebree, S. E. (2005). Relational spatial reasoning by a nonhuman: The example of capuchin monkeys. *Behavioral and Cognitive Neuroscience Reviews, 4,* 282–306.

Frank, M. C., Everett, D. L., Fedorenko, E., & Gibson, E. (in press). Number as a cognitive technology: Evidence from Pirahã language and cognition. *Cognition.*

Furlong, E. E., Boose, K. J., & Boysen, S. T. (2008). Raking it in: The impact of enculturation on chimpanzee tool use. *Animal Cognition, 11,* 83–97.

Gallese, V. & Lakoff, G. (2005). The brain's concepts: The role of the sensory–motor system in conceptual knowledge. *Cognitive Neuropsychology, 22,* 455–479.

Gallese, V., Fadiga, L., Fogassi, L., & Rizzolatti, G. (1996). Action recognition in the premotor cortex. *Brain, 119,* 593–609.

Gentilucci, M. & Corballis, M. C. (2006). From manual gesture to speech: A gradual transition. *Neuroscience and Biobehavioral Reviews, 30,* 949–960.

Gentilucci, M. & Dalla Volta, R. (2007). The motor system and the relationships between speech and gesture. *Gesture, 7,* 159–177.

Gentilucci, M., Stefanini, S., Roy, A. C., & Santunione, P. (2004). Action observation and speech production: Study on children and adults. *Neuropsychologia, 42,* 1554–1567.

Gibbons, A. (2007). Food for thought. *Science, 316,* 1558–1560.

Gick, M. L. & Holyoak, K. J. (1980). Analogical problem solving. *Cognitive Psychology, 12,* 306–355.

Gick, M. L. & Holyoak, K. J. (1983). Schema induction and analogical transfer. *Cognitive Psychology, 15,* 1–38.

Goldin-Meadow, S. (2006). Talking and thinking with our hands. *Current Directions in Psychological Science, 15,* 34–39.

Hare, B. (2007). From nonhuman to human mind: What changed and why? *Current Directions in Psychological Science, 16,* 60–64.

Hare, B., Call, J., & Tomasello, M. (2006). Chimpanzees deceive a human competitor by hiding. *Cognition, 101,* 495–514.

Hart, B. L., Hart, L. A., & Pinter-Wollman, N. (2008). Large brains and cognition: Where do elephants fit in?. *Neuroscience and Biobehavioral Reviews, 32,* 86–98.

Hattori, Y., Kuroshima, H., & Fujita, K. (2007). I know you are not looking at me: capuchin monkeys' (*Cebus apella*) sensitivity to human attentional states. *Animal Cognition, 10,* 141–148.

Haueisen, J. & Knösche, T. R. (2001). Involuntary motor activity in pianists evoked by music perception. *Journal of Cognitive Neuroscience, 13,* 786–792.

Hegarty, M. (1992). Mental animation: Inferring motion from static displays of mechanical systems. *Journal of Experimental Psychology: Learning, Memory, and Cognition, 18,* 1084–1102.

Holyoak, K. J. & Koh, K. (1987). Surface and structural similarity in analogical transfer. *Memory & Cognition, 15,* 332–340.

Iacoboni, M. & Dapretto, M. (2006). The mirror neuron system and the consequences of its dysfunction. *Nature Reviews Neuroscience, 7,* 942–951.

Jensen, K., Call, J., & Tomasello, M. (2007). Chimpanzees are rational maximizers in an ultimatum game. *Science, 318,* 107–109.

Johnson, K. E., Younger, B. A., & Furrer, S. D. (2005). Infants' symbolic comprehension of actions modeled with toy replicas. *Developmental Science, 8,* 299–318.

Johnson, M. & Lakoff, G. (2002). Why cognitive linguistics requires embodied realism. *Cognitive Linguistics, 13,* 245–263.

Jordan, K. E., Brannon, E. M., Logothetis, N. K., & Ghazanfar, A. A. (2005). Monkeys match the number of voices they hear to the number of faces they see. *Current Biology, 15,* 1034–1038.

Kenward, B., Weir, A. A. S., Rutz, C., & Kacelnik, A. (2005). Tool manufacture by naive juvenile crows. *Nature, 433,* 121.

Kilner, J. M., Friston, K. J., & Frith, C. D. (2007). Predictive coding: An account of the mirror neuron system. *Cognitive Processing, 8,* 159–166.

Kirsh, D. & Maglio, P. (1994). On distinguishing epistemic from pragmatic action. *Cognitive Science, 18,* 513–549.

Kohler, E., Keysers, C., Umilta, M. A., Fogassi, L., Gallese, V., & Rizzolatti, G. (2002). Hearing sounds, understanding actions: Action representation in mirror neurons. *Science, 297,* 846–848.

Kosslyn, S. M., Digirolamo, G. J., Thompson, W. L., & Alpert, N. M. (1998). Mental rotation of objects versus hands: Neural mechanisms revealed by positron emission tomography. *Psychophysiology, 35,* 151–161.

Kuczaj, S. A. & Walker, R. T. (2006). How do dolphins solve problems?. In E. A. Wasserman & T. R. Zentall (Eds.), *Comparative Cognition: Experimental Explorations of Animal Intelligence* (pp. 580–600). New York: Oxford University Press.

Kuhlmeier, V. A. & Boysen, S. T. (2002). Chimpanzees (*Pan troglodytes*) recognize spatial and object correspondences between a scale model and its referent. *Psychological Science, 13,* 60–63.

Kuhlmeier, V. A., Boysen, S. T., & Mukobi, K. L. (1999). Scale-model comprehension by chimpanzees (*Pan troglodytes*). *Journal of Comparative Psychology, 113,* 396–402.

Laden, G. & Wrangham, R. (2005). The rise of the hominids as an adaptive shift in fallback foods: Plant underground storage organs (USOs) and australopith origins. *Journal of Human Evolution, 49,* 482–498.

Lakoff, G. & Núñez, R. E. (2000). *Where Mathematics Comes from: How the Embodied Mind Brings Mathematics into Being*. New York: Basic Books.

Landy, D. & Goldstone, R. L. (2007). How abstract is symbolic thought?. *Journal of Experimental Psychology: Learning, Memory, and Cogntiion, 33,* 720–733.

Leavens, D. A., Hopkins, W. D., & Bard, K. A. (2005). Understanding the point of chimpanzee pointing. *Current Directions in Psychological Science, 14,* 185–189.

Lepage, J.-F. & Théoret, H. (2007). The mirror neuron system: Grasping others' actions from birth?. *Developmental Science, 10,* 513–529.

Lucas, P. W., Ang, K. Y., Sui, Z., Agrawal, K. R., Prinz, J. F., & Dominy, N. J. (2006). A brief review of the recent evolution of the human mouth in physiological and nutritional contexts. *Physiology & Behavior, 89,* 36–38.

MacNeilage, P. F. (1998). The frame/content theory of evolution of speech production. *Behavioral and Brain Sciences, 21,* 499–546.

Melis, A. P., Call, J., & Tomasello, M. (2006). Chimpanzees (*Pan troglodytes*) conceal visual and auditory information from others. *Journal of Comparative Psychology, 120,* 154–162.

Meltzoff, A. N. & Decety, J. (2003). What imitation tells us about social cognition: A rapprochement between developmental psychology and cognitive neuroscience. *Philosophical Transactions of the Royal Society of London B, 358,* 491–500.

Miklósi, A., Topál, J., & Csányi, V. (2007). Big thoughts in small brains? Dogs as a model for understanding human social cognition. *NeuroReport, 18,* 467–471.

de, A., Moura, A. C., & Lee, P. C. (2004). Capuchin stone tool use in Caatinga dry forest. *Science, 306,* 1909.

Mulcahy, N. J. & Call, J. (2008). Apes save tools for future use. *Science, 312,* 1038–1040.

Murai, C., Kosuig, D., Tomonaga, M., Tanaka, M., Matsuzawa, T., & Itakura, S. (2005). Can chimpanzee infants (*Pan troglodytes*) form categorical representations in the same manner as human infants (*Homo sapiens*)? *Developmental Science, 8,* 240–254.

Nieder, A. (2005). Counting on neurons: The neurobiology of numerical competence. *Nature Reviews Neuroscience, 6,* 177–190.

Nowak, M. A. & Sigmund, K. (2005). Evolution of indirect reciprocity. *Nature, 437,* 1291–1298.

Núñez, R. E. & Sweetser, E. (2006). With the future behind them: Convergent evidence from Aymara language and gesture in the crosslinguistic comparison of spatial construals of time. *Cognitive Science, 30,* 401–450.

O'Connor, M. (1996). The alphabet as a technology. In P. T. Daniels & W. Bright (Eds.), *The World's Writing Systems* (pp. 787–794). New York: Oxford University Press.

Pack, A. A. & Herman, L. M. (2007). The dolphin's (*tursiops truncatus*) understanding of human gazing and pointing: Knowing *what* and *where*. *Journal of Comparative Psychology, 121,* 34–45.

Pedone, R., Hummel, J. E., & Holyoak, K. J. (2001). The use of diagrams in analogical problem solving. *Memory & Cognition, 29,* 214–221.

Penn, D. C. & Povinelli, D. J. (2007). Causal cognition in human and nonhuman animals: A comparative, critical review. *Annual Review of Psychology, 58,* 97–118.

Pollick, A. S. & de Waal, F. B. M. (2007). Ape gestures and language evolution. *Proceedings of the National Academy of Sciences, 104,* 8184–8189.

Poole, J. H., Tyack, P. L., Stoeger-Horwath, A. S., & Watwood, S. (2005). Elephants are capable of vocal learning. *Nature, 434,* 455–456.

Raby, C. R., Alexis, D. M., Dickinson, A., & Clayton, N. S. (2007). *Nature, 445,* 919–921.

Rizzolatti, G. & Arbib, M. A. (1998). Language within our grasp. *Trends in Neurosciences, 21,* 188–194.

Russon, A. E. & Galdikas, B. M. (1993). Imitation in free-ranging rehab ilitant orangutans (*Pongo pygmaeus*). *Journal of Comparative Psychology, 107,* 147–161.

Scerif, G., Gomez, J.-C., & Byrne, R. W. (2004). What do Diana monkeys know about the focus of attention of a conspecific?. *Animal Behaviour, 68,* 1239–1247.

Schusterman, R. & Reichmuth, C. (2008). Novel sound production through contingency learning in the Pacific walrus (*Odobenus rosmarus divergens*). *Animal Cognition, 11,* 319–327.

Silk, J. B. (2006). Who are more helpful, humans or chimpanzees?. *Science, 311,* 1248–1249.

Skipper, J., Goldin-Meadow, S., Nusbaum, H. C., & Small, S. L. (2007). Speech-associated gestures, Broca's area, and the human mirror system. *Brain and Language, 101,* 260–277.

Suddendorf, T. & Corballis, M. C. (2007). The evolution of foresight: What is mental time travel, and is it unique to humans?. *Behavioral and Brain Sciences, 30,* 299–351.

Stevens, J. R. & Hauser, M. D. (2004). Why be nice? Psychological constraints on the evolution of cooperation. *Trends in Cognitive Sciences, 8,* 60–65.

Stoianov, I., Kramer, P., Umilta, C., & Zorzi, M. (2008). Visuospatial priming of the mental number line. *Cognition, 106,* 770–779.

Taub, S. F. (2001). *Language from the Body: Iconicity and Conceptual Metaphor in American Sign Language*. Cambridge: Cambridge University Press.

Tomasello, M. & Call, J. (2004). The role of humans in the cognitive development of apes revisited. *Animal Cognition, 7,* 213–215.

Troseth, G. L., Pickard, M. E. B., & DeLoache, J. S. (2007). Young children's use of scale models: testing an alternative to representational insight. *Developmental Science, 10,* 763–769.

van Schaik, C. P., Ancrenaz, M., Borgen, G., Galdikas, B., Knott, C. D., Singleton, I., Suzuki, A., Utami, S. S., & Merill, M. (2003). Orangutan cultures and the evolution of material culture. *Science, 299,* 102–105.

Ware, E. A., Uttal, D. H., Wetter, E. K., & DeLoache, J. S. (2006). Young children make scale errors when playing with dolls. *Developmental Science, 9,* 40–45.

Waga, I. C., Dacier, A. K., Pinha, P. S., & Tavares, M. C. H. (2006). Spontaneous tool use by wild capuchin monkeys (*Cebus libidinosus*) in the Cerrado. *Folia Primatologica, 77,* 337–344.

Warneken, F. & Tomasello, M. (2006). Altruistic helping in human infants and young chimpanzees. *Science, 311,* 1301–1303.

Wharton, C. M., Grafman, J., Flitman, S. S., Hansen, E. K., Brauner, J., Marks, A., & Honda, M. (2000). Toward neuroanatomical models of analogy: A positorn emission tomography study of analogical mapping. *Cognitive Psychology, 40,* 173–197.

Whiten, A. (2007). *Culture in primates*, December 11–14. Göttingen, Germany: Primate Behavior and Human Universals.

Whiten, A., Spiteri, A., Horner, V., Bonnie, K. E., Lambeth, S. P., Schapiro, S. J., & de Waal, F. B. (2007). Transmission of multiple traditions within and between chimpanzee groups. *Current Biology, 17,* 1038–1043.

Wilson, M. (2001). Perceiving imitatible stimuli: Consequences of isomorphism between input and output. *Psychological Bulletin, 127,* 543–553.

Wilson, M. (2001b). The case for sensorimotor coding in working memory. *Psychonomic Bulletin & Review, 8,* 44–57.

Wilson, M. (2002). Six views of embodied cognition. *Psychonomic Bulletin & Review, 9,* 625–636.

Wilson, M. & Fox, G. (2007). Working memory for language is not special: Evidence for an articulatory loop for novel stimuli. *Psychonomic Bulletin & Review, 14,* 470–473.

Wilson, M. & Knoblich, G. (2005). The case for motor involvement in perceiving conspecifics. *Psychological Bulletin, 131,* 460–473.

Wood, J. N., Glynn, D. D., Phillips, B. C., & Hauser, M. D. (2007). The perception of rational, goal-directed action in nonhuman primates. *Science, 317,* 1402–1405.

Yeakel, J. D., Bennett, N. C., Koch, P. L., & Dominy, N. J. (2007). The isotopic ecology of the African mole rats informs hypotheses on the evolution of the human diet. *Proceedings of the Royal Society B, 274,* 1723–1730.

Zentall, T. R. (2006a). Imitation: Definitions, evidence, and mechanisms. *Animal Cognition, 9,* 335–353.

Zimmer, C. (2005). *Smithsonian intimate guide to human origins.* New York: HarperCollins.

Zwaan, R. A. & Taylor, L. J. (2006). Seeing, acting, understanding: Motor resonance in language comprehension. *Journal of Experimental Psychology: General, 135,* 1–11.

20

THINKING WITH THE BODY: TOWARDS HIERARCHICAL, SCALABLE COGNITION

RICARDO SANZ, JAIME GÓMEZ, CARLOS HERNÁNDEZ AND IDOIA ALARCÓN

Autonomous Systems Laboratory, Technical University of Madrid, Jose Gutierrez Abascal 2, Madrid, Spain

INTRODUCTION

The reflection on the nature of mind has a long history. In our western tradition, this reflection has mostly taken place along the so-called dualist approach, where mind and body completely have different characteristics and even natures. In this dualistic context for understanding thought (and life!), the biggest problem has always been how these two realms—the physical and the mental—could possibly interact: the mind–body problem is the unavoidable sequel to this dualist standpoint.

The embodied cognition movement tries to reconcile this apparently multiple qualities (*duality* and *unity*) of human experience by means of analyzing the ways in which the body may affect cognition: supporting, sustaining, shaping, etc. Too many terms for a relation that is surely much simpler than generally thought or described in the literature. There is, in fact, nothing to reconcile. Mind and body cannot be separated because cognitive agents think with the body. Mind, as a separate entity, is in the eye of the beholder.

What we want to contend here is that the mind–body problem is not a problem of minds and bodies in the world—that is, a *physical* problem of interaction—but simply an artificial, conceptual problem for philosophers/scientists. The way to come out of the problem is to realize that minds and bodies are not separate entities, but what are separate are the mental concepts used in thinking about them; that is, what most thinkers use to think about minds and bodies as ontologically fully separate entities. However this way of thinking is misleading. Minds can be reduced to bodies because they are simply processes that run on them.

Explaining away the mind–body problem, then, does not require an explanation of the mind–body relation but the adoption of a particular systemic perspective: the mind as the control process of the body.

The approach that we propose may be considered to be very similar to conventional embodiment (Anderson, 2003) or may be viewed as completely divorced from embodied cognition and plagued with panpsychism. There are no minds without bodies and there are no bodies without minds—obviously ranging from the maximally complex to the practically inexistent. There are complexity ranges in all dimensions and cognitive agents go from the simple to the complex in both bodily and mental aspects.

SEPARATING MIND AND BODY

Our starting point involves some terminological changes. Following from the diagnosis above, we can say that the correct wording is not that *the mind emerges-from/is-supported-by/is-shaped-by bodily processes* but that indeed, those *bodily processes are the mind itself*. This may seem obvious for those processes going on inside the brain and it leads to the simple identification of the mind with whatever the brain is doing. With "bodily processes," however, we are not only referring to the processes occurring in the brain but perhaps in line with artificial life, to all the information-centric processes that constitute the very inner workings of the hierarchical structures of life.

Everything revolves around the notion of organic modularity (Callebaut & Rasskin-Gutman, 2005) and the provision of robust functions by the different subsystems that constitute a living being. Take for example the case of cardiac pace control (Rezek, 1997). The single purpose of the cardiovascular system is the transport of chemicals to be infused into cells across the whole body. If we consider, for instance, the heart, its function is to increase blood pressure to make the blood circulate against the resistance of elastic tissues. To provide the necessary cardiac robustness the heart can autonomously control its beating (Figure 20.1).

Obviously the autonomy of heart pacing is limited because the heart must also respond to the needs of other parts of the body (muscles, viscera, etc.) that are transmitted by different kinds of signals coming from different control levels—including the cortex-level mind. Indeed, the core system-integrated control of heart rate originates in the circulatory centers of the medulla oblongata and pons, in the brainstem. The control signals reach the heart through sympathetic and parasympathetic nerves which affect many aspects of cardiac operation: force of cardiac contraction (inotropism), rate of cardiac relaxation (lusitropism), heart rate (chronotropism), and impulse conduction (dromotropism) (Opie, 2003).

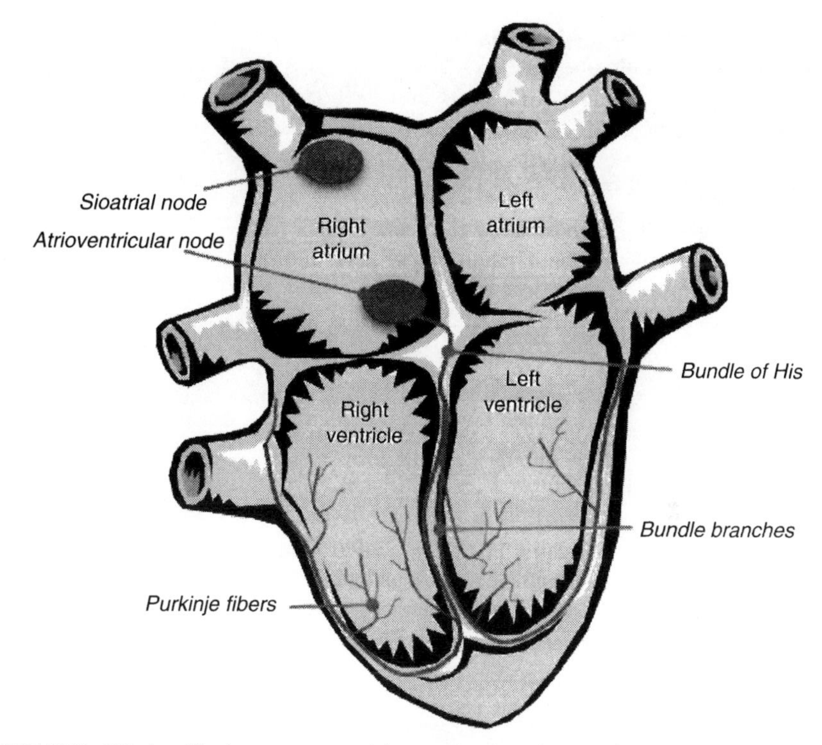

FIGURE 20.1 The heart can control its own beating—its core function—almost autonomously. The inherent cardiac rhythm originates in the pacemaker cells of the sinoatrial node to eventually reach the atrioventricular node, then through the bundle of His and its three branches—right, left anterior, left posterior—ultimately reaching the Purkinje cell network that interacts with myocardic cells. Obviously heart pace can also be controlled by upper control centers in the brain through sympathetic and parasympathetic nerves in be able to respond to demands from higher cognition processes.

The heart control example helps us to address the differences and mergers between two classes of functions that occur in biosystems: core functions and control functions. In the case of the heart, its core function is pumping—with adequate volume and pressure—and this function is performed by cardiac muscle and cardiac valves. The control function regulates pacing and pressure and is performed by the sinoatrial and atrioventricular nodes, the bundle of His and the Purkinje cells. However, cardiac muscle is very specialized, being the only type of muscle that is myogenic. This means that it can naturally contract and relax without receiving electrical impulses from nerves—that is, it is mostly autonomous, incorporating its own control function in part.

The observation that biological function is performed by a physical core process with an overlaid control process can be generalized down to the molecular biology of the cell and up to the psychophysics of the mind. Control is pervasive in biological function; according to Bayliss (1966), *we are living control systems*. There are numerous examples of analysis of physiological systems in terms of classical control theory (see e.g., Khoo, 1999). These control-theoretic analyses are not restricted to physical properties of biological systems, but are also applied to information-centric mental traits (Powers, 1973; Carver & Scheier, 1981; Nelson, 1993; Marken, 2002).

Notwithstanding the pervasiveness of the core process versus control process dichotomy, two important facts should not be overlooked:

- Control processes are necessarily realized as physical processes. Information, the very matter of control, does not exist in thin air and needs physical realization (Landauer, 1992).
- Sometimes the core and the control processes cannot be clearly separated because they are so intertwined in their physical realization or because the core process has intrinsic control capabilities; that is, it is a self-regulated process (as is the case with myocardic cells thanks to their myogenic capability).

This dichotomy—core processes versus control processes—is indeed the dichotomy of body and mind where minds are the control processes of physical bodies—or so we claim. However, while this separation can be completely clear cut in machines specifically designed in that way (process/control), it is not the case in biological systems. This is because evolution tends to operate in terms of modularization of function (Klingenberg, 2005) and not in terms of modularization of control and core processes; and also *because the very phenomenon of life is in itself a phenomenon of control* over complex physicochemical processes (Rosen, 1991). In any case, minds are informational processes controlling physical processes and realized on them. Agents are composed of *plants* and *controllers* that can usually be told apart in technical systems but not so easily in biological systems.

From an analytical perspective, therefore, the mind–body relation is really a relation between the physical and informational realms. The *information extraction frontier* is an interface where physical reality is transformed into informational reality. In biosystems this frontier is sometimes not very clear due to the bodily imbrication of control mechanisms but, in general, it is identified as sensorial system. The *information realization frontier* is an interface where informational reality is transformed into physical reality. Accordingly, this is generally identified in biosystems with the motor system.[1]

[1] A great deal of discussion has taken place around the notion of information in biological systems and the *extraction* versus *creation* of information by the agent from the environment. Obviously there is information, in the Shannon sense, in the environment. Only part of this is accessible to the agent through its senses and only part is relevant to its dwellings. We agree with the Batesonian view that what is informative for the agent is "a difference that makes a difference," that is, information from the world that is profitable for the agent (Bateson, 1979).

In technical systems these frontiers are usually well known and engineered for clear separability. They are composed of two sections: sensors and actuators. Sensors map the physical into the informational. Actuators work in reverse (informational into physical). In a sense, the existence of this frontier defines what a mind is. In the purest sense of systems analysis (Klir, 1969), this frontier specifies what should be considered *mental* and what *physical* (Figure 20.2).

Obviously, in purely psychological domains the firings of Purkinje cells (muscle cells) are not considered mental at all, but this specification of the informational/physical frontier is much clearer than many more arbitrary separations considered so far, and is much more in line with the separation of what is *system* and what is considered to be *environment*.

Additionally, minds are necessarily supported by a physical substrate. The realization of informational processes requires the use of physical body parts to deal with informational states. These parts can play a role in certain core physiological processes or can be fundamentally dedicated to informational tasks. The central nervous system (CNS) is a clear case of the latter. But we should never forget that, in general, the physical coupling of the physical substrate of mind and other bodily processes is maximal, for example, the state and competence of our mind totally depends on blood sugar content in strict connection with the functioning of liver and pancreas (Figure 20.3).

The need for a physical substrate entails that there are no disembodied minds. What we can find are minds whose physical substrate is partially independent of the physical body they control. Some may understand disembodiment in these terms. In fact it is a basic design objective of control engineers that the controller substrate is not affected by the physical events in the plant they control. However there is no possibility of total isolation for two reasons:

- Isolation is costly. Replacing copper wires with fiber optics to reduce electromagnetic interference with control signals costs a lot of money and

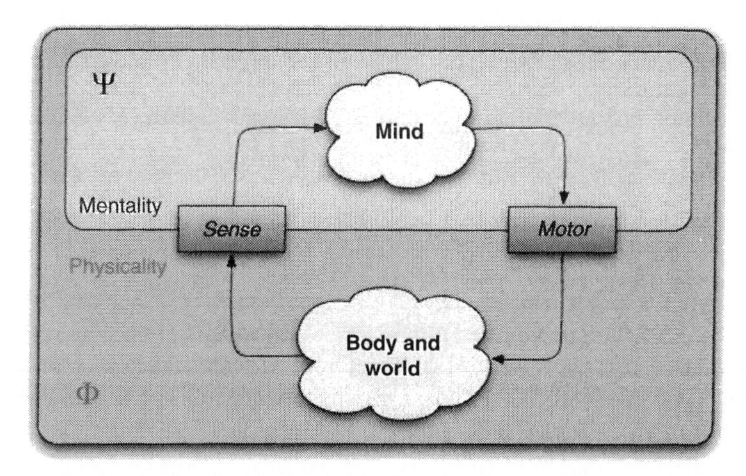

FIGURE 20.2 The physical-informational frontier defines the scope of mind.

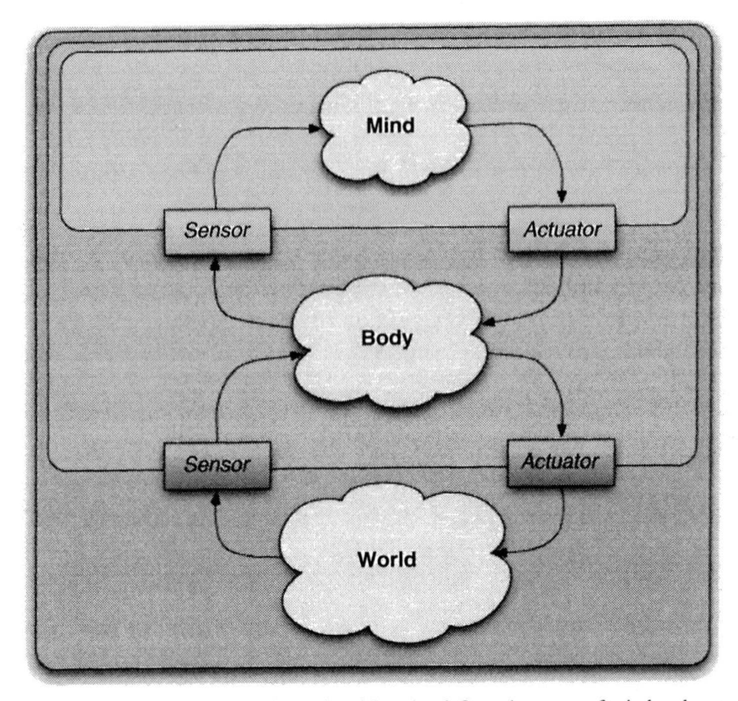

FIGURE 20.3 The physical-informational frontier defines the scope of mind and system. But this scope is arbitrary to the extent that—according to systems theory—what is the system and what is the environment is a contingent decision. In the analysis of cognitive systems we use three arbitrary parts to separate the universe—environment, body, and mind—relating states to one another. In some cases the specific physical realization of such states is important for the achievement of certain physical objectives. In other cases it is not.

a trade-off is mandatory. So, control engineers simply strive for sufficiently good isolation. Evolution must have worked similarly for biosystems for similar reasons.

- More importantly, information frontiers—where the physical becomes mental—are necessarily physical couplings between the mind's physical substrate and the rest of the body. While the previous reason for not having total isolation was purely economic and hence mostly contingent, this reason is absolutely necessary as the functional couplings are indispensable.

So, there are no disembodied minds, unless they are isolated from the body, in which case they are not minds. Therefore, "disembodied mind" is an oxymoron. "Embodied mind" is a truism. Brains in vats are just that: brain in vats; they could produce mentality but only to the extent of the enormously limited mind–body relation that such a limited body could sustain.

THE PHENOMENON OF CONTROL

Based on general analysis given in the previous section, we can now begin to characterize the phenomenon of mind as a phenomenon of control. The proper setting for such an analysis is within control systems theory and practice. This is a discipline based on dynamic systems theory and specifically focused on the dichotomy of body–mind, or plant controller, in our case.

Control theory (Ogata, 1990), as understood in the control world, is a deeply theoretical, mathematical endeavor. Control engineering is the engineering side where the theoretical results are put into practice in the form of controllers for machines and processes. There exists, however, a big gap between theory and practice—as is the case in all engineering disciplines:

- The theoretical results may turn out to be non-applicable for several reasons, among which there are: lack of understanding by practitioners, excessive constraints for their application, lack of plants matching the theoretical models, etc.
- There are domains of control technology lacking in theory. This lack may be due to missing interest on the part of control theorists (e.g., sensor drift problems) or may be due to a lack of an adequate formal model (e.g., human supervisory control).

In a sense, control theory has been driven by its mathematics, reaching a situation very similar to that of pure mathematics: disconnection from the real world. And, like most theoretical endeavors, it suffers from the "*Consider a spherical cow* ..." syndrome, producing solutions for yet-to-come problems. On the other side, control practice suffers from generalized under-training and lack of rigor in many of its activities. This is a purely economic management issue, because reasonably good solutions are enough for the real world.

In control systems analysis and design, the term "plant" is used to refer to the system we are interested in controlling and the term "environment" is to refer to the rest of the universe (Figure 20.4). Obviously, there are interactions between the plant and the environment which affect both the dynamics of the plant and the environment. We are interested only in isolated systems as degenerate theoretical cases of this interaction. The interaction can be classified into three categories:

Outputs: The quantities[2] coming from the plant that we are interested in. This could be, for instance, the production level in T/h in a cement plant or vehicle speed in a cruise control system.

Inputs: The quantities that we can manipulate to drive the plant to the operational point we are interested in. In the previous examples, these would be the coal burning rate or the position of the car throttle.

[2] The term *quantity* is used here in the precise sense proposed by Klir.

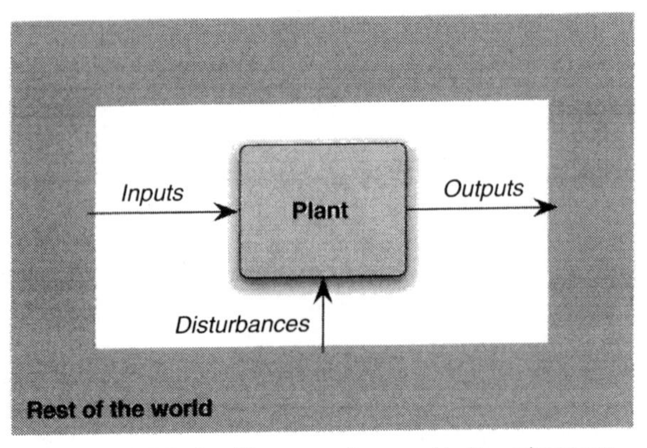

FIGURE 20.4 The system immersed in its environment.

Disturbances: They are material or energetic flows from the environment that cannot be controlled but nevertheless affect the plant's operation. Examples of this are the level of humidity in raw materials or the force of the wind in the road.[3]

The phenomenon of control can be simply stated (Figure 20.5) as:

> *If the dynamics of the interaction of a plant with its environment is not as desired—in terms of some observable quantities—it is possible in general to complement the system with an additional subsystem—a controller—so that the resultant dynamics of the system plant + environment + controller renders the desired dynamics at the target quantities.*

The task of devising the appropriate controller for a given plant and a given set of objectives is called *control design*. In the case of biosystems, the "designer" is evolution. This is apparently non-teleological but if analysed in detail is exactly equivalent to a teleological mechanism addressing *selfish gene* objectives (Dawkins, 1976). The control design problem is an inverse mathematical problem that can be exceedingly difficult to solve (and indeed is insolvable in many cases). The common strategy to achieve a solution in difficult cases is dual: it tries to simplify the mathematical problem by making approximations[4] and relaxing the target objectives.

We have said that it is possible "in general" but not "always" to complement a plant with a controller so as to reach a concrete global dynamics, because in some cases the necessary controller is physically unrealizable (e.g., would require non-causal behavior). The process of realization comes after the

[3] Strictly speaking there are also undesirable flows from the plant to the environment. In the past they were mostly ignored (if reasonable); now, the widely accepted *ecological* perspective forces control engineers to consider them as *outputs to be controlled*.

[4] The most common simplification is to linearize the models of the plant under control (because linear problems are easier to solve).

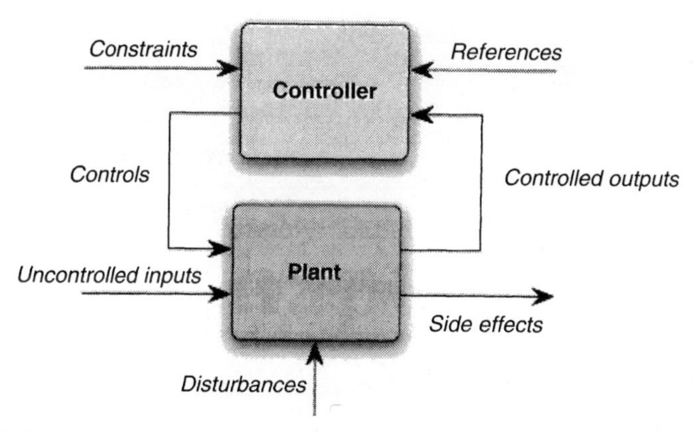

FIGURE 20.5 The controller is an additional subsystem so that the resultant dynamics of the system <plant + environment + controller> renders the desired values at the target output quantities.

design phase and is called *controller implementation*. Some of the implementational problems are directly addressed in the design phase (e.g., non-causal controllers are not considered acceptable designs) but other implementational problems cannot be handled in the design phase. The principal reason is that the construction process cannot normally be sufficiently formalized to be of use in the design process, which is deeply mathematical.

The most common implementational strategy today is the construction of controllers as software, as sets of interacting programs. So ultimately, the resulting mind controlling the physical body of the plant is a collection of interacting software *processes*—this being a precise term in computer science— running atop some computer and communications hardware. The discipline of control engineering has become a discipline of control + computing + communication. The computer metaphor for the mind is no longer a metaphor (Searle, 1990; Cisek, 1999); it is a technological asset.

CONTROL FROM BODY TO MIND

A particular topic that is very well addressed by control theory is that of linear feedback control.

A closed-loop feedback controller uses information coming from the actual outputs of the plant to determine the proper control actuations over it. Its name comes from the flow of information along the system: plant inputs (e.g., throttle in the cruise control system example) have an effect on the plant outputs (e.g., car speed), which is measured with sensors and processed by the controller; the result (the control signal) is used as input to the plant, closing the action loop (Figure 20.6).

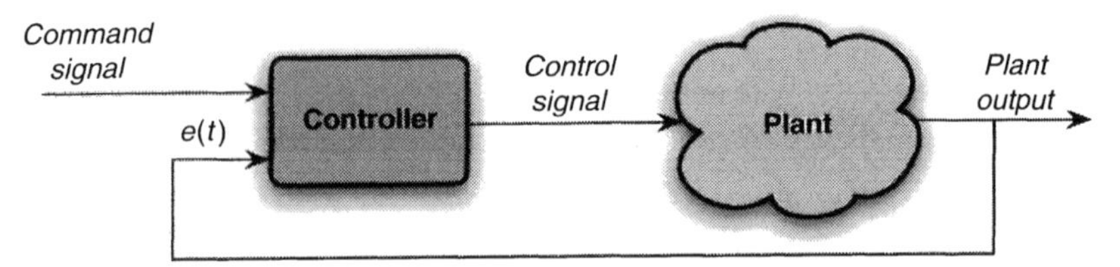

FIGURE 20.6 A closed-loop feedback controller uses information coming from the actual outputs of the plant to determine the appropriate control actuations over it.

In theory, all this complication is unnecessary because if we have a good model of the plant (that is, a formal mapping between its inputs and its outputs), the easiest and perfect control strategy is to invert such a model and use it as a controller (using the desired output as a reference to the controller):

$$desired\_output \rightarrow \textbf{PlantModel}^{-1} \rightarrow input \rightarrow \textbf{Plant} \rightarrow desired\_output$$

There are, however, two big problems with this strategy: (i) it is not easy to obtain a perfect model of a plant of non-minimal complexity and (ii) this model, if obtained, may not be invertible. So, in general, our controllers are based on plant models that differ from the real plant; that is the reason for using closed-loop controllers.

Closed-loop controllers have many advantages over open-loop controllers, mostly related to their capability in handling unmodeled dynamics:

- Disturbance rejection, that is, to make the system robust against unmodeled perturbations (such as unmeasured friction in a motor).
- Performance even with model discrepancies, when the model structure does not match the real plant perfectly.
- Unstable processes can be stabilized, that is, producing a qualitative change in the dynamics of the system.
- Robustness against plant drift, having reduced sensitivity to plant parameter variations.
- Improved reference tracking performance in the presence of noise.

The major drawback they have is that feedback control is, in a sense, necessarily slower because the controller only reacts when the plant departs from the desired behavior (i.e., after things go wrong). Alternative structures—like *feed-forward* control— are employed to compensate for these drawbacks and avoid going *behind the plant.*

PID CONTROLLERS

The most common control strategy uses a simple linear feedback to compensate for errors, speed of change, and accumulated error. It is called the PID controller—Proportional–Integral–Derivative—and refers to the three terms operating

$$u(t) = K_p \cdot e(t) + K_d T_d \frac{de(t)}{dt} + \frac{K_i}{T_i} \int e(t) \cdot dt$$

FIGURE 20.7 The PID controller is the most common control strategy, computing the control signal $u(t)$ from three terms on the tracking error $e(t) = r(t) - y(t)$.

on the error signal to produce a control signal (Åstrom & Hagglund, 1995). A PID controller has the general form as shown in Figure 20.7 where $u(t)$ is the control signal sent to the plant by the controller and $e(t)$ is the tracking error and can be shows as $e(t) = r(t) - y(t)$ (where $y(t)$ is the measured output and $r(t)$ is the desired output or reference). K_p, K_d, K_i, T_d, and T_i are the adjustment parameters of the controller.

MODE SWITCHING CONTROLLERS

In general the control capability of a PID controller is adequate to achieve good transitory and steady regime responses. In some cases, however, the simple linear response of the controller does not achieve the desired results and alternative strategies must be used. Even worse, the different strategies will only work adequately in some regions of the plant state space, where no single better alternative is available. In these cases the usual technique is called *mode switching control*, where the control system employs one single strategy from a variety of possible strategies, making the choice in terms of the present operational condition of the plant.

Figure 20.8 shows a simple mode switching control structure. The main difference between this mode and the basic feedback controller in Figure 20.6 is the existence of a battery of alternative control strategies:

- A *PID controller*, using the strategy described earlier
- A *bang-bang controller*, that provides increased responsiveness whilst sacrificing precision (this is used, e.g., in achieving minimum journey time in subway trains somewhat sacrificing passenger comfort)

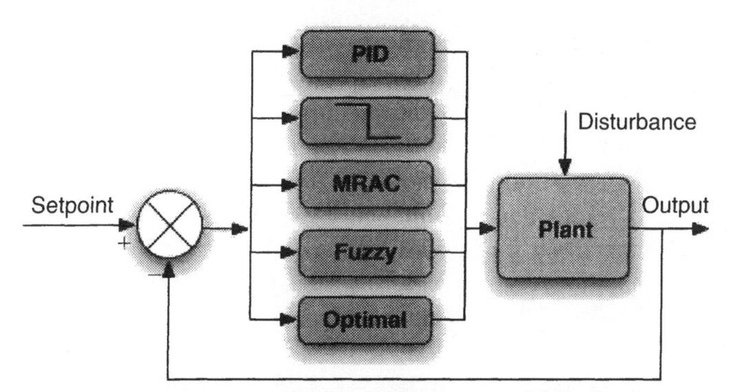

FIGURE 20.8 Mode switching controllers (MSC) select the best strategy to use—from a set of predefined strategies—for the current situation of the plant. The strategies in this case are (i) a PID controller, (ii) a bang-bang controller, (iii) a model reference adaptive controller (MRAC), (iv) a fuzzy controller, and (v) an optimal controller (Levine, 1996).

- A *model reference adaptive controller (MRAC)*, that will be described later in section "Model-Reference Adaptive Control"
- A *fuzzy controller*, that exploits control rules expressed in linguistic terms
- An *optimal controller*, that is, being deeply model based, and which tries to maximize certain relevant functions.

However, in the normal use of this control structure, the various controllers employed differ only in their parameterization and not in their strategy. Reference to such controllers can easily be found in the control systems' literature but a good, all-encompassing text, is that of Levine (1996).

MODEL-PREDICTIVE CONTROL

The foregoing discussion of feedback control has shown that the main problem for this strategy is that it is always *behind the plant*, and it can achieve the desired output values only in response to extant errors.

This is a big problem in certain kinds of systems, where reaching the desired output values at a specific moment in time is of maximum importance. Think for example of hitting a tennis ball with a racket. The racket must be in the exact position at the precise instant when the ball is passing. Error feedback control cannot achieve this. The only possibility is to have some form of anticipation.

Model-predictive control is a strategy based on predicting the future trajectory of a system by means of a *model* of it, and using this anticipatory capability to determine, at the present time, the control action necessary for taking the system to a certain state in a precise future instant (Figure 20.9).

The realization of the model-predictive controller usually takes the form of a two-level layered control. The outer layer—the MPC controller itself—uses the plant models and the current plant output measurements to calculate future trajectories of the manipulated variables that will result in an operation that fulfills all the constraints. The MPC layer then sends this set of manipulated variable changes to the inner layer regulatory controller as setpoints to be pursued in the process.

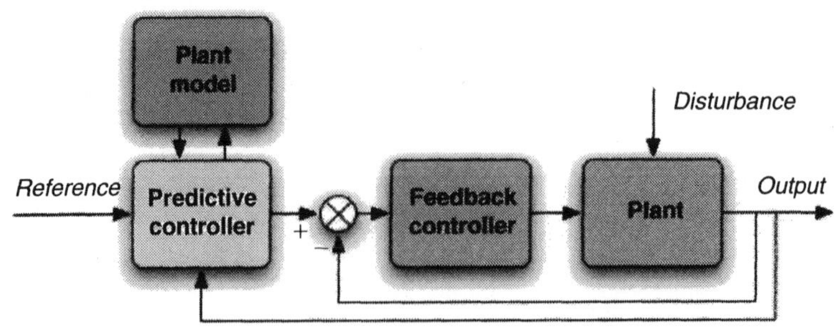

FIGURE 20.9 Model-predictive control is a strategy based on predicting the future trajectory of a system based on a model of it, and using this anticipatory capability to determine, at the present time, the control action necessary for taking the system to a certain state in a precise future instant.

There are two basic strategies for incorporating the models into the agent control structure:

Model learning: The agent builds the model of the plant by means of a controlled series of interactions. From a collection of action–effect pairs, the agent tunes the parameters of a specific model pattern. In the case of biological systems, the model pattern is a neural network realized in associative neural tissue that builds a concrete organization directed to generate action. The model-controller organization can be optimized (automated) to be faster by collapsing the model and the controller (in these systems, the separation of model and controller disappears). In the case of technical systems, model learning is carried out using a similar strategy called *system identification* (Ljung, 1987); this is used to tune the parameters of a certain class of dynamical models (hence presupposing a certain class of plant).

Model uploading: In cognitive agents with cognitive input/output capability, it is possible to externalize and internalize models from other agents that share a particular execution architecture. This is done using language to express the cognitive model in a way that can be used by other agents. In the case of technical systems this language is artificial and usually in close relation with the concrete implementation of the controller. This is currently changing, however, and a research trend now in control systems engineering is the exploitation of so-called neutral models; that is, models that are not built for a specific class of architectures or uses.

In both natural and artificial cases, the value of the model comes from its anticipatory capability in terms of agent goals (Rosen, 1985). The basic question here is the degree to which model and plant are isomorphic as this is what will make predictions reliable.

In general terms, any dynamical system can be infinitely approximated using a modeling formalism of the adequate capability. In the case of neural tissue or of artificial neural networks, it has been demonstrated that they offer universal approximation capability (i.e., they can approximate any system up to any required level of accuracy). The same can be said about the modeling formalisms used in control engineering, since differential equations—continuous, or difference equations, discrete and hence computable, are also universal approximators.

This means that a computer executing a difference equation model of adequate precision can capture any mental dynamics—up to any arbitrary level of accuracy. The exact level of accuracy is determined by a trade-off between the gains achieved by being more precise and the economic cost of doing so. This is the case for both natural and artificial beings.

MODEL-REFERENCE ADAPTIVE CONTROL

Adaptive control systems are reflective controllers able to modify their control law to cope with the fact that the parameters of the plant are uncertain or are

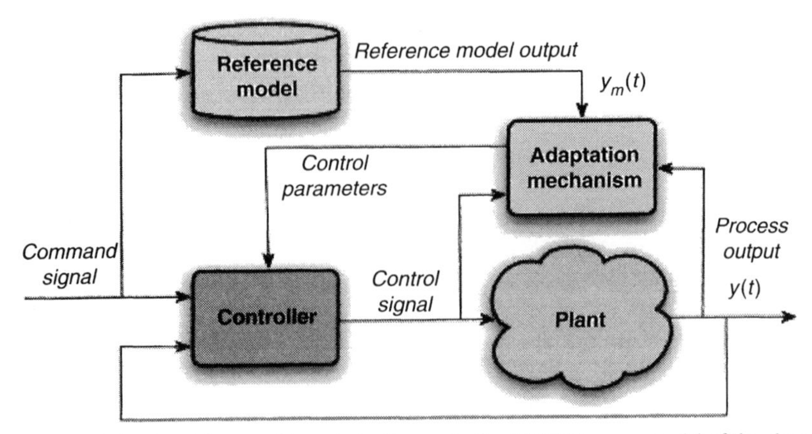

FIGURE 20.10 The model-based adaptive controller (MRAC) uses a model of the plant to determine the extent of deviation from the controller design conditions. This deviation is then used to re-tune the controller.

drifting from the initial conditions. For example, as a spacecraft flies, its mass will continuously decrease as a result of fuel consumption, and the control law must change accordingly. Another example is that of heavy production processes (e.g., cement manufacturing) where some sensors are constantly changing their measuring capability in a process of continuous grinding.

Figure 20.10 shows the architecture of a model-based adaptive controller. This controller has a law that is used to control the plant and at the same time it uses a model of the plant to determine to what extent the real plant is departing from what was envisaged. The behavioral differences between the real and the expected are then used by the adaptation mechanism to re-tune the control law parameters to increase its suitability for the real plant.

HIERARCHICAL CONTROL

A real plant can be very simple or can be extremely complex. Room thermostats—a favorite in philosophy of mind—are bang-bang controllers of extreme simplicity that are controlling a single magnitude in the plant room temperature. A real temperature control in a chemical industrial reactor can imply tens of sensors, actuators, and heterogeneous-nested control loops to achieve the desired performance.

A real industrial plant can have thousands of magnitudes under control and the organization of all these control loops is a major control system design challenge. This is so because not only must the different magnitudes be under control but they must be so in a *co-ordinated way* in order to achieve the global objectives of plant operation.

The strategy used to accomplish this is to organize the control loops in a hierarchy where low-level references for controllers are computed by upper-layer

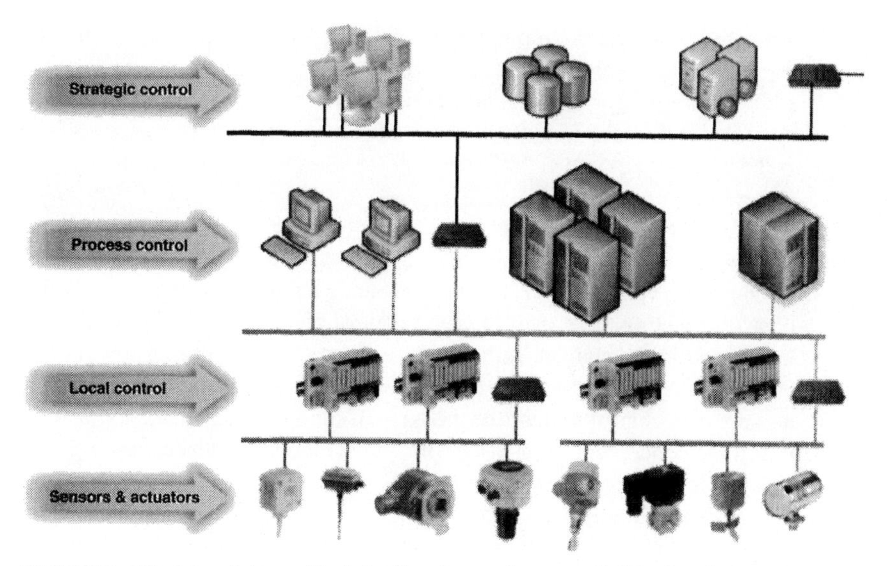

FIGURE 20.11 A hierarchical distributed control system (DCS) of an industrial plant is structured in many different control layers. Control objectives become more abstract at higher levels of control. The lower the level, the bigger the importance of temporal criticality and precision.

controllers that try to achieve more abstract and general setpoints. For example, in the reaction unit of a chemical plant, many low-level controllers control individual temperatures, pressures, flows, etc. to fulfill the higher level control objectives of the unit such as production and quality levels (Figure 20.11).

What is most interesting in studying the phenomenon of control in biological systems and large-scale process plant controllers are the striking similarities between both. While robot control systems in many cases try to mimic biosystems by exploiting what is known—or hypothesized—about their control systems, in the case of process plants, the bioinspired movement is yet to arrive (except possibly at the levels of expert process control, Åström et al., 1986).

Industrial control systems technology has developed following its own evolutionary path from the early analogical controllers of the mid-20th century to the fully computerized, zillion lines-of-code of today's whole plant controllers. Different types of organizations have appeared in the structuring of the core processes, in the structuring of control architectures, and, quite recently, in the co-structuring of process and control.

From the perspective of the control system we can observe an evolution that somewhat parallels the development of mental capabilities in biosystems:

1. The most simple control mechanism is a purely reactive mechanism that triggers some activity when certain conditions are met. Some examples of these are a large part of all protection and safety mechanisms in industrial systems. The overall behavior is similar to a multitude of safety reflexes in biosystems.

2. An additional level of complexity is achieved when the raw sensorial information is minimally processed to extract meaningful information for behavior triggering. This is done in elementary control and protection systems. In the case of biological systems, a well-known study in this field is the work of Lettwin et al. (1959)that involves retinal processing in frog's eyes.[5]
3. The next layer appears when it is possible to conceptualize the operation of the controller and feed its specific parametric values (e.g., setpoints or controller parameters). This layer is hence integrable with upper-level controls opening the possibility for a control hierarchy. It is also well known in biosystems that some motor actions coming from upstream in the CNS are executed by low-level controllers (core examples are the homeostatic control systems of the body, Cannon, 1932).
4. Using the conceptual openness of the control loop it is hence possible to layer control loop over control loop—this is called control loop nesting—so that upper-level behavior relies on the robust performance of lower level behavior—thanks to the integrated controller. In this way it is possible to use a production quality control in a chemical reactor with a plethora of lower level controllers underneath keeping flows, pressures, and temperatures at a suitable level. Following the homeostatic example of the previous case, we discover that large systemic processes—for example, digestion—rely on lower level processes to keep bodily magnitudes apace. Another interesting example is how the process of gait control relies on lower level muscular control (Grillner, 1985).
5. An interesting step forward occurs when engineers reach the conclusion that it is possible in general to separate controllers into two parts: a universal engine and data that specifies the particular control strategy to follow. This opens new possibilities for the reuse of engines. A clear example of this are the MPC controllers mentioned in section "Model-Predictive Control" and the controllers based on expert systems technology (Sanz et al., 1991).
6. The next and most interesting step in the development of complex control systems is the realization that a conceptualization of this separability (engine + knowledge) renders a new level of controller openness to metacognitive processes (Meystel & Sanz, 2002). In the case of human control systems this gives rise to introspection capabilities and the well known phenomenon of memetics and culture (Blackmore, 1999).

[5] "The output from the retina of the frog is a set of four distributed operations of the visual image. These operations are independent of the level of general illumination and express the image in terms of 1) local sharp edges and contrast, 2) the curvature of edge of a dark object, 3) the movement of edges, and 4) the local dimmings produced by movement or rapid general darkening. ... Could one better describe a system for detecting an accesible bug?" (Lettwin et al., 1959).

What is most interesting in the parallelism between technical industrial control systems and biological controllers is that they have come about in almost complete isolation. Certainly, the evolution of technical controllers has not substantially affected the evolution of control mechanisms in biosystems. However, the opposite is also true—with the possible exception of knowledge-based control where human expertize does figure in the technical system.

This could be interpreted in the sense that evolutionary pressure on control/cognition points toward the direction of layered metacognitive controllers; that is, it points toward the direction of consciousness (Sanz et al., 2002). In order to fully grasp this phenomenon a deeper analysis of the model-based nature of the control capability is needed.

MODEL-BASED COGNITION

As we have seen, models of the plant are critical assets in the construction of control systems. They are used to capture plant dynamics and serve as base information for the control design process. Indeed, they are explicitly used in some control schemata (e.g., model-predictive control).

Therefore, controller quality depends heavily on our capability to model the plant to be controlled adequately. However, modeling is not only essential as a supporting activity for humans performing a control system design process but also it is essential as a supporting activity because plant models become, in one form or another, parts of the controller. This leads to a crucially important conclusion: A controller is as good as its capability to exploit internalized models of the system it is controlling (Conant & Ashby, 1970).

At the end of the day, no matter what the controller architecture and the design process is, plant models become an integral part of effective controllers. Even the simplest controllers (e.g., the PID) have parameters that capture the plant's dynamics (the K_p, K_d, K_i, T_d, and T_i of Figure 20.7).

This assertion is important for cognitive science as will be seen later when *models* and *knowledge* will be equated to provide a formal grounding for an epistemological analysis of cognition-in-the-world. In doing so, we are moving the problem of knowledge from a cloudy philosophical context (Gettier, 1963) into the more precise and tractable waters of modeling (Zeigler & Tag Gon Kim, 2000).

STRUCTURE OF A MODEL-BASED COGNITIVE AGENT

The model-based control approach is very much in line with current trends in cognitive science that call for an understanding of minds in terms of internalized models. This trend began with Craik (1943) in the 1940s and gained popularity with the works of Johnson-Laird (1983) and Gentner and Stevens (1983), but still

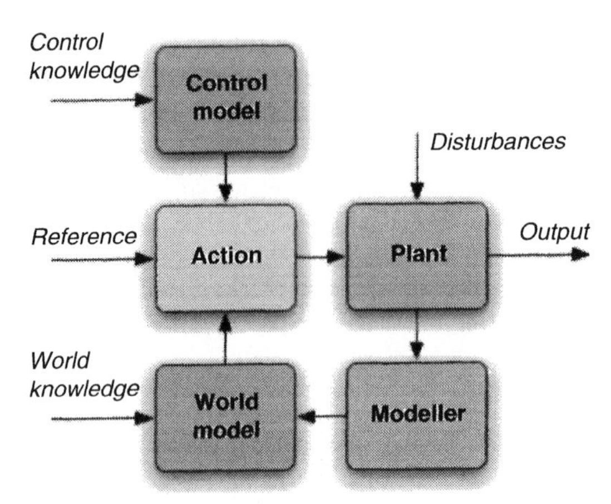

FIGURE 20.12 In generalized model-based control, the controller itself is able to generate the world models—a superset of the plant model—to be used in the performance of control actions. To increase generality the controller may employ a source of explicit knowledge about the control strategy to follow.

has to be cashed out as more than just a metaphor. Minds are to be understood as model-based control systems. According to the theoretical, system-level analysis of Conant and Ashby (1970), there is an evolutionary pressure to develop this type of architecture of cognition, and cognitive science will, sooner or later, necessarily take the model-based approach as a core disciplinary doctrine.

Figure 20.12 shows a basic structure for a generalized model-based controlled system that can both (i) generate its own plant models and (ii) accept external injection of explicit knowledge. In the interaction with the plant, the pair action/modeler performs both a change of plant state and a change in the controller itself. The *agent*—the name used to refer to the system—is able to create mental representations of the world it is interacting with and that includes both the plant and the world affecting it—its environment.

This kind of agent presents a maximal degree of cognitive autonomy but cannot however escape the constraints imposed upon it by the very realization of its core processes *action* and *modeler*.

Much has been said about processes of mental representation and cognitive dynamics. The model-based control approach to cognition, on its part, offers a consolidated perspective of the multiple aspects regarding situated, embodied cognition:

The relevance of the body: In section "Separating Mind and Body," we reached the conclusion that we cannot ignore the body as it provides the physical processes that sustain the informational processes of the mind. But beyond that obvious fact, the solution to the control problem in the context of autonomous systems requires an explicit consideration of body dynamics in relation to world dynamics. This is not the class of weak argument produced by the "embodied cognition" movement, but a purely mental argument: the solution of the entirely informational problem of control requires properly expressing its bodily components.

The relevance of the task: The age-old debate around the possibility of general intelligence dissipates somewhat in the context of model-based control. In Sanz, Matía and Galán (2000) we analyze the problem of *autonomy* in terms of three coupled entities: task, body, and world. Solving the control problem requires a maximal precision in the expression of the task; general statements like *achieving human-level problem solving* are not only imprecise but totally misleading because there is no such thing as *human-level competence* except at basic pre-cultural, almost strictly biological, capabilities of human bodies. Research on human-level cognitive competences has not yet properly addressed the question of systemic substrates for the provision of such competences (with some notable exceptions, as Sloman & Chrisley, 2003).

The relevance of the situation: In the very same sense, the relevant part of the dynamics of the reality surrounding the agent must be captured in the mind of the agent. However the necessary deepness of the representation of both agent and environment will be directly dependent on the nature of the control problem being solved and the specific control strategies followed by the agent. Using robust control schemata (Chen, 2000) the agent will be able to perform sufficiently well without deep representations if body and environment stay inside some boundary conditions. This is an effective and economic strategy based on maximally simple models (i.e., only in the case of sets of boundary conditions).

The relevance of epigenesis: The exact amount of epigenesis (Ziemke, 2002) necessary for the agent will depend on many factors: (i) the particular control problem; ii) The amount of a priori instantiated knowledge that the agent has in its components; and (iii) the level of change in body and environment dynamics that the agent must embrace. So, epigenesis is certainly on the side of adaptation but, nevertheless, building a complex mental structure from scratch may take aeons. Bootstrapping will be made easier by starting with cognitive engines that are both general and robust. See for example the work of Beer (2003)which presents a dynamical analysis of a minimal cognitive agent. The agent is implemented by means of a continuous time recurrent neural network that is adaptively configured.

OPERATION OF A MODEL-BASED COGNITIVE AGENT

The global operation of a cognitive system (Figure 20.13) is the concurrent activity of two major processes: the process called *action* that maps from the mental to the physical and the process called *perception* that maps from the physical to the mental. The classical cognitive paradigm, *sense–think–act*, maps

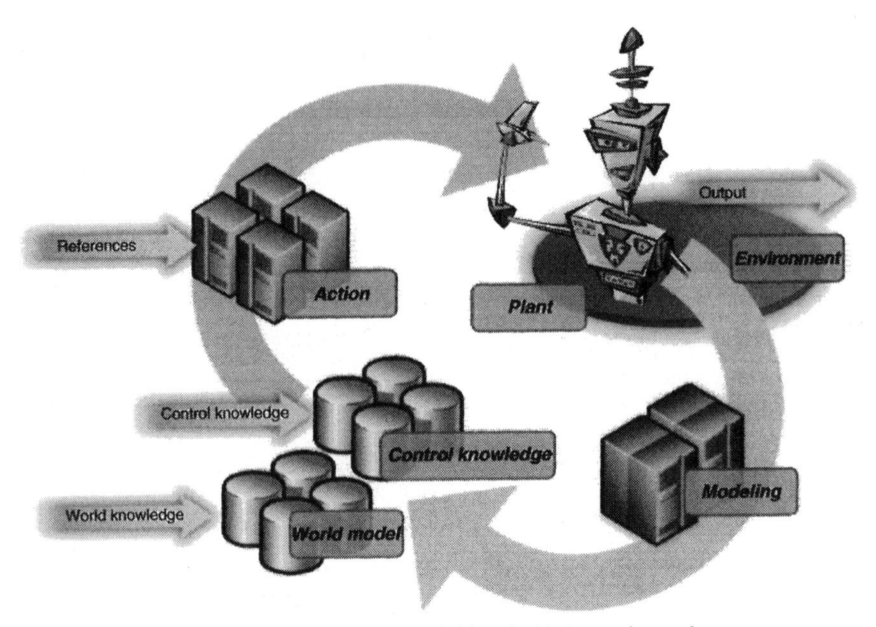

FIGURE 20.13 Body–mind in a double interaction cycle.

into a sense–think–act–behave cycle—much in line with embodied and dynamical systems approaches—that realizes the core cognitive pattern (Figure 20.14).

The operation of a cognitive agent realizing the model-based control architecture is very straightforward. In summary:

- The agent extracts relevant information from what Kuipers (2005)describes as *the fire hose of experience*. It must be borne in mind that the input comes in two stages, *sensing* and *perception* (López, 2007); the first is constrained by sensor capabilities and the second is limited to what is perceivable in terms of perceptual categories—that is, potential matches with the mental models.
- The different model-integrated percepts change the dynamic equilibrium of mind substrates, settling the many alternative potential operational modes of the agent on a reduced active mode set (Freeman, 2000). Hence effectively triggering the concurrent mental action dynamics that may produce externalized dynamics, motor outputs, and internalized dynamics—thought.
- A core basic internal dynamical process is the process of mental model construction and maintenance. This occurs upon detection of mismatches between perceptual flow and model flow—the product of model execution. The models are based either on (i) intrinsic biological capabilities—for example, our genetically implanted model of a human face,—or on (ii) experientially built models—a system identification process (Ljung, 1987),—and culturally transmitted (Aunger, 2000).

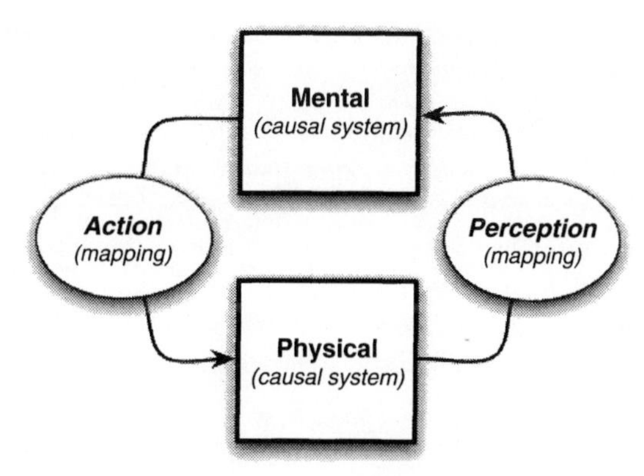

FIGURE 20.14 The cognitive pattern.

Notice that we sometimes use *modeling* instead of *perception* to stress the active modeling aspect underlying the perceptual process. In a more detailed architectural analysis, these two—*modeling* and *perception*—are resolved as different but related functions: perception is the process of external information injection into the models and to do this it exploits the functionality of the modeler—the model carer of the mind.

The approach depicted in Figure 20.12 is maximally general and this implies that in many cases we will only find degenerate versions of it. The most common degenerations are:

1. The collapse of the control knowledge representation and the action generation mechanisms into a single unit (faster, cheaper but less flexible).
2. The elimination of dynamicism in model construction rendering a somewhat classical feedback controller.
3. The collapse of mental and bodily subsystems in self-regulated processes (see also section "Separating Mind and Body").

What is missing to complete the "thinking bodies" picture we are presenting here is the fact that cognitive processes happen at all scales in a complex controlled system. Integration—the key to system-level cognition and consciousness—will be dealt with in the next section.

INTEGRATION IS KEY

This chapter began by analysing the control capabilities that bodily systems have at different levels. In the case of the heart, for instance, we showed that these capabilities range from the intrinsic self-organizing control capability of myogenic cells to upper brain control.

The structuring of all these control processes is driven by several heterogeneous forces:

- The first one calls for *simplicity* that provides robustness and evolvability.
- The second calls for *non-interference* that enables concurrence and modular evolution of the bodily subsystems.
- At the organism-level, efficient behavior requires that all these separate control loops must be *co-ordinated.*
- *Economic* reasons dictate that many of the functions carried out by bodily organs are shared across different subsystems.

Integration is the key issue for system-level behavior (Rossak & Ng, 1991; Grillner et al., 2005; Wheeler, 2007). Ultimately, all these forces result in an antithetical decomposition/integration organization process that renders a unified organism or system.

The concept of system-level integrated functionality is germane to the very concept of a biological organism. The idea that different parts of organisms are co-ordinated to form a functional whole was initially stated by Cuvier (1813) as the "principle of correlation" and it is currently termed "morphological integration" (Olson & Miller, 1999):

> This is because the number, direction, and shape of the bones that compose each part of an animal's body are always in a necessary relation to all the other parts, in such a way that-up to a point-one can infer the whole from any one of them and vice versa.
>
> Cuvier (1813)

In computer-based systems the issue of integration has been considered as simply a matter of proper interfacing at a primary level. However, when addressing issues of large-scale, enterprise-level systems, it is clear that sound integration not only calls for adequate integration mechanisms but also for a unified *integration architecture* (Rossak & Ng, 1991) (Figure 20.15).

The search for such a unified perspective on integrated control architecture in natural systems confronts the pervasive intricacy of biological function and control. However, some interesting work in theoretical biology has addressed this problem (Rosen, 1985).

In the case of complex technical control systems (Figure 20.15), our initial layered approach (Sanz, 1990) has been refined and extended (Alarcón et al., 1994; Sanz et al., 1999). We must stress, however, that from a model-based perspective of cognition, integration means control model federation across scales of the control hierarchy. This is an open research issue that is just beginning to receive adequate attention (Samad, 1998).

There are three aspects of critical importance in integrated control structures that affect the effectiveness of the integration and can be a source of emergent problems:

The question of ontologies: Models—and cognitive capabilities in general—are based on conceptualizations, that is, categorizations of information and

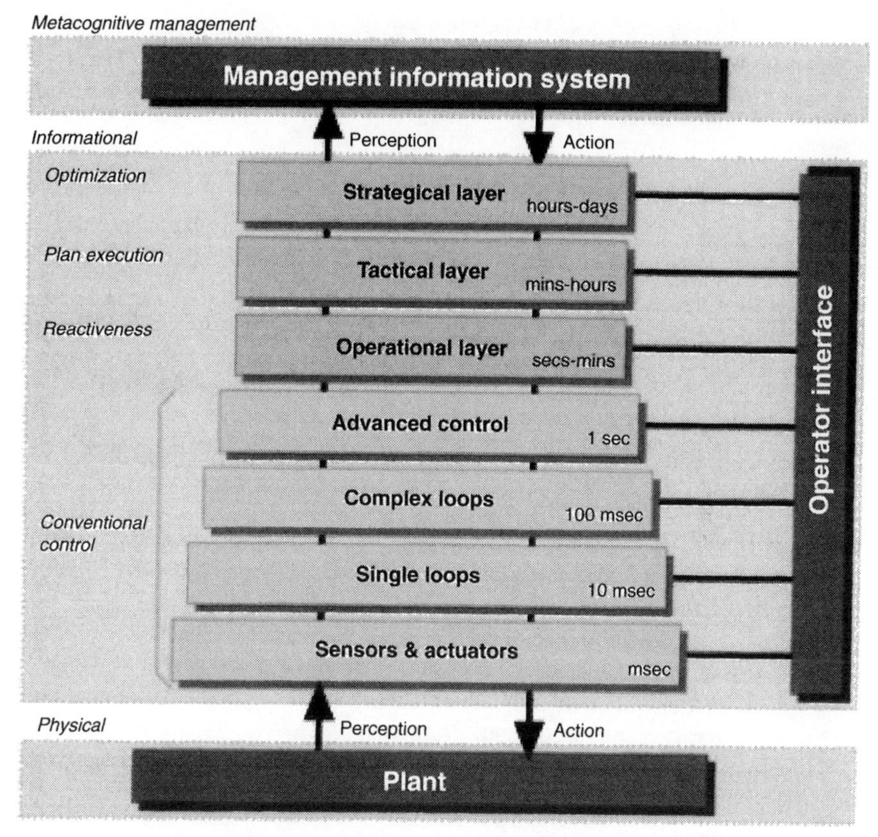

FIGURE 20.15 The integration landscape in a whole-plant integrated hierarchical control system. The lower layers provide homeostasis and fast responses to local goals. The upper layers implement highly cognitive multi-goal strategic control (including human-in-the-loop integration).

its flows. To integrate different systems it is necessary that they share a common ontology so they converge in the semantics of their interpretation and derived actions. In an effective, cohesive, integrated system they must necessarily converge because, ultimately, they will all be grounded on the physical, substratal ontologies of the layers in direct connection with physical aspects of the agent.

The question of model accuracy: The different models used in different layers can have different levels of precision. However, this is not a big problem as far as the control objectives of each layer match its precision. At the interface, the only net effect of this will be that of having quantized interactions (i.e., one side will see stepped changes in the other) that in general will be smoothed out.[6]

[6] But in some cases may produce instabiliy effects.

The question of temporality: The same can be said about time. Different layers will operate at different time scales. Some ripple effects will be produced at the interface that will be smoothed out in general but this interaction in some cases may produce unstable behavior.

CONCLUSIONS

We conclude this chapter with a summary of its main content in the form of a series of propositions:

- The mind–body relation is an informational–physical relation between a controller and the plant it is controlling.
- The mind—the controller—has a necessarily physical implementation.
- Cognition is the closed dynamical process of sense–think–act–behave. It is a systemic—emergent—phenomenon.
- Cognitive behavior is based on the exploitation of mental models of plant and environment in the determination of actions.
- Sensing is mapping physical states into informational states.
- Actuating is mapping informational states into physical states.
- Perception is model-integration of sensed information.
- Knowledge is executable dynamic models.
- Learning is model creation and caring.
- Cognitive loops organize in heterarchical/hierarchical integrated concurrent systems.

Figure 20.16 shows a summary depiction of the core fundamental cognitive organization as derived from the ideas presented so far. This elementary organization provides the core cognitive structuring that has to be implemented hierarchically and concurrently in each of the layers of the integrated hierarchy to achieve high-level cognitive capabilities.

Cognitive loop integration will ultimately render, at the end, a single unified cognitive architecture that, if properly provided with the necessary reflexive self-sensing and self-modeling, may allow for any level of cognitive capacity—including self-consciousness (Sanz et al., 2007).

To what extent does the epistemic control loop capture the nature of the cognitive organization of natural systems? It is obvious that not all systems with an embodied mind have this particular organization with all its details, but this structure must be seen as a maximal organization that may be realized in reduced ways for specific systems (see also the comments on model collapse in section "Model-Predictive Control").

Another question is related to the cognitive closure of the agent. Figure 20.16 shows a goal input channel but does so only to show the fundamental path of

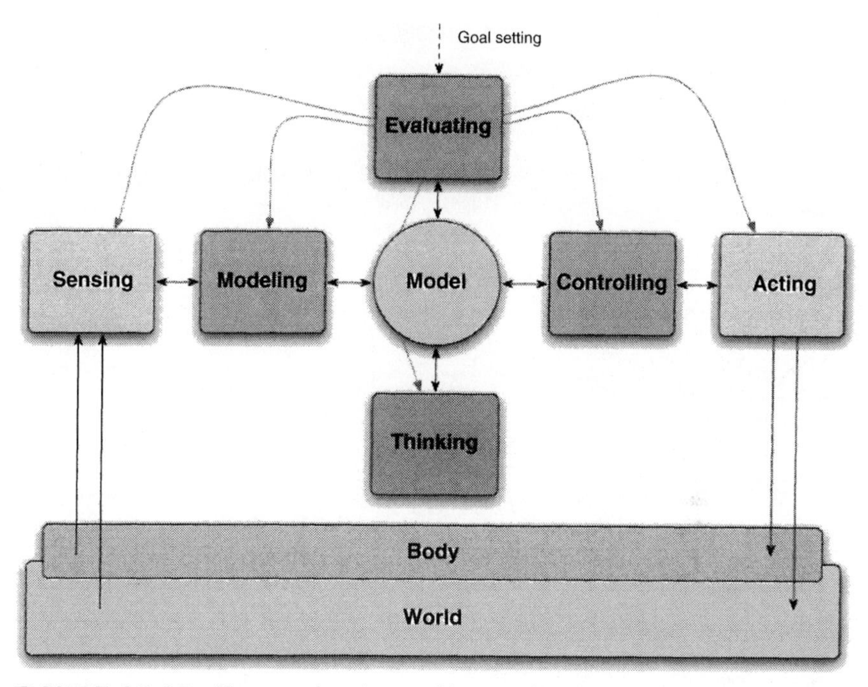

FIGURE 20.16 The core epistemic control loop provides the minimal structure for the provision of robust, model-based cognitive capabilities as described in this chapter. This is the minimal building block for general cognitive capabilities.

epistemic loop coupling. There are many other cognitive inputs—for example, model or perception sharing—that cognitive agents exploit.

The question of goal setting is particularly relevant from the perspective of autonomy. Those humans and other animals can accept explicit goal setting from the exterior is clear. That artificial machines do the same is also clear. In most cases of biological systems, though, goal setting is not explicit but unique and implicit: survival. Only biological systems that realize—embody—this goal in their very architecture will survive.

ACKNOWLEDGMENTS

The research leading to these results has received funding from multiple sources during years but we would specifically like to acknowledge the support received in the later stages from the European Community's *Sixth Framework Programme FP6/2004–2007* under grant agreement IST 027819 ICEA—*Integrating Cognition, Emotion and Autonomy*; the Spanish *Plan Nacional de I + D* under grant agreement DPI-2006–11798 C3—*Control Cognitivo Consciente*; and the Comunidad Autónoma de Madrid.

REFERENCES

Alarcón, M., Rodríguez, P., Almeida, L., Sanz, R., Fontaine, L., Gómez, P., Alamán, X., Nordin, P., Bejder, H., & de Pablo, E. (1994). Heterogeneous integration architecture for intelligent control. *Intelligent Systems Engineering*.

Anderson, M. L. (2003). Embodied cognition: A field guide. *Artificial Intelligence, 149*, 91–130.

Åström, K. J., Anton, J. J., & Årzén, K. E. (1986). Expert control. *Automatica, 22*(3), 277–286.

Åstrom, K. J. & Hagglund, T. (1995). *PID Controllers: Theory, Design, and Tuning*, 2nd ed. Research Triangle Park, NC: The Instrumentation, Systems, and Automation Society (ISA).

Aunger, R. (Ed.) (2000). *Darwinizing Culture: The Status of Memetics as a Science*. New York: Oxford University Press.

Bateson, G. (1979). *Mind and Nature. A Necessary Unity*. New York: E.P. Dutton, 2002 edition by Hampton Press.

Bayliss, L. (1966). *Living Control Systems*. London: The English Universities Press.

Beer, R. D. (2003). The dynamics of active categorical perception in an evolved model agent. *Adaptive Behavior, 11*(4).

Blackmore, S. J. (1999). *The Meme Machine*. Oxford: Oxford University Press.

Callebaut, W. & Rasskin-Gutman, D. (Eds.) (2005). *Modularity: Understanding the Development and Evolution of Natural Complex Systems*. Cambridge, MA: MIT Press.

Cannon, W. B. (1932). *The Wisdom of the Body*. New York: W. W. Norton.

Carver, C. S. & Scheier, M. F. (1981). *Attention and Self-Regulation: A Control-Theory Approach to Human Behavior*. New York: Springer-Verlag.

Chen, B. M. (2000). *Robust and H Control*. London: Springer.

Cisek, P. (1999). Beyond the computer metaphor: Behaviour as interaction. *Journal of Consciousness Studies, 6*(11–12), 125–142.

Conant, R. C. & Ashby, W. R. (1970). Every good regulator of a system must be a model of that system. *International Journal of Systems Science, 1*(2), 89–97.

Craik, K. J. (1943). *The Nature of Explanation*. New York: Cambridge University Press.

Cuvier, G. (1813). Essay on the theory of the earth.

Dawkins, R. (1976). *The Selfish Gene*. New York: Oxford University Press.

Freeman, W. J. (2000). The neurodynamics of intentionality in animal brains may provide a basis for constructing devices that are capable of intelligent behavior. *Performance Metrics for Intelligent Systems (PerMIS)*. NIST: Gaithersburg, MD, USA.

Gentner, D. & Stevens, A. L. (Eds.) (1983). *Mental Models*. Hillsdale, NJ: Lawrence Erlbaum Associates.

Gettier, E. L. (1963). Is justified true belief knowledge? *Analysis*(23), 121–123.

Grillner, S. (1985). Neurobiological bases of rhythmic motor acts in vertebrates. *Science, 228*(4696), 143–149.

Grillner, S., Kozlov, A., & Kotaleski, J. H. (2005). Integrative neuroscience: Linking levels of analyses. *Current Opinion in Neurobiology, 15*(5), 614–621.

Johnson-Laird, P. (1983). *Mental Models*. Cambridge, MA: Harvard University Press.

Khoo, M. C. K. (1999). *Physiological Control Systems: Analysis, Simulation, and Estimation*. IEEE Press Series on Biomedical Engineering. Wiley-IEEE Press: New York, USA.

Klingenberg, C. P. (2005). Developmental constraints, modules and evolvability. In B. Hallgrímsson & B. K. Hall (Eds.), *Variation: A Central Concept in Biology* (pp. 219–247). San Diego: Academic Press.

Klir, G. C. (1969). *An approach to General Systems Theory*. New York: Litton Educational Publishing, Inc.

Kuipers, B. (2005). Consciousness: Drinking from the firehose of experience. In *Proceedings of the AAAI Conference on Artificial Intelligence* (Vol. 20, pp. 1298–1305). Menlo Park: AAAI Press.

Landauer, R. (1992). Information is physical. In: *PhysComp '92, Workshop on Physics and Computation* (pp. 1–4): Dallas, TX, USA.

Lettwin, J. Y., Maturana, H. R., McCulloch, W. S., & Pitts, W. H. (1959). What the frog's eye tells the frog's brain. *Proceedings of the IRE, 47*(11), 1940–1951.

Levine, W. S. (1996). *The Control Handbook*. Boca Raton, FL: CRC Press.

Ljung, L. (1987). *System Identification. Theory for the User*. Upper Saddle River, NJ: Prentice Hall PTR.

López, I. (2007). *A Framework for Perception in Autonomous Systems*. Ph.D. thesis, Departamento de Automática, Universidad Politécnica de Madrid, Madrid, Spain.

Marken, R. S. (2002). Looking at behavior through control theory glasses. *Review of General Psychology, 6*(3), 260–270.

Meystel, A. & Sanz, R. (2002). Self identity in control systems. *Proceedings of the Workshop on Theoretical Fundamentals of Intelligent Systems*, Durham, NC.

Nelson, T. D. (1993). The hierarchical organization of behavior: A useful feedback model of self-regulation. *Current Directions in Psychological Research, 2*(4), 121–126.

Ogata, K. (1990). *Modern Control Engineering*. Upper Saddle River, NJ: Prentice-Hall.

Olson, E. C. & Miller, R. L. (1999). *Morphological Integration*. Chicago: University Of Chicago Press.

Opie, L. H. (2003). *Heart Physiology: From Cell to Circulation*, 4th ed. Philadelphia: Lippincott Williams & Wilkins.

Powers, W. T. (1973). *Behavior: The Control of Perception*. New York: Aldine/Walter de Gruyter.

Rezek, I. A. (1997). *Information Dynamics in Physiological Control Systems*. Ph.D. thesis, Imperial College of Science, Technology and Medicine, London, UK.

Rosen, R. (1985). *Anticipatory Systems*. New York: Pergamon Press.

Rosen, R. (1991). *Life Itself: A Comprehensive Inquiry into the Nature, Origin, and Fabrication of Life*. New York: Columbia University Press.

Rossak, W. & Ng, P. (1991). Some thought on systems integration—a conceptual framework. *International Journal on Systems Integration, 1*(1), 97–114.

Samad, T. (1998). Multi-modelling: Knowledge integration in complex systems. In T. Samad (Ed.), *Complexity Management: Multidisciplinary Perspectives on Automation and Control* (pp. 55–64). Honeywell Technology Center.

Sanz, R. (1990). *Arquitectura de Control Inteligente de Procesos*. Ph.D. thesis, Universidad Politécnica de Madrid, Madrid, Spain.

Sanz, R., Jiménez, A., & Galán, R. (1991). CONEX: A distributed architecture for intelligent process control. *Proceedings of the World Congress on Expert Systems*, Orlando, FL.

Sanz, R., López, I., & A. García, C. (2002). On the convergence of natural and artificial consciousness. *Proceedings of the conference Toward a Science of Consciousness*, Tucson, AZ.

Sanz, R., López, I., & Bermejo-Alonso, J. (2007). A rationale and vision for machine consciousness in complex controllers. In A. Chella & R. Manzotti (Eds.), *Artificial Consciousness*. New York: Imprint Academic.

Sanz, R., Matía, F., & Galán, S. (2000). Fridges, elephants and the meaning of autonomy and intelligence. *IEEE International Symposium on Intelligent Control, ISIC'2000*, Patras, Greece.

Sanz, R., Matía, F., & Puente, E. A. (1999). The ICa approach to intelligent autonomous systems. In S. Tzafestas (Ed.), *Advances in Autonomous Intelligent Systems, Microprocessor-Based and Intelligent Systems Engineering* (pp. 71–92). Dordretch, NL: Kluwer Academic Publishers, chapter 4.

Searle, J. R. (1990). Cognitive science and the computer metaphor. In B. Göranzon & M. Florin (Eds.), *Artificial Intelligence, Culture and Language: On Education and Work,* (pp. 23–34). New York: Springer-Verlag New York, Inc.

Sloman, A. & Chrisley, R. (2003). Virtual machines and consciousness. *Journal of Consciousness Studies, 10*(4–5), 133–172.

Wheeler, T. J. (2007). Analysis, modeling, emergence & integration in complex systems: A modeling and integration framework & system biology: Research article. *Complex, 13*(1), 60–75.

Zeigler, B. P. & Tag Gon Kim, H. P. (2000). *Theory of Modeling and Simulation*, 2nd ed. New York: Academic Press.

Ziemke, T. (2002). On the epigenesis of meaning in robots and organisms. *Sign Systems Studies, 30,* 101–111.

21

On the Grounds of (X)-Grounded Cognition

Michael L. Anderson

Department of Psychology, Franklin & Marshall College, Lancaster, PA; Institute for Advanced Computer Studies, Program in Neuroscience and Cognitive Science, University of Maryland, College Park, MD, USA

For the least the last 10 years, there has been growing interest in, and growing evidence for, the intimate relations between more abstract or higher order cognition—such as reasoning, planning, and language use—and the more concrete, immediate, or lower order operations of the perceptual and motor systems that support seeing, feeling, moving, and manipulating. A sub-field of the larger research program in *embodied* cognition (Clark, 1997, 1998; Wilson, 2001; Anderson, 2003, 2007d, 2008; Gibbs, 2006), this work has generally proceeded under the banner of *grounded* cognition, and works to support the claim that thinking is inherently tied to—grounded in—perceiving and acting. Thus, Glenberg and Kaschak (2002) discuss "grounding language in action"; Gallese and Lakoff (2005) argue that concepts are "grounded in the sensory–motor system;" and Barsalou (1999) at various times talks of "grounding cognition in perception," "grounding conceptual knowledge in modality-specific systems" (Barsalou et al., 2003), and most recently simply of "grounded cognition" (Barsalou, 2008).

Yet despite a great deal of terminological consensus, in fact there are nearly as many theories of grounding—what it is, and what it means—as there are theorists. Some, like Glenberg and Kaschak stress the origin (and continuing importance) of cognitive structures in controlling bodily action (Glenberg, 1997; Glenberg & Kaschak, 2002; see also Anderson & Rosenberg, 2008). The view seems to be: cognition is grounded in x (e.g., action) if some of x's elements are deployed in guiding it. Thus, their theory of language comprehension emphasizes the capacity to process the affordances associated with sentence elements to generate representations of possible actions. Insofar as these affordances both guide action and guide comprehension, comprehension is action-grounded. Others, like Barsalou and his many collaborators, suggest that the relation is one of *simulation*: cognition is grounded in x if it requires, depends upon or otherwise involves simulations of x experiences. "As people represent *TREES*, for example, they [engage] … a partial reenactment of the perceptual, motor, and introspective states that occur as people actually experience trees."[1] (Barsalou et al., 2005, p. 251; see also Prinz, 2002) And finally, there are many advocates for *conceptual metaphor* theories, which hold that cognition is grounded in x just in case it inherits from domain x an inferential structure that limits and guides one's thinking (Lakoff & Johnson, 1980, 1999; Lakoff & Núñez, 2000; Fauconnier & Turner, 2002). For instance, there are some apparent similarities between our notions of a *purpose* and of a *destination*—we imagine a goal as being at some place ahead of us, we plan a route, we imagine obstacles, and we set benchmarks to track our progress. Thus, the argument goes, our thinking about purposes is grounded in our experience of moving through space.

One thing that is especially interesting about the current state of affairs is that despite significant differences in the underlying theoretical frameworks, there is little overt disagreement between the various camps. Each approvingly cites the work of the others, and often even casts their own theories in the others' terms. For instance, Glenberg and Kaschak (2002) cite Barsalou (1999), and explain that their theory also "proposes that language is made meaningful by cognitively simulating actions implied by sentences" (p. 559)—this even though their theory is entirely prospective, rooted in what actions are *possible*, and requires no re-enactment of any specific experience. Moreover, they *also* indicate that their findings are compatible with the more metaphorical-structuring-friendly framework of Talmy (1988), whereby causal sentences are understood by analogy with contrasts between agonists and antagonists. Similarly, in the course of developing a version of conceptual grounding that attempts to combine elements of both conceptual metaphor theory and simulation-based accounts, Gallese and Lakoff (2005) cite Glenberg and Kaschak (2002) as providing evidence for their view. Indeed, even theorists who are suspicious of grounded cognition—such as Lera Boroditsky, who is explicitly critical of the inferential move from evidence for

[1]In the essay being quoted, words in italics and all capital letters indicate concepts. Thus *TREES* is the categorical concept to which all and only trees belong. Given this, the quote should probably actually read: "When people represent trees with *TREES* …" but the meaning is nevertheless clear.

[2]See, Boroditsky and Ramscar (2002, p. 188): "[C]ontrary to the very strong, embodied view (that abstract thought is based directly on sensory motor representations), we found that actual motion was neither necessary nor sufficient to influence people's thinking about time. Rather it is thinking about spatial motion that seems to influence thinking about time. It appears that thinking about abstract domains is built on representations of more experience-based domains that are functionally separable from representations directly involved in sensorimotor experience itself." This is a significant dissent in the context of Boroditsky's other work, since she is arguing in part that this functional separation allows room for specifically cultural influences to shape cognitive structures. Abstract thought is not grounded in concrete, embodied experience, but in abstract, culturally influenced *thinking* about embodied experience. This line of reasoning appears to point in precisely the opposite direction from that advocated by the theorists of grounded cognition.

metaphorical structuring to evidence for experiential grounding[2]—are typically cited as providing further evidence in favor of grounded cognition, without mention of their dissent.[3] As supportive and collegial as this makes conferences on embodied and grounded cognition, I would like to submit that it is not the most scientifically productive situation. The theories noted earlier (and these represent only a small fraction of the work in this area) are different enough that they ought to make differentiating predictions. The failure to highlight and test for the different predictions made by competing theories of cognitive grounding represents a missed opportunity to challenge and improve those theories. And if indeed all the scientific evidence gathered so far supports all the various theories—and one can sometimes get this impression, reading the literature—that suggests there is something wrong with the evidence, with the theories, or with both.

One particular domain of evidence that has been over-generously interpreted in support of various theories of grounded cognition is neuro-imaging data. Consider the interesting fact that mental planning can activate higher motor areas even when the planning itself involves no motor activity (Dagher et al., 1999). Anderson (2007d) cites this finding as support for the view that "many, if not all, higher-level cognitive processes are body-based in the sense that they make use of (partial) simulations or emulations of sensorimotor processes through the re-activation of neural circuitry that is also active in bodily perception and action" (Svensson et al., 2007)—but then adds a footnote noting it also supports Lakoff and Johnson (1999). Similarly, each of the following findings has been cited in support of conceptual metaphor theory (Gallese & Lakoff, 2005), *and* in support of simulation-based views (Barsalou et al., 2003; Barsalou, 2008):

- Evidence that watching actions, imagining actions, and doing actions all activate similar networks of brain regions (Decety et al., 1990; Jeannerod, 1994; Decety et al., 1997; Decety & Grèzes, 1999).
- Evidence that brain areas involved in motor control functions are also activated in verb retrieval tasks, while naming colors and animals activated brain regions associated with visual processing (Damasio & Tranel, 1993; Martin et al., 1995, 1996).
- Evidence that perceiving objects and object names activates brain regions associated with grasping (Chao & Martin, 2000).

If all these findings do indeed support the various theories, it suggests that brain imaging data isn't all that useful a tool for distinguishing between theories of grounded cognition; and if they do not, it suggests that very many cognitive scientists—myself most certainly included—have been less than careful in evaluating the available evidence. So, what's going on? My own impression of the situation is that most tests of embodied/grounded cognition are not targeted

[3]For instance, Barsalou (2008, p. 621) writes: "Lakoff and Johnson (1980, 1999) proposed that abstract concepts are grounded metaphorically in embodied and situated knowledge. . . Increasing evidence suggests that these metaphors play central roles in thought (Boroditsky & Ramscar, 2002; Gibbs, 2006)."

tests of the specific theory under consideration; rather they are designed to differentiate predictions made by a generic theory of grounded cognition from a generic theory of abstract, computational, amodal, or otherwise largely cognitivist theory of cognition. At this, they are effective. Even critics of embodied cognition have largely ceded the cognitivist high ground, and are fighting rear guard actions against further extensions of the basic claims (see e.g., Rupert, 2004; Weiskopf, 2007). But we theorists of embodied cognition ought by now to be in a position where the embodied, situated, and distributed approaches to the study of the mind are seen not primarily as criticisms of some prevailing paradigm, but as established, vibrant and fruitful research programs in their own right, needing no justification other than their own success. The proliferation of presumably incompatible, but apparently equally well-supported models of grounded cognition is one concrete and (at least partly) harmful result of the general failure to move beyond the idea of providing an alternative to cognitivism, and toward building a general, unified, and specifically supported theory of cognition on embodied first principles.

Allow me to suggest that the best way forward is first to take a step back. For in the course of developing my own version of x-grounded cognition (where x = "action"), I came to a somewhat surprising—and certainly sobering—conclusion: most (if not all) of the brain imaging results, and much (if not most) of the behavioral data taken to support specific theories of grounded or embodied cognition can in fact be accounted for by a generic theory of the evolution of the cortex. It's called the massive redeployment hypothesis (MRH), and its fundamental tenet is that the human cortex evolved by neural exaptation, whereby circuits originally developed to serve some specific purpose are used for new purposes and combined to support new capacities, without disrupting their participation in existing programs.[4] Such redeployment of existing neural circuits is favored by straightforward considerations of efficiency, and MRH need not (and indeed, cannot) make any general assumptions about the functional implications of reuse. Whether a given instance of neuronal reuse results in a circuit that implements simulation, or supports metaphorical structuring, or reproduces the process of affordance meshing in another domain, or simply indicates that some low-level computational function is being borrowed is not something the mere fact of redeployment can adjudicate.

All this will be made clearer by a more detailed account of MRH and the evidence that supports it—something I will briefly offer in the next section—but I want to say at the outset that I do *not* believe that MRH undermines any theory of embodied or grounded cognition. In fact, I think MRH has three very important things to offer: first, it identifies what might be characterized as the generic grounds for x-grounded cognition. The discovery of frequent redeployment of

[4]MRH is related to (but more general and, I think, more empirically supported than) both the theory of neural exploitation proposed by Gallese and Lakoff (2005) and the neuronal recycling hypothesis developed by Dehaene and Cohen (2007).

neural circuits across many different domains—action, perception, language, reasoning, and the like—is indeed an important fact that needs to be more fully understood and assimilated into prevailing theories of cognition.[5] Second—and precisely because MRH makes clear that the mere fact of redeployment is not by itself evidence for embodied cognition, or for any more specific relation of grounding, but is only a starting point for further inference and investigation—it focuses attention on uncovering the precise nature of the inheritance that redeployment enables. That is, if a later-developing cognitive function—say, verb retrieval—reuses cortical circuits originally developed for motor control, then clearly there will be some sort of functional inheritance as a result. But *what* sort, exactly? Only answering this question, and not just noticing the overlap, will provide specific evidence for any more substantive account of grounded cognition (or tell us whether grounding is even the best metaphor to be using in a particular case). Third, MRH suggests a method for actually answering such questions, by leveraging cross-domain cognitive modeling to attribute functional roles to redeployed circuits. This has the potential to tell us interesting things not just about the newer cognitive function(s) in which a given circuit was redeployed, but about the older function in which it was originally developed. That is, discovery of specific functional inheritances between language and motor control, or categorization and perception, or any such similar x and y, will tell us something interesting about *both* domains.

THE MASSIVE REDEPLOYMENT HYPOTHESIS

As indicated earlier, MRH is both a theory about the functional topography of the cortex, and an account of how it got that way. According to MRH, neural circuits evolved for one use are frequently *exapted* for later uses, while retaining their original functional role. That is, the process of cognitive evolution is analogous to component reuse in software engineering (Heineman & Councill, 2001). Components originally developed to serve a specific purpose are frequently reused in later software packages. The new software may serve a purpose very different from the software for which the component was originally designed, but may nevertheless require some of the same low-level computational functions (e.g., sorting). Thus, efficient development dictates reuse of existing components where possible. Note that in such reuse, the component just does whatever it does (e.g., sorts lists) for all the software packages into which it has been integrated, even if that computational function serves a very different high-level purpose in each individual case. This important aspect of component reuse in software engineering is also part of the hypothesis as it applies to neural redeployment.

The end result of such reuse in the brain is a structure in which brain areas are typically recruited to support many different functions across cognitive domains.

[5]See Stewart and West (2007) for some work along these lines.

Such a story about the organization and development of the cortex has some interesting implications for its overall functional architecture. For instance, on this theory, brain areas are not domain-restricted entities. If this were not the case, if a typical brain region in fact served a very limited set of cognitive functions, then this would suggest instead a localization-based development, whereby the brain evolved by generating new, dedicated regions for each new purpose. Moreover, according to MRH we should expect that differences in domain functions will be accounted for primarily by differences in the way brain areas cooperate with one another, rather than by differences in which brain areas are used in each domain. If neural circuits do not change their roles when they are exapted, and if they are used in many different cognitive domains, then the only way to get different functions while using the same components is to put them together in different ways. Another straightforward consequence of MRH is that more recently evolved cognitive functions will utilize more, and more widely scattered brain areas than phlyogenetically older functions. Again, the reason is simple: the more established neural components there are when a given cognitive capacity is evolving, the more likely that one of them will already serve some purpose useful for the emerging capacity, and there is little reason to suppose that the most useful areas will be grouped together (and less reason to suppose this as evolutionary time passes, making available more functions supported by more areas). Finally, MRH predicts that evolutionarily older brain areas will be deployed in more cognitive functions. This is presumably because the longer an area has been around, the more likely it will have proved useful to some evolving cognitive capacity, and be incorporated into the functional network of brain regions supporting the new task.

Preliminary investigations have uncovered evidence for all four of these predictions (see Anderson, 2007a, b, c; Anderson et al., in press for details of the methods and results). For instance, in my lab we recently performed a coactivation analysis of 472 brain imaging experiments (representing about 10 years worth of studies from *Journal of Cognitive Neuroscience*) from 8 different cognitive domains (action, attention, emotion, language, memory, mental imagery, reasoning, and visual perception). We coded the results of each experiment in terms of which Brodmann areas were activated (using only post-subtraction activations), and determined the baseline chance of activation for each area (and for each possible pairing) by dividing the number of experiments in which it was activated by the total number of experiments in the database. Then, for each pair of Brodmann areas, we used a chi-squared measure to see if their observed degree of coactivation (in a given domain) was significantly different from what would be predicted by chance. We also performed a binomial analysis, since a binomial measure can provide directional information. [It is sometimes the case that, while area A and area B are coactive more (or less) often than would be predicted by chance, the effect is asymmetric, such that area B is more active when area A is active, but not the reverse.]

The idea is that if co-activation indicates functional cooperation, such an analysis should reveal the cortical networks supporting cognitive functions in

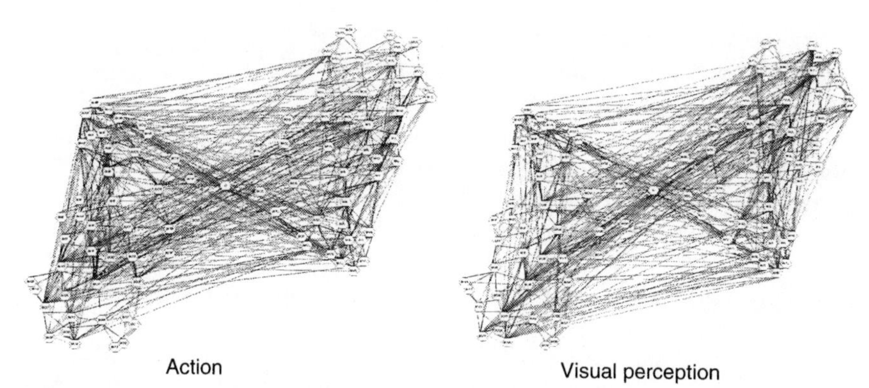

FIGURE 21.1 Cortex represented as adjacency + co-activation graphs. Here the Brodmann areas are nodes, with black lines between adjacent areas and orange lines between areas showing significant co-activation. The graph on the left shows co-activations from 56 action tasks, and the graph on the right shows co-activations from 57 visual perception tasks. Graphs rendered with aiSee v. 2.2. (See color plate)

the different domains. A graph offers a very nice representational format for the results. In the graphs we build, the nodes represent Brodmann areas, and the lines between nodes indicate significant co-activation (i.e., apparent functional cooperation). Figure 21.1 represents the co-activation graphs for the action and visual perception domains. The co-activation graph has been superimposed on an adjacency graph (where lines indicate that the connected areas share a physical border in the cortex) for ease of cross-domain visual comparison.

I mentioned earlier that MRH predicts that brain regions should support tasks across many different domains, but that the pattern of cooperation between the areas should be different in different domains. There is an obvious analog for these features in our co-activation graphs: comparing the graphs from different domains, node overlaps indicate Brodmann areas that support tasks in both domains, whereas edge overlaps would indicate a similar pattern of cooperation between Brodmann areas. Thus, MRH predicts a great deal of node overlap between co-activation graphs, but little edge overlap.

Using Dice's coefficient as our measure [$D = 2(o_{1,2})/(n_1 + n_2)$ where o is the number of overlapping elements, and n is the total number of elements in each set] and doing a pairwise comparison between all eight domains bears this prediction out. On average, there is very little edge overlap between the domains (Mean(D) = 0.15, SD 0.04), but a great deal of node overlap (Mean(D) = 0.81, SD 0.04); the difference is significant (2-tailed t-test, $P << 0.001$). Other specific results from the study include the fact that between action and visual perception the node overlap is 0.85 and the edge overlap is 0.14; between action and language we found a similarly high node overlap of 0.82 with an edge overlap of 0.06; and between visual perception and language the node overlap is 0.77 with an edge overlap of 0.17.

IMPLICATIONS OF MRH FOR X-GROUNDED COGNITION

The results reported above are interesting on their own, and certainly offer some support for MRH. But consider the following: insofar as MRH predicts high node overlap in this case, it *also* (by the same token) predicts a great deal of overlap of activations as seen in fMRI images of cognitive functions from different domains. That is to say, the fact that both language and action, and both language and perception, activate some of the same regions of the brain is an unsurprising consequence of the way the brain evolved; it does not represent, in and of itself, an anomaly that can only be explained by some version of x-grounded or embodied cognition.

And yet, there *is* a great deal of overlap in the networks supporting these functions—nearly as much as there is between perception and action, which one would expect to be intimately entwined indeed. This is by no means an insignificant fact. The question is, what shall we make of it? MRH itself has no answer to this question; it does, however, suggest a method for answering it. As noted earlier, the question of whether and how a given cognitive domain is x-grounded comes down to figuring out what that domain inherited from x; that is, it means knowing both what the shared neural components do in and of themselves (identifying their role), *and* what they are being redeployed to do (identifying their cognitive use). Given widespread overlap in the networks supporting different cognitive domains, MRH suggests that to determine the functional role of a given brain region it is better to focus on the brain region and consider its participation across multiple task categories. This is roughly the opposite of current practice, which generally involves choosing a given cognitive task (or task category) and identifying the various brain regions implicated in those tasks. Thus, rather than thinking about and modeling language functions in isolation from perception, attention, motor control, and other high-level cognitive domains, instead one needs to consider what sorts of components (and/or sub-functions) could serve functionality across domains. Finding the role of a given brain area will be something like finding the right letter to go into a box on a (multidimensional) crossword puzzle, determined not just by the answer to a single clue, but by all the clues whose answers cross that box. This makes the task both harder, because it is multiply constrained, but also easier, because it offers the possibility of leveraging information from several sources to make the attribution. For instance, the overlaps should suggest more fine-grained predictions about such matters as priming and cognitive interference, and this opens the possibility of designing experiments leveraging these overlaps, for example in further imaging, cross-domain priming, and interference studies.[6]

[6]Note the other implication, however: where there are shared neural resources, there will be interference. Thus, just as with MRH's implications for fMRI images, so too with its implications for cognitive interference; the mere existence of interference between cognitive domains like language and motor control is not an anomalous finding explainable only in terms of grounded or embodied cognition. Any such claims must rest instead on the exact nature of the interference in question.

This approach will have the effect of not just encouraging more creative hypotheses for the roles of brain areas, but also result in more integrated models of cognitive uses and domains. I have mentioned earlier the various theories supporting the idea that language is grounded in motor control. One common theme shared by all the accounts is that the action grounding of language implies that language is somehow *like* motor control—on Glenberg and Kaschak's affordance-mesh account, for instance, putting linguistic elements together in a meaningful sentence is like putting motor primitives together in an executable motor plan. But since this hypothesized functional inheritance is the result of shared neural circuits, and also these neural circuits are presumably playing the same role for each domain, this relation helps highlight a reverse implication that is worth considering: motor control should be in some way *like* language understanding. An affordance-based account suggests the following intriguing possibility: since affordances, the perceived availability of objects for certain kinds of interaction, aren't just motor programs, but features of the environment with specific significance for the organism, this opens the possibility that the motor control system is also, already, a primitive meaning processor (Gorniak & Roy, 2006). This would offer one explanation of how it is even possible to leverage motor control to support and constrain higher order processes like language understanding. After all, on a more mechanistic understanding of the nature of motor control, it would be hard to say why a motor control system would have *any* of the right basic elements for building a language-understanding system.

Of course, whether this is indeed the nature of the functional inheritance between language and motor control remains to be established; and the mere fact of overlapping neural circuits between the domains in no way does so. It nevertheless serves as a good example of the kinds of re-thinking that become possible when taking a more integrative approach. Much of this re-thinking will generate models that bear out the tenets of the embodied/grounded view of cognition—but not all of it. A case in point is some recent work on the relationship between finger gnosis and mathematical ability. Finger gnosis—the awareness of one's fingers—is commonly assessed via the ability to distinguish, without visual feedback, which fingers have been lightly touched. Developmentally, finger gnosis has been found to predict children's mathematics performance (for a review see Penner-Wilger et al., 2007), and studies have suggested that these two capacities are supported by some shared brain regions. For instance, Zago et al. (2001) found that a region associated with the representation of fingers (left parieto-premotor circuit) was activated during adults' arithmetic performance, and rTMS applied to the left angular gyrus that has been found to disrupt adults' performance on both finger gnosis and number magnitude tasks (Rusconi et al., 2005).

Now, any theorist with sympathies for embodied accounts of abstract cognition will be inclined to interpret this relation as an evidence for the grounding of mathematics in embodied experience, perhaps following Butterworth (1999) in the claim that using one's fingers to count causes one's finger representations and number representations to become intertwined. But in this case, considerations of cross-domain modeling seem to point in a different direction. As Anderson and Penner-Wilger (2007; Penner-Wilger & Anderson, 2008) note, one foundational element in any calculating circuit is a register for storing the number(s) to be manipulated. Such a register is typically implemented as a series of switches that can be independently activated. Likewise, at least one way to implement the ability to know whether and which fingers have been touched (and other aspects of a general "finger sense") would be with such a register of independent switches. Such a finger register—one part of the functional complex supporting finger gnosis—would be a candidate for redeployment in any later developing complex with functional elements able to take advantage of a component with this abstract functional structure. This, Anderson and Penner-Wilger suggest, is exactly what the number representation complex did.

There is some interesting evidence that seems to favor the view. Note, for instance, that Butterworth's position makes the experience of using the fingers to count a necessary condition for the observed intertwining, whereas the redeployment requires only that there be an intact finger register (i.e., intact finger gnosis) that can be put to various uses. Thus, the fact that children with Spina Bifida have poor finger agility co-morbid with significant mathematical difficulties (Banister & Tew, 1991; Barnes et al., 2005) has generally been taken as support for Butterworth's position. But since these children also have finger agnosia, the finding is equally compatible with the redeployment view. In contrast, children with developmental coordination disorder (DCD) have poor finger agility, but most have preserved finger gnosis, and do not generally evidence significant mathematical difficulties (Cermak & Larkin, 2001; Hamilton, 2002). This finding is consistent with the redeployment view, but appears to present some difficulties for Butterworth's position. Of course, these are just preliminary results, and whether the suggestion will be borne out by future investigations is an open question (for a discussion see Penner-Wilger & Anderson, 2008). But for current purposes, the point is twofold: (1) Such a proposal for one of the components of finger gnosis is unlikely to have occurred to researchers focusing only on results from their own domain; this may suggest the fruitfulness of the approach to modeling advocated here. (2) Not every overlap between cognitive domains is evidence that one is grounded in the other—at least not in the robust sense required by the various theories of embodied cognition.

Does any of this mean that cognition is not embodied, is not grounded? That mathematical understanding does not involve sensory–motor simulation? No. But it does mean that in many cases much more work needs to be done to establish the claim, whether gathering new and better empirical evidence, or reworking existing evidence to more clearly support a specific position. We must ask: what is it about *this* overlap of neural circuitry that suggests simulation in particular, rather than metaphorical mapping, or something else entirely? Which details in the general finding of cognitive interference indicate more than just a resource bottleneck? Is there some directionality, some selectivity to the interference that may give us insight into the nature of the functional inheritance that the overlap enables? A few researchers have started to focus on such specifics, as a way to decide *not* between amodal and modal theories of cognition (for which such details are often beside the point), but precisely between competing theories of grounded and embodied cognition (Casasanto & Lozano, 2007). It is time for more of us to follow their lead.

CONCLUSION

In this chapter I have laid down a challenge to the field of embodied and grounded cognition. Should the field see fit to pick it up, there is good reason to believe that the results will be very positive. To do that, we must move beyond the too simple task of finding evidence against abstract, amodal, and cognitivist theories of cognition and focus on detailing and supporting specific accounts of the functional inheritances that abstract higher order cognition has received from the substrates on which it is built. This will mean being a bit more critical of each other's work—though hopefully not less friendly toward one another. And indeed, I hope that this chapter is taken in just such a friendly, but constructively critical, spirit.

ACKNOWLEDGMENTS

Discussions with Daniel Casasanto helped me to clarify certain aspects of the current state of the field. Thanks to Tony Chemero for her helpful comments on an earlier draft. Thanks are also due to Paco Calvo and Toni Gomila for inviting this contribution, and for their patience while it was prepared.

REFERENCES

Anderson, M. L. (2003). Embodied cognition: A field guide. *Artificial Intelligence, 149*(1), 91–103.

Anderson, M. L. (2007a). Evolution of cognitive function via redeployment of brain areas. *The Neuroscientist, 13*(1), 13–21.

Anderson, M. L. (2007b). Massive redeployment, exaptation, and the functional integration of cognitive operations. *Synthese, 159*(3), 329–345.

Anderson, M. L. (2007c). The massive redeployment hypothesis and the functional topography of the brain. *Philosophical Psychology, 21*(2), 143–174.

Anderson, M. L. (2007d). How to study the mind: An introduction to embodied cognition. In F. Santoianni & C. Sabatano (Eds.), *Brain Development in Learning Environments: Embodied and Perceptual Advancements*. Cambridge Scholars Publishing. Newcastle, UK; 65–82.

Anderson, M. L. (2008). Embodiment, evolution, and the nature of the mind. In B. Hardy-Vallée & N. Payette (Eds.), *Beyond the Brain: Embodied, Situated and Distributed Cognition*. Cambridge Scholars Publishing. Newcastle, UK; 15–28.

Anderson, M. L. & Penner-Wilger, M. (2007). Do redeployed finger representations underlie math ability? In D. S. McNamara & J. G. Trafton (Eds.), *Proceedings of the 29th Annual Cognitive Science Society* (p. 1703). Austin, TX: Cognitive Science Society.

Anderson, M. L. & Rosenberg, G. (in press). Content and action: The guidance theory of representation. *Journal of Mind and Behavior*.

Anderson, M. L., Brumbaugh, J., & Suben, A. (in press). Investigating functional cooperation in the human brain using simple graph-theoretic methods. In: P. M. Pardalos, V. Boginski, and P. Xanthopoulos, (eds.). Computational Neuroscience. Springer.

Banister, C. M. & Tew, B. (1991). *Current Concepts in Spina Bifida and Hydrocephalus*. Cambridge: Cambridge Press.

Barnes, M. A., Smith-Chant, B. L., & Landry, S. (2005). Number processing in neurodevelopmental disorders: Spina bifida myelomenigocele. In J. I. D. Campbell (Ed.), *Handbook of Mathematical Cognition*. New York, NY: Psychology Press.

Barsalou, L. W. (1999). Perceptual symbol systems. *Behavioral and Brain Sciences, 22,* 577–660.

Barsalou, L. W. (2008). Grounded cognition. *Annual Review of Psychology, 59,* 617–645.

Barsalou, L. W., Simmons, W. K., Barbey, A. K., & Wilson, C. D. (2003). Grounding conceptual knowledge in modality-specific systems. *Trends in Cognitive Sciences, 7*(2), 84–91.

Barsalou, L. W., Pecher, D., Zeelenberg, R., Simmons, W. K., & Hamann, S. B. (2005). Multimodal simulation in conceptual processing. In W. Ahn, R. Goldstone, B. Love, A. Markman, & P. Wolff (Eds.), *Categorization Inside and Outside the Lab: Essays in Honor of Douglas L. Medin* (pp. 249–270). Washington, DC: American Psychological Association.

Boroditsky, L. & Ramscar, M. (2002). The roles of body and mind in abstract thought. *Psychological Science, 13*(2), 185–188.

Butterworth, B. (1999). *The Mathematical Brain*. London: Nelson.

Casasanto, D. & Lozano, S. (2007). Meaning and motor action, In D. S. McNamara & J. G. Trafton (Eds.), *Proceedings of the 29th Annual Cognitive Science Society* (p. 149–154). Austin, TX: Cognitive Science Society.

Cermak, S. A. & Larkin, D. (2001). *Developmental Coordination Disorder*. Albany, NY: Delmar.

Chao, L. L. & Martin, A. (2000). Representation of manipulable man-made objects in the dorsal stream. *Neuroimage, 12,* 478–484.

Clark, A. (1997). *Being There: Putting Brain, Body, and World Together Again*. Cambridge, MA: MIT Press.

Clark, A. (1998). Embodied, situated, and distributed cognition. In W. Bechtel & G. Graham (Eds.), *A Companion to Cognitive Science*. Malden, MA: Blackwell Publishers. 506–517.

Dagher, A., Owen, A., Boecker, H., & Brooks, D. (1999). Mapping the network for planning. *Brain, 122,* 1973–1987.

Damasio, A. & Tranel, D. (1993). Nouns and verbs are retrieved with differently distributed neural systems. *Proceedings of the National Academy of Science USA, 90,* 4957–4960.

Decety, J. & Grèzes, J. (1999). Neural mechanisms subserving the perception of human actions. *Trends in Cognitive Sciences, 3,* 172–178.

Decety, J., Sjoholm, H., Ryding, E., Stenberg, G., & Ingvar, D. (1990). The cerebellum participates in cognitive activity: Tomographic measurements of regional cerebral blood flow. *Brain Research, 535,* 313–317.

Decety, J., Grezes, J., Costes, N., Perani, D., Jeannerod, M., Procyk, E., Grassi, F., & Fazio, F. (1997). Brain activity during observation of actions. Influence of action content and subject's strategy. *Brain, 120,* 1763–1777.

Dehaene, S. & Cohen, L. (2007). Cultural recycling of cortical maps. *Neuron, 56,* 384–398.

Fauconnier, G. & Turner, M. (2002). *The Way We Think: Conceptual Blending and the Mind's Hidden Complexities*. New York: Basic Books.

Gallese, V. & Lakoff, G. (2005). The brain's concepts: The role of the sensory-motor system in conceptual knowledge. *Cognitive Neuropsychology, 22*(3–4), 455–479.

Gibbs, R. W. Jr., (2006). *Embodiment and Cognitive Science*. New York: Cambridge University Press.

Glenberg, A. (1997). What memory is for. *Behavior and Brain Sciences, 20,* 1–55.

Glenberg, A. & Kaschak, M. (2002). Grounding language in action. *Psychonomic Bulletin and Review, 9,* 558–565.

Gorniak, P. & Roy, D. (2006). Perceived affordances as a substrate for linguistic concepts. In R. Sun (Ed.), *Proceedings of the 29th Annual Cognitive Science Society* (p. 279–284). Austin, TX: Cognitive Science Society.

Hamilton, S. S. (2002). Evaluation of clumsiness in children. *American Family Physician, 66,* 1435–1440.

Heineman, G. T. & Councill, W. T. (2001). *Component-Based Software Engineering: Putting the Pieces Together.* New York: Addison-Wesley.

Jeannerod, M. (1994). The representing brain: Neural correlates of motor intention and imagery. *Behavioral Brain Science, 17,* 187–245.

Lakoff, G. & Johnson, M. (1980). *Metaphors We Live by.* Chicago, IL: University of Chicago Press.

Lakoff, G. & Johnson, M. (1999). *Philosophy in the Flesh: The Embodied Mind and its Challenge to Western Thought.* New York: Basic Books.

Lakoff, G. & Núñez, R. (2000). *Where Mathematics Comes from: How the Embodied Mind Brings Mathematics into Being.* New York: Basic Books.

Martin, A., Haxby, J. V., Lalonde, F. M., Wiggs, C. L., & Ungerleider, L. G. (1995). Discrete cortical regions associated with knowledge of color and knowledge of action. *Science, 270,* 102–105.

Martin, A., Wiggs, C. L., Ungerleider, L. G., & Haxby, J. V. (1996). Neural correlates of category-specific knowledge. *Nature, 379,* 649–652.

Penner-Wilger, M., Fast, L., LeFevre, J., Smith-Chant, B. L., Skwarchuk, S., Kamawar, D., & Bisanz, J. (2007). The foundations of numeracy: Subitizing, finger gnosia, and fine-motor ability. In D. S. McNamara & J. G. Trafton (Eds.), *Proceedings of the 29th Annual Cognitive Science Society* (p. 1385–1390. Austin, TX: Cognitive Science Society.

Penner-Wilger, M. & Anderson, M. L. (2008). An alternative view of the relation between finger gnosis and math ability: Redeployment of finger representations for the representation of number. In V. Sloutsky, B. Love, & K. McRae (Eds.), *Proceedings of the 30th Annual Conference of the Cognitive Science Society.* Austin, TX: Cognitive Science Society.

Prinz, J. (2002). *Furnishing the Mind: Concepts and Their Perceptual Basis.* Cambridge, MA: MIT Press.

Rupert, R. (2004). Challenges to the hypothesis of extended cognition. *Journal of Philosophy, 101,* 389–428.

Rusconi, E., Walsh, V., & Butterworth, B. (2005). Dexterity with numbers: rTMS over left angular gyrus disrupts finger gnosis and number processing. *Neuropsychologica, 43,* 1609–1624.

Stewart, T. & West, R. (2007). Cognitive redeployment in ACT-R: Salience, vision, and memory, *Proceedings of ICCM-2007-Eighth International Conference on Cognitive Modeling,* 313–318. Ann Arbor, MI.

Svensson, H., Lindblom, J., & Ziemke, T. (2007). Making sense of embodied cognition: Simulation theories of shared neural mechanisms for sensorimotor and cognitive processes. In J. Zlatev, T. Ziemke, & R. Frank (Eds.), *Body, Language and Mind* (vol. 1). Berlin: Mouton de Gruyter.

Talmy, L. (1988). Force dynamics in language and cognition. *Cognitive Science, 12,* 49–100.

Weiskopf, D. (2007). Concept empiricism and the vehicles of thought. *The Journal of Consciousness Studies, 14,* 156–183.

Wilson, M. (2001). The case for sensorimotor coding in working memory. *Psychonomic Bulletin and Review, 8,* 44–57.

Zago, L., Pesenti, M., Mellet, E., Crivello, F., Mazoyer, B., & Tzourio-Mazoyer, N. (2001). Neural correlates of simple and complex mental calculation. *NeuroImage, 13,* 314–327.

SECTION VII

Emotion and Social Interaction

22

UNDERSTANDING OTHERS: EMBODIED SOCIAL COGNITION

SHAUN GALLAGHER

Philosophy and Cognitive Sciences, University of Central Florida, Orlando, FL, USA and University of Hertfordshire Hatfield, GB

The standard and dominant approaches to social cognition rarely emphasize intersubjective interaction, and even when they do mention interaction they frame the problem in terms of two minds that have to communicate across the seemingly thin air of an unbridgeable gap. From this viewpoint, interaction is not a solution but simply an another way to state the problem of other minds. Consider, for example, the following formulation:

> … the study of social interaction … is concerned with the question of how two minds shape each other mutually through reciprocal interactions. To understand interactive minds we have to understand how thoughts, feelings, intentions, and beliefs can be transmitted from one mind to the other.
>
> (Singer et al., 2004, p. xvii)

On standard accounts of theory of mind (ToM) this gap between minds is bridged by some kind of cognitive processes in one mind providing the means to infer what is going on in the mind of the other, since the mind of the other is imperceptible. What one needs to bridge this gap is either theory (folk psychology), or simulation, or a combination of theory and simulation that will permit an inferential form of mind-reading or "mentalizing."

In this chapter, after reviewing some of the traditional ToM models of social cognition, I outline an alternative model on the basis of evidence from developmental psychology and phenomenology. In this alternative model, embodied, second-person interaction plays a central (although not an exclusive) role in our ability to understand other people. Finally, I discuss a recent development of simulation theory (ST) that champions an embodied simulationist approach.

Traditional ToM accounts make little mention of how the body might fit into the process of understanding others. At best, we take an observational stance toward the other's body and treat it as the source of evidence for constructing an inference. Proponents of theory theory (TT) contend that inference formation happens as the result of a mental consultation with a theory or a set of folk-psychological rules that will allow one to deduce an explanation of the observed behavior in terms of beliefs and desires understood as the other's mental states. ST eschews theory and opts for simulation routines that are run on the mechanisms of one's own mind. Here, for example, is a clear statement of how an explicit simulation works:

> First, the attributor creates in herself pretend states intended to match those of the target. In other words, the attributor attempts to put herself in the target's 'mental shoes'. The second step is to feed these initial pretend states [e.g., beliefs] into some mechanism of the attributor's own psychology … and allow that mechanism to operate on the pretend states so as to generate one or more new states [e.g., decisions]. Third, the attributor assigns the output state to the target … [e.g., we infer or project the decision to the other's mind].
>
> (Goldman, 2005, pp. 80–81)

Both approaches share certain fundamental assumptions. First, they assume that the problem is best posed as one that involves lack of access to other minds. Minds are hidden away behind or beyond the behavior that may be manifested. The task, then, is to explain or predict the behavior in terms of mental states that can only be inferred.

A second assumption taken up by both TT and ST is that theory use or simulation use, respectively, constitute the primary and pervasive means for social cognition. Thus we find proponents of these ToM approaches making universalistic claims, of which the following are good examples.

> [H]umans everywhere interpret the behavior of others in … mentalistic terms because we all come equipped with a "theory of mind" module (ToMM) that is compelled to interpret others this way, with mentalistic terms as its natural language.
>
> (Tooby & Cosmides, 1995, p. xvii)

> It is hard for us to make sense of behavior in any other way than via the mentalistic framework.—'attribution of mental states is to humans as echolocation is to the bat. It is our natural way of understanding the social environment'.
>
> (Baron-Cohen, 1995, pp. 3–4; see also Leslie, 2000; Currie & Sterelny, 2000; Frith & Happé, 1999; Wellman, 1993; Karmiloff-Smith, 1992; Malle, 2002, for similar statements).

> The strongest form of ST would say that all cases of (third-person) mentalization employ simulation. A moderate version would say, for example, that simulation is the *default* method of mentalization … I am attracted to the moderate version … Simulation is the primitive, root form of interpersonal mentalization.
>
> (Goldman, 2002, pp. 7–8)

Third, they assume that our relations with others are always from an observational stance. Perception is characterized as a third-person process where one person is observing the behavior of the other person rather than interacting with him/her in a second-person fashion. This observational stance is very clear in almost all false-belief tests, which TT appeals to as scientific evidence about the development of our mentalizing ability. For example, a subject (often a child) is asked to observe the behavior of two other children (or sometimes puppets). Sally puts a marble in a basket and leaves the room; another child, Anne, moves the marble from the basket to a box. Sally comes back in the room and the subject is asked where Sally will look for the marble. Four year olds tend to answer correctly that Sally will look in the basket; 3 year olds tend to answer incorrectly that Sally will look in the box, where the marble actually is. This is taken as evidence that the 3-year-old subjects (and some autistic subjects) are unable to appreciate that having a different perspective could lead to Sally's false belief; 4-year-old children apparently have developed a ToM that can deal with false beliefs (Wimmer & Perner, 1983; Leslie & Frith, 1988). Such experiments are designed so that the subject is simply a third-person observer of events; the subject never participates in the events or interacts with Sally or Anne. Theory theorists also fail to point out that even the youngest of the non-autistic children tested interact with the experimenter, and tend to understand what the experimenter wants them to do.

ST also takes observation to be the starting point, and inferential judgment to be the ending point of the intersubjective process. To put ourselves in the other's shoes, we need to first observe where those shoes are; that is, we need to observe the behavior of the other person before retreating into our own mind to run the simulation. The entire description of the simulation process is governed by the observational stance.

There are various debates within TT and ST, one of which concerns whether the processes involved are explicit (consciously controlled) or implicit. The strongest version of the implicit model is discussed below, but it should be noted that we could raise a simple phenomenological objection against explicit models that evoke conscious or introspective forms of theorizing or simulation. Simply put, if we carefully consult our everyday ordinary experiences of encountering others, we do not find ourselves taking observational stances in the third person; we do not find ourselves always trying to explain or predict their behavior, or attempting to get into their heads to ascertain what their beliefs or desires are. Most of our encounters are second person, interactive encounters, and most of what we need for understanding others is often readily available.

AN EMBODIED APPROACH

What we are calling an embodied or interactive approach involves a complex set of practices that can be found from infancy onward. From this viewpoint, much of what we call the mind is not something hidden away, but is something that is more directly accessible. Consider the phenomenologist Max Scheler's characterization of intersubjective perception as a direct perception.

> For we certainly believe ourselves to be directly acquainted with another person's joy in his laughter, with his sorrow and pain in his tears, with his shame in his blushing, with his entreaty in his outstretched hands … And with the tenor of this thoughts in the sound of his words. If anyone tells me that this is not 'perception', for it cannot be so, in view of the fact that a perception is simply a 'complex of physical sensations … I would beg him to turn aside from such questionable theories and address himself to the phenomenological facts'.
>
> (Scheler, 1954, pp. 260–261)

The idea is that there is a good amount of information that we can pick up in an ongoing direct perception of the other person's body that will give us a sense of what is going on with them. We can perceive their feelings and intentions in their postures, movements, facial expressions, gestures, vocal intonations, and actions. Scheler is not alone on this issue. We can find in Wittgenstein a number of similar statements.

> Look into someone else's face, and see the consciousness in it, and a particular shade of consciousness. You see on it, in it, joy, indifference, interest, excitement, torpor, and so on. … Do you look into yourself in order to recognize the fury in *his* face?
>
> (Wittgenstein, 1967, p. 229)

> In general I do not surmise fear in him—I *see* it. I do not feel that I am deducing the probable existence of something inside from something outside; rather it is as if the human face were in a way translucent and that I were seeing it not in reflected light but rather in its own.
>
> (Wittgenstein, 1980, p. 170)

That we do not look into ourselves to see what the other person is experiencing suggests that it is not a simulation process. And to say that I am not surmising or deducing the other's experience means that it is not through a theoretical inference that I gain access to the other.

Although, from this viewpoint, access is not regarded as a problem, this is not to say that the other person is entirely transparent, or that the meaning of all behavior can be perceptually grasped; behavior is often ambiguous, people are not always revealing of their emotions and thoughts. The claim here is not that direct perception can penetrate to the soul of the other person and discover his/her innermost emotional states. Nor is the claim that we can never be misled by what we perceive. The claim is rather that for the most part, in most of our encounters in everyday life, direct perception delivers a significant amount of important information for understanding others. In addition, it would only be something that I discover through these means that would lead me to the idea that perhaps something more is going on with the other person.

Moreover, in ordinary everyday encounters with others, I am not taking an observational stance; I am not off to the side thinking or trying to figure out what they are doing. Rather, I am responding to them in an embodied way, and I am part of the situation. As we will see shortly, our own motor and emotional systems are intricately involved in our perception of others, and we should think of perception here as an enactive rather than a passive process. What we call social cognition is first of all social interaction. What I perceive in these cases does not constitute something short of understanding. Rather my understanding of the other person is constituted within the perception–action loops that define the various things that I am doing with or in response to others.

Evidence for this can be found in many developmental studies, and generally falls under what the developmental psychologist Colwyn Trevarthen calls "primary intersubjectivity" (Trevarthen, 1979). We do not arrive in the world as a *tabula rasa*—and our slate starts to fill up very quickly. Developmental studies consistently tell us that neonate perception is already relatively smart. The newborn infant can pick out a human face from the crowd of objects in its environment, with sufficient detail that will enable it to imitate the gesture it sees on that face (Meltzoff & Moore, 1977, 1994). There is an increasing evidence that infants automatically attune to smiles (and other facial gestures) with an enactive, mimetic, response (Schilbach et al., 2008). The young infant is visually attracted to movement and in specific ways to biological movement, and auditorily attracted to certain kinds of sounds, such as its mother's voice. Infants "vocalize and gesture in a way that seems [affectively and temporally] 'tuned' to the vocalizations and gestures of the other person" (Gopnik & Meltzoff, 1997, p. 131). Human infants show a wide range of facial expressions, such as complex emotional, gestural, prosodic, and tactile face-to-face interaction patterns which are absent or rare in non-human primates (Falk, 2004; Herrmann et al., 2007), but notably without the intervention of theory or simulation. Moreover and in a non-mentalizing way, they are able to see bodily movement as expressive of emotion, goal-directed intentional movement, and they are able to perceive other persons as agents. This does not require advanced cognitive abilities, inference, or simulation skills; rather, it is a perceptual capacity that is "fast, automatic, irresistible and highly stimulus-driven" (Scholl & Tremoulet, 2000, p. 299).

Infants are able to detect correspondences between visual and auditory information that specify the expression of emotions as early as 5–7 months (Walker, 1982; also Hobson, 1993, 2002). At 9 months, infants follow the other person's eyes (Senju, Johnson & Csibra, 2006), and start to perceive various movements of the head, the mouth, the hands, and more general body movements as meaningful, goal-directed movements. Baldwin and colleagues, for example, have shown that infants at 10–11 months are able to parse some kinds of continuous action according to intentional boundaries (Baldwin & Baird, 2001; Baird & Baldwin, 2001). Such perceptions give the infant, by the end of the first year of life, a non-mentalistic, perceptually based embodied understanding of the intentions and dispositions of other persons (Baldwin, 1993; Johnson et al., 1998; Allison et al., 2000; Johnson, 2000). These capabilities do not disappear in adulthood but they mature and become more sophisticated (see Dittrich et al., 1996). This can be clearly shown in a micro-analysis of the postures, movements, gestures, gazes, and facial expressions of people as they engage in a novel task and where communication among them is intrinsic to the actions that they take (see Niedenthal et al., 2005; Lindblom, 2007).

This initial set of direct perceptual practices does not give us the full account of social cognition, and the information we pick up directly from the other person's embodied comportments is far from sufficient for the often rich and nuanced understanding that we can have of the other person. This primary intersubjectivity, however, is immediately supplemented and enhanced by a secondary intersubjectivity (Trevarthen & Hubley, 1978). Expressions, intonations, gestures, and movements, along with the bodies that manifest them, do not float freely in the air; we find them in the world, and infants soon start to notice how others interact with the world. Infants begin to tie actions to pragmatic contexts around the age of 1 year; they enter into *contexts* of shared attention—shared situations—in which they learn what things mean and what they are for. Behavior representative of joint attention begins to develop around 9–14 months (Phillips et al., 1992). In such interactions, the child looks to the body and the expressive movement of the other to discern the intention of the person or to find the meaning of some object. The child can understand that the other person *wants* food or *intends* to open the door; that the other can *see* him (the child) or is *looking at* the door.[1] They begin to see that another's movements and expressions often depend on meaningful and pragmatic contexts and are mediated by the surrounding world. Others are not given (and never were given) primarily as objects that we encounter cognitively, or in need of explanation. We perceive them as agents whose actions are framed in pragmatic and socially defined contexts. It follows that there is not one uniform way in which we relate to others, but that our relations are mediated through the various pragmatic (and ultimately, institutional) circumstances of our encounters. Indeed, we are caught up in such pragmatic circumstances, and are already existing in reference to others, from the very beginning (consider for example the infant's dependency on others for nourishment), even if it takes some time to sort out which agents provide sustenance, and which ones are engaged in other kinds of activities.

As we noted, children do not simply observe others; they are not passive observers. Rather they interact with others and in doing so they develop further capabilities in the contexts of those interactions. If the capacities of primary intersubjectivity, like the detection of intentions in expressive movement and eye direction, are sufficient to enable the child to recognize dyadic relations between the other and the self, or between the other and the world, something more is added to this in secondary intersubjectivity. As noted, in joint attention, beginning around 9–14 months, the child alternates between monitoring the gaze of the other and what the other is gazing at, checking to verify that they are continuing to look at the same thing. Indeed, the child also learns to point at approximately this same time. At 18 months, children comprehend what another person intends

[1]This is not taking an intentional stance, that is, treating the other *as if* they had desires or beliefs hidden away in their minds; rather, the intentionality is perceived in the embodied actions of others.

to do with an instrument in a specific context. They are able to re-enact to completion the goal-directed behavior that someone else fails to complete. Thus, the child, on seeing an adult who tries to manipulate a toy and who appears frustrated about being unable to do so, quite readily picks up the toy and shows the adult how to do it (Meltzoff, 1995; Meltzoff & Brooks, 2001; see also Herrmann et al., 2007).[2]

Our understanding of the actions of others occurs on the highest, most appropriate pragmatic level possible. That is, we understand actions at the most relevant pragmatic (intentional, goal-oriented) level, ignoring possible subpersonal or lower level descriptions, but in most cases also ignoring interpretations in terms of beliefs, desires, or hidden mental states. Rather than *making an inference* to what the other person is intending by starting with bodily movements, and moving from there to the level of mental events, we see actions as meaningful in the context of the physical and intersubjective environment. If, in the vicinity of a locked door, I see you reach for a set of keys, I would know your intentions as much from the door and the keys, your bodily posture and expression as from anything that I postulate in your mind. We interpret the actions of others in terms of their goals and intentions set in contextualized situations, rather than abstractly in terms of either their muscular performance or their beliefs. The environment, which is not only a physical location, but also a pragmatic context and a social situation, is never perceived neutrally (without meaning), either in regard to our own possible actions, or in regard to the actions and possibilities of others. In this regard, the world itself does much of the work involved in social cognition. As Gibson's theory of affordances (Gibson, 1979) suggests, we see things in relation to their possible uses, and therefore never as a disembodied observer. Likewise, our perception of the other person, as another agent, is never of an entity existing outside of a situation, but rather of an agent in a pragmatic context that throws light on the intentions (or possible intentions) of that agent.

There is much more to say about the role of socially defined situations and the roles that people play in them. As children develop, and precisely because they have the embodied capabilities defined by primary and secondary intersubjectivity,

[2]Onishi and Baillargeon (2005) have recently shown that infants at 15 months apparently mentalize the false beliefs of others. The data from their experiments suggest that infants see what the other person intends to do and is surprised (or at least notices) when the behavior of the other violates what the infant knows about the context (specifically about who has seen or not seen certain events). Although Onishi and Baillargeon interpret the data entirely in a ToM framework of mentalizing the other's beliefs, an alternative interpretation in terms of perceived meaningful (contextualized) behavior, actions, and intentions is clearly available. See Woodward & Sommerville (2000): "[...] 12-month-old infants interpreted action in context in two senses: They used both the other actions performed by the actor and the causal constraints in the situation to interpret an ambiguous action .... infants as young as 6 months construe grasping as goal directed, infants under 12 months may be able to interpret the goal of an action on the basis of sequences [of actions in context]" (pp. 76–77). Appeal to hidden beliefs or mental states is not required. See also Király et al. (2003) and Biro et al. (2007).

they easily learn what to expect of other people in such situations, and these expectations define the default cultural framework for understanding others. When I enter a classroom or a grocery store, I can immediately see who the teacher is or who the cashier is, and I can intuitively understand what they are doing, and for my particular purposes that may be sufficient for my interactions. We have no need for theories or simulations; most of our social understanding is shaped by scripts and short narratives that we learn as children (Hutto, 2007). We do not ordinarily need to go further than the already rich and complex comprehension that we gain through the perception of a situated agent—that is, of an agent who is situated in an environment which also tells us something about what that person is doing and thinking. If I see the situation and what the agent is doing in it, and how the agent is doing it, and what the agent is expressing (e.g., through his/her gestures and style of movement), and this perception is already informed by my own interaction with them and others, as well as by my previous situated experiences, my habitual ways of understanding, and by cultural norms and established practices, and so forth, then in cases which we encounter in our normal ordinary engagements the work of understanding is already sufficiently accomplished and I do not have to go any further. I do not have to start thinking about what might be going on in the other person's mind since everything I need for understanding him/her is there in his/her action and in our shared world.

Again, there is more to be said about the role of narratives in fine-tuning our social understandings. We gain narrative competency as young children, and along with it comes the ability to employ a folk-psychological practice in those rare cases where we may be entirely puzzled about someone's actions (Hutto, 2007; Gallagher & Hutto, 2008). If the cashier is dancing on the counter, or the teacher starts to throw water balloons in class, then we may adopt an observational stance (as we duck) and start to theorize or simulate about what the state of his/her mind might be. This kind of practice, however, is the exception rather than the primary or pervasive way by which we come to understand others.

IMPLICIT SIMULATION OR EMBODIED PRACTICES

The embodied practices of primary and secondary intersubjectivity, involving direct perception and pragmatic contextualizations, clearly contrast with the claims made by theory theorists and simulation theorists who conceive of social cognition as a purely mentalistic or cognitive process. Recently, however, ST has appealed to the neuroscience of resonance systems and mirror neurons (MNs) as offering scientific evidence for a form of implicit simulation. This, of course, depends on a specific interpretation of the scientific data.

We know that the perceiver's motor system is activated when he/she perceives another person performing an intentional action. The same or overlapping neural areas in parts of the frontal and parietal cortices, and specifically, MNs in the pre-motor cortex, in Broca's area, and in the parietal cortex of the human brain are activated both when the subject engages in specific instrumental actions, and when the subject observes someone else engage in those actions (Rizzolatti et al., 1996, 2000; Grèzes & Decety, 2001). Some simulation theorists claim that these processes underpin (or are the neural correlates) of explicit acts of simulation (Jeannerod & Pacherie, 2004, p. 129). *Implicit* simulation theorists, however, contend that these subpersonal processes themselves just are a simulation of the other's intentions. Vittorio Gallese, for example, claims that activation of MNs involves "automatic, implicit, and nonreflexive simulation mechanisms ..." (2005, p. 117; see also Gallese, 2007). According to Gallese, one's empathic experience of the other person at the phenomenological level is underpinned by the activity of "mirror matching neural circuits" at the brain level, which he interprets as "simulation routines, *as if* processes enabling models of others to be created" at the functional level (2001, p. 45). On this hypothesis, at the explicit, phenomenological level, one is not explicitly (consciously) simulating; rather the simulation process remains entirely at the subpersonal level.

There is a growing consensus forming around this implicit simulation idea. Decety & Grèzes (2006, p. 6) summarize Rizzolatti's position in this way:

> By automatically matching the agent's observed action onto its own motor repertoire without executing it, the firing of mirror neurons in the observer brain simulates the agent's observed action and thereby contributes to the understanding of the perceived action

Goldman (2006) distinguishes between simulation as a form of high-level (explicit) mind-reading and simulation as a low-level (implicit) mind-reading where the latter is "simple, primitive, automatic, and largely below the level of consciousness" (p. 113), and the prototype for which is "the mirroring type of simulation process" (p. 147). Research suggests that MN activation is a simulation not only of the goal of the observed action but of the intention of the acting individual, and therefore a form of mind-reading. MNs discriminate identical movements according to the intentional action and contexts in which these movements are embedded (Fogassi et al., 2005; Iacoboni et al., 2005; Kaplan & Iacoboni, 2006). Neural simulation has also been extended as an explanation of how we grasp emotions and pain in others (Avenanti & Aglioti, 2006; Minio-Paluello et al., 2006; Gallese et al., 2007). Oberman & Ramachandran (2007), who amass evidence that the MN system as an internal simulation mechanism is dysfunctional in cases of autism, reinforce the idea that "simulator neurons" are responsible for understanding actions, thoughts, and emotions.

There are, however, several conceptual problems involved in calling subpersonal mirror resonance processes "simulations" (Gallagher, 2007a, b). There are good reasons to think that subpersonal processes, such as MN activation, fail to meet the definition of simulation as it is developed in ST. In that definition, simulation involves two essential aspects: first, simulation involves instrumental control of a model as we use it to understand something that we cannot understand

directly. Second, simulation involves pretense—the idea that we use our own mental states "as if" they were the mental states of others. In contrast, however, subpersonal mirroring processes do not have an instrumental character, nor are they under our control. Rather, they are automatic and, indeed, they are elicited by the actions of others. The perceiver does not launch an MN activation as a means for making sense of the other's action; rather, the process is one of perceptual elicitation where the perceived action calls forth the activation of these neurons. Furthermore, because MNs are activated both when I act and when I see someone else act, they are neutral with respect to who the agent is (deVignemont, 2004; Jeannerod & Pacherie, 2004; Gallese, 2005; Hurley, 2005). As a result, MNs do not involve pretense, which requires distinguishing one agent (me) from another (you). There is no I or you registered in MNs, per se (see Georgieff & Jeannerod, 1998).

These kinds of issues motivate a weakened or minimal definition of simulation which jettisons the instrumental and pretense aspects and defines simulation as simply a form of matching (Goldman & Sripada, 2005; Goldman, 2006). This strategy, however, fails to explain how we understand others who are engaged in very different activities from us, or who are experiencing very different emotions. For example, I may see someone acting in a certain way (picking up an insect, for instance) and clearly enjoying it, while at the same time I feel disgust about that very action and make a pushing away gesture. Neither my emotional state nor my motoric state matches up with the relevant states of the other person, yet I clearly understand his/her emotional and motor states—they are in fact motivating my own. Furthermore, there is neuroscientific evidence that shows that MN activation does not necessarily involve a precise match between motor system execution and observed action, but may be involved in "logically related" actions (e.g., complementary actions) or in anticipating future action (Csibra, 2005; Iacoboni et al., 2005). All of this goes against the idea that MNs are simulating anything.

To deny that mirror resonance processes constitute simulations, however, is not to deny that MNs may play an important role in our interactions with others, possibly contributing to our ability to understand others, or to keep track of ongoing intersubjective relations. Rather, the alternative and more parsimonious interpretation of MN activation is that it constitutes part of the neuronal correlates of direct intersubjective perception. That is, the articulated neuronal processes that include activation in various sensory areas, but also resonating activation of MNs in the motor system, are part of what underpins a non-articulated immediate perception of the other person's intentional actions, rather than a distinct process of simulating their intentions (Gallagher, 2007a, b, 2008).[3] On this view, we need to think of perception as an enactive process (Hurley, 1998; Noë, 2004; Varela et al., 1991), as involving sensory–motor skills rather than as just sensory input/processing, as

[3]Note that MN activation is only part of the story and likely not sufficient for social perception of intentions. MNs, for example, were first discovered in monkeys, but this does not mean that monkeys are capable of social perception in the same way that humans are.

an active, skillful, embodied engagement with the world rather than as the passive reception of information from the environment. In the context of social cognition, it seems appropriate to think of mirror resonance processes as part of the structure of the perceptual process when it is a perception of another person's actions. Accordingly, mirror activation is not the initiation of simulation; it subtends a direct intersubjective perception of what the other is doing. On this interpretation, MN activation fits properly with the direct perception account of intersubjective understanding and interaction, and helps to explain such capacities already operative in infancy in certain embodied practices—practices that are emotional, sensory–motor, non-conceptual, and directly perceptual—practices that involve a perceptual sense of others and that constitute a common bodily intentionality shared by both the perceiving subject and the perceived other (Gallagher, 2001, 2005).

CONCLUSION

On the embodied view of social cognition, the mind of the other person is not something that is hidden away and inaccessible. In perceiving the actions and expressive movements of the other person in the interactive contexts of the surrounding world, one already grasps their meaning; no inference to a hidden set of mental states (beliefs, desires, etc.) is necessary. When I see the other's action or gesture, I see (I *immediately perceive*) the meaning in the action or gesture; and when I am in a process of interacting with the other, my own actions and reactions help to constitute that meaning. I not only see, but I resonate with (or against), and react to the joy or the anger, or the intention that is in the face or in the posture or in the gesture or action of the other.

The alternative, non-simulationist interpretation of the neuroscience of MNs coheres with the larger non-ToMistic, interaction view of social cognition. This view, supported by evidence from developmental and neuroscientific studies, suggests that before we are in a position to theorize, simulate, explain, or predict mental states in others, we are already in a position to interact with and to understand others in terms of their contextualized expressions, gestures, and purposive movements, reflecting their intentions and emotions. We already have specific perception-based understandings about what others feel, whether they are attending to us or not, how they are acting toward us and others, whether their intentions are friendly or not, and so forth; and in most cases, we have this without the need for personal-level theorizing or simulating about what the other person believes or desires. Moreover, we understand this without the benefit of anything that on the subpersonal level could be considered an extra cognitive step, a simulation, or inference.

REFERENCES

Allison, T., Puce, Q., & McCarthy, G. (2000). Social perception from visual cues: Role of the STS region. *Trends in Cognitive Science*, *4*(7), 267–278.

Avenanti, A. & Aglioti, S. M. (2006). The sensorimotor side of empathy for pain. In M. Mancia, (Ed.), *Psychoanalysis and Neuroscience* (pp. 235–256). Milan: Springer.

Baird, J. A. & Baldwin, D. A. (2001). Making sense of human behavior: Action parsing and intentional inference. In B. F. Malle, L. J. Moses & D. A. Baldwin (Eds.), *Intentions and Intentionality: Foundations of Social Cognition* (pp. 193–206). Cambridge, MA: MIT Press.

Baldwin, D. A. (1993). Infants' ability to consult the speaker for clues to word reference. *Journal of Child Language*, *20*, 395–418.

Baldwin, D. A. & Baird, J. A. (2001). Discerning intentions in dynamic human action. *Trends in Cognitive Science*, *5*(4), 171–178.

Baldwin, D. A., Baird, J. A., Saylor, M. M., & Clark, M. A. (2001). Infants parse dynamic action. *Child Development*, *72*, 708–717.

Baron-Cohen, S. (1995). *Mindblindness: An Essay on Autism and Theory of Mind*. Cambridge, MA: MIT Press.

Biro, S., Csibra, G., & Gergely, G. (2007). The role of behavioral cues in understanding goal-directed actions in infancy. *Progress in Brain Research*, *164*, 303–322.

Csibra, G. (2005). Mirror neurons and action observation. Is simulation involved? ESF Interdisciplines. http://www.interdisciplines.org/mirror/papers/.

Currie, G. & Sterelny, K. (2000). How to think about the modularity of mind-reading. *The Philosophical Quarterly*, *50*(199), 145–160.

Decety, J. & Grèzes, J. (2006). The power of simulation: Imagining one's own and other's behavior. *Brain Research*, *1079*, 4–14.

deVignemont, F. (2004). The co-consciousness hypothesis. *Phenomenology and the Cognitive Sciences*, *3*(1), 97–114.

Dittrich, W. H., Troscianko, T., Lea, S. E. G., & Morgan, D. (1996). Perception of emotion from dynamic point-light displays represented in dance. *Perception*, *25*, 727–738.

Falk, D. (2004). Prelinguistic evolution in early hominids: Whence motherese? *Behavioral and Brain Sciences*, *27*(4), 491–503.

Fogassi, L., Ferrari, P. F., Gesierich, B., Rozzi, S., Chersi, F., & Rizzolatti, G. (2005). Parietal lobe: From action organization to intention understanding. *Science*, *308*, 662–667.

Frith, U. & Happé, F. (1999). Theory of mind and self-consciousness: What is it like to be autistic? *Mind and Language*, *14*(1), 1–22.

Gallagher, S. (2001). The practice of mind: Theory, simulation, or interaction? *Journal of Consciousness Studies*, *8*(5–7), 83–107.

Gallagher, S. (2005). *How the Body Shapes the Mind*. Oxford: Oxford University Press.

Gallagher, S. (2007a). Logical and phenomenological arguments against simulation theory. In D. Hutto, & M. Ratcliffe, (Eds.), *Folk Psychology Re-assessed* (pp. 63–78). Dordrecht: Springer Publishers.

Gallagher, S. (2007b). Simulation trouble. *Social Neuroscience*, *2*(3–4), 353–365.

Gallagher, S. (2008). Direct perception in the social context. *Consciousness and Cognition*, *17*, 535–543.

Gallagher, S. & Hutto, D. (2008). Primary interaction and narrative practice. In: Zlatev, Racine, Sinha and Itkonen (Eds). *The Shared Mind: Perspectives on Intersubjectivity* (pp. 17–38). Amsterdam: John Benjamins.

Gallese, V. (2001). The shared manifold' hypothesis: From mirror neurons to empathy. *Journal of Consciousness Studies*, *8*, 33–50.

Gallese, V. (2005). Being like me: Self-other identity, mirror neurons and empathy. In S. Hurley & N. Chater (Eds.), *Perspectives on imitation I* (pp. 101–118). Cambridge, MA: MIT Press.

Gallese, V. (2007). Before and below "theory of mind": Embodied simulation and the neural correlates of social cognition. *Philosophical Transactions of the Royal Society, B-Biological Sciences*, *362*(1480), 659–669.

Gallese, V., Eagle, M. N. & Migone, P. (2007). Intentional attunement: Mirror neurons and the neural underpinnings of interpersonal relations. *Journal of the American Psychoanalytic Association*, *55*(1), 131–176.

Georgieff, N. & Jeannerod, M. (1998). Beyond consciousness of external events: A "Who" system for consciousness of action and self-consciousness. *Consciousness and Cognition*, *7*, 465–477.

Gibson, J. J. (1979). *The Ecological Approach to Visual Perception*. Boston, MA: Houghton-Mifflin.
Goldman, A. (2006). *Simulating minds: The philosophy, psychology and neuroscience of mindreading*. Oxford, England: Oxford University Press.
Goldman, A. (2005). Imitation, mind reading, and simulation. In Hurley & Chater (Eds.), *Perspectives on Imitation II* (pp. 79–93). Cambridge, MA: MIT Press.
Goldman, A. I. (2002). Simulation theory and mental concepts. In J. Dokic & J. Proust (Eds.), *Simulation and Knowledge of Action* (pp. 1–19). Amsterdam: John Benjamins.
Goldman, A. I. & Sripada, C. S. (2005). Simulationist models of face-based emotion recognition. *Cognition, 94*, 193–213.
Gopnik, A. & Meltzoff, A. N. (1997). *Words, Thoughts, and Theories*. Cambridge, MA: MIT Press.
Grèzes, J. & Decety, J. (2001). Functional anatomy of execution, mental simulation, and verb generation of actions: A meta-analysis. *Human Brain Mapping, 12*, 1–19.
Herrmann, E., Call, J., Hare, B., & Tomasello, M. (2007). Humans evolved specialized skills of social cognition: The cultural intelligence hypothesis. *Science, 317*(5843), 1360–1366.
Hobson, P. (1993). The emotional origins of social understanding. *Philosophical Psychology, 6*, 227–249.
Hobson, P. (2002). *The Cradle of Thought*. London: Macmillan.
Hurley, S. L. (1998). *Consciousness in Action*. Cambridge, MA: Harvard University Press.
Hurley, S. L. (2005). Active perception and perceiving action: The shared circuits model. In T. Gendler, & J. Hawthorne, (Eds.), *Perceptual Experience*. New York: Oxford University Press.
Hutto, D. (2007). *Folk Psychological Narratives*. Cambridge, MA: MIT Press.
Iacoboni, M., Molnar-Szakacs, I., Gallese, V., Buccino, G., Mazziotta, J. C. & Giacomo Rizzolatti, G. (2005). Grasping the intentions of others with one's own mirror neuron system. *PLoS Biology, 3*(3), 529–535.
Jeannerod & Pacherie, (2004). Agency, simulation, and self-identification. *Mind and Language, 19*(2), 113–146.
Johnson, S. et al. (1998). Whose gaze will infants follow? The elicitation of gaze-following in 12-month-old infants. *Developmental Science, 1*, 233–238.
Johnson, S. C. (2000). The recognition of mentalistic agents in infancy. *Trends in Cognitive Science, 4*, 22–28.
Kaplan, J. T. & Iacoboni, M. (2006). Getting a grip on other minds: Mirror neurons, intention understanding, and cognitive empathy. *Social Neuroscience, 1*(3–4), 175–183.
Karmiloff-Smith, A. (1992). *Beyond Modularity: A Developmental Perspective on Cognitive Science*. Cambridge, MA: MIT Press.
Király, I., Jovanovic, B., Prinz, W., Aschersleben, G. & Gergely, G. (2003). The early origins of goal attribution in infancy. *Consciousness and Cognition, 12*(4), 752–769.
Leslie, A. (2000). Theory of mind as a mechanism of selective attention. In M. Gazzaniga (Ed.), *The New Cognitive Neurosciences* (pp. 1235–1247). Cambridge, MA: MIT Press.
Leslie, A. & Frith, U. (1988). Autistic children's understanding of seeing, knowing and believing. *British Journal of Developmental Psychology, 6*, 315–324.
Lindblom, J. (2007). *Minding the Body: Interacting Socially through Embodied Action*. Linköping: Linköping Studies in Science and Technology, Dissertation No. 1112.
Malle, B. F. (2002). The relation between language and theory of mind in development and evolution. In T. Givón & B. F. Malle (Eds.), *The Evolution of Language out of Pre-Language* (pp. 265–284). Amsterdam: John Benjamins.
Meltzoff, A. N. (1995). Understanding the intentions of others: Re-enactment of intended acts by 18-month-old children. *Developmental Psychology, 31*, 838–850.
Meltzoff, A. N. & Brooks, R. (2001). Like me as a building block for understanding other minds: Bodily acts, attention, and intention. In B. F. Malle, L. J. Moses & D. A. Baldwin (Eds.), *Intentions and Intentionality: Foundations of Social Cognition* (pp. 171–191). Cambridge, MA: MIT Press.
Meltzoff, A. & Moore, M. K. (1977). Imitation of facial and manual gestures by human neonates. *Science, 198*, 75–78.
Meltzoff, A. & Moore, M. K. (1994). Imitation, memory, and the representation of persons. *Infant Behavior and Development, 17*, 83–99.
Minio-Paluello, I., Avenanti, A., & Aglioti, S. M. (2006). *Social Neuroscience, 1*(3–4), 320–333.

Niedenthal, P. M., Barsalou, L. M., Winkielman, P., Krauth-Gruber, S., & Ric, F. (2005). Embodiment in attitudes, social perception, and emotion. *Personality and Social Psychology Review*, *9*(3), 184–211.

Noë, A. (2004). *Action in Perception*. Cambridge, MA: MIT Press.

Oberman, L. M. & Ramachandran, V. S. (2007). The simulating social mind: The role of the mirror neuron system and simulation in the social and communicative deficits of autism spectrum disorders. *Psychological Bulletin*, *133*(2), 310–327.

Onishi, K. H. & Baillargeon, R. (2005). Do 15-month-old infants understand false beliefs? *Science*, *308*(5719), 255–258.

Phillips, W., Baron-Cohen, S., & Rutter, M. (1992). The role of eye-contact in the detection of goals: Evidence from normal toddlers, and children with autism or mental handicap. *Development and Psychopathology*, *4*, 375–383.

Rizzolatti, G., Fadiga, L., Gallese, V., & Fogassi, L. (1996). Premotor cortex and the recognition of motor actions. *Cognitive Brain Research*, *3*, 131–141.

Rizzolatti, G., Fogassi, L., & Gallese, V., (2000). Cortical mechanisms subserving object grasping and action recognition: A new view on the cortical motor functions. In M. S. Gazzaniga (Ed.), *The New Cognitive Neurosciences* (pp. 539–552). Cambridge, MA: MIT Press.

Scheler, M. (1954). *The Nature of Sympathy*. Trans. P. Heath. London: Routledge and Kegan Paul. Original: *Wesen und Formen der Sympathie*. Bonn: Verlag Friedrich Cohen, 1923.

Schilbach, L., Eickhoff, S. B., Mojzisch, A., & Vogeley, K. (2008). What's in a smile? Neural correlates of facial embodiment during social interaction. *Social Neuroscience*, *3*(1), 37–50.

Scholl, B. J. & Tremoulet, P. D. (2000). Perceptual causality and animacy. *Trends in Cognitive Sciences*, *4*(8), 299–309.

Senju, A., Johnson, M. H., & Csibra, G. (2006). The development and neural basis of referential gaze perception. *Social Neuroscience*, *1*(3–4), 220–234.

Singer, W., Wolpert, D., & Frith, C. (2004). Introduction: The study of social interactions. In C. Frith & D. Wolpert (Eds.), *The Neuroscience of Social Interaction* (pp. xii–xxvii). Oxford: Oxford University Press.

Tooby, J. & Cosmides, L. (1995). Foreword to S. Baron-Cohen, *Mindblindness: An Essay on Autism and Theory of Mind* (pp. xi–xviii). Cambridge, MA: MIT Press.

Trevarthen, C. B. (1979). Communication and cooperation in early infancy: A description of primary intersubjectivity. In M. Bullowa (Ed.), *Before Speech* (pp. 321–347). Cambridge: Cambridge University Press.

Trevarthen, C. & Hubley, P. (1978). Secondary intersubjectivity: Confidence, confiding and acts of meaning in the first year. In A. Lock (Ed.), *Action, Gesture and Symbol: The Emergence of Language* (pp. 183–229). London: Academic Press.

Varela, F. J., Thompson, E. & Rosch, E. (1991). *The Embodied Mind: Cognitive Science and Human Experience*. Cambridge: MIT Press.

Walker, A. S. (1982). Intermodal perception of expressive behaviors by human infants. *Journal of Experimental Child Psychology*, *33*, 514–535.

Wellman, H. M. (1993). Early understanding of mind: The normal case. In S. Baron-Cohen, H. Tager-Flusberg & D. J. Cohen (Eds.), *Understanding Other Minds: Perspectives from Autism* (pp. 10–39). Oxford: Oxford University Press.

Wimmer, H. & Perner, J. (1983). Beliefs about beliefs: Representation and constraining function of wrong beliefs in young children's understanding of deception. *Cognition*, *13*, 103–128.

Wittgenstein, L. (1967). *Zettel*. Eds. G. E. M. Anscombe & G. H. von Wright, trans. G. E. M. Anscombe. Berkeley: University of California Press.

Wittgenstein, L. (1980). *Remarks on the Philosophy of Psychology*, Vol. II. Eds. G. H. von Wright and H. Nyman, trans. C. G. Luckhardt & M. A. E. Aue. Oxford: Blackwell.

Woodward, A. L. & Sommerville, J. A. (2000). Twelve-month-old infants interpret action in context. *Psychological Science*, *11*, 73–77.

23

Getting to the Heart of Emotions and Consciousness

Maxine Sheets-Johnstone

University of Oregon, Yachats, OR, USA

INTRODUCTION: DESCRIPTIVE FOUNDATIONS AND ANIMATION

The aim of this chapter is to open a path that penetrates to the core of affective experience—to the heart of emotion and consciousness—the topic of this section of the Handbook. Opening such a path requires a consistent view of the integral wholeness of life; that is, it requires foundational experiential understandings of commonly separated aspects, precisely such aspects as emotion and consciousness, and more finely, empathy and altruism, for example, aggression and submission, disgust and joy. In light of this requirement, phylogeny and ontogeny are of signal importance as are phenomenological analyses. These facts and analyses provide the descriptive foundations necessary for foundational experiential understandings.

Common topics in investigations of social emotions such as empathy and altruism rarely provide descriptive foundations. Such topics thus fall outside the aim of this chapter as does a specific focus on a particular kind of behavior or emotion. Current theoretically fashionable perspectives that attempt to capture overlooked or essential features of life, namely, the enactive approach and embodiment theses, similarly fall outside the purview of this chapter. Phylogenetic and ontogenetic facts and phenomenological analyses will in fact ultimately suggest that these perspectives are lexical band-aids covering over long-lingering ignorances of the realities of life itself; they are descriptively as well as linguistically deflective.

The realities of life itself are of fundamental significance. Because they are, they warrant serious investigation and study from the beginning, the beginning in precisely a phylogenetic, ontogenetic, and phenomenological sense. To take phylogeny as an example, this clearly does not mean that present-day scientists and philosophers are responsible for concentrated investigations and studies *en par* with those of Darwin, approximating to the span of creatures he studied or to the years he devoted to their study. On the other hand, specializing in one animal or species of animal—including *Homo sapiens sapiens*—and generalizing phylogenetically in one way and another from there has its hazards. A striking instance is E. O. Wilson's monumentally thorough study of ants, which blossomed into the theoretical science known as sociobiology (Wilson, 1975), which in turn blossomed into theoretical entities such as selfish genes and brain modules. When theory overtakes real-life observations, or when ready-made categories of behavior triumph over finely detailed descriptive accounts of the social interactions of wolves, chimpanzees, or humans, for example, or of beavers building dams or birds building nests, the basic realities of life itself are elided. To be noted too, however, is that descriptive accounts of the lives of nonhuman animals may be faulted for their anthropomorphism or for offering merely anecdotal rather than laboratory-controlled data. Jane Goodall's lifelong studies of chimpanzees (Goodall, 1968, 1971, 1990) are testimonial to the impropriety of many such charges as are the studies of other researchers from a variety of perspectives (for a breadth of perspectives, see ethologists Mitchell et al., 1997; see also philosopher John Fisher, 1996, and biologist Stephen Jay Gould, 2000). Moreover the studies of Goodall, Strum (1987), Hall and DeVore (1972), and Schaller (1963), along with those of many other primate researchers, attest the importance of recognizing descriptive foundations: descriptive foundations are the bedrock of phylogenetic matters of fact, and thus, of evolutionary continuities (Sheets-Johnstone, 2002; see also, Sheets-Johnstone, 1986, 1994, 1996, 1999a).

The realities of life itself are implicit in the original descriptive foundations set forth by Darwin. Indeed, the centerpin of Darwin's extensive as well as lifelong studies of animals was precisely living forms that move themselves. From his first studies as a biologist on The Beagle (Darwin, 1958 [1839]) to his last studies of earthworms (Darwin, 1976 [1881]), his morphological concerns were consistently tied to animation; that is, to how animate forms make a living, given the animate forms they are. His emphasis was thus not on a static morphology, but on what we might term *morphology-in-motion.* Morphology-in-motion—animation—is first and foremost a subject–world relationship. Perception, emotion, cognition, and imagination all derive from the basic fact that whatever the animate form, it lives not in a vacuum but in a world particularized by its being the animate form it is. Precisely because it does not live in a vacuum, it is unnecessary to "embed" its perceptions, cognitions, or affective experiences in the world. Its interest, curiosity, hesitation, fright, and so on, its turning toward or turning away, and its approach or avoidance are emblematic of its affective motivations to move in distinctive ways with respect to the world in which it lives.

Animation is actually theoretically of a piece with the biological concept of "responsivity": "Plant seedlings bend toward light; meal-worms congregate in dampness; cats pounce on small moving objects; even certain bacteria move toward or away from certain chemicals.... [T]the capacity to respond is a fundamental and almost universal characteristic of life" (Curtis, 1975, p. 28). It is notable that we find just such observations throughout Darwin's writings, specifically with respect to emotions. He writes, for example, "Terror acts in the same manner on them [the lower animals] as on us, causing the muscles to tremble, the heart to palpitate, the sphincters to be relaxed, and the hair to stand on end" (Darwin, 1981 [1871], vol. 1, p. 39). He goes on to write of suspicion in "most wild animals," of courage and timidity being "variable qualities in ... individuals of the same species," of some animals of a species being good-tempered and others ill-tempered, and of maternal affection in non-human animal life (ibid., pp. 39–40). In short, and even before he examines emotions at length in his well-known book *The Expression of the Emotions in Man and Animals* (Darwin, 1965 [1872]), Darwin dwells at length on the responsivity of living creatures—on the primordial *animation* that is at the heart of life and across virtually the whole of the animal kingdom.

Precisely in this context, it is of moment to note Darwin's estimation of the vexing relationship of mind and body, of the challenge it presents, and of the proper mode of conceiving and approaching the challenge of understanding that relationship. He writes, "Experience shows the problem of the mind cannot be solved by attacking the citadel itself—the mind is function of body—we must bring some *stable* foundation to argue from" (Darwin, 1987 [1838], p. 564). While further comment will be made below on this insightful observation, the point of moment here is that *animation* is indeed the *stable* foundation from which to argue, for *animation* is inclusive of the whole of life, and for this reason is integral to all-inclusive and penetrating understandings of emotion and consciousness. In particular, animation tells us why distinguishing between behavior and movement is of vital significance; it tells us why concepts emanating from movement are of vital significance to animate life; it tells us why emotions too are descriptively declinable in terms of force, space, and time, why they too are manifestations of dynamic bodily feelings, in this instance, not just kinetic but affective dynamic bodily feelings; and finally, it tells us why emotions and movement are dynamically congruent. We will examine each of these four essentially enlightening aspects of animation in turn.

ON THE DISTINCTION BETWEEN BEHAVIOR AND MOVEMENT

Descriptions of behavior rely first of all on pre-assigned categorical placements. An individual is eating, for example, or mating, fighting, or exploring; in other words, the individual is doing something that carries a ready-made label.

Identified within a ready-made category, the doing thereby falls also within a ready-made category of knowledge. The observer already knows what is involved on the basis of his or her own experience and applies that knowledge to his or her object of study. In short, the observer's first-person knowledge is the basis of his or her third-person behavioral ascriptions. What such ascriptions elide are the kinetic dynamics of any particular behavior. This fact was recognized by ethologists such as Ilan Golani who, by utilizing a movement notation system, was able to analyze the qualitative kinetic structure of movement, thereby coming to understand the actual dynamics of life itself. What these dynamics show is that "cognition is not separated from perception, perception is not separated from movement, and movement is not separated from an environment nor from a larger category designated as a behavior; on the contrary, the movement–perceptual system *is* behavior in the sense that it is the actual 'real-time', 'real-life' event as it unfolds" (Sheets-Johnstone, 1999a, p. 218). Ethologist John Fentress's intricate studies of "how mice scratch their faces" (Fentress, 1989, pp. 45–46) and his and other's combined study of ritualized fighting in wolves (Moran et al., 1981) demonstrate the significance of analyzing and understanding the kinetic dynamics of life itself. In his early explorations with automatons, neuroscientist Gerald Edelman arrived at similar conclusions. He found that *movement* was instrumental in gaining knowledge of the world, that the automaton Darwin III, for example, "categorizes only on the basis of experience" (Edelman, 1992, p. 93); Darwin III could "decide," for example, "that something is an object, that the object is striped, and that the object is bumpy" (ibid.) only on the basis of freely varied movement. Edelman's findings testify to the fact that animation is first and foremost a subject–world relationship and that life is grounded in animation: animals are impelled to move on the basis of their interest or aversion to what they perceive, what they recognize, and so on.

Though they have been largely overtaken by an attentive fixation on *the brain*, the above-mentioned studies should not be dismissed; they are not passé but curatively topical to the ills that plague a reductionist-leaning neuroscience. The point is succinctly made by Foolen et al. (2007) when they rightfully question whether mirror neurons are cause or effect of experience. If the latter, then morphology-in-motion—animation—is the core phenomenon of life, the core from which animal faculties and capacities arise. In other words, movement is our mother tongue (Sheets-Johnstone, 1999a). Behavior is no match for this core phenomenon. Moreover certain methodological correspondences are evident between the above-mentioned studies and the studies of Edmund Husserl, the founder of phenomenology, and Hermann von Helmholtz, the noted 19th-century physicist–physiologist. Both Husserl and von Helmholtz emphasized the centrality of movement to perception and both made use of the free variation of movement in their epistemologically tethered research pursuits. For example, in the context of describing aspects of infant-child play with objects, von Helmholtz concludes that "the child learns to recognize the different views which the same object can afford in correlation with the movements which he is constantly giving it" (von Helmholtz, 1971a [1878],

p. 214). In another text, he states that "our body's movement sets us in varying spatial relations to the objects we perceive, so that the impressions which these objects make upon us change as we move (von Helmholtz, 1971b [1868], p. 373). He furthermore devises not thought experiments but real-life *movement* experiments such as the following in quest of understanding what he calls "judgment[s] of relief in the floor-plane":

> This [judgment] can be tested by standing in a level meadow and first observing the relief of the ground in the ordinary way. There may be little irregularities here and there, but still the surface appears to be distinctly horizontal for a long way off. Then bend the head over and look at it from underneath the arm; or stand on a stump or a little elevation in the ground, and stoop down and look between the legs, without changing much the vertical distance of the head above the level ground. The farther portions of the meadow will then cease to appear level and will look more like a wall painted on the sky. I have frequently made observations of this kind as I was walking along the road between Heidelberg and Mannheim (von Helmholtz, 1962 [1910], pp. 433–34).

Clearly, when experience is meticulously examined and meticulously varied, we can make self-evident the fact that movement quintessentially informs perception. The practice of phenomenology rigorously testifies to this fact. As Husserl states, "We constantly find here [in moving and perceiving] this two-fold articulation: kinesthetic sensations on the one side, the motivating; and the sensations of features on the other, the motivated" (Husserl, 1989, p. 63). He furthermore underscores the consequential nature of the articulation, stating not only that "[*I*]*f* the eye turns in a certain way, *then* so does the 'image'" (ibid.); but more broadly that "'exhibitings of' are related back to correlative multiplicities of kinesthetic processes having the peculiar character of the 'I do', 'I move' (to which even the 'I hold still' must be added).... [A] hidden intentional 'if-then' relation is at work here" (Husserl, 1970, p. 161).

It is of prime importance to recognize that these movement–perceptual relationships—in reality, movement–perceptual–cognitional relationships—are informed by and articulated within an affective experience of some kind. *Animate beings are moved to move*. To be moved by and move with interest toward something is different from being moved by and move with apprehension toward it; to be moved by and move with delight toward something is different from being moved by and move with disgust away from it. Animation opens the path toward just such holistic understandings of life. Behavior as it is commonly spoken of and written about does not approximate to the deeper and more complex facets of movement that are at the heart of animation. It does not uncover the consequential dynamic relationship of perception to movement, the significance of freely varied movement, or the integral role of emotion in perception and cognition. Indeed, from the viewpoint of affective experience, if I draw close to something or run away from it, if I am determined to solve a problem, convinced of the truth of my findings, or dumbstruck, for example, my experience is clearly not simply perceptual or cognitional but affective and as such, a testimonial to the reality of my being an animate being, a being that is moved to move, whether to explore, avoid, persist in my efforts, stand my ground, or, as Husserl points out, even to hold still. In sum, recognition of, and attention to movement opens our eyes to the complex reality that is animation, a reality that defines the multiple facets of a subject–world relationship.

CONCEPTS EMANATING FROM MOVEMENT

As indicated earlier, and as suggested by Darwin's observation that "the mind is function of body," primary attention to movement is the key to understanding the complex realities of animation, whether a matter of tigers, bats, humans, lizards, langurs, bees, or even bacteria. With respect to humans, we may first note that movement is the primary object of attention of human infants (Spitz, 1983); it is the primary mode of social communication in infancy (Stern, 1985); and it is the primary source of nonlinguistic concepts (Sheets-Johnstone, 1990). The latter concepts are not poor relatives of later linguistic concepts; quite the reverse in that naming where something is—*inside* or *far away*, for example—or naming the temporal span of something—*sudden* or *prolonged*, for example—is rooted not in the words themselves but in experience. In short, nonlinguistic kinetic concepts ground fundamental concepts of space, time, and force. This fact is validated by a diversity of research reports and conclusions about infants and young children: for example, by Clark's finding that *in* is the first locative to be learned (Clark, 1979; see also Grieve et al., 1977; Cook, 1978; see too Piaget, 1967, 1968, and Bower, 1974 on "being inside"; see also Sheets-Johnstone, 1990 for a phenomenological explication of an infant's experiences of *in* and *being inside*); by Bower's observations that not only is an infant fascinated by the opening and closing of its hand, but that it is fascinated by putting something inside someone else's hand, then closing it, then opening it, and so on (Bower, 1979); and by Stern's observation that an infant's *hunger storm* is not a sensation but an ongoing dynamic (Stern, 1990). We indeed begin life by thinking in movement (Sheets-Johnstone, 1999a; see also Bruner, 1991; Bloom, 1993). This form of thinking does not disappear, but remains at the core of our capacities in the world, including our capacity to think in language, that is, in words. Infants are indeed not *prelinguistic*; language is *post-kinetic* (Sheets-Johnstone, 1999a). Not only this but movement forms the I that moves before the I that moves forms movement (ibid.). In other words, movement is there from birth and before (Furuhjelm et al., 1976), feeding the faculties and concepts that mature precisely in the course of moving oneself.

To appreciate the concepts generated by movement, consider the experience of walking. The qualitative character of our experience depends on the inherently qualitative nature of movement. Our walk may be jaunty, for example, or slow and labored, determined, hesitant, rushed, or relaxed, and so on. Moreover we may follow a straight path across a parking lot, zig-zag erratically in walking down a street to avoid collisions with others, or follow regular cut backs along a mountain trail, our movement in each instance creating a distinctive linear pattern. If it is windy, we may be tilted forward or if in pain, we may be twisted, our body in each instance having a distinctive linear design. Movement indeed has four fundamental qualities: tensional, linear, areal, and projectional quality. These qualities are of a piece in any movement sequence, that is, not simply in walking, but in reaching for a glass, kicking a ball, marching out of a room, drawing figure-eights in the air, and so on. They are inherent in any movement, but they can be analyzed phenomenologically, meaning that the nature of each quality can be spelled out. As intimated earlier, the intensity of our walk may vary and change, as when we realize that we are late and must hurry, a shift that increases our originally relaxed gait. Also, our body has a certain linear design and creates certain linear patterns in walking, its design and pattern being directional lines. Consider too that our body creates areal designs and patterns as well as linear ones. In walking, our body may be anywhere from contracted to expanded, depending, for example, on whether we are trying to hide our presence as we walk into or out of a room or whether we are walking open-armed to greet a friend; the movement itself may be anywhere from intensive to extensive, depending upon the spatial amplitude of our walk. We might, for example, stop short in an otherwise extensive walk through a forest, hunch over, slowly rise up to look ahead at what we thought was a bear, then take quite small steps backward in retreat. As with all qualities of movement, areal design and pattern too may vary and thus change the character of our walk. Consider finally the projectional quality of our walk, that is, whether our movement is abrupt, sustained, ballistic, or a combination of these qualities: a goose-stepping walk is notably different from a smooth, ongoing, relatively unaccented gait, for example. To be noted is that any particular projectional designation is actually inclusive of infinite degrees of shading and that any particular movement may be a combination of basic projectional qualities, as when, for example, we turn our head abruptly to the side while continuing in our even-footed gait, or when we begin swinging our arms back and forth, giving them an initial impulse upward, then letting them follow through on their own momentum and gravitational pulls.

However brief the above analysis, it should be evident that kinesthesia is at the source of fundamental concepts of force, time, and space—of intensity, direction, amplitude, duration, and so on—and at the source as well of the originally qualitative character of these concepts, that is, not only of what is experienced as strong, moderate, or weak in the course of self-movement, for example, but also of what is fading and what is growing in intensity, what is sudden and what is attenuated, what is continuing to move on the basis of an initial impulse or powerfully driven at every moment, what is moving resolutely forward or diverging erratically onto different paths, what is loose and open-ended or tight and constricted, and so on. These qualitatively-laden concepts are integral to our understandings of emotion as well as movement. In particular, recognition of these concepts as deriving from our earliest experiences of movement onward is integral not only to our understanding of the subtle and intricate complexities of movement and thus of our capacity to learn to move ourselves efficiently and effectively in the world in the first place, but also of our understanding of affective feelings and thus ultimately of the dynamic congruency of movement and emotion.

AFFECTIVE FEELINGS

Emotions too are declinable in terms of space, time, and force. To appreciate this fact at its core, we need to look closely again at infancy and acknowledge what it teaches us about our humanness. Adult thoughts and theories about our human ways risk validity by dismissing—or denigrating (Dennett 1983, p. 384)—that period of our lives we all lived through as infants. We clearly see the truth of this claim in the fact that, for an infant, an emotional experience is a *whole-body experience*; an infant does not feel distress simply in its abdomen or pleasure only on its face. Accordingly, why is whole-body affective experience transformed into partial-body affective experience in many adult estimations of emotion? More finely put, why is misery or joy or sadness or anger conceived as felt only in one's head or face, or hand or belly? An adult example well illustrates the point at issue.

When one has missed attending a meeting and wants to catch up on happenings by reading the minutes, one may feel by turns interested, indifferent, doubtful, surprised, disappointed, and angered in the course of reading. The feelings are likely fleeting as well as in the background in relation to the reading itself. Yet they inform one's experience of reading the minutes and may in fact come strongly to the fore depending on their intensity. If they do, one finds oneself immersed in an affective dynamics, feeling them not only here and there, as if anger were simply a matter of clenched teeth and fists, or surprise simply a matter of a gasp and a pounding heart—though one may certainly feel specifically located tensions, a rush of air in one's throat, or sudden hammering heartbeats. One is, on the contrary, immersed throughout; the affective dynamics envelop one in a whole-body sense.

Infant psychiatrist and clinical psychologist George Downing asks, "When we think about emotion, should our focus be on the face or on the body as a whole?" (Downing, 2005, p. 429). When he attempts to show how "[r]esearch strategies limited to the face," while having had "considerable success … may marginalize phenomena that, in fact, are deserving of more attention," he surely points us in the direction of reconsidering not just a partialized-body view of emotion but a view that is developmentally disjoint. That is, if, as Downing notes, "Affectively speaking, infants would seem to be full-body creatures from the start" (ibid.), how is it that adult humans commonly lose not sight but the *felt bodily sense* of their full-body affectivity? Having lost touch with their emotions in this experiential sense, adult humans thereby commonly fail to recognize the dynamics of emotion, conceiving emotions as *states* of being or as nothing more than discrete bodily *sensations*.

Affective dynamics are finely delineated by infant psychiatrist and clinical psychologist Daniel Stern in his illustrations of "affect attunement," situations in which a mother creates a qualitatively felt dynamic with her infant. Consider the following examples.

> A nine-month-old girl becomes very excited about a toy and reaches for it. As she grabs it, she lets out an exuberant "aaaah!" and looks at her mother. Her mother looks back, scrunches up her shoulders, and performs a terrific shimmy with her upper body, like a go-go dancer. The shimmy lasts only about as long as her daughter's "aaaah!" but is equally excited, joyful, and intense (Stern, 1985, p. 140).
>
> A nine-month-old boy is sitting facing his mother. He has a rattle in his hand and is shaking it up and down with a display of interest and mild amusement. As mother watches, she begins to nod her head up and down, keeping a tight beat with her son's arm motions. (ibid., p. 141).

Stern analyzes such instances of affect attunement in terms closely analogous to the qualities of movement specified in the previous sections, namely, in terms of *intensity*, *timing*, and *shape*. He breaks these dimensions down in greater detail, describing how a mother matches the "absolute intensity," "intensity contour," "temporal beat," "rhythm," "duration," and "shape" of her infant's dynamics (ibid., p. 146). The immediate point of note here is that whatever the mode of attunement, whether aural or kinetic, it is not a question of imitation but of dynamics, dynamics created by infant and mother together through some mode of bodily movement. The further point is that although the dynamics are clearly created through distinct bodily movements, including voice-producing movements, they are *whole-body experiences*, and this because they are experienced not objectively as an arm moving or a head nodding, or as someone shaking a rattle or banging on a toy (ibid., pp. 140–141), but as a wholly qualitative phenomenon having a certain spatio-temporal-energic character, such as when a mother matches her infant's kinetic dynamics with a vocalized "'kaaaaa-*bam*, kaaaaa-*bam*" (ibid., p. 140).

In sum, affects, like movement, are whole-body spatio-temporal-energic phenomena; precisely as Stern indicates, they have distinct spatial contours, intensities, and temporalities. When we explode in anger, burst into song, begin to doubt, nurse a grudge, hesitate to speak out, continue to grieve, turn away in disgust, are seized by fear, and so on, it is *experientially* evident not only that emotions are manifestations of feelings but that emotions are distinctive in both a bodily-felt and bodily-observable sense and are therefore descriptively declinable.

DYNAMIC CONGRUENCY

In his article "Action and Emotion in Development of Cultural Intelligence: Why Infants Have Feelings Like Ours," infant psychologist Colwyn Trevarthen defines emotions "as manners of moving, and of responding to movement"

(Trevarthen, 2005, p. 63). He emphasizes the sensitivity of infants to "animacy" (ibid., p. 80), and more broadly, the way in which animal bodies are "motivated with intrinsic rhythm and intensity in the 'vitality' or 'sentic forms' of emotions," stressing in this context the dynamic temporal dimension of emotion and movement (ibid., p. 64).

Trevarthen's conception of emotion and his emphasis on animacy substantively echo Darwin's observation of "the intimate relation which exists between almost all the emotions and their outward manifestations." It is puzzling as well as notable that many present-day scientists and philosophers investigating emotions seem oblivious not only of Darwin's original insight into the intimate relation between emotion and movement but of ontogenetical studies of emotion such as those of Trevarthen and Stern. Their thinking is instead tied to behavior or action. In effect, they fail to realize that, to paraphrase anthropologist Claude Levi-Strauss, "movement is good to think." As previous sections of this chapter have shown, movement opens paths to multiple dimensions of animation. A discussional comment in a recent book on emotion suggests that a sense of the import of movement hovers at the edge of awareness. Answering a question about dynamics with respect to the ability of chimpanzees to recognize the facial expressions of conspecifics, ethologist Lisa Parr states, "Normally, chimpanzees don't really see facial expressions in a static way.... Maybe movement in and of itself produces a lot of information about individual identity and maybe even about the type of expression that's being made" (Parr, 2003, p. 80). Earlier, in her article in the book, she comments, "When facial expressions were presented as dynamic stimuli using video, subjects showed no preference for trials in which distinctive features were present. Therefore, the addition of movement, vocalizations and context significantly changed the manner by which chimpanzees discriminated some facial expressions" (ibid., p. 73). Most interestingly too, she notes, "The perception of facial expressions ... produces a low-level motor mimicry in the perceiver that can be measured using electromyographic recordings. These subtle movements correlate to the self-perception of emotion, suggesting an integral link between facial action and emotional experience" (ibid., p. 72).

Clearly, thinking in terms of *action*—like thinking in terms of *behavior*—deflects us from recognizing the rich and subtle spatio-temporal-energic dynamics of movement, and in turn from recognizing the rich and subtle dynamics of emotions and the intimate relation between movement and emotions. In short, once we realize that movement is indeed good to think, we begin thinking in terms of *dynamics*. Not only are Stern's studies of affect attunement testimony to these dynamics but so also is his identification of "vitality affects," which he describes precisely in kinetic terms such as "'surging', 'fading away', 'fleeting', 'explosive', 'crescendo', 'decrescendo', 'bursting', 'drawn out', and so on" (Stern, 1985, p. 54). What ontogenetical studies of emotion clearly point to and elucidate through "vitality affects," "animacy," "affect attunement," and the like, is precisely movement and the intimate relation between emotion and movement.

The term *dynamic congruency* was introduced and the phenomenon of affective/kinetic concordance was originally analyzed in a 1999 article that combined an empirical approach with a phenomenological one (Sheets-Johnstone, 1999b). The empirical analysis is based not on ontogenetical research but on a variety of studies of emotions, studies that, in a different but no less trenchant way, carry forward Darwin's basic insight that movement and emotion go hand in hand (Sheets-Johnstone, 1999b). In particular, the article begins by showing how the research of medical doctor and neuropsychiatrist Edmund Jacobson grounds emotions in a neuromuscular dynamic; how that of neuropsychiatrist Nina Bull grounds emotions in "motor attitudes," that is, in bodily postures and in a readiness to move; and how that of psychologist Joseph de Rivera grounds emotions in our experience of being literally "moved" by emotions. The article proceeds to demonstrate the phenomenological import of these studies, precisely by identifying the *formal* congruency of emotion and movement, that is, their concordantly experienced qualitative dynamic. The *concept* of dynamic congruency is thus rooted in experience and is descriptively analyzable in terms of experience. To be noted is that a prime and in fact sterling value of actually *doing phenomenology* is to describe experience, which is to say, to meet the challenge of languaging experience. A careful, exacting, and evidentially supportable vocabulary is indeed essential to the attainment of veritable descriptive foundations. We can readily appreciate this fact by noting that it is the *natural* dynamic congruity of emotions and movement that allows us successfully both to mime feelings we do not actually feel and to inhibit the expression of those that we do. Indeed, the natural dynamic congruency of emotions and movement attests to the rationality of animation. If animals—human ones included—were not moved to move in ways they actually move or if they did not move in ways they were moved to move, there would be no possibility of moving efficiently and effectively in the world. Animation would literally have no inherent rhyme or reason.

In sum, animation is the "*stable* foundation" for understandings of consciousness in its entire multiple and varied forms. No more than any other animal do human animals need to be "embodied" or "embedded" in order to be fully accounted for and understood. To comprehend their foundational animation requires meeting the challenge of examining and describing experience in a rigorous and methodologically enlightened way, finding in the process a language commensurate with the realities of animation.

REFERENCES

Bloom, L. (1993). *The Transition from Infancy to Language: Acquiring the Power of Expression.* New York: Cambridge University Press.

Bower, T. G. R. (1974). *Development in Infancy.* San Francisco: W. H. Freeman.

Bower, T. G. R. (1979). *Human Development.* San Francisco: W. H. Freeman and Co.

Bruner, J. (1991). *Acts of Meaning.* Cambridge, MA: Harvard University Press.

Clark, E. V. (1979). Building a vocabulary: Words for objects, actions and relations. In P. Fletcher & M. Garman (Eds.), *Language Acquisition* (pp. 149–160). Cambridge: Cambridge University Press.

Cook, N. (1978). In, on and under revisited again. *Papers and Reports on Child Language Development 15* (pp. 38–45). Stanford, CA: Stanford University Press.

Curtis, H. (1975). *Biology*, 2nd ed. New York: Worth Publishers.

Darwin, C. 1958 [1839]. *The Voyage of the Beagle*. New York: Bantam Books.

Darwin, C. 1965 [1872]. *The Expression of the Emotions in Man and Animals*. Chicago: University of Chicago Press.

Darwin, C. 1976 [1881]. *The Formation of Vegetable Mould Through the Action of Worms with Observations on Their Habits*. Ontario, CA: Bookworm Publishing Co.

Darwin, C. 1981 [1871]. *The Descent of Man and Selection in Relation to Sex*. Princeton: Princeton University Press.

Darwin, C. 1987 [1838]. *Charles Darwin's Notebooks, 1836–1844*. P. H. Barrett, P. J. Gautrey, S. Herbert, D. Kohn, & S. Smith (Eds.). Ithaca: Cornell University Press.

Dennett, D. (1983). Intentional systems in cognitive ethology: The "Panglossian" paradigm defended. *The Behavioral and Brain Sciences, 6*, 343–390.

Downing, G. (2005). Discussion: Emotion, body, and parent-infant interaction. In J. Nadel & D. Muir (Eds.), *Emotional Development* (pp. 429–449). Oxford: Oxford University Press.

Edelman, G. M. (1992). *Bright Air, Brilliant Fire: On the Matter of the Mind*. New York: Basic Books.

Fentress, J. C. (1989). Developmental roots of behavioral order: Systemic approaches to the examination of core developmental issues. In M. R. Gunnar & E. Thelen (Eds.), *Systems and Development* (pp. 35–76). Hillsdale, NJ: Lawrence Erlbaum Associates.

Fisher, J. A. (1996). The myth of anthropomorphism. In M. Bekoff & D. Jamieson (Eds.), *Readings in Animal Cognition* (pp. 3–16). Cambridge, MA: Bradford Book/MIT Press.

Foolen, A., Lüdtke, U., Zlatev, J., & Racine, T. (2007). *Moving Ourselves, Moving Others: The Role of E(motion) for Intersubjectivity, Consciousness and Language*. Book proposal to John Benjamins. Forthcoming 2009 from same.

Furuhjelm, M., Ingelman-Sundbert, A., & Wirsén, C. (1976). *A Child Is Born*, rev. ed. New York: Delacourte Press.

Golani, I. (1976). Homeostatic motor processes in mammalian interactions: A choreography of display. In P. G. Bateson & P. H. Klopfer (Eds.), *Perspectives in Ethology* (Vol. 2, pp. 69–134). New York: Plenum Publishing.

Goodall, J. (1968). The behaviour of free-living chimpanzees in the Gombe Stream Area. *Animal Behaviour Monographs, 1*(Part 3), 161–311.

Goodall, J. (1971). *In the Shadow of Man*. New York: Dell Publishing.

Goodall, J. (1990). *Through a Window: My Thirty Years with the Chimpanzees of Gombe*. Boston: Houghton Mifflin Company.

Gould, S. J. (2000). Foreword: A lover's quarrel. In M. Bekoff (Ed.), *The Smile of a Dolphin: Remarkable Accounts of Animal Emotions* (pp. 13–17). New York: Discovery Books.

Grieve, R., Hoogenraad, R., & Murray, D. (1977). On the young child's use of lexis and syntax in understanding locative instructions. *Cognition, 5*, 235–250.

Hall, K. R. L. & DeVore, I. (1972). Baboon social behavior. In P. Dolhinow (Ed.), *Primate Patterns* (pp. 125–180). New York: Holt, Rinehart and Winston.

Husserl, E. (1970). *The Crisis of European Sciences and Transcendental Phenomenology*. D. Carr. (trans.), Evanston, IL: Northwestern University Press.

Husserl, E. (1989). *Ideas Pertaining to a Pure Phenomenology and to a Phenomenological Philosophy: Book 2 (Ideas II)*. R. Rojcewicz & A. Schuwer (trans.), Boston: Kluwer Academic.

Mitchell, R. W., Nicholas, T. S., & Miles, H. L. (Eds.) (1997). *Anthropomorphism, Anecdotes, and Animals*. New York: State University of New York Press.

Moran, G., Fentress, J. C., & Golani, I. (1981). A description of relational patterns of movement during 'ritualized fighting' in wolves. *Animal Behavior, 29*, 1146–1165.

Paar, L. A. (2003). The discrimination of faces and their emotional content by chimpanzees (*Pan troglodytes*). In P. Ekman, J. J. Campos, R. J. Davidson, & F. B. M. de Waal (Eds.), *Emotions Inside Out: 130 Years after Darwin's Expression of the Emotions in Man and Animals* (pp. 56–78). New York: New York Academy of Sciences.

Piaget, J. (1967). *La construction du réel chez l'enfant*. Neuchatel: Delachaux et Niestlé.
Piaget, J. (1968). *La naissance de l'intelligence chez l'enfant*, 6th ed. Neuchatel: Delachaux et Niestlé.
Schaller, G. B. (1963). *The Mountain Gorilla: Ecology and Behavior*. Chicago: University of Chicago Press.
Sheets-Johnstone, M. (1986). Existential fit and evolutionary continuities. *Synthese, 66,* 219–248.
Sheets-Johnstone, M. (1990). *The Roots of Thinking*. Philadelphia: Temple University Press.
Sheets-Johnstone, M. (1994). *The Roots of Power: Animate Form and Gendered Bodies*. Chicago: Open Court Publishing.
Sheets-Johnstone, M. (1996). Taking evolution seriously: A matter of primate intelligence. *Etica & Animali, 8,* 115–130.
Sheets-Johnstone, M. (1999a). *The Primacy of Movement*. Amsterdam/Philadelphia: John Benjamins Publishing.
Sheets-Johnstone, M. (1999b). Emotions and movement: A beginning empirical-phenomenological analysis of their relationship. *Journal of Consciousness Studies, 6*(11–12), 259–277.
Sheets-Johnstone, M. (2002). Descriptive foundations. *Interdisciplinary Studies in Literature and Environment, 9*(1), 165–179.
Spitz, R. A. (1983). *Dialogues from Infancy*. R. M. Emde (Ed.), New York: International Universities Press.
Stern, D. N. (1985). *The Interpersonal World of the Infant: A View from Psychoanalysis and Developmental Psychology*. New York: Basic Books.
Stern, D. N. (1990). *Diary of a Baby*. New York: Basic Books.
Strum, S. C. (1987). *Almost Human: A Journey into the World of Baboons*. New York: W. W. Norton.
Trevarthen, C. (2005). Action and emotion in development of cultural intelligence: Why infants have feelings like ours. In J. Nadel & D. Muir (Eds.), *Emotional Development* (pp. 61–91). Oxford: Oxford University Press.
von Helmholtz, H. 1962 [1910]. *Physiological Optics* (vol. III). ed. and trans. J. P. C. Southall. New York: Dover Publications.
von Helmholtz, H. 1971a [1878]. The facts of perception. In *Selected Writings of Hermann von Helmholtz*, ed. and trans. R. Kahl. Middletown, CT: Wesleyan University Press, pp. 366–408.
von Helmholtz, H. 1971b [1868]. Recent progress in the theory of vision. In *Selected Writings of Hermann von Helmholtz*, ed. and trans. R. Kahl. Middletown, CT: Wesleyan University Press, pp. 144–222.
Wilson, E. O. (1975). *Sociobiology: The New Synthesis*. Cambridge, MA: Harvard University Press.

INDEX